Lecture Notes in Computer Science

Lecture Notes in Artificial Intelligence 16371

The series Lecture Notes in Artificial Intelligence (LNAI) was established in 1988 as a topical subseries of LNCS devoted to artificial intelligence.

The series publishes state-of-the-art research results at a high level. As with the LNCS mother series, the mission of the series is to serve the international R & D community by providing an invaluable service, mainly focused on the publication of conference and workshop proceedings and postproceedings.

Miaomiao Liu · Xin Yu · Chang Xu · Yiliao Song
Editors

AI 2025: Advances in Artificial Intelligence

38th Australasian Joint Conference on Artificial Intelligence, AI 2025
Canberra, ACT, Australia, December 1–5, 2025
Proceedings, Part II

Editors
Miaomiao Liu
Australian National University
Canberra, ACT, Australia

Xin Yu
University of Queensland
St Lucia, QLD, Australia

Chang Xu
University of Sydney
Sydney, NSW, Australia

Yiliao Song
Adelaide University
Adelaide, SA, Australia

ISSN 0302-9743 ISSN 1611-3349 (electronic)
Lecture Notes in Artificial Intelligence
ISBN 978-981-95-4971-9 ISBN 978-981-95-4972-6 (eBook)
https://doi.org/10.1007/978-981-95-4972-6

LNCS Sublibrary: SL7 – Artificial Intelligence

This Springer imprint is published by the registered company Springer Nature Singapore Pte Ltd.
The registered company address is: 152 Beach Road, #21-01/04 Gateway East, Singapore 189721, Singapore

Preface

This volume contains the papers presented at the 38th Australasian Joint Conference on Artificial Intelligence (AJCAI 2025). The conference was held during December 1–5, 2025, and was hosted by the Australian National University. This annual conference is one of the longest-running conferences in artificial intelligence, with the first conference held in Sydney in 1987. The conference remains the premier event for artificial intelligence in Australasia, offering a forum for researchers and practitioners across all subfields of artificial intelligence to meet and discuss recent advances.

AJCAI 2025 received 152 submissions, and each submission was reviewed by at least two Program Committee (PC) members or external reviewers in a double-blind process. After a thorough discussion and rigorous scrutiny by the reviewers, 73 papers were accepted as full papers and 1 paper was accepted as a short paper. These 74 submissions were accepted for publication in these proceedings with an acceptance rate of 49%. AJCAI 2025 had seven keynote talks by the following distinguished scientists: Toby Walsh, Jing Jiang, Ling Chen, Mengjie Zhang, Nicholas Mattei, Geoff Webb and Marcus Hutter.

The following are notable aspects of the AJCAI 2024 conference:

- AJCAI 2025 was jointly held with the Defence Artificial Intelligence 2025 Symposium (December 1, 2025). The Defence Artificial Intelligence Symposium is an exciting opportunity for Defence and AI researchers to come together and explore priorities, opportunities and commonalities.
- The conference included a Government and Industry Day on December 4. The focus of the Government and Industry Day was on cutting-edge AI advancements in addressing regulatory and ethical challenges.
- The AJCAI 2025 program included two workshops held on December 2, Federation of Agentic Foundation Models (FAFMs), and AIML in Computational and Experimental Biology, and another invited panel discussion from The ARC Centre for Information Resilience (CIRES).
- The AJCAI 2025 program included five tutorials on December 2: Immersive AI Avatar Simulations for Education and Training: Designing Engaging, Ethical and Scalable Solutions; Demystifying Optimization and Generalization of Deep Learning via Feature Learning Theory; Reinforcement Learning: Single and Multi-Agent Foundations and Applications; Applied AI for Complex Environmental Sensing: Marine Case Studies; AI Alignment with Human Values: Foundations, Methods, and Challenges for Foundation Models.
- The AJCAI 2025 program included a PhD Forum, held on December 2, to mentor and assist postgraduate students in developing their research, with mentorship provided by research leaders.

We especially appreciate the work of the members of the Program Committee and the external reviewers for their expertise and tireless effort in assessing the papers within a strict timeline. We are also very grateful to the members of the Organizing Committee for their efforts in the preparation, promotion and organisation of the conference, especially the general chairs, Stephen Gould, Guodong Long and Helen Huang, for coordinating the whole event.

Lastly, we thank the National Committee for Artificial Intelligence of the Australian Computer Society, Springer, for the professional service provided by the Lecture Notes in Artificial Intelligence editorial and publishing teams, and our conference sponsors: Australian National University, Google, Monash University, CSIRO-Data61, Pioneer, Yep AI, Australian Computer Society, Defence Artificial Intelligence Research Network (DAIRNET), Follow Me AI, Computing Research and Education Association of Australasia, UNSW AI Institute and Springer.

October 2025

Miaomiao Liu
Xin Yu
Chang Xu
Yiliao Song

Organization

General Chairs

Stephen Gould	Australian National University, Australia
Guodong Long	University of Technology Sydney, Australia
Helen Huang	University of Queensland, Australia

Program Committee Chairs

Miaomiao Liu	Australian National University, Australia
Xin Yu	University of Queensland, Australia

Advisory Board

Chengqi Zhang	University of Technology Sydney, Australia
Geoff Webb	Monash University, Australia
Jie Lu	University of Technology Sydney, Australia
Abdul Sattar	Griffith University, Australia
Mengjie Zhang	Victoria University of Wellington, New Zealand
Andy Song	RMIT University, Australia

Senior Program Committee

Shan Xue	Macquarie University, Australia
Liyuan Pan	Beijing Institute of Technology, China
Kaihao Zhao	Harbin Institute of Technology, China
Miao Xu	University of Queensland, Australia
Xin Yu	University of Queensland, Australia
Haytham M. Fayek	RMIT University, Australia
Junyu Xuan	University of Technology Sydney, Australia
Yanbin Liu	Auckland University of Technology, New Zealand
Lizhen Qu	Monash University, Australia
Xuequan Lu	La Trobe University, Australia
Susan Wei	University of Melbourne, Australia
Yuxia Wu	Singapore Management University, Singapore

Jingfeng Zhang	University of Auckland, New Zealand
Miaomiao Liu	Australian National University, Australia
Felipe Trevizan	Australian National University, Australia
Runnan Chen	University of Sydney, Australia

Program Committee

Yuan Liu	Hong Kong University of Science and Technology, China
Zehong Cao	University of South Australia, Australia
Jun Wang	University of Sydney, Australia
Yi Mei	Victoria University of Wellington, New Zealand
Brendon J. Woodford	University of Otago, New Zealand
Maurice Pagnucco	University of New South Wales, Australia
Xin Guo	University of Queensland, Australia
Cong Lei	University of Sydney, Australia
Changqin Huang	Zhejiang Normal University, China
Jiale Liu	Pennsylvania State University, USA
Guangyan Huang	Deakin University, Australia
Xueping Peng	University of Technology Sydney, Australia
Jean Lee	University of Sydney, Australia
Zihe Liu	University of Technology Sydney, Australia
Yawen Zhao	University of Queensland, Australia
Abhi Desai	New England College, USA
Youquan Liu	Hochschule Bremerhaven, Germany
Kun Han	University of Queensland, Australia
Jianglin Qiao	University of South Australia, Australia
Kun Wang	University of Technology Sydney, Australia
Wei Duan	University of Technology Sydney, Australia
Huaxi Huang	Shanghai Artificial Intelligence Laboratory, China
Zesheng Ye	University of Melbourne, Australia
Liangwei Nathan Zheng	University of Adelaide, Australia
Jiahao Ma	Australian National University, Australia
Mehmet Bilgehan Bezcioglu	Australian National University, Australia
Huan Huo	University of Technology Sydney, Australia
Zhuo Huang	University of Sydney, Australia
Shaofei Shen	University of Queensland, Australia
Leah Gerrard	University of Technology Sydney, Australia
Yuanyuan Wang	University of Melbourne, Australia
Chenhao Zhang	University of Queensland, Australia
Ran Wang	University of Technology Sydney, Australia

Huiqiang Chen	University of Technology Sydney, Australia
Nicolas Bievre	Facebook, USA
Dhivya Nagasubramanian	University of Minnesota - Twin Cities, USA
Dongting Hu	University of Melbourne, Australia
Xunye Tian	University of Melbourne, Australia
Marcus Gallagher	University of Queensland, Australia
Muyang Li	University of Sydney, Australia
Yeliz Yesilada	University of Manchester, UK
Wenjie Wang	University of Melbourne, Australia
Xiangyu Sun	University of Sydney, Australia
Yexiong Lin	University of Sydney, Australia
Rina Carines Cabral	University of Sydney, Australia
Nayyar Zaidi	Deakin University, Australia
Miaomiao Liu	Australian National University, Australia
Dharmender Salian	Avco Consulting, USA
Ningyuan Zhang	University of Technology Sydney, Australia
Hoang Chuong Nguyen	Australian National University, Australia
Renjie Wu	Australian National University, Australia
Qingtao Yu	Australian National University, Australia
Changlin Song	Australian National University, Australia
Yue Yin	Australian National University, Australia
Yexin Mao	Australian National University, Australia
Zhiwei Li	University of Technology Sydney, Australia
Shutong Chen	University of Technology Sydney, Australia
Yiyuan Yang	University of Technology Sydney, Australia
Shengchao Chen	University of Technology Sydney, Australia
Jinhong Ni	Australian National University, Australia
Chaoyue Xing	Australian National University, Australia
Tianchi Sha	University of Technology Sydney, Australia
Guohang Zeng	University of Technology Sydney, Australia
Peter Baumgartner	CSIRO, Australia
Yixuan Qiu	University of Queensland, Australia
Tong Wu	Auckland University of Technology, New Zealand
Stephen Chen	York University, Canada
Yuhao Wu	University of Sydney, Australia
Tianchen Guo	University of Queensland, Australia
Xin Shen	University of Queensland, Australia
Sam Bahrami	Australian National University, Australia
M. A. Hakim Newton	University of Newcastle, Australia
Jianhua Yang	University of Western Sydney, Australia
Matthew Damigos	Ionian University, Greece
Harith Al-Sahaf	Victoria University of Wellington, New Zealand

Yanjun Shu	Harbin Institute of Technology, China
Chen Liu	University of Queensland, Australia

Volume Editors

Chang Xu	University of Sydney, Australia
Yiliao Song	University of Adelaide, Australia

Sponsors

Contents

Reinforcement Learning and Robotics

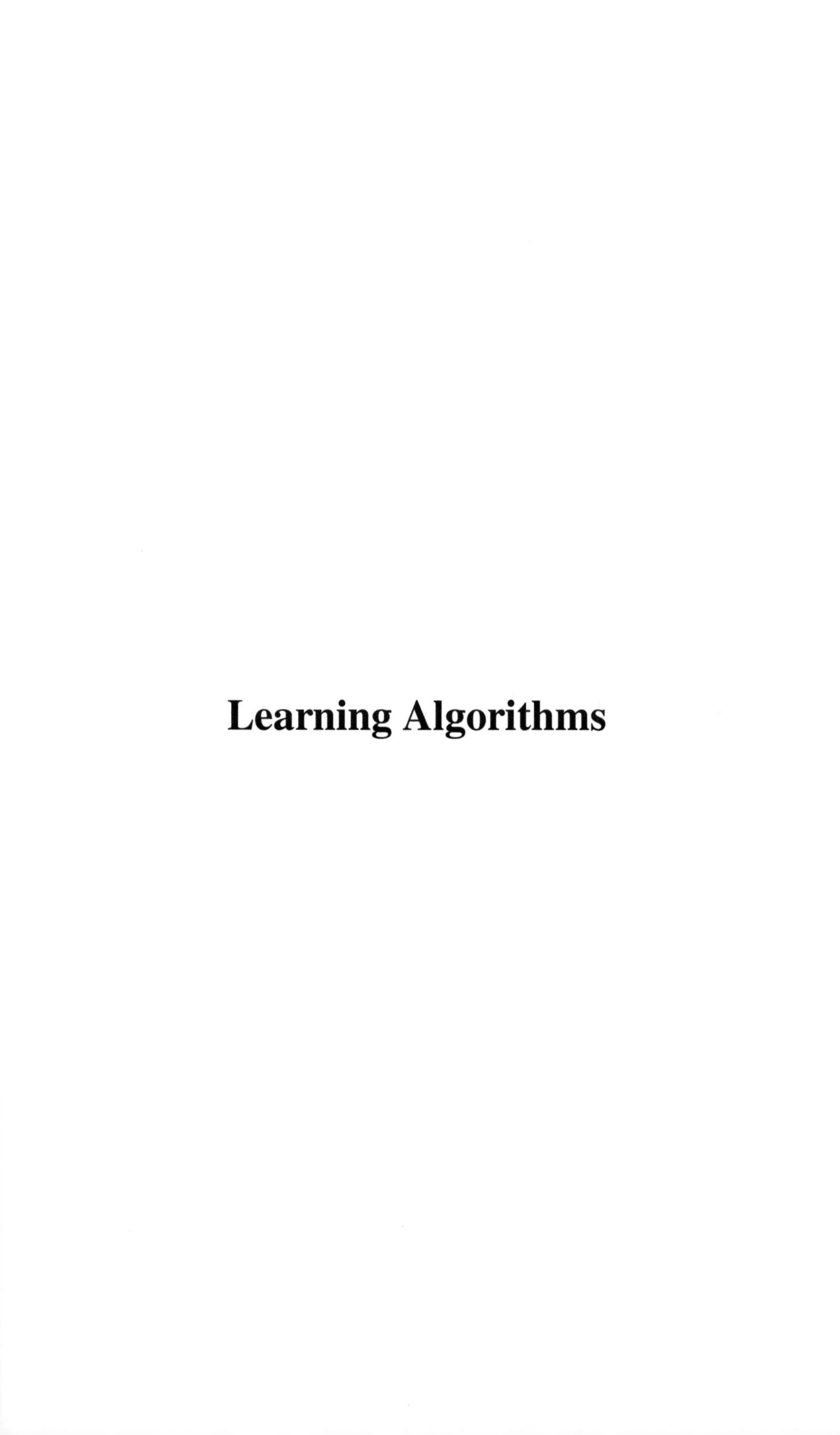

Learning Algorithms

Price Equilibrium Routing: A Lightweight Framework for Expert Selection in Mixture-of-Experts

Yiyang Zhao, Liangwei Nathan Zheng, Wei Emma Zhang, and Weitong Chen(✉)

The University of Adelaide, Adelaide, SA 5005, Australia
{yiyang.zhao,liangwei.zheng,wei.e.zhang,weitong.chen}@adelaide.edu.au

Abstract. Expert collapse, where a few experts dominate token routing while others remain underutilized, represents a fundamental challenge in Mixture-of-Experts (MoE) models for multimodal learning. Current load balancing approaches rely on auxiliary losses that may conflict with primary task objectives and require careful hyperparameter tuning. We propose Price Equilibrium Routing (PER), a market-inspired framework that addresses expert imbalance through dynamic pricing. Each expert maintains a price that adjusts based on its utilization: overused experts become more expensive and less likely to be selected, while underused ones become cheaper and more attractive. This creates self-regulating behavior that balances expert utilization without explicit constraints. PER is applied as a preprocessing step that adjusts utility scores prior to routing, making it compatible with a wide range of MoE architectures. We evaluate PER across clinical prediction tasks (MIMIC-III/IV) and sentiment analysis (CMU-MOSI), demonstrating improved expert utilization balance with coefficient of variation reductions. PER achieves consistent performance gains across diverse tasks, demonstrating its effectiveness in balancing expert utilization while preserving task accuracy. Overall, this work highlights the promise of economic principles for stable and interpretable MoE routing.

Keywords: Mixture-of-Experts · Multimodal Learning · Load Balancing

1 Introduction

Multimodal learning has recently advanced a range of applications, including computer vision [9], language processing, time series analysis [19,20], clinical practice [18], and financial forecasting [10]. Mixture-of-Experts (MoE) models offer an efficient way to scale neural networks in such scenarios by routing multimodal tokens to specialized expert networks. This allows increased model capacity without proportional inference cost. However, MoE models often suffer from expert collapse, where a few experts dominate routing while others are underutilized [6,21], reducing overall efficiency and performance. This issue is often

M. Liu et al. (Eds.): AI 2025, LNAI 16371, pp. 3–16, 2026.
https://doi.org/10.1007/978-981-95-4972-6_1

exacerbated by the sharp distribution of softmax, which amplifies extreme scores and favors a single expert during training.

Current approaches to address load balancing in MoE models typically rely on auxiliary balancing losses, such as entropy regularization or load balancing losses, which force more uniform token distribution across experts [4,6,16]. While effective in many settings, these losses often require careful weight tuning and may conflict with the primary task objectives [21], leading to slower convergence and sub-optimal solutions. In contrast, Price Equilibrium Routing (PER) introduces a few interpretable hyperparameters that directly govern its price adjustment behavior and are decoupled from the primary loss, avoiding the intricate weight-balancing often required by auxiliary objectives.

Inspired by the analysis in [21], the main reason of expert collapse is the usage of extreme routing distribution from softmax distribution, whereas the replacement of routing function can effectively mitigate this phenomenon. We propose PER, a dynamic routing mechanism that complements stochastic regularization with an economically grounded feedback process, reducing randomness and improving interpretability in expert selection. By adjusting each expert's selection probability based on historical usage, PER establishes a self-regulating equilibrium that naturally balances expert utilization and mitigates collapse. Furthermore, PER can operate in conjunction with existing balancing losses and can be integrated into a wide range of MoE frameworks with minor architectural changes. This design provides a principled, interpretable, and modular framework for expert selection, improving both training stability and final model performance across diverse multimodal benchmarks. We summarize our key contributions as follows:

- A novel routing mechanism that dynamically adjusts expert selection probabilities through a market-inspired price mechanism, enabling balanced and interpretable expert usage.
- Compatible with both traditional and structure-based gating strategies, enabling broad adoption across various MoE frameworks.
- Improves training stability by preventing expert collapse through real-time price feedback.
- Demonstrates consistent performance gains on three multimodal benchmarks over baselines.

2 Related Works

2.1 Sparse Mixture-of-Experts

Sparse MoE models, beginning with the Switch Transformer [4], use top-k routing and expert capacity limits to address expert imbalance. Subsequent extensions have focused on improving routing quality and load balance. For instance, contrastive learning has been integrated for multimodal alignment in LlMoE [12], while alternative routing mechanisms such as "Expert Choice" [22] and

Laplacian-based selection in Fuse-MoE [6] have been proposed to enhance expert-token affinity. Flex-MoE [16] further generalizes these ideas with additional auxiliary losses for balancing arbitrary multimodal inputs, and ReMoE [15] employs a differentiable gate with $\mathcal{L}_1$ regularization to promote sparsity.

To ensure expert balance, recent models commonly adopt auxiliary losses such as load variance [16] and entropy regularization [6], but these techniques often introduce instability, manifesting as sharp sinusoidal oscillations during training [21]. Moreover, methods like Fuse-MoE and Flex-MoE retain tight coupling between softmax routing and optimization objectives, leading to conflicts in expert selection and reduced specialization.

2.2 Structure-Based Gating Mechanisms

Structure-based gating mechanisms seek to balance expert load without explicit loss terms by encoding regularization into the gating architecture. A representative example is ConfNet [21], which introduces a confidence-based gating strategy where each expert computes a confidence score for each input token. These scores directly guide expert selection through either top-k selection or weighted aggregation. However, these methods typically lack dynamic feedback mechanisms, making them susceptible to routing instability. Structure-based methods often suffer from routing instability, where small fluctuations in confidence scores can lead to abrupt changes in expert selection, especially under top-k gating.

2.3 Economic Frameworks in AI

While economic principles have been applied to various AI problems, including resource allocation in distributed systems [5] and auction-based task assignment [3], their application to neural network routing remains relatively unexplored. Market-based approaches have shown promise in multi-agent systems and resource scheduling, providing inspiration for our price-based routing mechanism.

3 Method

3.1 Overview of Price Equilibrium Routing

Price Equilibrium Routing (PER) is a novel routing regularization mechanism inspired by economic equilibrium theory. It introduces expert-specific prices that dynamically adjust based on utilization, treating overused experts as "expensive" and underused ones as "cheap" to promote balanced expert assignment. Unlike prior methods, PER adjusts selection probabilities dynamically and intrinsically, without relying on gradient-based constraints or randomness within the mechanism itself, thereby encouraging more balanced expert assignment by design.

To achieve this, our implementation incorporates PER as a complementary mechanism for traditional routing objectives. PER provides a real-time balancing signal through dynamic pricing, which helps stabilize expert assignment and mitigate routing volatility during training. This mechanism is particularly beneficial for structure-based gating methods, which embed expert selection into the architecture itself and typically lack an explicit loss-based feedback signal. By incorporating price-based adjustments into the routing probabilities, PER makes the outputs compatible with standard gate loss formulations.

In practice, we apply PER in conjunction with the CV^2 gate loss, which is computed on the price-adjusted routing distributions. This hybrid design combines the short-term adaptability of PER with the long-term optimization benefits of CV^2 loss. While PER dynamically reacts to load imbalances at each forward pass, CV^2 loss encourages the gating network to learn smoother and more balanced routing distributions across training epochs. The two mechanisms work together, reinforcing one another to improve training stability and expert utilization.

In conclusion, PER is implemented as a lightweight, plug-in mechanism that is compatible with both traditional softmax-based and structure-based gating strategies in our implementation, and can be integrated into existing MoE frameworks. This makes PER a flexible and promising addition to the MoE design space.

3.2 Economic Intuition

The core insight behind PER is to view expert selection as a resource allocation problem where tokens represent demand and experts represent capacity-constrained service providers. In traditional MoE routing, tokens are assigned to experts based solely on utility scores computed by a router network. This can lead to expert collapse when multiple tokens find the same expert most suitable, creating severe demand imbalances.

PER introduces a pricing mechanism to regulate this demand. Each expert maintains a dynamic price that reflects its current utilization level. When experts become popular (high-demand), their price increases, making them less attractive for future token assignments. At the same time, the price of less popular (low-demand) experts decreases, leading to an increase in their likelihood of being selected in subsequent routing steps. This dynamic pricing mechanism introduces a more balanced token distribution across experts.

The key advantage is that optimal resource allocation emerges from local price adjustments without requiring global coordination or explicit constraint enforcement. Tokens naturally distribute across experts as they seek the best "value for money", which is the highest utility per unit price (Fig. 1).

3.3 PER Framework: Price-Based Expert Selection

The framework operates through the following key stages (1) **Price Adjustment**: The expert logits used for PER routing are computed as a linear projection over token embeddings. PER modifies each logit by dividing it by the corresponding expert price, producing a price-adjusted score. This operation

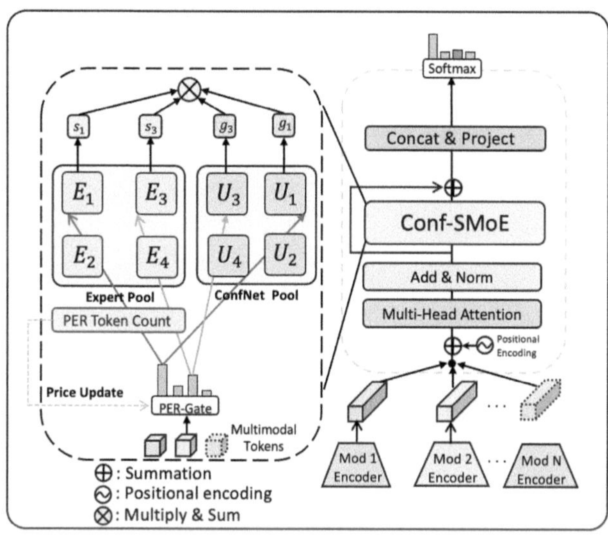

Fig. 1. Price Equilibrium Routing Framework.

balances both token-expert affinity and relative availability (cost) in the routing decision. (2) **Top-k Selection and Sparse Softmax**: For each token, the top-k experts with the highest adjusted logits are selected. A sparse softmax is then applied over these top-k adjusted logits, producing gating weights that determine the probability with which each selected expert will process the token. (3) **Token Routing and Expert Processing**: Tokens are routed to selected experts based on gating weights, and each expert processes them independently. (4) **Loss Computation**: After the forward pass, the model computes the primary task loss and gate loss, with optional auxiliary losses such as confidence regularization. (5) **Expert Utilization Measurement and Price Update**: Expert utilization is measured by token counts per batch, and prices are updated accordingly: overused experts become more expensive, underused ones cheaper, closing the feedback loop. (6) **Iterative Adaptation**: The updated prices are used in the next batch, allowing the system to dynamically adapt expert selection based on demand.

This pipeline establishes a decoupled control loop for expert balancing: the gating network learns to model token-expert affinities, while the price mechanism autonomously regulates expert availability through utilization-driven feedback. In practice, the gating network is further regularized by a CV^2 gate loss to promote balanced utilization and improve training stability. For a utilization vector $x \in \mathbb{R}^E$, representing the actual load per expert, the CV^2 loss is defined as:

$$\mathcal{L}_{\text{CV}^2}(x) = \frac{\text{Var}(x)}{(\text{Mean}(x))^2 + \epsilon}$$

where Var(x) is the variance of x, Mean(x) is the mean of x and ϵ is a small constant (e.g., 10^{-10}) added for numerical stability.

The key advantages of PER lie in its training stability, compatibility, and interpretability. This mechanism introduces a dynamic, price-based routing mechanism designed to enhance expert utilization in MoE models. It improves training stability by discouraging expert overuse and promoting underutilized experts through adaptive pricing, naturally mitigating expert collapse without requiring additional constraints. The mechanism is theoretically compatible with both traditional and structure-based gating strategies, and can be integrated into existing MoE frameworks with minimal changes. Furthermore, PER introduces interpretability into expert routing: frequently selected experts become more "expensive," while underused ones become "cheaper," enabling a controllable, market-inspired routing process.

3.4 Price Update Mechanism

The heart of PER is a multiplicative update rule that adjusts expert prices based on their utilization relative to the system average. We first present the basic form and then introduce temporal smoothing for training stability.

Basic Price Update. The fundamental price update rule is:

$$p_j(t+1) = p_j(t) \cdot \left(1 + \eta \cdot \left(\frac{c_j}{c_{\text{avg}}} - 1\right)\right) \tag{1}$$

where $p_j(t)$ is the price of expert j at time step t, c_j is the count of tokens routed to expert j in the current batch, $c_{\text{avg}} = \frac{M}{N}$ is the average token count per expert (with M total tokens and N experts), and η is the price learning rate.

This update rule implements three distinct behavioral regimes:

- **Overutilization** ($c_j > c_{\text{avg}}$): Price increases, reducing future selection probability
- **Underutilization** ($c_j < c_{\text{avg}}$): Price decreases, increasing future selection probability
- **Balanced utilization** ($c_j = c_{\text{avg}}$): Price remains stable

Price-Adjusted Routing. Expert selection incorporates these prices through a temperature-scaled softmax routing probability:

$$P(\text{expert}_j|\text{token}_i) = \frac{\exp(u_{ij}/p_j^\alpha)}{\sum_{k=1}^{N} \exp(u_{ik}/p_k^\alpha)} \tag{2}$$

where u_{ij} is the utility of expert j for token i computed by the router network, and α is a temperature parameter that controls price sensitivity. The temperature parameter allows fine-tuning of how strongly prices influence routing decisions: $\alpha = 0$ ignores prices entirely, while larger α values make routing more sensitive to price differences. As expert price increases, the routing probability decreases, implementing the economic intuition that expensive experts become less attractive.

3.5 Temporal Smoothing for Training Stability

To ensure training stability and prevent oscillatory behavior, we implement PER with temporal smoothing that dampens rapid price fluctuations:

$$p_j(t+1) = p_j(t) \cdot \left(1 + (1-\beta) \cdot \eta \cdot \left(\frac{c_j}{c_{\text{avg}}} - 1\right)\right) \tag{3}$$

where $\beta \in [0,1]$ is a smoothing factor that controls the rate of price adjustment, and η is the price learning rate. When $\beta = 0$, this reduces to the basic update rule (Eq. 1). As β approaches 1, price updates become more conservative, with $(1-\beta)$ acting as an effective learning rate multiplier. This temporal smoothing prevents sharp price oscillations that could destabilize training while maintaining the core balancing behavior.

3.6 Hyperparameter Configuration

The PER mechanism introduces three key hyperparameters that control the dynamics of price-based routing: the price learning rate (η), price sensitivity factor (α), and temporal smoothing coefficient (β).

Price Learning Rate (η): This parameter controls the speed of price adjustments in response to utilization imbalances. It appears in the price update equation (Eq. 1) and determines how aggressively prices react to expert overutilization or underutilization. A higher η leads to faster price adjustments, causing prices to respond more rapidly to demand changes, while a lower η results in more gradual price changes that may provide greater stability but slower adaptation to routing imbalances. In practice, we found that $\eta = 0.15$ provides stable performance across different experimental datasets, balancing responsiveness with training stability.

Price Sensitivity Factor (α): This parameter controls how strongly current prices influence routing decisions and appears in the routing probability equation (Eq. 2). When $\alpha = 0$, prices have no effect on routing, reducing PER to standard utility-based routing. Higher α values make routing decisions more sensitive to price differences, while lower α values reduce the influence of prices on expert selection. We empirically found that optimal α values vary across different model scales and task complexities, typically ranging from 0.5 to 2.0.

Temporal Smoothing Coefficient (β): This parameter controls the rate of price adjustment and prevents oscillatory behavior during training. It appears in the temporally smoothed price update equation (Eq. 3), where $\beta \in [0,1]$ acts as a smoothing factor. When $\beta = 0$, the mechanism reduces to the basic update rule, while β approaching 1 makes price updates increasingly conservative. Our experiments demonstrate that $\beta = 0.5$ provides robust performance across various experimental settings by preventing sharp price oscillations while maintaining sufficient adaptability.

4 Experiment

To ensure a fair comparison, we adopt the best-performing hyperparameter configuration for each baseline, as reported in its original paper or official implementation, while standardising the hidden dimension to **128** for every model. To eliminate confounding factors introduced by heterogeneous encoder architectures, *all* methods share the same modality-specific encoders:

- **Text (MIMIC-III/IV):** pretrained *ClinicalBERT* [1].
- **Irregular time series & ECG:** a single-layer patch-embedding module.
- **Chest X-ray (CXR):** pretrained *DenseNet* from *TorchXRayVision* [2].

For **CMU-MOSI**, only privacy-protected, pre-extracted features are available. Consequently, we append a single linear mapping layer that projects each modality into the common 128-dimensional embedding space (Table 1).

4.1 Dataset

Table 1. Dataset statistics

Dataset	Task	#Samples	Modality	#Classes	Is Imbalanced?
MIMIC3	IHM	6621	Time Series, Text	2	Yes
	LOS	6621	Time Series, Text	2	Yes
	PHE	12,278	Time Series, Text	2	Yes
MIMIC4	IHM	22,733	Time Series, Text, ECG, Image	2	Yes
	LOS	22,733	Time Series, Text, ECG, Image	2	Yes
	PHE	47,027	Time Series, Text, ECG, Image	2	Yes
CMU-MOSI	Sentiment Analysis	1870	Audio, Video, Text	3	Yes

We adopt 3 benchmarking datasets MIMIC3, MIMIC4 and CMU-MOSI in multimodal learning scenarios and validate our method's ability on multimodal fusion. We include a short introduction to each task below:

IHM is a binary classification task in which each patient's time series is truncated to the first 48 h of ICU readings; the model then predicts whether in-hospital mortality will occur during that stay.

LOS is likewise framed as a binary classification task. Following the 48-IHM setup, we retain only patients who remain in the ICU for at least 48 h and predict whether their subsequent length of stay will exceed the median for the cohort.

PHE is a multilabel classification task that predicts the presence of any of 25 acute-care conditions during a given ICU stay. These conditions are defined in the MIMIC benchmark [7].

CMU-MOSI is a multimodal sentiment-analysis benchmark comprising short video clips annotated with sentiment scores. Each sample provides synchronized audio, visual, and textual modalities.

4.2 Baseline Model

To ensure a fair comparison with state-of-the-art methods, we include representative baselines from multimodal fusion and MoE literature:

- **SMIL** [11] adopts Bayesian meta-learning for multimodal modeling.
- **ShaSpec** [14] disentangles modality-shared and modality-specific representations.
- **TF** [17] performs tensor-based multimodal fusion.
- **mmFormer** [13] applies Transformer-based fusion across modalities.
- **LIMoE** [12] uses contrastive learning to align modality-specific embeddings.
- **FuseMoE** [6] selects experts through softmax or Laplacian-based gating.
- **FlexMoE** [16] routes different modality combinations to specialized experts.

5 Experiment Setting

In Table 2, we present comprehensive experimental results for our proposed Price Equilibrium Routing (PER) framework across three multimodal benchmarking datasets. Our experiment setting focuses on evaluating the effectiveness of the economic-inspired routing mechanism in achieving balanced expert utilization while maintaining superior performance compared to existing MoE approaches.

We conduct experiments on three diverse multimodal learning scenarios: clinical prediction tasks using MIMIC-III and MIMIC-IV datasets, and sentiment analysis using the CMU-MOSI dataset. For MIMIC datasets, we evaluate on three critical ICU prediction tasks: in-hospital mortality (IHM), length-of-stay (LOS), and phenotyping (PHE). The multimodal inputs include time series data, clinical notes, ECG signals, and chest X-ray images where available. For CMU-MOSI, we utilize audio, visual, and textual modalities for sentiment classification. All experiments are repeated with different random seeds. We report mean and standard deviation across runs to assess improvements over baselines.

6 Primary Results

Our experimental results demonstrate several key findings: (1) PER-MoE consistently outperforms state-of-the-art baseline methods across all tasks and datasets, showing particularly strong improvements over existing MoE approaches such as FuseMoE and FlexMoE. In MIMIC-IV tasks, PER-MoE achieves F1 score improvements ranging from 1.10% to 10.34% and AUC improvements ranging from 1.15% to 4.79% compared to the best-performing baselines. (2) The performance gains are consistent across different dataset scales and domains, with MIMIC-III showing F1 improvements of 0.52% to 3.13% and AUC improvements of 1.03% to 2.77%. (3) On the CMU-MOSI sentiment analysis task, our method achieves notable improvements of 4.50% in F1 score and 2.17% in AUC compared to FlexMoE, demonstrating the generalizability of our approach beyond clinical domains and the effectiveness of our approach in addressing routing instability issues commonly observed in multimodal sentiment analysis.

These empirical results validate our core hypothesis that price-based equilibrium routing effectively balances expert utilization without sacrificing model performance, leading to more efficient and robust multimodal learning compared to auxiliary loss-based load balancing strategies. The improvements are particularly pronounced in scenarios where routing stability is crucial, such as sentiment analysis tasks with high input variability.

Table 2. Main results with three datasets

Task	Metric	SMIL	ShaSpec	mmFormer	TF	LIMoE	FuseMoE-S	FuseMoE-L	FlexMoE	**PER-MoE**
MIMIC4-IHM	F1	40.12 ± 1.22	33.51 ± 1.44	44.52 ± 1.04	11.32 ± 0.63	43.88 ± 0.44	30.14 ± 1.15	40.21 ± 1.29	35.28 ± 0.33	**45.62 ± 1.19**
	AUC	76.63 ± 0.98	78.21 ± 0.61	80.18 ± 0.53	67.21 ± 0.69	82.31 ± 0.82	72.28 ± 0.33	78.05 ± 0.72	80.35 ± 0.44	**84.98 ± 0.51**
MIMIC4-LOS	F1	57.38 ± 0.13	56.19 ± 0.13	58.44 ± 0.31	43.06 ± 1.33	59.36 ± 0.94	57.59 ± 0.76	58.31 ± 0.46	56.28 ± 0.53	**60.69 ± 0.69**
	AUC	76.23 ± 0.38	73.11 ± 0.10	74.61 ± 0.32	45.21 ± 0.41	76.33 ± 0.17	72.59 ± 1.07	72.59 ± 0.61	74.71 ± 0.37	**77.38 ± 0.59**
MIMIC4-PHE	F1	**29.41 ± 0.31**	26.84 ± 0.13	26.31 ± 0.25	26.76 ± 0.53	25.38 ± 0.52	12.45 ± 0.54	12.25 ± 0.55	24.63 ± 0.71	28.79 ± 1.05
	AUC	64.68 ± 1.23	62.41 ± 0.44	70.38 ± 0.20	54.20 ± 0.44	72.50 ± 0.67	55.48 ± 0.39	58.52 ± 0.27	71.57 ± 0.47	**74.65 ± 0.18**
MIMIC3-IHM	F1	48.56 ± 0.35	34.36 ± 0.73	43.80 ± 0.67	23.09 ± 1.13	47.62 ± 0.21	48.62 ± 0.27	**51.63 ± 0.39**	47.81 ± 0.31	49.70 ± 0.98
	AUC	84.00 ± 1.10	79.86 ± 0.53	81.27 ± 0.76	78.47 ± 0.65	83.30 ± 0.64	85.81 ± 0.13	86.22 ± 0.15	85.10 ± 0.61	**87.15 ± 0.42**
MIMIC3-LOS	F1	62.01 ± 0.54	57.27 ± 0.36	61.10 ± 0.65	56.23 ± 0.79	62.75 ± 0.10	62.26 ± 0.36	62.27 ± 1.61	64.18 ± 0.34	**65.06 ± 0.92**
	AUC	80.08 ± 0.35	74.71 ± 0.07	77.94 ± 0.12	72.60 ± 0.79	79.80 ± 0.27	80.05 ± 0.51	79.70 ± 0.43	80.49 ± 0.48	**81.75 ± 0.51**
MIMIC3-PHE	F1	33.01 ± 0.30	19.52 ± 0.39	35.10 ± 0.38	30.38 ± 0.38	36.22 ± 0.39	33.48 ± 0.22	34.07 ± 0.51	35.31 ± 0.57	**36.41 ± 0.68**
	AUC	75.17 ± 1.84	64.61 ± 0.25	71.79 ± 0.26	66.56 ± 0.17	77.93 ± 0.44	77.22 ± 0.29	77.19 ± 0.33	77.37 ± 0.10	**79.43 ± 0.09**
CMU-MOSI	F1	54.56 ± 0.43	41.78 ± 0.64	43.28 ± 0.36	50.22 ± 0.20	50.91 ± 0.19	47.69 ± 0.39	48.15 ± 0.57	51.50 ± 0.55	**56.00 ± 1.66**
	AUC	77.25 ± 0.24	65.08 ± 0.39	64.48 ± 0.51	74.62 ± 0.49	78.12 ± 0.31	75.26 ± 0.20	75.96 ± 0.29	79.38 ± 0.23	**81.55 ± 0.44**

6.1 Expert Utilization Analysis

To understand the superior performance of our Price Equilibrium Routing framework, we analyze the expert selection patterns through comprehensive utilization metrics and visualization of routing decisions during training. We use the same CMU-MOSI dataset and experimental setup as reported in Table 2, with additional expert utilization metrics calculated specifically for this analysis. The experimental setup optionally includes an auxiliary scoring module [21] for output fusion, which is not involved in routing decisions and does not affect the PER mechanism.

Figure 2 presents a comparative analysis of expert utilization patterns between ConfNet and our PER routing mechanism across training epochs on the CMU-MOSI dataset.

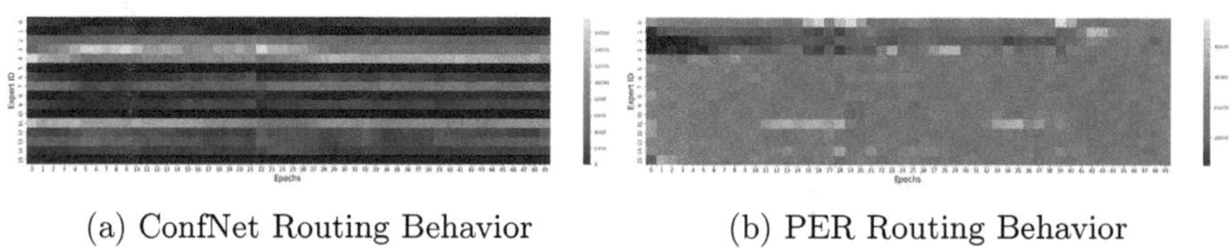

(a) ConfNet Routing Behavior (b) PER Routing Behavior

Fig. 2. Expert utilization heatmaps comparing routing patterns across 50 training epochs with 16 experts on CMU-MOSI. Each row represents an expert, and each column represents an epoch. Color intensity indicates usage frequency. Figure (a) shows the expert routing pattern under ConfNet, where a few experts dominate the computation. In contrast, Figure (b) demonstrates the more balanced utilization achieved by PER routing.

The visualization reveals fundamental differences in expert utilization patterns that explain the performance improvements of PER. In Fig. 2a, ConfNet exhibits uneven expert utilization patterns, where expert usage is concentrated among a subset of experts while others receive significantly fewer tokens. This utilization imbalance is quantified by a high Coefficient of Variation (CV) of 1.15 and CV^2 of 1.33, indicating substantial disparity in expert usage patterns and departure from uniform resource allocation.

In contrast, Fig. 2b demonstrates how PER achieves significantly more balanced expert utilization. The price-based mechanism dynamically adjusts the input logits through the multiplicative price update rule:

$$p_j(t+1) = p_j(t) \cdot \left(1 + (1-\beta) \cdot \eta \cdot \left(\frac{c_j}{c_{\text{avg}}} - 1\right)\right)$$

where β is the smoothing factor, η is the price learning rate, and c_j/c_{avg} is the utilization ratio. This prevents any single expert from monopolizing the routing decisions. The balanced utilization is reflected in dramatically improved balance metrics: CV reduced to 0.24.

6.2 Quantitative Analysis of Load Balancing

The effectiveness of PER in addressing potential expert collapse is further validated through comprehensive utilization metrics comparing ConfNet and PER

Table 3. Expert Usage Balance Metrics (CMU-MOSI, 16 experts, 50 epochs)

Metric	ConfNet	PER Routing
Coefficient of Variation (CV)	1.15 ± 0.07	0.24 ± 0.11
CV^2 (Balance Score)	1.33 ± 0.16	0.07 ± 0.06
Gini Coefficient	0.59 ± 0.02	0.12 ± 0.05

routing on the CMU-MOSI dataset over 50 training epochs with 16 experts (Table 3).

PER leads to substantially more balanced expert utilization, as evidenced by much lower values of Coefficient of Variation (CV), Balance Score (CV^2), and Gini coefficient. These metrics respectively capture variability, imbalance, and inequality in load distribution. CV quantifies the relative dispersion of expert usage, with lower values indicating more uniform routing. Its squared form, CV^2, is commonly used to evaluate expert imbalance; lower CV^2 implies better load balance. CV^2 is implicitly optimized in GShard [8], effectively reducing expert load variance relative to the mean. The Gini coefficient, a standard inequality metric originating from economics, measures distributional disparity, where 0 denotes perfect equality and 1 indicates maximum imbalance.

6.3 Ablation Study on PER's Contribution to Load Balancing

To isolate the contribution of PER, we compare the full model (PER + CV^2) against a variant where the PER mechanism is disabled (CV^2-only), setting all price adjustment parameters (η, α, β) to 0. As shown in Table X, PER leads to consistently lower Coefficient of Variation (CV), CV^2, and Gini coefficient across tasks. This suggests that PER plays a key role in enhancing expert utilization balance, which is critical for training stability and mitigating expert collapse in large-scale or sparse settings (Table 4).

Table 4. Impact of PER on Expert Routing Balance across MIMIC-III Tasks

Model	IHM			LOS			PHE		
	CV	CV^2	Gini	CV	CV^2	Gini	CV	CV^2	Gini
PER + CV^2	0.225 ± 0.161	0.0677 ± 0.0746	0.119 ± 0.082	0.380 ± 0.026	0.1477 ± 0.0195	0.208 ± 0.012	0.375 ± 0.064	0.1436 ± 0.0459	0.200 ± 0.039
CV^2-only	0.366 ± 0.150	0.1491 ± 0.0971	0.197 ± 0.077	0.451 ± 0.066	0.2066 ± 0.0592	0.249 ± 0.037	0.450 ± 0.036	0.2036 ± 0.0317	0.233 ± 0.020

7 Conclusion

This work introduces Price Equilibrium Routing (PER), a market-inspired approach to expert load balancing in Mixture-of-Experts models. Unlike auxiliary loss methods that create optimization conflicts, PER operates through dynamic price adjustments that naturally regulate expert utilization without interfering with primary task objectives. Our experimental evaluation across clinical prediction and sentiment analysis tasks demonstrates that PER consistently improves expert utilization balance, with coefficient of variation reductions compared to structure-based baselines.

Acknowledgment. This work was supported by Australian Research Council ARC Early Career Industry Fellowship (Grant No.IE240100275), Australian Research Council Discovery Projects (Grant No.DP240103070), Australian Research Council Linkage(Grant No.LP230200821), University of Adelaide, Global Partnership Engagement Fund, 2025.

References

1. Alsentzer, E., et al.: Publicly available clinical BERT embeddings. arXiv preprint arXiv:1904.03323 (2019)
2. Cohen, J.P., et al.: Torchxrayvision: a library of chest x-ray datasets and models. In: International Conference on Medical Imaging with Deep Learning, pp. 231–249. PMLR (2022)
3. Dias, M.B., Zlot, R., Kalra, N., Stentz, A.: Market-based multirobot coordination: a survey and analysis. Proc. IEEE **94**(7), 1257–1270 (2006)
4. Fedus, W., Zoph, B., Shazeer, N.: Switch transformers: scaling to trillion parameter models with simple and efficient sparsity. J. Mach. Learn. Res. **23**(120), 1–39 (2022)
5. Ghodsi, A., Zaharia, M., Hindman, B., Konwinski, A., Shenker, S., Stoica, I.: Dominant resource fairness: fair allocation of multiple resource types. In: 8th USENIX Symposium on Networked Systems Design and Implementation (NSDI 11) (2011)
6. Han, X., Nguyen, H., Harris, C., Ho, N., Saria, S.: Fusemoe: mixture-of-experts transformers for fleximodal fusion. arXiv preprint arXiv:2402.03226 (2024)
7. Johnson, A.E., et al.: Mimic-iii, a freely accessible critical care database. Sci. Data **3**(1), 1–9 (2016)
8. Lepikhin, D., et al.: Gshard: scaling giant models with conditional computation and automatic sharding. In: International Conference on Learning Representations (ICLR) (2021). https://openreview.net/forum?id=qrwe7XHTmYb
9. Liang, W., Dong, C., Zheng, L., Li, Z., Zhang, W., Chen, W.: Calibrating deep neural network using euclidean distance. arXiv preprint arXiv:2410.18321 (2024)
10. Liang, W., Li, Z., Chen, W.: Enhancing financial market predictions: causality-driven feature selection. In: International Conference on Advanced Data Mining and Applications, pp. 149–163. Springer (2024)
11. Ma, M., Ren, J., Zhao, L., Tulyakov, S., Wu, C., Peng, X.: Smil: multimodal learning with severely missing modality. In: Proceedings of the AAAI Conference on Artificial Intelligence, vol. 35, pp. 2302–2310 (2021)
12. Mustafa, B., Riquelme, C., Puigcerver, J., Jenatton, R., Houlsby, N.: Multimodal contrastive learning with limoe: the language-image mixture of experts. In: Advances in Neural Information Processing Systems, vol. 35, pp. 9564–9576 (2022)
13. Tsai, Y.H.H., Bai, S., Liang, P.P., Kolter, J.Z., Morency, L.P., Salakhutdinov, R.: Multimodal transformer for unaligned multimodal language sequences. In: Proceedings of the conference. Association for computational linguistics. Meeting, vol. 2019, p. 6558 (2019)

14. Wang, H., Chen, Y., Ma, C., Avery, J., Hull, L., Carneiro, G.: Multi-modal learning with missing modality via shared-specific feature modelling. In: Proceedings of the IEEE/CVF Conference on Computer Vision and Pattern Recognition, pp. 15878–15887 (2023)
15. Wang, Z., Zhu, J., Chen, J.: Remoe: fully differentiable mixture-of-experts with relu routing. arXiv preprint arXiv:2412.14711 (2024)
16. Yun, S., et al.: Flex-moe: modeling arbitrary modality combination via the flexible mixture-of-experts. arXiv preprint arXiv:2410.08245 (2024)
17. Zadeh, A., Chen, M., Poria, S., Cambria, E., Morency, L.P.: Tensor fusion network for multimodal sentiment analysis. arXiv preprint arXiv:1707.07250 (2017)
18. Zheng, L.N., Dong, C.G., Zhang, W.E., Chen, X., Yue, L., Chen, W.: Devil in the tail: a multi-modal framework for drug-drug interaction prediction in long tail distinction. In: Proceedings of the 33rd ACM International Conference on Information and Knowledge Management, pp. 3395–3404 (2024)
19. Zheng, L.N., et al.: Revisited large language model for time series analysis through modality alignment. arXiv preprint arXiv:2410.12326 (2024)
20. Zheng, L.N., et al.: Irregularity-informed time series analysis: adaptive modelling of spatial and temporal dynamics. In: Proceedings of the 33rd ACM International Conference on Information and Knowledge Management, pp. 3405–3414 (2024)
21. Zheng, L.N., Zhang, W.E., Guo, M., Xu, M., Maennel, O., Chen, W.: Rethinking gating mechanism in sparse MOE: handling arbitrary modality inputs with confidence-guided gate. arXiv preprint arXiv:2505.19525 (2025)
22. Zhou, Y., et al.: Mixture-of-experts with expert choice routing. In: Advances in Neural Information Processing Systems, vol. 35, pp. 7103–7114 (2022)

Minimum Message Length t-Test

Enes Makalic(✉) and Daniel F. Schmidt

Faculty of Information Technology, Monash University, Melbourne, VIC 3800, Australia
enes.makalic@monash.edu

Abstract. The frequentist two-sample t-test for comparison of two means is one of the most widely used hypothesis tests in practice. Recently, several Bayesian alternatives to the t-test were derived and shown to have some advantages, such as incorporation of prior expert knowledge, over the standard methodology. This manuscript introduces a new Bayesian t-test based on the minimum message length (MML) principle of inductive inference. The aim of the manuscript is to introduce the general reader to MML and demonstrate how MML can be applied to a problem of central importance in statistics.

Keywords: t-test · Bayesian statistics · minimum message length

1 Introduction

Perhaps the most popular hypothesis test in use today is the frequentist two-sample t-test for comparison of two means. As an example of the t-test's popularity, a search for the term 'paired t tests' returns approximately 12,000 articles from 1990 onwards in PubMed. Similarly, articles published in peer reviewed psychology journals feature, on average, more than three t-tests per article [23]. Formally, the t-test assumes that we observe normally distributed data from two, possibly different, populations:

$$Y_{1i} \sim \mathcal{N}\left(\mu + \frac{\sigma\delta}{2}, \sigma^2\right), \quad Y_{2j} \sim \mathcal{N}\left(\mu - \frac{\sigma\delta}{2}, \sigma^2\right), \tag{1}$$

for $i = 1, \ldots, n_1$ and $j = 1, \ldots, n_2$. Here, the unknown parameter $\mu \in \mathbb{R}$ is the overall (grand) mean, $\sigma > 0$ is a standard deviation that is common to both groups, $\delta \geq 0$ is the standardised effect size and (n_1, n_2) are the sample sizes of the two groups. This formulation of the t-test in terms of the overall mean and standardised effect size is due to [6]. The usual null hypothesis says that the mean is the same in both groups while the alternative hypothesis specifies a different mean for each group; formally, we have

$$\mathcal{H}_0 : \delta = 0, \quad \mathcal{H}_1 : \delta > 0. \tag{2}$$

Without loss of generality, we assume that the population mean in group 1 is greater than the population mean in group 2, if a difference exists. Let $\boldsymbol{\theta}_0 =$

M. Liu et al. (Eds.): AI 2025, LNAI 16371, pp. 17–29, 2026.
https://doi.org/10.1007/978-981-95-4972-6_2

$\{\mu, \sigma\} \in (\mathbb{R} \times \mathbb{R}_+) \equiv \Theta_0$ and $\boldsymbol{\theta}_1 = \{\mu, \sigma, \delta\} \in (\mathbb{R} \times \mathbb{R}_+^2) \equiv \Theta_1$ denote the parameter set under the two models. Given data $\mathbf{y} = \{\mathbf{y}_1, \mathbf{y}_2\}$ where $\mathbf{y}_1 = (y_{1,1}, \ldots, y_{n_1,1})'$ and $\mathbf{y}_2 = (y_{2,1}, \ldots, y_{n_2,1})'$, the task is to estimate the effect size $\delta > 0$ and thus determine whether the population means of the two groups are equal ($\delta = 0$).

Recently, there has been a great deal of interest in Bayesian alternatives to the frequentist t-test [6,7,9,12,22]. A key quantity in Bayesian analysis is the Bayes factor

$$\underbrace{\frac{P(\mathcal{H}_1|\mathbf{y})}{P(\mathcal{H}_0|\mathbf{y})}}_{\text{Posterior odds}} = \underbrace{\frac{p(\mathbf{y}|\mathcal{H}_1)}{p(\mathbf{y}|\mathcal{H}_0)}}_{\text{Bayes factor}} \underbrace{\frac{P(\mathcal{H}_1)}{P(\mathcal{H}_0)}}_{\text{Prior odds}}, \tag{3}$$

which is the ratio of the marginal likelihoods of the data obtained by integrating the parameters out with respect to their prior distribution; that is,

$$p(\mathbf{y}|\mathcal{H}_i) = \int p(\mathbf{y}|\boldsymbol{\theta}_i)\pi_i(\boldsymbol{\theta}_i)\, d\boldsymbol{\theta}_i, \quad i = 0, 1, \tag{4}$$

where $\pi_i(\boldsymbol{\theta}_i)$ is the prior distribution of the parameters under model i. The Bayes factor, henceforth denoted by BF_{10}, measures the amount of evidence in support of hypothesis $\mathcal{H}_1$ over hypothesis $\mathcal{H}_0$, where $\mathrm{BF}_{10} > 3$ indicates substantial evidence in favour of $\mathcal{H}_1$ [8]. Given a proper prior $\pi(\delta)$ on the effect size δ, Gronau et al. [7], generalizing the work of [6,12], showed that the Bayes factor for the t-test problem can be expressed as

$$\mathrm{BF}_{10} = \frac{\int T_\nu(t|\sqrt{n_\delta}\delta)\pi(\delta)\, d\delta}{T_\nu(t)} \tag{5}$$

where t is the observed t-statistic

$$t = \sqrt{n_\delta}(\bar{\mathbf{y}}_1 - \bar{\mathbf{y}}_2)/s_p, \quad \nu s_p = (n_1 - 1)s_1^2 + (n_2 - 1)s_2^2, \tag{6}$$

and s_j ($j = 1, 2$) are the usual unbiased sample variance estimates. Here, $\nu = n_1 + n_2 - 2$ denotes the degrees of freedom, $n_\delta = (1/n_1 + 1/n_2)^{-1}$ is the effective sample size, and $T_\nu(\cdot|a)$ is the non-central Student t distribution with degrees of freedom ν and non-centrality parameter a. For most prior distributions $\pi(\delta)$, the integral in the Bayes factor must be computed by numerical integration.

In this article, we introduce a new Bayesian version of the t-test that follows the same setup as Gronau et al. [7], but diverges from their work by instead applying the minimum message length principle of inductive inference to the t-test problem. Minimum message length (MML) is a Bayesian principle rooted in information theory that allows for model selection and parameter estimation within the same framework. The aim of this manuscript is thus two-fold: (1) to introduce the MML principle of inductive inference and, as an example MML application, (2) to demonstrate how to perform MML analysis of the two-sample t-test problem.

2 Minimum Message Length

The minimum message length (MML) principle [14,15,17,19] is a framework for inductive inference rooted in information theory. The main idea behind the minimum message length (MML) principle is that parameter estimation and model selection are instances of data compression. It is well known that random data sequences are inherently incompressible. By contrast, compressing a dataset successfully reveals its underlying structure. MML provides a method for computing the minimum length of a message that describes some data $\mathbf{y} \in \mathbb{R}^n$, with the requirement that the message contains the data as well as a model describing the data. Formally, the two parts of the message are:

1. the *assertion*: code for the model describing the model structure and model parameters $\boldsymbol{\theta} \in \boldsymbol{\Theta} \in \mathbb{R}^p$. We use $I(\boldsymbol{\theta})$ to denote the codelength of the assertion; and
2. the *detail*: code for the data $\mathbf{y}$ using the model $p(\mathbf{y}|\boldsymbol{\theta})$ from the assertion. We use $I(D|\boldsymbol{\theta})$ to denote the codelength of the detail.

Measured in information units (e.g., bits), the total MML message length $I(D, \boldsymbol{\theta})$

$$I(D, \boldsymbol{\theta}) = \underbrace{I(\boldsymbol{\theta})}_{\text{assertion}} + \underbrace{I(D|\boldsymbol{\theta})}_{\text{detail}} \tag{7}$$

is the sum of the codelengths for the assertion and detail. The assertion length reflects the complexity of the model: longer assertions allow for precise specification of model parameters and model structure. Conversely, shorter assertions imply simpler models with less precise parameter encoding. The detail length measures how well the asserted model fits the observed data. A complex model that is characterised by a long assertion, offers greater explanatory capacity and is able to encode more data strings than models of lesser complexity. The model

$$\hat{\boldsymbol{\theta}}(D) = \underset{\boldsymbol{\theta} \in \boldsymbol{\Theta}}{\arg\min} \{I(D, \boldsymbol{\theta})\} \tag{8}$$

that minimises the length of this two-part message message is deemed an optimal description of the data. Minimizing the combined message length requires striking a balance between the model's complexity (assertion) and its ability to fit the data (detail).

2.1 A Practical Approximation to the Strict MML Codelength

The strict MML codelength (SMML) approximation [14,16] to (7) is the gold standard codelength approximation in the MML literature. However, practically computing the SMML codelength is NP-Hard in general and only possible for simple models [3,4]. As such, SMML remains predominantly of theoretical interest only. Various approximation schemes have been developed for the SMML codelength and among these, the MML87 approximation [14,19] is the most widely adopted. Under some regularity conditions [14] (pp. 226), the MML87 codelength for data D is

$$\mathcal{I}_{87}(D,\boldsymbol{\theta}) = \underbrace{-\log \pi_{\boldsymbol{\theta}}(\boldsymbol{\theta}) + \frac{1}{2}\log|J_{\boldsymbol{\theta}}(\boldsymbol{\theta})| + \frac{p}{2}\log \kappa_p + }_{\text{assertion}}\underbrace{\frac{p}{2} - \log p(D|\boldsymbol{\theta})}_{\text{detail}} \tag{9}$$

where $\pi_{\boldsymbol{\theta}}(\boldsymbol{\theta})$ is the prior distribution for the parameters $\boldsymbol{\theta}$, $|J_{\boldsymbol{\theta}}(\boldsymbol{\theta})|$ is the determinant of the expected Fisher information matrix, $p(D|\boldsymbol{\theta})$ is the likelihood function of the model and κ_p is a lattice quantization constant [1,2]; for small p we have

$$\kappa_1 = \frac{1}{12}, \quad \kappa_2 = \frac{5}{36\sqrt{3}}, \quad \kappa_3 = \frac{19}{192 \times 2^{1/3}}.$$

The best known lattice quantizers for dimensions $p \leq 48$ are available in [1]. For large p, κ_p is well-approximated by [14]:

$$\frac{p}{2}(\log \kappa_p + 1) \approx -\frac{p}{2}\log 2\pi + \frac{1}{2}\log p\pi - \gamma,$$

where $\gamma \approx 0.5772$ is the Euler–Mascheroni constant.

For a broad class of sufficiently regular models, the MML87 codelength coincides with the exact MML codelength to within negligible error, yet is considerably simpler to evaluate, as it depends only on the prior distribution of the model parameters and the determinant of the expected Fisher information matrix. Unlike the strict MML codelength which is computed as an expectation over all possible data with respect to the data marginal distribution, the MML87 codelength is only computed for the observed data. For large sample sizes $n \to \infty$, it is easy to show that the MML87 codelength is identical to the well-known Bayesian information criterion (BIC) [13]

$$\mathcal{I}_{87}(D,\boldsymbol{\theta}) = -\log p(D|\boldsymbol{\theta}) + \frac{p}{2} + O(1), \tag{10}$$

where the $O(1)$ term depends on the prior distribution, the Fisher information and the number of parameters p. Estimates minimising the MML87 codelength are invariant under one-to-one reparameterizations, mirroring the invariance of the maximum likelihood estimator. MML87 has been applied to a wide range of statistical models including decision trees [21], factor analysis [20] and mixture models [18], among others. An introduction and short tutorial on MML can be found in [24].

3 Minimum Message Length *t*-Test

In order to perform the two-sample t test under the MML framework, we require codelengths of the data under the null distribution ($\delta = 0$) and under the alternative hypothesis ($\delta > 0$). Under the null model, the negative log-likelihood of the data $\mathbf{y}$ is

$$\begin{aligned} -\log \ell_0(\boldsymbol{\theta}_0) &= \frac{n_1}{2}\log(2\pi\sigma^2) + \frac{1}{2\sigma^2}\sum_{i=1}^{n_1}(y_{1i}-\mu)^2 + \frac{n_2}{2}\log(2\pi\sigma^2) + \frac{1}{2\sigma^2}\sum_{j=1}^{n_2}(y_{2j}-\mu)^2 \\ &= \frac{n}{2}\log(2\pi\sigma^2) + \frac{1}{2\sigma^2}\sum_{i=1}^{n}(y_i-\mu)^2 \end{aligned}$$

where $\mathbf{y} = (y_1, \ldots, y_n)$ denotes the two-samples stacked into a single data vector of length $n = (n_1 + n_2)$. Let $\pi_0(\boldsymbol{\theta}_0)$ denote the prior distributions for the parameters under the null hypothesis. We use the usual right Haar prior for the mean and standard deviation:

$$\pi_0(\boldsymbol{\theta}_0) = (\Omega\sigma)^{-1} \tag{11}$$

defined over some suitable parameter range Ω implying a location and scale invariant distribution of the mean and standard deviation, respectively. The determinant of the expected Fisher information matrix for the normal distribution is $|J(\boldsymbol{\theta}_0)| = 2n^2/\sigma^4$ leading to the MML87 codelength:

$$\begin{aligned} \mathcal{I}_0(\mathbf{y}, \boldsymbol{\theta}_0) &= -\log \ell_0(\boldsymbol{\theta}_0) + \frac{1}{2}\log|J(\boldsymbol{\theta}_0)| - \log \pi_0(\boldsymbol{\theta}_0) + \log \kappa_2 + 1 \\ &= \frac{1}{2\sigma^2}\sum_{i=1}^{n}(y_i-\mu)^2 + (n-1)\log(\sigma) + \frac{1}{2}\log\left(2^{n+1}\pi^n\left(ne\kappa_2\Omega\right)^2\right) \end{aligned}$$

where $\kappa_2 = 5/36/\sqrt{3} \approx 0.0802$. MML estimates that minimize the codelength $\mathcal{I}_0(\mathbf{y}, \boldsymbol{\theta}_0)$ correspond to the usual sample mean and the unbiased sample variance estimates

$$\hat{\mu}(\mathbf{y}) = \frac{1}{n}\sum_{i=1}^{n} y_i = \bar{\mathbf{y}}, \quad \hat{\sigma}^2(\mathbf{y}) = \frac{1}{n-1}\sum_{i=1}^{n}(y_i - \bar{\mathbf{y}})^2. \tag{12}$$

The minimum MML87 codelength for the data under the null hypothesis is

$$\mathcal{I}_0(\mathbf{y}) = \frac{n-1}{2}\left(1 + \log\hat{\sigma}^2(\mathbf{y})\right) + \frac{1}{2}\log\left(2^{n+1}\pi^n\left(ne\kappa_2\Omega\right)^2\right). \tag{13}$$

We observe that the MML principle allows for automatic parameter estimation and model selection, as long as we can compute codelengths of the candidate models. Next, we derive the MML87 codelength for the alternative hypothesis where the two means are different at population level ($\delta > 0$).

Recall that $\boldsymbol{\theta}_1 = \{\mu, \sigma, \delta\} \in (\mathbb{R} \times \mathbb{R}_+^2) \equiv \Theta_1$ denotes the parameter set of the alternative hypothesis, which compared to the null model also includes the effect size δ. The negative log-likelihood of the data $\mathbf{y}$ under the alternative hypothesis, $-\log \ell_1(\mathbf{y}|\boldsymbol{\theta}_1)$, is

$$\frac{n}{2}\log(2\pi\sigma^2) + \frac{1}{2\sigma^2}\sum_{i=1}^{n_1}\left(y_{1i} - \mu - \frac{\sigma\delta}{2}\right)^2 + \frac{1}{2\sigma^2}\sum_{j=1}^{n_2}\left(y_{2j} - \mu + \frac{\sigma\delta}{2}\right)^2 \tag{14}$$

where, as before, $n = (n_1 + n_2)$ is the total sample size. Let

$$S_j = \sum_{i=1}^{n_j} y_{ji}, \quad S^2 = \sum_{j=1}^{2}\sum_{i=1}^{n_j} y_{ji}^2 = \mathbf{y}'\mathbf{y}, \quad j = 1, 2 \tag{15}$$

denote the sufficient statistics. Maximum likelihood estimates of the model parameters are

$$\begin{aligned}
\hat{\mu}_{\mathrm{ML}}(\mathbf{y}) &= \frac{1}{2}\left(\frac{S_1}{n_1} + \frac{S_2}{n_2}\right) = \frac{1}{2}\left(\bar{\mathbf{y}}_1 + \bar{\mathbf{y}}_2\right) \\
\hat{\sigma}^2_{\mathrm{ML}}(\mathbf{y}) &= \frac{1}{n}\left(S^2 - \left(\frac{S_1^2}{n_1} + \frac{S_2^2}{n_2}\right)\right) = \frac{1}{n}\left(S^2 - (n_1\bar{\mathbf{y}}_1^2 + n_2\bar{\mathbf{y}}_2^2)\right) \\
\hat{\delta}_{\mathrm{ML}}(\mathbf{y}) &= \frac{1}{\hat{\sigma}_{\mathrm{ML}}}\left(\frac{S_1}{n_1} - \frac{S_2}{n_2}\right) = \frac{1}{\hat{\sigma}_{\mathrm{ML}}}\left(\bar{\mathbf{y}}_1 - \bar{\mathbf{y}}_2\right).
\end{aligned} \tag{16}$$

Next we derive the MML87 codelength for the model and the corresponding MML87 estimates. For the MML codelength of the data under the alternative hypothesis, we again require a prior distribution for all the model parameters and the determinant of the expected Fisher information matrix. Following [7], the prior distribution for the parameters $\boldsymbol{\theta}_1$ is chosen to be

$$\pi_1(\boldsymbol{\theta}_1) = \pi_0(\boldsymbol{\theta}_0)\pi(\delta), \quad \pi(\delta) = \frac{2}{\pi(1 + \delta^2)}, \tag{17}$$

where $\pi(\delta)$ is the standard half-Cauchy distribution. The half-Cauchy distribution has a simple analytical form and heavy tails that prevent the prior from being overly informative about large effect sizes. This prior distribution is also recommended by Gelman [5] for scale parameters as a reasonable default when little prior information is available. If prior knowledge is available, the codelength may be trivially extended to include other prior distributions for the effect size, such as the t distribution with location and scale hyperparameters μ_δ and $\gamma_\delta > 0$, respectively. Note that the prior distribution for the mean and standard deviation parameters is the same under the null and alternative hypotheses, which implies that the range of parameters Ω will have no effect on MML inference as it equally contributes to codelengths of both hypotheses. The determinant of the expected Fisher information matrix under the alternative hypothesis is

$$|J(\boldsymbol{\theta}_1)| = \frac{2n_1 n_2(n_1 + n_2)}{\sigma^4}. \tag{18}$$

Substituting (14), (17) and (18) into (9) yields the MML87 codelength $\mathcal{I}(\mathbf{y}, \boldsymbol{\theta}_1)$ of the data under the alternative hypothesis model. Under this choice of the prior distribution, MML87 estimates of the parameters $\boldsymbol{\theta}_1$ are unavailable in closed form and must be obtained via numerical optimisation

$$\hat{\boldsymbol{\theta}}_1(\mathbf{y}) = \underset{\boldsymbol{\theta}_1 \in \Theta_1}{\arg\min}\, \mathcal{I}(\mathbf{y}, \boldsymbol{\theta}_1). \tag{19}$$

To perform a MML t-test, we compute the codelength of the null model $\mathcal{I}(\mathbf{y}, \hat{\boldsymbol{\theta}}_0)$ and the alternative model $\mathcal{I}(\mathbf{y}, \hat{\boldsymbol{\theta}}_1)$, with the preferred model being the one with the shorter codelength. As with standard Bayesian analysis, a codelength difference of about 3 nits or more indicates substantial preference of the model with the shorter codelength. Recall that we can also interpret the difference in MML codelengths as the log posterior probability in favour of the model with the shorter codelength, similar to the Bayes factor.

4 Discussion

A brief simulation was performed to compare the behaviour of the MML87 estimate of the standardised effect size δ to the maximum likelihood (ML) estimate (16). Data with sample sizes $(n_1, n_2) \in \{5, 10, 20, 50\}$ was generated from the model $(\mu = 0, \sigma = 1)$ with $0.1 \leq \delta \leq 5$ and both ML and MML87 were asked to nominate an estimate of δ. The experiment was repeated for 10^5 iterations and the estimates were compared using the normalized mean squared error metric

$$\text{NMSE} = \frac{(\delta - \hat{\delta}(\mathbf{y}))^2}{\delta^2}, \tag{20}$$

averaged over all 10^5 iterations for each experiment. Results for sample sizes $n_1 = n_2 = 5$ and $n_1 = n_2 = 10$ are shown in Fig. 1. Behaviour of the ML and MML87 estimates for moderate to large sample sizes was indistinguishable. When the sample size was small, ML tends to overestimate δ and underestimate σ. This is in contrast to MML87 which tends to underestimate δ and is therefore more conservative.

In terms of model selection, unlike the frequentist methodology, neither the Bayes factor nor MML are designed to explicitly control the type I error. Empirically, to control the average type I error rate at 5% when $0.1 \leq \delta \leq 2$ and $n = 15$ for the MML criterion, we select the alternative model whenever $(\mathcal{I}_1(y) + \log(1.62)) < \mathcal{I}_0(y)$. Figure 2 shows the statistical power of the MML test versus the standard frequentist t-test when $n = 15$ and when we control the average type I error rate for both approaches at 5%. We observe a slightly higher power for the MML t-test for all effect sizes tested. As expected, for larger sample sizes $(n_1, n_2 \geq 30)$, both approaches are virtually identical in terms of power.

An advantage of MML over some alternative approaches is that it allows comparison of models with different structures and parameters as long as we can compute their codelengths. The codelength is measured in the unit of information (e.g., bits) and is therefore a universal yardstick for model comparison. It is easy to extend the MML t-test to, for example, handle the case of unequal variances in the two groups, a version of the so-called Behrens–Fisher problem. MML87 codelengths for a similar model were derived in [11] and can be compared to the codelengths of the null and alternative hypotheses under the equal variance assumption derived in Sect. 3 to discriminate between the candidate models. We could instead consider the more general setup with four candidate models:

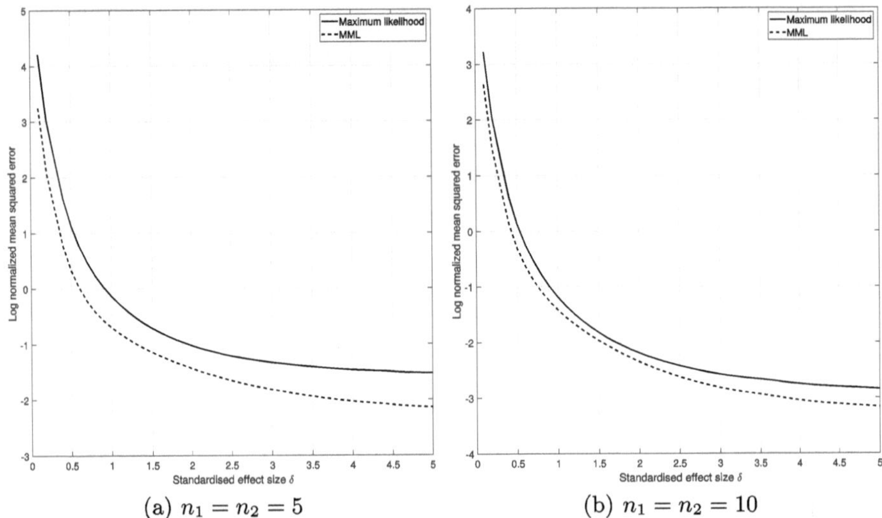

(a) $n_1 = n_2 = 5$ (b) $n_1 = n_2 = 10$

Fig. 1. Normalized mean squared error for the frequentist, Bayesian and MML t-test computed for $0.1 \leq \delta \leq 2$ with $n_1 = n_2 = 5$ and $n_1 = n_2 = 10$ data samples averaged over 10^5 simulations. In all cases, the mean and standard deviation of the data generating model were $\mu = 0, \sigma = 1$.

$$\begin{aligned}
\mathcal{M}_1 :\quad & Y_{1i} \sim \mathcal{N}\left(\mu, \sigma^2\right), && Y_{2j} \sim \mathcal{N}\left(\mu, \sigma^2\right) \\
\mathcal{M}_2 :\quad & Y_{1i} \sim \mathcal{N}\left(\mu, \sigma_1^2\right), && Y_{2j} \sim \mathcal{N}\left(\mu, \sigma_2^2\right) \\
\mathcal{M}_3 :\quad & Y_{1i} \sim \mathcal{N}\left(\mu + \frac{\sigma\delta}{2}, \sigma^2\right), && Y_{2j} \sim \mathcal{N}\left(\mu - \frac{\sigma\delta}{2}, \sigma^2\right) \\
\mathcal{M}_4 :\quad & Y_{1i} \sim \mathcal{N}\left(\mu + \frac{\sqrt{\sigma_1\sigma_2}\,\delta}{2}, \sigma_1^2\right), && Y_{2j} \sim \mathcal{N}\left(\mu - \frac{\sqrt{\sigma_1\sigma_2}\,\delta}{2}, \sigma_2^2\right)
\end{aligned}$$

Describing samples from two groups of size $i = 1, \ldots, n_1$ and $j = 1, \ldots, n_2$, respectively. Given data $(\mathbf{y}_1, \mathbf{y}_2)$, the task is to determine which of the four models generated the data. Within the MML framework, this may be accomplished by computing the codelength of the data under each of the four models and selecting the model with the shortest codelength as optimal.

Model $\mathcal{M}_1$ represents two data sets from one normal distribution, the null hypothesis examined in this paper. Model $\mathcal{M}_2$ says that the two groups have a common mean but different variances. Model $\mathcal{M}_3$ describes two groups with different means and a common variance, the alternative hypothesis examined in this paper, while model $\mathcal{M}_4$ describes groups with different means and different variances. Model $\mathcal{M}_4$ is another way of expressing the Behrens–Fisher problem inspired by [7]. Here, the effect size δ is scaled by the geometric mean of the

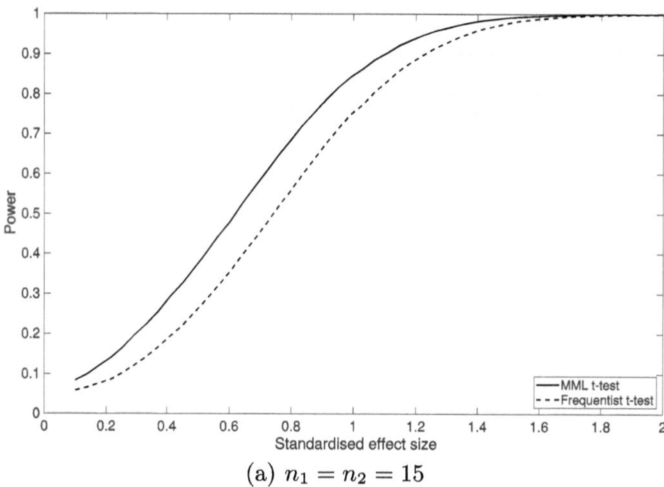

(a) $n_1 = n_2 = 15$

Fig. 2. Statistical power for the frequentist and MML t-test computed for $0.1 \leq \delta \leq 2$ with $n_1 = n_2 = 15$ data samples averaged over 10^5 simulations. In all cases, the mean and standard deviation of the data generating model were $\mu = 0, \sigma = 1$.

standard deviations of each group, which is a symmetric normalisation of heteroscedasticity that makes δ scale-invariant. In the special case that the standard deviations of the two groups are identical, $\mathcal{M}_4$ reduces to $\mathcal{M}_3$, as expected.

The MML87 codelengths for models $\mathcal{M}_1$ and $\mathcal{M}_3$ are derived in Sect. 3. The negative log likelihood under Model $\mathcal{M}_2$, $-\log \ell_2(\boldsymbol{\theta}_2)$, is

$$\frac{n_1}{2}\log(2\pi\sigma_1^2) + \frac{1}{2\sigma_1^2}\sum_{i=1}^{n_1}(y_{1i}-\mu)^2 + \frac{n_2}{2}\log(2\pi\sigma_2^2) + \frac{1}{2\sigma_2^2}\sum_{j=1}^{n_2}(y_{2j}-\mu)^2 .$$

Similarly for $\mathcal{M}_4$, $-\log \ell_4(\mathbf{y}|\boldsymbol{\theta}_4)$, we have

$$\frac{n_1}{2}\log(2\pi\sigma_1^2) + \frac{1}{2\sigma_1^2}\sum_{i=1}^{n_1}\left(y_{1i}-\mu-\frac{(\sigma_1\sigma_2)^{\frac{1}{2}}}{2}\delta\right)^2 + \frac{n_2}{2}\log(2\pi\sigma_2^2)$$
$$+\frac{1}{2\sigma_2^2}\sum_{j=1}^{n_2}\left(y_{2j}-\mu+\frac{(\sigma_1\sigma_2)^{\frac{1}{2}}}{2}\delta\right)^2 .$$

In the absence of suitable prior information, we adopt half-Cauchy priors for all scale parameters $(\sigma, \sigma_1, \sigma_2, \delta)$ and a uniform, location invariant prior for the grand mean μ, similar to Sect. 3. Further, it is not difficult to show that the determinant of the expected Fisher information matrix for each model is

$$|J(\boldsymbol{\theta}_1)| = \frac{2n^2}{\sigma^4}, \qquad |J(\boldsymbol{\theta}_2)| = \frac{4n_1n_2\left(n_2\sigma_1^2 + n_1\sigma_2^2\right)}{\sigma_1^4\sigma_2^4},$$

$$|J(\boldsymbol{\theta}_3)| = \frac{2n_1n_2(n_1+n_2)}{\sigma^4}, \qquad |J(\boldsymbol{\theta}_4)| = \frac{4n_1^2n_2^2}{\sigma_1^3\sigma_2^3}.$$

A brief simulation was carried out to compare the minimum message length criterion against the popular Bayesian information criterion (BIC) on how well they can discriminate between the four competing models. For each simulation run, data was generated from each model with the following population parameters $\mu = 0$, $\sigma = \sigma_1 = 1$, $\sigma_2 = \{0.5, 2, 4\}$ and $\delta = \{0.5, 1, 2\}$. In each experiment, sample sizes were $n_1 = n_2 = 15$. BIC and MML87 were then asked to select the model that best describes the generated data. Model classification accuracy and Kullback–Leibler (KL) divergence [10] from the data generating model were used to measure the performance of the two criteria. For completeness, the KL divergence between two univariate Gaussian distributions $p \sim N(\mu_0, \sigma_0^2)$ and $q \sim N(\mu_1, \sigma_1^2)$ is

$$\mathrm{KL}(p \,\|\, q) = \frac{1}{2}\left[\frac{\sigma_0^2}{\sigma_1^2} + \frac{(\mu_1-\mu_0)^2}{\sigma_1^2} - 1 + \ln\left(\frac{\sigma_1^2}{\sigma_0^2}\right)\right].$$

Table 1 shows the averaged test results over 10^5 simulation runs. Models selected by MML87 have smaller KL divergence compared to the BIC selected models for each parameter set tested. This is unsurprising given that BIC may be viewed as an asymptotic approximation to the MML87 codelength (see (10)). As expected, the two criteria were virtually indistinguishable in experiments with larger sample sizes $(n_1, n_2 > 100)$. Lastly, we note that as long as we can compute MML codelengths, this experiment could be extended to include more than two groups or to compare more complex models, such as a Gaussian linear regression with multiple explanatory variables or decision trees with Gaussian distributions in the leaves. Matlab code implementing the MML t-test and the experiments in this paper is available at https://github.com/EnesMakalic/MML-t-test.git.

Table 1. Model selection accuracy and Kullback–Leibler divergence for the Bayesian information criterion (BIC) and MML87 with sample sizes $n_1 = n_2 = 15$, averaged over 10^5 simulations. The data generating model is described in the first three columns with $\mu = 0, \sigma = \sigma_1 = 1$ in all experiments.

Model	δ	σ_2	Accuracy (%)		KL Divergence	
			BIC	MML	BIC	MML
	0.5	–	85.6 ± 35.1	83.9 ± 36.8	0.063 ± 0.092	**0.053 ± 0.073**
		–	85.5 ± 35.2	83.8 ± 36.9	0.063 ± 0.092	**0.053 ± 0.074**
		–	85.8 ± 34.9	84.0 ± 36.7	0.063 ± 0.093	**0.053 ± 0.075**
$\mathcal{M}_1$	1.0	–	85.7 ± 35.0	84.1 ± 36.6	0.063 ± 0.093	**0.053 ± 0.074**
		–	85.6 ± 35.1	83.8 ± 36.8	0.063 ± 0.093	**0.053 ± 0.074**
		–	85.6 ± 35.1	83.9 ± 36.7	0.063 ± 0.092	**0.053 ± 0.074**
	2.0	–	85.5 ± 35.2	83.9 ± 36.7	0.063 ± 0.093	**0.053 ± 0.074**
		–	85.9 ± 34.8	84.0 ± 36.7	0.062 ± 0.092	**0.052 ± 0.073**
		–	85.6 ± 35.1	84.1 ± 36.6	0.063 ± 0.094	**0.053 ± 0.075**
	0.5	0.5	61.6 ± 48.6	63.8 ± 48.1	0.112 ± 0.111	**0.092 ± 0.092**
		2.0	61.4 ± 48.7	64.3 ± 47.9	0.112 ± 0.113	**0.095 ± 0.097**
		4.0	58.0 ± 49.4	77.9 ± 41.5	0.102 ± 0.106	**0.073 ± 0.087**
$\mathcal{M}_2$	1.0	0.5	61.8 ± 48.6	64.1 ± 48.0	0.111 ± 0.112	**0.092 ± 0.093**
		2.0	61.6 ± 48.6	64.6 ± 47.8	0.112 ± 0.111	**0.094 ± 0.095**
		4.0	58.3 ± 49.3	78.0 ± 41.4	0.101 ± 0.108	**0.073 ± 0.090**
	2.0	0.5	61.9 ± 48.6	64.2 ± 47.9	0.112 ± 0.113	**0.092 ± 0.094**
		2.0	61.6 ± 48.6	64.2 ± 47.9	0.113 ± 0.112	**0.095 ± 0.096**
		4.0	57.9 ± 49.4	77.8 ± 41.6	0.102 ± 0.108	**0.074 ± 0.089**
	0.5	–	30.5 ± 46.1	49.7 ± 50.0	0.088 ± 0.092	**0.069 ± 0.076**
		–	30.4 ± 46.0	49.9 ± 50.0	0.087 ± 0.090	**0.069 ± 0.075**
		–	30.6 ± 46.1	49.9 ± 50.0	0.087 ± 0.090	**0.069 ± 0.075**
$\mathcal{M}_3$	1.0	–	75.0 ± 43.3	86.4 ± 34.2	0.092 ± 0.100	**0.070 ± 0.081**
		–	75.0 ± 43.3	86.4 ± 34.3	0.092 ± 0.103	**0.070 ± 0.083**
		–	75.1 ± 43.2	86.5 ± 34.2	0.091 ± 0.099	**0.070 ± 0.080**
	2.0	–	92.1 ± 27.0	93.8 ± 24.2	0.077 ± 0.095	**0.066 ± 0.076**
		–	92.0 ± 27.2	93.7 ± 24.2	0.077 ± 0.094	**0.066 ± 0.075**
		–	92.1 ± 26.9	93.8 ± 24.1	0.077 ± 0.094	**0.066 ± 0.075**
	0.5	0.5	35.3 ± 47.8	38.5 ± 48.7	0.127 ± 0.111	**0.106 ± 0.095**
		2.0	35.1 ± 47.7	39.9 ± 49.0	0.128 ± 0.112	**0.108 ± 0.099**
		4.0	58.7 ± 49.2	54.9 ± 49.8	0.103 ± 0.107	**0.082 ± 0.091**
$\mathcal{M}_4$	1.0	0.5	66.5 ± 47.2	66.0 ± 47.4	0.129 ± 0.119	**0.109 ± 0.100**
		2.0	66.6 ± 47.2	67.3 ± 46.9	0.128 ± 0.120	**0.112 ± 0.104**
		4.0	85.6 ± 35.1	85.3 ± 35.5	0.101 ± 0.105	**0.089 ± 0.092**
	2.0	0.5	77.5 ± 41.8	72.7 ± 44.6	0.117 ± 0.115	**0.107 ± 0.097**
		2.0	77.5 ± 41.7	73.7 ± 44.0	0.117 ± 0.115	**0.109 ± 0.099**
		4.0	99.7 ± 5.3	99.7 ± 5.8	0.096 ± 0.105	**0.090 ± 0.093**

References

1. Agrell, E., Allen, B.: On the best lattice quantizers. IEEE Trans. Inf. Theory **69**(12), 7650–7658 (2023). https://doi.org/10.1109/TIT.2023.3291313
2. Conway, J.H., Sloane, N.J.A.: Sphere Packing, Lattices and Groups, 3rd edn. Springer, Cham (1998)
3. Dowty, J.G.: SMML estimators for 1-dimensional continuous data. Comput. J. **58**(1), 126–133 (2013). https://doi.org/10.1093/comjnl/bxt145
4. Farr, G.E., Wallace, C.S.: The complexity of strict minimum message length inference. Comput. J. **45**(3), 285–292 (2002)
5. Gelman, A.: Prior distributions for variance parameters in hierarchical models. Bayesian Anal. **1**(3), 515–533 (2006)
6. Gönen, M., Johnson, W.O., Lu, Y., Westfall, P.H.: The Bayesian two-sample t test. Am. Stat. **59**(3), 252–257 (2005). https://doi.org/10.1198/000313005X55233
7. Gronau, Q.F., Ly, A., Wagenmakers, E.J.: Informed bayesian *t*-tests. Am. Stat. **74**(2), 137–143 (2019). https://doi.org/10.1080/00031305.2018.1562983
8. Kass, R.E., Raftery, A.E.: Bayes factors. J. Am. Stat. Assoc. **90**(430), 773–795 (1995)
9. Kelter, R.: Analysis of type I and II error rates of Bayesian and frequentist parametric and nonparametric two-sample hypothesis tests under preliminary assessment of normality. Comput. Stat. **36**(2), 1263–1288 (2020). https://doi.org/10.1007/s00180-020-01034-7
10. Kullback, S., Leibler, R.A.: On information and sufficiency. Ann. Math. Stat. **22**(1), 79–86 (1951)
11. Makalic, E., Schmidt, D.F.: Minimum message length analysis of the Behrens–Fisher problem. In: Algorithmic Probability and Friends, Bayesian Prediction and Artificial Intelligence, pp. 250–260. Springer, Heidelberg (2013). https://doi.org/10.1007/978-3-642-44958-1_19
12. Rouder, J.N., Speckman, P.L., Sun, D., Morey, R.D., Iverson, G.: Bayesian t tests for accepting and rejecting the null hypothesis. Psychon. Bull. Rev. **16**(2), 225–237 (2009). https://doi.org/10.3758/PBR.16.2.225
13. Schwarz, G.: Estimating the dimension of a model. Ann. Stat. **6**(2), 461–464 (1978)
14. Wallace, C.S.: Statistical and Inductive Inference by Minimum Message Length. Information Science and Statistics, 1st edn. Springer, Heidelberg (2005)
15. Wallace, C.S., Boulton, D.M.: An information measure for classification. Comput. J. **11**(2), 185–194 (1968). http://www.allisons.org/ll/MML/Structured/1968-WB-CJ/
16. Wallace, C.S., Boulton, D.M.: An invariant Bayes method for point estimation. Classificat. Soc. Bull. **3**(3), 11–34 (1975)
17. Wallace, C.S., Dowe, D.L.: Refinements of MDL and MML coding. Comput. J. **42**(4), 330–337 (1999)
18. Wallace, C.S., Dowe, D.L.: MML clustering of multi-state, Poisson, von Mises circular and Gaussian distributions. Stat. Comput. **10**(1), 73–83 (2000)
19. Wallace, C.S., Freeman, P.R.: Estimation and inference by compact coding. J. R. Stat. Soc. (Ser. B) **49**(3), 240–252 (1987)
20. Wallace, C.S., Freeman, P.R.: Single-factor analysis by minimum message length estimation. J. R. Stat. Soc. (Ser. B) **54**(1), 195–209 (1992)
21. Wallace, C.S., Patrick, J.D.: Coding decision trees. Mach. Learn. **11**(1), 7–22 (1993)
22. Wang, M., Liu, G.: A simple two-sample Bayesian t-test for hypothesis testing. Am. Stat. **70**(2), 195–201 (2016). https://doi.org/10.1080/00031305.2015.1093027

23. Wetzels, R., Matzke, D., Lee, M.D., Rouder, J.N., Iverson, G.J., Wagenmakers, E.J.: Statistical evidence in experimental psychology. Perspect. Psychol. Sci. **6**(3), 291–298 (2011). https://doi.org/10.1177/1745691611406923
24. Wong, C.K., Makalic, E., Schmidt, D.F.: Minimum message length inference of the poisson and geometric models using heavy-tailed prior distributions. J. Math. Psychol. **83**, 1–11 (2018). https://doi.org/10.1016/j.jmp.2018.01.001

Towards Scalable Backpropagation-Free Gradient Estimation

Daniel Wang(✉), Evan Markou, and Dylan Campbell

Australian National University, Canberra, Australia
{u7918232,Evan.Markou,Dylan.Campbell}@anu.edu.au

Abstract. While backpropagation—reverse-mode automatic differentiation—has been extraordinarily successful in deep learning, it requires two passes (forward and backward) through the neural network and the storage of intermediate activations. Existing gradient estimation methods that instead use forward-mode automatic differentiation struggle to scale beyond small networks due to the high variance of the estimates. Efforts to mitigate this have so far introduced significant bias to the estimates, reducing their utility. We introduce a gradient estimation approach that reduces both bias and variance by manipulating upstream Jacobian matrices when computing guess directions. It shows promising results and has the potential to scale to larger networks, indeed performing better as the network width is increased. Our understanding of this method is facilitated by analyses of bias and variance, and their connection to the low-dimensional structure of neural network gradients.

Keywords: Optimisation · Low-rank gradient subspace · Forward-mode differentiation · Automatic differentiation · Deep learning

1 Introduction

While backpropagation [1] is ubiquitous in deep learning, there has been recent interest in alternative memory-efficient methods of optimising a neural network without using backpropagation [2–5]. One such class of methods is based on estimating the gradient using forward-mode automatic differentiation (AD). Forward-mode AD efficiently evaluates the Jacobian–vector product in a single forward pass and avoids the memory cost of storing intermediate activations required by backpropagation. Additional motivation for backpropagation alternatives arises from the fact that backpropagation is biologically implausible [6].

Using forward-mode AD, an unbiased estimator of the gradient can be constructed by randomly sampling a guess direction and scaling it by the directional derivative in the guess direction. This method is defined as the forward gradient [7]. However, because this estimator requires sampling in a high-dimensional guessing space equal to the number of parameters, high variance prohibits this method from scaling beyond training small networks on toy datasets. For a network with N parameters, the cosine similarity between the estimated and true

M. Liu et al. (Eds.): AI 2025, LNAI 16371, pp. 30–42, 2026.
https://doi.org/10.1007/978-981-95-4972-6_3

gradient is $O(\frac{1}{\sqrt{N}})$, which means that the guess direction becomes orthogonal to the true gradient as network size increases. A number of works based on the forward gradient, including state-of-the-art gradient guessing method "$\tilde{W}^\top$" [2], have focused on decreasing its variance by reducing the guessing space [2,7,8]. Neural network gradients have also been shown to exhibit a low-dimensional structure [9,10], which if properly exploited has the potential to significantly improve gradient guessing methods.

In this paper, we introduce a novel method $\tilde{W}^\perp$ that extends the $\tilde{W}^\top$ approach, reducing both its bias and its variance by considering the low-rank structure of the guessing space. Our experiments show that the $\tilde{W}^\perp$ method performs considerably better than $\tilde{W}^\top$ for larger networks. We also observe a possible connection to the low-dimensional structure of the gradient in that it lies in a low-dimensional subspace of the image of the upstream Jacobian matrix, providing additional insight into our gradient guessing methods.

2 Related Work and Technical Background

Forward-Mode Automatic Differentiation (AD). Given a function $f : \mathbb{R}^n \to \mathbb{R}^m$, forward-mode AD computes the Jacobian–vector product Jv where $J \in \mathbb{R}^{m\times n}$ is the Jacobian matrix and $v \in \mathbb{R}^n$ is an arbitrary vector. In the context of optimising a neural network where the loss function $L : \mathbb{R}^n \to \mathbb{R}$ typically has a scalar output, the Jacobian–vector product is simply the directional derivative in the direction v. In practice, automatic differentiation frameworks such as PyTorch and JAX provide functions to compute this efficiently.

Forward Gradient (Weight Perturbation [3]). For the remainder of this paper we will consider an l-layer MLP with weights W and biases b with layer widths of $d_1, d_2, \ldots, d_l$, ReLU activations, and loss function L. The ith linear layer can be written as

$$s_i = W_i x_i + b_i, \quad s_i \in \mathbb{R}^{d_{i+1}}, \tag{1}$$

where $W_i \in \mathbb{R}^{d_{i+1}\times d_i}$, $b_i \in \mathbb{R}^{d_{i+1}}$, and

$$x_{i+1} = \mathrm{ReLU}(s_i), \quad x_{i+1} \in \mathbb{R}^{d_{i+1}}. \tag{2}$$

We will only consider optimising the weights W_i, but these methods can be easily extended to the biases too. At layer i, the vanilla forward gradient method generates a guess of $\partial L/\partial W_i$ as follows. First, a guess direction y_i of the same dimension as W_i is sampled from the standard normal distribution $\mathcal{N}(\mathbf{0}, \mathbf{I})$. Then, the Jacobian–vector product $y^\top \frac{\partial L}{\partial W}$—the directional derivative of $\partial L/\partial W$ in the direction y, where W denotes the concatenated W_is and y denotes the concatenated y_is—is computed in the forward pass. The guess for $\partial L/\partial W_i$ is defined as $\frac{\hat{\partial L}}{\partial W_i} = (y_i^\top \frac{\partial L}{\partial W_i}) y_i$: the guess direction scaled by the directional derivative in that direction. With this approach, weight updates are performed by replacing the true gradient with the guess $\hat{\partial L/\partial W_i}$. In general, sampling the guess direction from a zero-mean unit-variance distribution with independent components yields an unbiased but high variance estimator of the true gradient.

Activation Perturbation [7]. This method guesses the gradient with respect to the activations instead of the weights. In deep neural networks, the weight matrix $W_i \in \mathbb{R}^{d_{i+1} \times d_i}$ contains many more entries than the activation vector $s_i \in \mathbb{R}^{d_{i+1}}$. Therefore, guessing the gradient with respect to the activations $\partial L/\partial s_i$ reduces the dimension of guessing space and hence decreases variance. Similar to before, the guess $\frac{\hat{\partial L}}{\partial s_i} = (y^\top \frac{\partial L}{\partial s})y_i$, where y denotes the concatenated guess directions y_i and s denotes the concatenated activations s_i at each layer, is computed by sampling a random vector $y_i \in \mathbb{R}^{d_{i+1}}$ and scaling it by the Jacobian–vector product $y^\top \frac{\partial L}{\partial s}$. Given the guess $\frac{\hat{\partial L}}{\partial s_i}$, the gradient guess with respect to the weight matrix can be obtained via the chain rule,

$$\frac{\hat{\partial L}}{\partial W_i} = \frac{\hat{\partial L}}{\partial s_i}\frac{\partial s_i}{\partial W_i} = \frac{\hat{\partial L}}{\partial s_i} x_i^\top \,. \tag{3}$$

As well as reducing variance, Ren et al. [7] show that activation perturbation is also unbiased. Consequently, it performs better than weight perturbation.

$\tilde{W}^\top$ **method** [2]. The $\tilde{W}^\top$ method extends the activation perturbation approach by incorporating terms along the chain rule gradient decomposition up to the next layer's activations into the guess for $\partial L/\partial s_i$. For a network with linear layers and ReLU activations, the chain rule gives

$$\frac{\hat{\partial L}}{\partial W_i} = \frac{\hat{\partial L}}{\partial s_{i+1}}\frac{\partial s_{i+1}}{\partial x_{i+1}}\frac{\partial x_{i+1}}{\partial s_i}\frac{\partial s_i}{\partial W_i} = \epsilon_{i+1}^\top W_{i+1} M_i x_i^\top \,, \tag{4}$$

where ϵ_{i+1} is a sample from a standard normal distribution that approximates the unknown $\partial L/\partial s_{i+1}$, W_{i+1} is the weight matrix of the next layer, M_i is the diagonal matrix corresponding to the ReLU derivative mask, and x_i is the input to layer i. Using the first three terms in Eq. (4), a guess direction $y_i^\top = \epsilon_{i+1}^\top W_{i+1} M_i$ for $\partial L/\partial s_i$ is generated by sampling ϵ_{i+1} and multiplying with the next layer weight matrix $W_{i+1}^\top$ and the ReLU mask M_i. As before, the Jacobian–vector product is $\frac{\partial L}{\partial s}^\top y$, and for each layer the guess for $\frac{\partial L}{\partial s_i}$ is $\frac{\hat{\partial L}}{\partial s_i} = (y^\top \frac{\partial L}{\partial s})y_i$. The weight update for W_i is obtained from the last term of Eq. (4) by taking the outer product $\frac{\hat{\partial L}}{\partial s_i} x_i^\top$.

The intuition behind this approach is that it leverages information from the network architecture and narrows the guessing space with each additional term in the chain rule. The next layer weight matrix is typically low-rank and the ReLU mask is sparse, leading to a large variance reduction. Although we focus on an MLP with ReLU activations, in a more general setting the guess direction can be simply written as $y_i = \tilde{W}_{i+1}^\top \epsilon_{i+1}$, where $\tilde{W}_{i+1}$ represents the composition of each term in the chain rule from the current layer's activations s_i up to the next layer's activations s_{i+1}. For a ReLU MLP, $\tilde{W}_{i+1} = W_{i+1} M_i$.

However, because $y_i = \tilde{W}_{i+1}^\top \epsilon_{i+1}$ is no longer a zero-mean unit-variance random variable with independent components, this estimator is no longer unbiased.

Singhal et al. [2] show that the bias at layer i is given by

$$\mathbb{E}\left[\frac{\partial \hat{L}}{\partial s_i}\right] - \frac{\partial L}{\partial s_i} = (\mathrm{Cov}(y_i) - I)\frac{\partial L}{\partial s_i}, \tag{5}$$

where $\partial L/\partial s_i$ is the true gradient and y_i is the guess direction with

$$\mathrm{Cov}(y_i) = \tilde{W}_{i+1}^{\top}\tilde{W}_{i+1}. \tag{6}$$

Despite this incurred bias, the $\tilde{W}^{\top}$ method significantly outperforms activation perturbation due to a large decrease in variance.

Gradients Lies in a Low-Dimensional Subspace. The gradient guessing methods described above are based on the principle of reducing variance by reducing the dimensionality of the guessing space. Related to the idea of gradients within a low-dimensional subspace is the work of Gur-Ari et al. [9], which observes that for a deep neural network trained on a k-class classification task, the gradient predominantly lies in the low-dimensional subspace spanned by the eigenvectors corresponding to the top-k largest eigenvalues of the Hessian. Interestingly, Song et al. [10] found that performing SGD in the small top-k subspace fails to decrease the loss, but performing SGD in the orthogonal complement of this subspace recovers the original SGD performance, leading to the hypothesis that this phenomenon is caused by noise in stochastic gradient descent as opposed to full-batch gradient descent. In the context of forward gradient guessing methods, this phenomenon reveals the underlying structure of gradients which can be exploited by, as we show in the next section, guessing within a small subspace that aligns with the true gradient.

3 A Low-Bias, Low-Variance Guess Direction

We propose the $\tilde{W}^{\perp}$ method, based on the $\tilde{W}^{\top}$ method, that reduces both the bias and variance of the guess direction by orthogonalising the upstream Jacobian matrix. A variant, $\tilde{W}^{\perp}$-NS, accelerates the orthogonalisation using Newton–Schulz iterations. Finally, to better understand the effect of bias and variance on the performance of gradient guessing methods, we also introduce a preconditioning method for comparison that is unbiased but has higher variance.

3.1 $\tilde{W}^{\perp}$: Orthogonalising the Guess Projection Matrix

In this section, we consider layer i in isolation and drop the layer index i, so the general upstream Jacobian matrix up to the next layer is denoted by $\tilde{W}^{\top}$ and the guess direction is $y = \tilde{W}^{\top}\epsilon$. That is, $\tilde{W}^{\top}$ projects the Gaussian sample ϵ to the guess direction y. Recall that for a ReLU MLP, $\tilde{W}_{i+1} = W_{i+1}M_i$, where W_{i+1} is the weight matrix of the next layer and M_i is the ReLU derivative mask, a diagonal binary matrix. The central idea is to manipulate $\tilde{W}$ into an orthonormal, low-rank matrix $\tilde{W}'$ that further narrows the guessing space while bringing the covariance matrix $\mathrm{Cov}(y) = \tilde{W}'^{\mathsf{T}}\tilde{W}'$ closer to the identity, thus reducing both bias and variance. We expect $\mathrm{rank}(\tilde{W})$ to be relatively low, since for a ReLU network it is at most the number of non-zero activations. Let the reduced SVD of $\tilde{W}$ be

$$\tilde{W} = U_r \Lambda_r V_r^{\mathsf{T}}, \tag{7}$$

where $r = \mathrm{rank}(\tilde{W})$. Then, we manipulate $\tilde{W}$ into the orthonormal matrix

$$\tilde{W}' = U_r V_r^{\mathsf{T}}, \tag{8}$$

so the guess direction is $y = \tilde{W}'^{\mathsf{T}}\epsilon$ with covariance matrix $\mathrm{Cov}(y) = \tilde{W}'^{\mathsf{T}}\tilde{W}' = V_r U_r^{\mathsf{T}} U_r V_r^{\mathsf{T}} = I_r$, being the identity matrix in the first r diagonal elements and 0 everywhere else. To control the bias–variance trade-off, integer values of k from 1 to r can be chosen as a hyperparameter to select only the rows and columns in U_r and V_r^{T} corresponding to the k largest singular values so that $\tilde{W}'_k = U_k V_k^{\mathsf{T}}$. Smaller values of k reduce the dimensionality of the guessing space at the cost of pushing the covariance matrix I_k further from the identity and thus incurring higher bias.

3.2 $\tilde{W}^{\perp}$-NS: Fast Orthogonalisation via Newton–Schulz Iterations

The Newton–Schulz iteration method [11–13] can be used instead of SVD to significantly speed up our $\tilde{W}^{\perp}$ method. The idea is that, given an odd polynomial, applying it to a matrix $A = USV^{\mathsf{T}}$ is the same as applying it element-wise to the singular values S. That is,

$$p(A) = \sum_{k=0}^{K} a_k A(A^{\mathsf{T}}A)^k = U\left(\sum_{k=0}^{K} a_k S^{2k+1}\right)V^{\mathsf{T}} = Up(S)V^{\mathsf{T}}. \tag{9}$$

Applying this N times yields

$$p^N(A) = Up^N(S)V^{\mathsf{T}}, \tag{10}$$

where on the left hand side $p^N(A)$ is the matrix polynomial applied N times to A and on the right hand side it is the real-valued polynomial applied N times element-wise to the singular values. With carefully chosen coefficients, repeated composition of the polynomial $p^N(A)$ can be made to converge to various functions as $N \to \infty$. Jordan et al. [13] use this method to orthogonalise $p^N(A) \approx UV^{\mathsf{T}}$ by mapping all singular values to 1. They find a degree five polynomial with $N = 5$ to approximate the function $f(x) = 1$ on $[0, 1]$.

The Newton–Schulz iteration can be easily adapted to our $\tilde{W}^{\perp}$ method, labelled $\tilde{W}^{\perp}$-NS. We wish to obtain $p^N(\tilde{W}) \approx U_k V_k^{\mathsf{T}}$ by mapping the singular values larger than σ_k to 1 and the ones smaller than σ_k to 0, where $\sigma_k \in [0, 1]$ is the kth largest singular value. That is, we need to approximate the step function defined by $f_{\sigma_k}(x) = 0$ for $x \in [0, \sigma_k]$ and $f(x) = 1$ for $x \in (\sigma_k, 1]$, where σ_k is constrained to lie in $[0, 1]$ by normalising $\tilde{W}$ to have unit spectral norm. Since the step function is slightly more complex than $f(x) = 1$ on $[0, 1]$, we fit a degree seven polynomial with coefficients obtained via gradient descent.

3.3 $\tilde{W}^{\mathsf{P}}$: Preconditioning the Guess Projection Matrix

Both the $\tilde{W}^{\top}$ and $\tilde{W}^{\perp}$ methods reduce variance and increase accuracy over the unbiased activation perturbation approach, but at the cost of introducing bias. In this section, we develop an unbiased estimator, based on the $\tilde{W}^{\top}$ method, at the cost of increasing the variance. This allows us to investigate the bias–variance trade-off. Specifically, we pre-multiply $\tilde{W}^{\top}$ by $(\tilde{W}^{\top}\tilde{W})^{-1/2}$ so that $\tilde{W}'^{\top} \approx (\tilde{W}^{\top}\tilde{W})^{-1/2}\tilde{W}^{\top}$. Then $\tilde{W}'^{\top}\tilde{W}' = I$, and $y = \tilde{W}'^{\top}\epsilon$ has identity covariance. To be precise, let the SVD of $\tilde{W}$ be $\tilde{W} = USV^{\top}$. Since $\tilde{W}$ is not guaranteed to be full-rank, we introduce a small constant $\sigma = 10^{-5}$ and modify $\tilde{W}$ to be $\tilde{W}_{\sigma} = U(S^2 + \sigma I)^{1/2}V^{\top}$, so that

$$\tilde{W}_{\sigma}^{\top}\tilde{W}_{\sigma} = V(S^2 + \sigma I)V^{\top}\,, \tag{11}$$

can be diagonalised while keeping $\tilde{W}_{\sigma}$ close to $\tilde{W}$. Setting

$$\tilde{W}_{\sigma}'^{\top} = (\tilde{W}_{\sigma}^{\top}\tilde{W}_{\sigma})^{-1/2}\tilde{W}_{\sigma}^{\top}\,, \tag{12}$$

yields an unbiased estimator, because the covariance matrix is identity:

$$\begin{aligned}
\mathrm{Cov}(y) &= \tilde{W}_{\sigma}'^{\top}\tilde{W}_{\sigma}' && (13)\\
&= (\tilde{W}_{\sigma}^{\top}\tilde{W}_{\sigma})^{-1/2}\tilde{W}_{\sigma}^{\top}\tilde{W}_{\sigma}(\tilde{W}_{\sigma}^{\top}\tilde{W}_{\sigma})^{-1/2} && (14)\\
&= V(S^2 + \sigma I)^{-1/2}V^{\top}V(S^2 + \sigma I)V^{\top}V(S^2 + \sigma I)^{-1/2}V^{\top} && (15)\\
&= I\,. && (16)
\end{aligned}$$

This method has high variance because $(\tilde{W}_{\sigma}^{\top}\tilde{W}_{\sigma})^{-1/2}$ contains large singular values, due to σ being small. Moreover, unlike the $W^{\top}$ and $W^{\perp}$ methods, the dimension of the guessing space is not reduced, because $\tilde{W}_{\sigma}'$ is a full rank matrix.

4 Experiments

4.1 Experiment Setup

Implementation Details and Dataset. Following Singhal et al. [2], the experiments in this section involve optimising an MLP with three hidden layers of size 128 and ReLU activations. Our methods and existing methods from previous works are implemented in PyTorch, using `functorch.jvp` for forward mode automatic differentiation[1]. The experiments were run on an Apple M1 Pro Macbook with an 8-core CPU and 14-core GPU with 16GB of unified memory on the MNIST-1D dataset [14] for 300 epochs. Note that 300 epochs is not enough for convergence. The choice of optimiser, AdamW [15] with learning rate 10^{-4} and batch size 512, is taken from the best performing configuration in Singhal et al. [2] with no additional hyperparameter tuning.

[1] The code is available at https://github.com/danielwang0452/Forward-W-Perp.

Metrics: Bias. For a gradient guess direction $y_i^b = \tilde{W}_{i+1}^{b\top}\epsilon_i^b$ at layer i where b indexes over a batch of size B, the bias magnitude is given by

$$\text{Bias}\left(\frac{\partial \hat{L}}{\partial s_i}\right) = \left\|\frac{1}{B}\sum_{b=1}^{B}(\text{Cov}(y_i^b) - I)\frac{\partial L}{\partial s_i^b}\right\|_2. \tag{17}$$

Metrics: Variance. We compute the variance magnitude at layer i via the decomposition of MSE into bias and variance using the empirical mean squared error (MSE) and squared bias, given by

$$\widehat{\text{Var}}\left(\frac{\partial \hat{L}}{\partial s_i}\right) = \left\|\underbrace{\frac{1}{B}\sum_{b=1}^{B}\left(\frac{\partial L}{\partial s_i^b} - \frac{\partial \hat{L}}{\partial s_i^b}\right)^{\circ 2}}_{\text{MSE}} - \underbrace{\left(\frac{1}{B}\sum_{b=1}^{B}\left(\text{Cov}(y_i^b) - I\right)\frac{\partial L}{\partial s_i^b}\right)^{\circ 2}}_{\text{Squared Bias}}\right\|_2, \tag{18}$$

where $\circ 2$ is the element-wise square.

Metrics: Covariance Frobenius Norm. Another way to measure bias is to consider $\text{Cov}(y_i) - I$ independent of the true gradient $\frac{\partial L}{\partial s}$.

We quantify this with the Frobenius norm $||(\text{Cov}(y) - I)||_F$, averaged over the batch dimension. Intuitively, a smaller Covariance Frobenius Norm means that the covariance matrix is closer to identity, resulting in lower bias.

4.2 Results

$W^\perp$: Optimal Choice of k. We evaluate $W^\perp$ with different values of k, which controls the bias–variance trade-off, against the baseline $W^\top$ method to examine its effect on training accuracy, shown in Table 1. Smaller values of k reduce variance by reducing the dimensionality of the guessing space. However, reducing k also increases the bias since the covariance matrix I_k is further from the identity. For $k = \text{rank}(W)$ the final training accuracy is almost identical to $W^\top$, and as smaller values of k are chosen the accuracy increases. This result shows that a lower-dimensional guessing space with reduced variance counteracts the additional bias, except for the extreme low variance, high bias $k = 1$ case where there is a sharp decrease in accuracy.

Why Small Values of k Perform Best. The phenomenon that gradients lie in a low-dimensional subspace offers an explanation as to why small values of k attain the highest accuracy. Given the idea behind the $\tilde{W}^\top$ method, that multiplying by the upstream Jacobian matrix $\tilde{W}^\top$ constrains the gradient guess to lie in its image, denoted by $\text{Im}(\tilde{W}^\top)$, it is reasonable to speculate that an analogous phenomenon occurs here where the true gradient lies not only in $\text{Im}(\tilde{W}^\top)$ but in a low-dimensional subspace of $\text{Im}(\tilde{W}^\top)$. This can be empirically verified by projecting the true gradient $\partial L/\partial s$ onto $\text{Im}(\tilde{W})$, similar to Gurari et al. [9]. In the reduced SVD of $\tilde{W} = U_k\Lambda_k V_k^\top$, U_k contains an orthonormal

Table 1. Training accuracies after 300 epochs for different values of k in the $\tilde{W}^{\perp}$ method. The $\tilde{W}^{\perp}$ accuracies are comparable to $\tilde{W}^{\top}$, with smaller values $k = 10$ and $k = 25$ reaching the highest accuracy. In general, the trend is that smaller k perform better due to larger variance reduction, despite increased bias.

Method	k	Accuracy
$\tilde{W}^{\top}$		0.312
$\tilde{W}^{\perp}$	rank($\tilde{W}$)	0.311
$\tilde{W}^{\perp}$	25	0.323
$\tilde{W}^{\perp}$	10	**0.324**
$\tilde{W}^{\perp}$	1	0.275

basis of $\mathrm{Im}(\tilde{W}^{\top})$, and the overlap o_{top} of the true gradient in $\mathrm{Im}(\tilde{W}^{\top})$ can be computed by projecting $\frac{\partial L}{\partial s}$ onto $\mathrm{Im}(\tilde{W}^{\top})$, given by

$$o_{\text{top}} = \frac{\left\| U_k U_k^{\top} \frac{\partial L}{\partial s} \right\|}{\left\| \frac{\partial L}{\partial s} \right\|}. \tag{19}$$

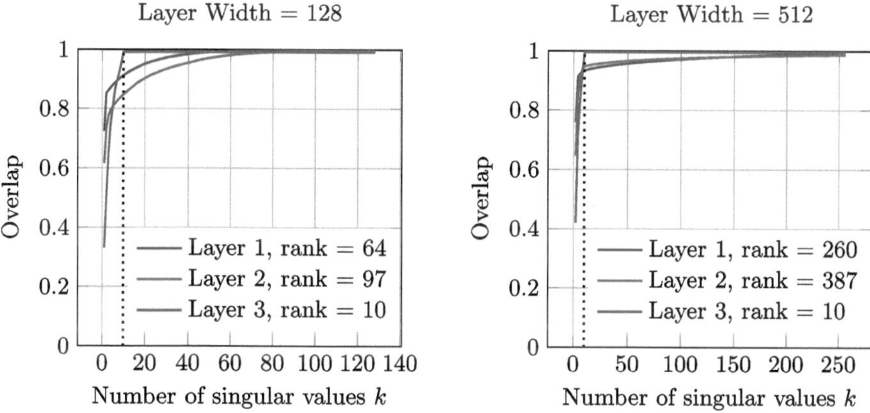

Fig. 1. Overlap of the true gradient $\frac{\partial L}{\partial s}$ onto the k-dimensional subspace of $\mathrm{Im}(\tilde{W}^{\top})$ corresponding to the k largest singular values of W, for layer widths of 128 and 512. In both cases, $\frac{\partial L}{\partial s}$ predominantly lies in a subspace of much lower dimension than rank(W); around $k = 10$ captures most of the gradient, reducing the guessing space while introducing minimal bias. Note that the Layer 3 weight matrix is low rank since it is connected to the output layer of width 10.

Figure 1 shows that the true gradient does in fact mostly lie in in the small k-dimensional subspace spanned by V_k. Intuitively, choosing the smallest value of k that is just large enough for the k-dimensional subspace to contain most of the true gradient will incur minimal additional bias while greatly reducing variance by narrowing the guessing space. For the network architecture and dataset in our case, $k = 10$ is a good choice.

Table 2. Variance, bias, and covariance Frobenius norm for different methods.

Method	Variance (18)	Bias (17)	Cov. Norm
$\tilde{W}^\top$	1.50×10^{-4}	2.75×10^{-4}	**0.084**
$\tilde{W}^\perp$-bottom	1.30×10^{-5}	2.83×10^{-4}	0.085
$\tilde{W}^\perp$ ($k = 10$)	$\mathbf{1.03 \times 10^{-5}}$	$\mathbf{2.86 \times 10^{-5}}$	0.085

$\tilde{W}^\perp$: Understanding the Bias Reduction. Next, we observe that for appropriately chosen k, our $\tilde{W}^\perp$ method outperforms $\tilde{W}^\top$ in both variance *and* bias. This is shown in Table 2. The variance reduction is achieved by constraining the guesses to a low-dimensional subspace, while the bias reduction is achieved by aligning the guesses with the true gradient, which is less obvious. Recall that the bias is zero when the covariance matrix is identity. For $\tilde{W}^\perp$, the purpose of orthogonalisation is to zero out off-diagonal entries in the covariance matrix I_k to bring it closer to identity. However, for small values of k, I_k ends up being equally distant from identity (in the Frobenius norm) as $\tilde{W}^\top \tilde{W}$, the covariance matrix of the $\tilde{W}^\top$ method. This might suggest that we have not actually decreased the bias in $\tilde{W}^\perp$. However, Table 2 shows that this is not the case.

To illustrate that the bias reduction comes from aligning the guessing space with the true gradient, consider the following variant of $\tilde{W}^\perp$ which we refer to as $\tilde{W}^\perp$-bottom. In the SVD of $\tilde{W} = USV^\top$, we take $\tilde{W}' = U_k V_k^\top$ corresponding to the $k = 10$ *smallest* (in this case 0) singular values, where we know the true gradient does not lie. Both $\tilde{W}^\perp$ and $\tilde{W}^\perp$-bottom have the same covariance matrix I_k, so according to Eq. (5), the difference in bias comes from the difference in the true gradients $\frac{\partial L}{\partial s}$. In other words, the optimisation trajectory of $\tilde{W}^\perp$ has smaller gradient norm $\left\|\frac{\partial L}{\partial s}\right\|_2$ than that of $\tilde{W}^\perp$-bottom, and hence lower bias. The conclusion is that $\tilde{W}^\perp$ reduces bias by aligning the guessing space with the underlying low-rank structure of the true gradient.

$\tilde{W}^\perp$: Scaling to Larger Networks. Compared to previous methods, the $\tilde{W}^\perp$ method shows promising results for larger networks. Figure 2 and Fig. 3 show that the gap in training accuracy between $\tilde{W}^\perp$ with $k = 10$ and $\tilde{W}^\top$ increases as the hidden layer width increases from 64 to 128 and 512. This behaviour can be explained by the variance reduction from narrowing the guessing space. Recall the idea behind $\tilde{W}^\top$, that multiplying a randomly sampled ϵ with $\tilde{W}^\top$ restricts the guess direction $y = \tilde{W}^\top \epsilon$ to lie in a lower dimensional $\mathrm{Im}(\tilde{W}^\top)$. However, as layer width increases, so does the rank of the next layer weight matrix and hence the dimensionality of $\mathrm{Im}(\tilde{W}^\top)$. $\tilde{W}^\perp$ does not suffer from this problem, because the dimensionality of the guessing space can be controlled through the hyperparameter k. In our experiments, the value of $k = 10$ which was observed to work well in the width 128 experiments still performs well for width 512 because the 10-dimensional subspace still contains a large portion of the true gradient. In general, these results indicate that for a layer of width n, most of the true gradient will be contained in a k-dimensional subspace with $k << n$, and that k grows much slower than n. In practice, this means that the guessing space and hence variance of the $\tilde{W}^{\perp}$ method can be kept almost constant as layer width increases, enabling $\tilde{W}^{\perp}$ to be scaled to larger networks than previous methods.

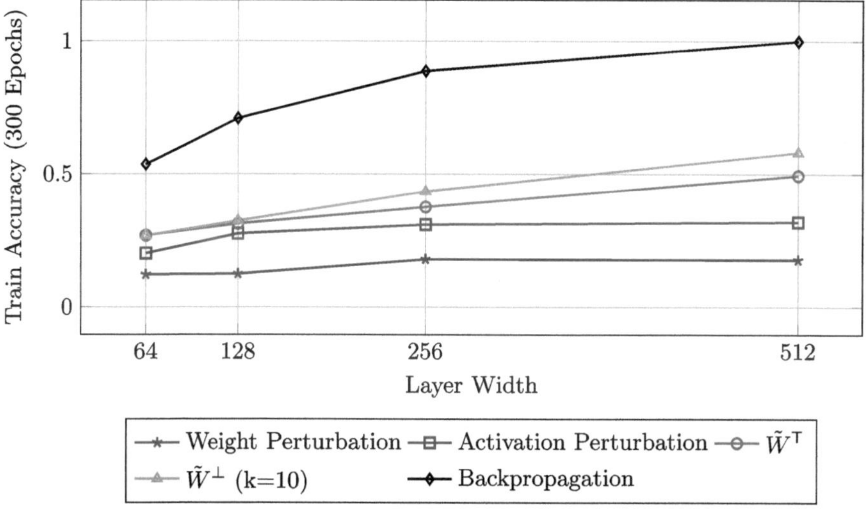

Fig. 2. Training accuracy as the hidden layer width increases.

Table 3. Train accuracies across methods and hidden layer widths. As layer width increases from 64 to 512, the accuracy gap between $\tilde{W}^{\top}$ and $W^{\perp}$ increases from 0% up to almost 10%. This result indicates that $\tilde{W}^{\perp}$ will scale to larger networks much better than previous methods. However, there is still a significant gap between gradient guessing methods and backpropagation.

Layer Width	Weight Perturbation	Activation Perturbation	$\tilde{\mathbf{W}}^{\top}$	$\tilde{\mathbf{W}}^{\perp}(\mathbf{k} = \mathbf{10})$	Backprop
64	0.121	0.201	0.268	0.268	0.536
128	0.124	0.275	0.312	0.323	0.709
256	0.179	0.309	0.376	0.434	0.887
512	0.177	0.319	0.494	0.581	1.000

Table 4. Train and test accuracies and training times (300 epochs) for our unoptimised† implementations of different methods across layer sizes of 128 and 512. †PyTorch MPS backend on MacOS laptop; SVD on the CPU.

Method	Layer Size	Train Time	Train Accuracy	Test Accuracy
$\tilde{W}^{\perp} - NS$	512	6h	**0.594**	0.477
$\tilde{W}^{\perp}, k = 10$	512	40h	0.581	**0.498**
$\tilde{W}^{\top}$	512	26 m	0.486	0.438
$\tilde{W}^{\perp} - NS$	128	9 m	**0.326**	**0.291**
$\tilde{W}^{\perp}, k = 10$	128	1h	0.324	0.283
$\tilde{W}^{\top}$	128	5 m	0.312	0.277

$\tilde{W}^{\perp}$-NS Results. Here, we verify that the $\tilde{W}^{\perp}$-NS approximation to $\tilde{W}^{\perp}$ with $k = 10$ does not harm performance. First, we describe a few implementation details that further speed up $\tilde{W}^{\perp}$-NS. To avoid having to repeatedly determine the polynomial coefficients that approximate f_{σ_k} (since σ_k varies across layers and over the course of training), we pre-determine a set of coefficients that approximate f_{σ_k} at intervals $\sigma_k = 0.1, 0.2, ..., 0.9$ via gradient descent and use the set of coefficients that σ_k is closest to at the current training iteration. Computing the spectral norm and σ_k can be done efficiently via the Implicitly Restarted Lanczos Method [16]. This can be further sped up by observing that the singular values change slowly over the course of training, allowing us to get away with computing σ_k once every 50 iterations. Table 4 shows that after 300 epochs $\tilde{W}^{\perp}$-NS attains comparable accuracies to $\tilde{W}^{\perp}$, and in some cases even surpasses it while being significantly faster than $W^{\perp}$. Our $\tilde{W}^{\perp}$ method and its $\tilde{W}^{\perp}$-NS variant surpasses the baseline $\tilde{W}^{\top}$ in all cases. For the MLP with a layer width of 512, $W^{\perp}$-NS has around 10× smaller memory usage than $W^{\perp}$.

Table 5. Layer 1 variance and bias, and accuracy at 300 epochs for an MLP with layer width of 256 trained on our methods and the baseline $\tilde{W}^{\top}$. $\tilde{W}^{\perp}$, with lower variance and bias than $\tilde{W}^{\top}$, reaches the highest accuracy by a wide margin while $\tilde{W}^{\mathrm{P}}$ (preconditioning), with the lowest bias but highest variance, reaches lower accuracy than $\tilde{W}^{\top}$.

Method	Variance (18)	Bias (17)	Train Accuracy	Test Accuracy
$\tilde{W}^{\perp}$ $(k = 10)$	$\mathbf{6.33 \times 10^{-6}}$	3.63×10^{-5}	**0.433**	**0.383**
$\tilde{W}^{\mathrm{P}}$	1.75×10^{-4}	$\mathbf{1.35 \times 10^{-8}}$	0.367	0.314
$\tilde{W}^{\top}$	7.04×10^{-5}	1.68×10^{-4}	0.375	0.316

$\tilde{W}^{\mathrm{P}}$ Preconditioning Results. While the earlier weight perturbation and activation perturbation methods emphasise unbiasedness, the baseline $\tilde{W}^{\top}$ and our $\tilde{W}^{\perp}$ methods emphasise reducing variance but at the cost of incurring bias. To gauge the effectiveness of lowering variance versus lowering bias, we compare our $\tilde{W}^{\mathrm{P}}$ preconditioning method with $\tilde{W}^{\perp}$ and the $\tilde{W}^{\top}$ baseline in Table 5. We see that lower variance correlates with higher accuracy, while the high variance preconditioning method, despite being unbiased, achieves the lowest accuracy. This result shows that removing bias at the expense of increasing variance is ineffective, and therefore it is better to have low variance even if it incurs bias. Based on the accuracy gains of $\tilde{W}^{\perp}$ as layer widths increase (Fig. 2), we hypothesise that this principle becomes even more relevant for larger networks.

5 Conclusion

In this paper, we introduced the $\tilde{W}^{\perp}$ method to addresses the central problem of high variance in gradient estimation methods. Our $\tilde{W}^{\perp}$ method exploits the low-dimensional structure of gradients by orthogonalising the upstream Jacobian matrix, thereby constraining the guessing space and reducing bias and variance. Experiments show promising results with $\tilde{W}^{\perp}$ as network width increases, while the $\tilde{W}^{\mathrm{P}}$ preconditioning method demonstrates that removing bias at the cost of increased variance is ineffective. Furthermore, we greatly speed up $\tilde{W}^{\perp}$ by using Newton–Schulz iterations for approximate orthogonalisation. In future work, these ideas can be expanded to more practical architectures, such as networks with normalisation layers or residual connections, or approaches that better consider the underlying structure of gradients in deep neural networks.

References

1. Rumelhart, D.E., Hinton, G.E., Williams, R.J.: Learning representations by back-propagating errors. Nature **323**(6088), 533–536 (1986)
2. Singhal, U., et al.: How to guess a gradient. arXiv preprint arXiv:2312.04709 (2023)
3. Baydin, A.G., Pearlmutter, B.A., Syme, D., Wood, F., Torr, P.: Gradients without backpropagation. arXiv preprint arXiv:2202.08587 (2022)
4. Silver, D., Goyal, A., Danihelka, I., Hessel, M., van Hasselt, H.: Learning by directional gradient descent. In: International Conference on Learning Representations (2022)
5. Chandra, K.: An unexpected challenge in using forward-mode automatic differentiation for low-memory deep learning. Stanford Digital Repository (2021)
6. Crick, F.: The recent excitement about neural networks. Nature **337**(6203), 129–132 (1989)
7. Ren, M., Kornblith, S., Liao, R., Hinton, G.: Scaling forward gradient with local losses. In: The Eleventh International Conference on Learning Representations (2023)
8. Fournier, L., Rivaud, S., Belilovsky, E., Eickenberg, M., Oyallon, E.: Can forward gradient match backpropagation? In: Proceedings of the 40th International Conference on Machine Learning, ICML'23 (2023)
9. Gur-Ari, G., Roberts, D.A., Dyer, E.: Gradient descent happens in a tiny subspace. arXiv preprint arXiv:1812.04754 (2018)
10. Song, M., Ahn, K., Yun, C.: Does SGD really happen in tiny subspaces? In: The Thirteenth International Conference on Learning Representations (2025)

11. Kovarik, Z.: Some iterative methods for improving orthonormality. SIAM J. Numer. Anal. **7**(3), 386–389 (1970)
12. Björck, Å., Bowie, C.: An iterative algorithm for computing the best estimate of an orthogonal matrix. SIAM J. Numer. Anal. **8**(2), 358–364 (1971)
13. Jordan, K., et al.: Muon: an optimizer for hidden layers in neural networks (2024)
14. Greydanus, S., Kobak, D.: Scaling down deep learning with MNIST-1D. In: Proceedings of the 41st International Conference on Machine Learning (2024)
15. Loshchilov, I., Hutter, F.: Decoupled weight decay regularization. In: International Conference on Learning Representations (2019)
16. Calvetti, D., Reichel, L., Sorensen, A.: An implicitly restarted lanczos method for large symmetric eigenvalue problems. Electron. Trans. Numer. Anal. **2**, 1–21 (1994)

Coordinate-Free k-Means Clustering

Dan C. Popescu(✉) and Philip Kilby

Canberra, ACT, Australia
danny.c.popescu@gmail.com, philip@kilby.id.au

Abstract. In this paper we present generalizations of the k-means clustering algorithm, with guaranteed monotonic convergence, which do not require that data points be localized in a vector space. Instead, only a dissimilarity matrix between data points is needed. This generalization allows k-means-like clustering to be applied to data-sets that cannot be embedded exactly in an Euclidean space, such as road networks, and data-sets with categorical data. If the data points have localizations in an Euclidean space and the dissimilarity matrix is the matrix of mutual squared Euclidean distances between points, one of our algorithms perfectly mirrors the classical k-means. If the dissimilarity matrix does not correspond to an Euclidean scenario, our experimental results using both real-world and synthetic data sets show that our proposed generalizations of k-means generally lead to more compact clusters than k-medoids clustering.

Keywords: clustering · k-means · k-medoids

1 Introduction

Clustering is a technique of partitioning a set P of objects, or data points, into a number of disjoint subsets called clusters, such that all objects belonging to the same cluster are very similar to each other. This assumes a predefined distance, or dissimilarity function on the set $P \times P$, which quantifies how close, or similar, are any two objects in P. It also requires the definition of an objective function to be minimized, which is derived from the distance defined on $P \times P$, to measure the spread of the clusters. Clustering techniques are very relevant to many areas of science and engineering, such as data compression, data mining, information retrieval, pattern recognition, and machine learning.

Among the many clustering algorithms, k-means is certainly one of the most popular. Originally proposed in Lloyd's 1982 paper [1], over the last 40 years it also turned out to be very useful in many other types of applications. These include document categorization from databases, optimization of transport logistics, segmentations of market data, to name just a few. The popularity of k-means is mainly due to its main strengths of simplicity and robustness. k-means also has some weaknesses; the most notable, that it it essentially an iterative local search optimization algorithm, and as such ends up in local minima. Therefore, several

M. Liu et al. (Eds.): AI 2025, LNAI 16371, pp. 43–55, 2026.
https://doi.org/10.1007/978-981-95-4972-6_4

initializations and runs of the algorithm are generally required to find a good solution. Modified versions of the original k-means formulation can mitigate the need for a good initialisation [2].

Pivotal to the k-means algorithm is the concept of cluster means, around which data clustering is performed. It requires that data points have localizations in a Euclidean vector space, the distance function is induced by the corresponding Euclidean norm, and cluster means are calculated as vector means of the position vectors of the cluster points. While this paradigm covers a large number of practical applications, there are nevertheless scenarios where it may not be possible to provide a localization for the data points that is compatible with the distance function on the set $P \times P$. For example, a network of geographic locations, linked by roads over rugged terrain, would be represented by a distance matrix between locations which exhibits the properties of a metric space, but is not Euclidean. If data clustering is required in such scenarios, then k-means can not be directly used. In [3] an alternative approach is proposed to avoid some explicit computations of the cluster centers, however the framework is still limited to the Euclidean context. In [4] the paradigm is shifted towards calculating proximity based on a local neighbourhood search. While this may allow for more flexible cluster shapes, the lack of a well-defined global objective function to minimise can lead to less compact clusters. A well-known classical alternative for the non-Euclidean clustering context is to use k-medoids [5], which can be regarded as a more robust version of k-means, relying only on the existing data points to create cluster centers. The k-medoids approach eliminates the need for computing new cluster centers, possibly at the expense of some loss in cluster quality.

The aim of our communication is not to revisit the merits, or propose new applications, for the k-means clustering methodology. This has already received ample coverage in the literature. References [6] [7] [8], to cite just a few, contain good surveys. It is remarkable to note, for example, that the survey in [9] identifies "k-means" as the most significant keyword found in modern power stations publications. Our aim is to provide an alternative cast and interpretation of the k-means methodology, that relies exclusively on the dissimilarity function and does not require the explicit computation of cluster means. This allows for a natural generalization of our proposed methodology for situations where the dissimilarity matrix is not a Euclidean distance matrix (EDM). We then show, through experimental results, that the proposed generalization of k-means generally outperforms k-medoids in terms of cluster quality, if a more balanced measure is used to quantify cluster spreads.

Our paper is organized as follows. In Sect. 2 we briefly review the basics of the k-means and k-medoids algorithms. In Sect. 3 we introduce our first generalization of the k-means algorithm. In Sect. 4 we introduce our generalized notions of cluster spread and cluster inflation, and further generalize our algorithm for the case of general pre-distance matrices. We show several experimental results in Sect. 5 and then present our conclusions in Sect. 6.

2 Preliminaries on k-Means and k-Medoids

Let $P = \{p_1, p_2, \cdots p_n\}$ be a finite set of n objects, or data points. With the set P is associated an $n \times n$ dissimilarity matrix $D = (d_{ij})$ $(1 \leq i, j \leq n)$ where entry d_{ij} quantifies how close, or similar, the data point p_i is to data point p_j, with larger values corresponding to less similarity. In the most general case, we only assume that D is a pre-distance matrix; that is, D has zero diagonal and that $d_{ij} = d_{ji} > 0$ for all $(1 \leq i < j \leq n)$. This means that any object or data point is most similar to itself, and that the dissimilarity between any pair of data points does not depend on the order in which we consider the points. Note that these are relatively mild assumptions, likely to cover a large number of practical scenarios.

Given integer $k < n$ the aim to find an optimal k-partion of the set P, $\mathcal{C} = \{C_1, C_2, \cdots, C_k\}$ of the set P into k non-empty "good" clusters $C_i (1 \leq i \leq k)$ with $\bigcup_{i=1}^{n} C_i = P$ and $C_i \bigcap C_j = \emptyset$ for $1 \leq i < j \leq k$. The quality of any particular clustering solution is measured by computing an objective function $\mathcal{F}$ which depends on P, D and $\mathcal{C}$. Generally stated, the aim is to minimize the objective function $\mathcal{F}$ by placing the data points that are most similar to each other into the same cluster.

The k-means and k-medoids algorithms are well-known. Here we only briefly remind of their basics. The k-means clustering algorithm assumes that each data point has an associated position, in a Euclidean space of a given dimension m. The dissimilarity between any pair of data points is given by the squared Euclidean distance between their position vectors. k-means starts with an initially given k-partition of the data points, computes cluster centers as the vector mean of all members in the cluster, and then reassigns data points to the nearest cluster center. k-medoids utilses a similar strategy, except that cluster centers are constrained to be exactly at the data points; therefore eliminating the need for computing new cluster centers, but also possibly leading to a less precise evaluation of the cluster spreads. For both algorithms the objective function $\mathcal{F}$ to minimize is the sum of all intra-cluster distances. Both algorithms have monotonic convergence to local minima, which depend on the initialisation of the algorithm.

3 The Basic k-Clusters Algorithm

In this section we introduce a few notations and then we present our first basic generalization of the k-means algorithm, which we call the k-clusters algorithm. Our approach is gradual. Initially we show how one can recast the k-means algorithm, such that it is no longer required to have explicit localization of data points in a Euclidean space. Then, in the following sections, we present some refinements that allows for our proposed algorithm to be further generalized, ensuring its convergence even for situations where the dissimilarity matrix is a very general pre-distance matrix.

We start with the remark that for the k-means algorithm, the dissimilarity matrix D is a Euclidean Distance Matrix (EDM), which is a matrix of *squared*

Euclidean distances between data points. We also note that cluster spreads are defined using the squared distances from cluster centers. The key observation is the following identities from [10], involving the squared distances from points in a Euclidean space to the geometric center of any finite set of points:

$$\sum_{i=1}^{s} \|\bar{c} - \bar{x}_i\|_2^2 = \frac{1}{s} \sum_{1 \le i < j \le s} \|\bar{x}_i - \bar{x}_j\|_2^2 \tag{1}$$

$$\|\bar{y} - \bar{c}\|_2^2 = \frac{1}{s} \sum_{i=1}^{s} \|\bar{y} - \bar{x}_i\|_2^2 - \frac{1}{s^2} \sum_{1 \le i < j \le s} \|\bar{x}_i - \bar{x}_j\|_2^2 \tag{2}$$

where $\bar{y}$ and all $\bar{x}_i$ $(1 \le i \le s)$ in the set $S = \{\bar{x}_1, \bar{x}_2, ..., \bar{x}_s\}$ are any vectors in a Euclidean space $\mathbb{R}^m$, and $\bar{c}$ is the geometric center of the set S: $\bar{c} = \mu(S)$. Equations (1) and (2) provide a basis for reformulating the k-means algorithm, only by using the mutual squared distances corresponding to the dissimilarity matrix and without having to explicitly compute any cluster center. Given a data set P with an associated dissimilarity matrix $D = (d_{ij})$, for any point $p_i \in P$ and any non-empty subset $S \subset P$ of size $|S| \ge 1$, we introduce the following notations:

$$\|S\| = \frac{1}{|S|} \sum_{p_i, p_j \in S, (i<j)} d_{ij} \tag{3}$$

$$< p_i, S > = \sum_{p_j \in S} d_{ij} \tag{4}$$

$$[p_i, S] = \frac{1}{|S|}(< p_i, S > - \|S\|) \tag{5}$$

We call $\|S\|$ the *spread of cluster* S, we call $<p_i,S>$ the *span of point* p_i *to cluster* S and we call $[p_i,S]$ the *inflation of point* p_i *to cluster* S. Note that in order to calculate any of the functions defined by Eqs. (3), (4) and (5) only the values in the matrix D are needed. For the case when the dissimilarity matrix D is an EDM, the value of the span given by Eq. (4) is an exact match for the right side of Eq. (1) and thus corresponds exactly to the sum of squared distances, for all points in the cluster, to the cluster center. Similarly, Eq. (5) is just a rewrite of Eq. (2) and therefore $[p_i,S]$, the inflation of point p_i to cluster S, corresponds to its squared distance to the center of S. Using this terminology, we would aim for having cluster configurations where every point has minimal inflation to the cluster it belongs, compared to all other clusters. This can be assessed by computing a $k \times n$ inflation matrix $\mathcal{I}$, of points' inflations to clusters, such that $\mathcal{I}_{ij} = [p_j,C_i]$ for all $(1 \le i \le k)$ and $(1 \le j \le n)$.

For a given k-cluster partition $\mathcal{C} = \{C_1, C_2, \cdots, C_k\}$ of the set P we now define the sum of the cluster spreads $\mathcal{F}$ as:

$$\mathcal{F} = \mathcal{F}(P, D, \mathcal{C}) = \sum_{i=1}^{k} \|C_i\| \tag{6}$$

We outline our proposed k-clusters algorithm in the pseudocode of Algorithm 1.

Algorithm 1 k-clusters algorithm

Input: n-set P, dissimilarity matrix D, initial k-partition $\mathcal{C}^0$ of P
Output: $\mathcal{C}^0_{\text{opt}}$: local optimum k-cluster partition of P
procedure k-CLUSTERS(P,S,$\mathcal{C}^0$)
    ▷ Initialize objective function based on initial partition
    $\mathcal{F}^0 = \sum_{i=1}^{k} \sum_{p_j \in C_i^0} \|C_i^0\|$
    $t = 0$ ▷ initialize iteration counter t
    repeat
        $t = t + 1$
        for all $1 \leq i \leq k$ **do**
            Initialize $C_i^t = \emptyset$
        end for
        ▷ Create new clusters at iteration t
        Compute $k \times n$ inflation matrix $\mathcal{I}$: $\mathcal{I}_{ij} = [p_j, C_i^{t-1}]$
        for all $1 \leq j \leq n$ **do**
            $i^* = \operatorname*{argmin}_{1 \leq i \leq k} \mathcal{I}_{ij}$
            $C_{i^*}^t = C_{i^*}^t \bigcup \{p_j\}$
        end for
        ▷ Update objective function
        $\mathcal{F}^t = \sum_{i=1}^{k} \sum_{p_j \in C_i^t} \|C_i^t\|$
    until $\mathcal{F}^t \geq \mathcal{F}^{t-1}$
    $\mathcal{C}^0_{\text{opt}} = \mathcal{C}^{t-1}$
end procedure

With reference strictly to cluster evolution, the outlined k-clustering algorithm is a step by step, exact replica, of the classical k-means clustering algorithm, and as such will have guaranteed monotonic convergence if the dissimilarity matrix is Euclidean. The main novelty of k-clusters is that it removes the need for having an explicit localization for data points in a Euclidean space $\mathbb{R}^m$. This is a very significant advantage, for two main reasons. Firstly, many dissimilarity matrices are EDMs, even if that may not be immediately obvious. For example, Schoenberg's results regarding the isometric embedding of certain metric spaces into Hilbert spaces [11] imply that some matrices of Minkovski norm powers are automatically EDMs. As remarkable examples, any matrix of (non-squared) Euclidean distances, corresponding to the ℓ_2^1 norm (instead of the commonly used ℓ_2^2 norm) or any matrix corresponding to the ℓ_1^1 norm (also sometimes referred to as the Manhattan norm), in any $\mathbb{R}^m$, would always be an EDM, albeit in general leading to embeddings into Euclidean spaces of significantly higher dimension than the original measurement space. Secondly, it will allow for generalizing the methodology to cope with more general dissimilarity matrices. If the dissimilarity matrix is not Euclidean, it turns out that the monotonic covergence of the k-clusters algorithm is no longer guaranteed. To further guarantee local monotonic convergence for the case of a general dissimilarity matrix, another refinement is need.

4 A Refinement

To refine our algorithm, such that it could cope with dissimilarity matrices that are arbitrary pre-distance matrices, we need to have a closer look at the basic exchange mechanism of one data point between clusters. Directly from Eqs. (3) and (5) we generally have, for any given cluster S:

$$\sum_{p_i \in S} [p_i, S] = \|S\| \tag{7}$$

which simply shows that the spread of a cluster is the sum of all inflations into it by all member points. When a data point joins or leaves a cluster, it affects the cluster spread in proportion to its inflation factor, but the proportionality factor is not 1. To quantify this more precisely, consider a cluster S of size $|S| = s$ and a data point p that does not belong to S. Using again Eqs. (3), (4) and (5):

$$\|S \cup \{p\}\| = \frac{1}{s+1}\Big(\sum_{p_i, p_j \in S, (i<j)} d_{ij} + < p, S >\Big) = \frac{1}{s+1}(s\|S\| + < p, S >)$$
$$= \frac{1}{s+1}(s\|S\| + \|S\| + s[p, S]) = \|S\| + \frac{s}{s+1}[p, S]$$

and therefore:

$$\|S \cup \{p\}\| - \|S\| = \frac{s}{s+1}[p, S] \tag{8}$$

We also have that:

$$[p, S \cup \{p\}] = \frac{1}{s+1}(< p, S \cup \{p\} > - \|S \cup \{p\}\|)$$
$$= \frac{1}{s+1}(< p, S > - \|S\| - \frac{s}{s+1}[p, S]) = \frac{s^2}{(s+1)^2}[p, S]$$

and from Eqs. (8) and (9) we get:

$$\|S \cup \{p\}\| - \|S\| = \frac{s+1}{s}[p, S \cup \{p\}] \tag{9}$$

Based on Eqs. (8) and (9) we can now compute the exact loss/decrease Δ^p_{ij} in the objective function $\mathcal{F}$ of Eq. (6), if a data point p moves from its cluster S_i of size $|S_i|$ into a new cluster S_j of size $|S_j|$:

$$\Delta^p_{ij} = \frac{|S_i|}{|S_i| - 1}[p, S_i] - \frac{|S_j|}{|S_j| + 1}[p, S_j] \tag{10}$$

Eq. (10) is a sharper version of the somewhat more relaxed rule of assessing cluster proximity used by the basic k-clusters algorithm, which relies on just comparing cluster inflation values. It shows that cluster inflation values need to be slightly adjusted by certain factors, which depend strictly on cluster sizes. Equation (10) represents a natural generalization of the cluster switch rule proposed in [12] and [13] for the Euclidean scenario.

If there is any switch value Δ_{ij}^p that turns out to be strictly positive, the current clustering solution can still be improved by switching point p from cluster S_i into cluster S_j. If there is no positive Δ_{ij}^p value, a local minimum of the objective function $\mathcal{F}$ has been reached and the clustering iteration will stop. The good news with using Δ_{ij}^p of Eq. (10) to assess cluster switch values is that it can be used in the in the most general context. It does not require any assumptions about the signs of the inflation factors involved, or about the dissimilarity matrix, other than it is a pre-distance matrix. The less good news is that while Δ_{ij}^p is a perfect indicator on the effect of *one switch* for a point p between clusters S_i and S_j, Eq. (10) is based on the assumption that there are no other changes to either cluster S_i or cluster S_j. It is therefore not easy to assess the cumulative effect of multiple such switches, if they have in common either the source or the destination cluster. We will call two or more switches *disjoint*, if no cluster index appears more than once across all sources and destinations of the considered switches.

It is clear that several disjoint switches will act independently of each other in adjusting the value of the objective function $\mathcal{F}$, and can be performed simultaneously, with cumulative effect in decreasing the value of $\mathcal{F}$. This naturally leads to the question: if several positive switches are available, which ones should be chosen? We propose a greedy approach strategy, based on choosing positive switches in the order of decreasing Δ_{ij}^p values, and such that they remain all disjoint. This can be done by building a vector of positive switch tuples (Δ_{ij}^p, p, i, j) sorted by the Δ-value field in decreasing order. This sorted vector is then scanned in natural order, to gradually perform cluster switches that are disjoint with previously selected ones. Note that because each switch choice eliminates two clusters from further consideration, no more than $\lfloor k/2 \rfloor$ switches can be considered simultaneously. A pseudocode for this generalized clustering procedure, which we call k-clustersDis is given in Algorithm 2. The most important property of Algorithm 2 is that it converges monotonically, for any given pre-distance matrix and for any initialization. Its main weakness is the slower convergence, primarily due to the limitation that no more than $\lfloor k/2 \rfloor$ data points can be exchanged between clusters at any iteration.

Finally, we designate by *k-clustersGen* our general, synthesis algorithm. It starts by applying k-clusters, for as long as the objective function decreases, then applies k-clustersDis until convergence. Therefore k-clustersGen combines the efficient initial quick descent of k-clusters with the final refinement (at slower rate, but guaranteed monotonic convergence) of k-clustersDis.

In the end of this section we mention that some preliminary suggestions about the possibility of generalizing k-means have appeared in preprints on the arXiv open access repository. In [14] a derivation of the distance to the cluster

Algorithm 2 k-clustersDis algorithm

Input: n-set P, dissimilarity matrix D, initial k-partition $\mathcal{C}^0$ of P
Output: $\mathcal{C}^0_{\text{opt}}$: local optimum k-cluster partition of P
procedure k-CLUSTERSDIS(P,S,$\mathcal{C}^0$)
  ▷ Initialize objective function based on initial partition
  $\mathcal{F}^0 = \sum_{i=1}^{k} \sum_{p_j \in C_i^0} \|C_i^0\|$
  $t = 0$ ▷ initialize iteration counter t
  repeat
    $t = t + 1$
    ▷ Create new clusters at iteration t
    for all $1 \leq i \leq k$ **do**
      Initialize $C_i^t = C_i^{t-1}$
    end for
    $\mathcal{V} = \emptyset$ ▷ Initialize vector of cluster switches
    for all $1 \leq j \leq n$ **do**
      Let i_1 be the cluster index of point p_j: $p_j \in C_{i_1}$
      for all $1 \leq i_2 \leq k$ **do**
        Compute switch value $\Delta_{i_1 i_2}^{p_j}$
        if $\Delta_{i_1 i_2}^{p_j} > 0$ **then**
          Append ($\Delta_{i_1 i_2}^{p_j}$,j,i_1,i_2) to vector $\mathcal{V}$
        end if
      end for
    end for
    $\mathcal{T} = \emptyset$ ▷ Initialize set of ticked-off clusters
    $\mathcal{V}^* = \text{sort}(\mathcal{V})$ ▷ Sort $\mathcal{V}$ decreasing by Δ-field
    for all $1 \leq i \leq$ size($\mathcal{V}^*$) **do**
      $(\Delta,j,i_s,i_d) = \mathcal{V}^*(i)$
      if $\Delta > 0$ **and** $\mathcal{T} \cap \{i_s, i_d\} = \emptyset$ **then**
        $\mathcal{T} = \mathcal{T} \cup \{i_s, i_d\}$
        $C_{i_s}^t = C_{i_s}^t \setminus \{j\}$
        $C_{i_d}^t = C_{i_d}^t \cup \{j\}$
        $\mathcal{F}^t = \mathcal{F}^t - \Delta$
      end if
    end for
  until $\mathcal{F}^t == \mathcal{F}^{t-1}$
  $\mathcal{C}^0_{\text{opt}} = \mathcal{C}^{t-1}$
end procedure

center in terms of the entries of the distance matrix is given. This is equivalent to one of the results presented earlier more clearly and comprehensively in [10], with the authors of [14] apparently unaware of this prior work. In [15] an implementation of a generalized k-means algorithm is suggested, however in a rather crude form, that cannot guarantee monotonic convergence to local minima. The main strength of our approach consists in the clear articulation of the fundamental underlying concepts. The formal definitions of cluster inflation and cluster spread described in our present work generalize the concept of distance to the cluster center, and represent a significant difference with, and improvement to the previous works.

5 Experimental Results

Figure 1 displays the results of applying both our k-clusters algorithm and the k-medoids clustering algorithm, on a dataset of 160 data points located in a 2-dimensional Euclidean space, which make the shape of four equal size circles. Both algorithms have been run with $k = 4$ as the number of clusters and the ℓ_1^1 norm was used as dissimilarity measure. Both algorithms have been run 100 times, initialized with the same set of random seeds, which have been generated with random uniform distribution. The left image in Fig. 1 displays the best clustering result for k-clusters, which has a cluster spread value of 1256.52, while the right image corresponds to the best result of k-medoids algorithm, corresponding to a spread value of 1431.53. While the k-medoids never finds the minimal clustering solution of spread 1256.52 with any of the 100 random initial seeds, the k-clusters algorithm finds the minimal solution 76 times out of the overall 100 initial random seedings.

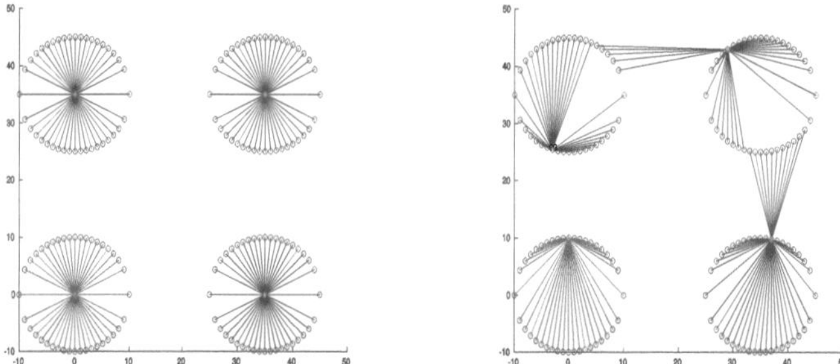

Fig. 1. Clustering results: by k-clusters (left image) and k-medoids (right image). The cluster spreads are 1256.52 and 1431.53, respectively.

Tables 1 and 2 summarize comparisons of clustering tests performed on real-world datasets, using several types of non-Euclidean dissimilarity measures, with external labeling, using both the k-clustersGen and k-medoids. We have generated 50 random seeds for each of the datasets, which were used by both k-clustersGen and k-medoids. The first column of the tables gives the cluster spread value of the externally labeled partition, while the second and the third column display the spread values of the best clusters obtained by the k-clustersGen and k-means algorithms, respectively. In addition, the degrees of correlation of the clustering results with the structure of the externally labeled (benchmark) partition are also presented. The Adjusted Rand Index (ARI) [16,17] and the Normalized Mutual Information (NMI) values [18] between the best cluster and the benchmark partition are displayed in columns 4 to 7, with sub-index 1 corresponding to the k-clustersGen method results and sub-index 2 corresponding to the k-medoids results.

Table 1 displays a comparison of clustering test results on text document selected from the database collection presented in [19], where in each data-set documents have been labeled according to their source type. For each dataset of documents, a predefined set of representative keywords has been selected, and each document has been represented by a vector of keyword frequencies within the respective document. In our tests we used a dissimilarity measure derived from the Goodman-Kruskal correlation coefficient GK(x,y) [20], a rank-based correlation coefficient resulting from counting the numbers of concordant and discordant pairs of coordinates in both vectors x and y. It therefore assesses the relative importance of likewise terms, which leads to a suitable measure of similarity for the text data described. The Goodman-Kruskal correlation coefficient has values in the range $[-1, 1]$, with the values close to 1 corresponding to high similarity. Thus the matrix of values $1 - GK(x, y)$ with values in the range $[0, 2]$ is a pre-distance matrix. This matrix, which in general is not an EDM, was used.

Table 2 presents a comparison of clustering tests on geographic locations, obtained from the Australian Post data described in detail in [21]. The dissimilarity between locations is given by the actual road distance, and the resulting dissimilarity matrix is non-Euclidean. The labeled clusters are obtained by a vehicle routing algorithm, as described in [22].

Table 1. Statistics of relative performance of k-clustersGen (index 1) and k-medoids (index 2) over document datasets with labeled clusters.

Data set	||labeled||	||k-clustersGen||	||k-medoids||	ARI_1	ARI_2	NMI_1	NMI_2
Articles-1442-5	30.05	29.97	30.12	0.990	0.972	0.987	0.967
Cbr-ilp-son-ir	92.52	91.36	109.96	0.896	0.220	0.860	0.273
Pubmed2000sel	220.20	220.08	262.99	0.996	0.744	0.992	0.687
Scopus2008	263.57	257.63	432.78	0.890	0.187	0.887	0.241

Table 2. Statistics of relative performance of k-clustersGen (index 1) and k-medoids (index 2) over geographic locations with labeled clusters.

Data set	||labeled||	||k-clustersGen||	||k-medoids||	ARI_1	ARI_2	NMI_1	NMI_2
AP100-17	220.62	203.41	228.24	0.673	0.519	0.872	0.803
AP100-18	205.16	193.60	220.89	0.783	0.629	0.915	0.854
AP100-19	199.36	184.90	209.29	0.687	0.655	0.887	0.871
AP100-20	191.77	177.96	201.03	0.714	0.608	0.896	0.855

The most important remark for the results of these tests is that k-clustersGen ends up finding clusters of lower spreads than the k-medoids clustering algorithm. Additionally, the ARI and NMI indexes also show that k-clustersGen finds clusters which are better aligned with externally labeled partition.

Table 3 draws a similar comparison of clustering results based on the publicly available dataset of high-dimensional synthetic Gaussian clusters described in [23]. All high-dimensional datasets have 1024 items grouped into 16 clusters, are externally labeled, have very good cluster separation and the external labeling is very well aligned with the minimal clusters.

Table 3. Statistics of relative performance of k-clustersGen and k-medoids over synthetic labelled dataset with 16 Gaussian clusters, for several dissimilarity measures.

Dissimilarity	||labeled||	||k-clustersGen||	||k-medoids||	ARI_1	ARI_2	NMI_1	NMI_2
ℓ_2^1	9.57×10^3	1.86×10^5	3.17×10^5	0.924	0.859	0.978	0.957
ℓ_2^2	2.32×10^6	4.66×10^6	7.39×10^6	0.935	0.867	0.983	0.960
COS	0.194	3.19	3.22	0.935	0.919	0.983	0.976
Mahalanobis	81.86	81.86	90.57	1.00	0.791	1.00	0.847

We show a comparative analysis of clustering results on the dataset of dimension 32, which has the most challenging separation of clusters of all six high-dimensional datasets. The tests were again performed using the same set of 50 random initial seeds, using four dissimilarity measures: Euclidean distance, squared Euclidean distance, Cosine (or chordal) distance and the Mahalanobis distance. The k-clustersGen has again performed better than k-medoids in all experiments, both in terms of finding a cluster of lower spread and of best correlation with the standard labeled partition. Remarkably, k-clustersGen with the Mahalanobis distance seemed to best exploit the intrinsic correlations of the dataset and finds a solution of perfect alignment with the labeled standard partition.

6 Conclusions

We have proposed a new clustering methodology that represents a generalization of Lloyd's classical k-means clustering algorithm. The key element of novelty with our proposed methodology is that it requires only the dissimilarity matrix between datapoints, and does not rely on having position vectors associated with the datapoints, to compute cluster centers. Therefore the proposed methodology will work for both Euclidean and non-Euclidean datasets.

We have defined generalized notions of cluster inflation and cluster spread, which eliminate the need of calculating explicit cluster centers. We have shown how both values can be computed directly from the dissimilarity matrix, and we have outlined monotonically convergent algorithms.

The relaxed requirement of only needing a pre-distance type dissimilarity matrix allows for using of our proposed methods in application areas where the classical k-means clustering method can not be used, and where one may have to rely on k-medoids clustering, in the absence of a similar competing alternative.

Our tests show that our proposed methodology generally produces more compact clusters than k-medoids, for both synthetic and real-world data, and we believe it will become useful in a variety of practical application areas.

References

1. Lloyd, S.P.: Least squares quantization in PCM. IEEE Trans. Inf. Theory **28**(2), 129–137 (1982)
2. Sinaga, K.P., Yang, M.S.: Unsupervised K-means clustering algorithm. IEEE Access **8**, 80716–80727 (2023)
3. Nie, F., Li, Z., Wang, R., Li, X.: An effective and efficient algorithm for K-means clustering with new formulation. IEEE Trans. Knowl. Data Eng. **35**(4), 3433–3433 (2023)
4. Cheng, D., Huang, J., Zhang, S., Xia, S., Wang, G., Xie, J.: K-means clustering with natural density peaks for discovering arbitrary-shaped clusters. IEEE Trans. Neural Netw. Learn. Sys. **35**(8), 11077–11090 (2024)
5. Kaufman, L., Rousseeuw, P.J.: Finding Groups in Data: An Introduction to Cluster Analysis. Wiley-Interscience. John Wiley & Sons, Hoboken (1990)
6. Aggarwal, C.C., Reddy, C.K.: Data Clustering: Algorithms and Application. Chapman & Hall, CRC Press, Boca Raton (2013)
7. Meila, M., Murtagh, F., Henig, C., Rocci, R.: Handbook of Cluster Analysis. Chapman and Hall, CRC Press, New York (2015)
8. Hastie, T., Tibshirani, R., Friedman, J.: The Elements of Statistical Learning: Data Mining, Inference, and Prediction. Springer Series in Statistics. Springer, Heidelberg (2017)
9. Miraftabzadeh, S.M., Colombo, C.G., Longo, M., Foiadelli, F.: K-means and alternative clustering methods in modern power systems. IEEE Access **11**, 119596–119633 (2023)
10. Apostol, T.M., Mnatsakanian, M.A.: Sums of squares of distances in m-space. Amer. Math. Monthly **110**, 516–526 (2003)
11. Schoenberg, I.J.: On certain metric spaces arising from Euclidean spaces by a change of metric and their imbedding in Hilbert spaces. Ann. Math. **30**, 787–793 (1937)
12. Hartigan, J.A., Wong, M.A.: Algorithm AS 136: a K-means clustering algorithm. J. R. Stat. Soc. Ser. C (Appl. Stat.) **28**(1), 100–108 (1979)
13. Telgarski, M., Vattani, A.: Hartigan's method: k-means clustering without voronoi. J. Mach. Learn. Res. **9**, 820–827 (2010)
14. Szalkai, B.: Generalizing k-means for an arbitrary distance matrix. arXiv preprint (2013)
15. Szalkai, B.: An implementation of the relational k-means algorithm. arXiv preprint (2013)
16. Hubert, L., Arabie, P.: Comparing partitions. J. Classif. **2**, 193–218 (1985)
17. Santos, J.M., Embrechts, M.: On the use of the adjusted rand index as a metric for evaluating supervised classification. In: Alippi, C., Polycarpou, M., Panayiotou, C., Ellinas, G. (eds.) ICANN 2009. LNCS, vol. 5769, pp. 175–184. Springer, Heidelberg (2009). https://doi.org/10.1007/978-3-642-04277-5_18
18. Kvalseth, T.O.: Entropy and correlation: some comments. IEEE Trans. Syst. Man Cybern. **17**, 517–519 (1987)

19. Soares, V.H.A., Campello, R.J.G.B., Nourashrafeddin, S., Milios, E., Naldi, M.C.: Combining semantic and term frequency similarities for text clustering. Knowl. Inf. Syst. **61**(3), 1485–1516 (2019). https://doi.org/10.1007/s10115-018-1278-7
20. Goodman, L.A., Kruskal, W.H.: Measures of association for cross classifications. J. Am. Stat. Assoc. **49**, 732–764 (1954)
21. Ernst, A., Krishnamoorthy, M.: Efficient algorithm for the uncapacitated single allocation p-hub median problem. Locat. Sci. **4**, 139–154 (1996)
22. Ropke, S., Pisinger, D.: An adaptive large neighbourhood search for the pickup and delivery problem with time windows. Transp. Sci. **40**(4), 455–472 (2006)
23. Franti, P., Virmajoki, O., Hautamaki, V.: Fast agglomerative clustering using a k-nearest neighbor graph. IEEE Trans. Pattern Anal. Mach. Intell. **28**, 1875–1881 (2006)

The Ensemble Kalman Update Is an Empirical Matheron Update

Dan MacKinlay(✉)

CSIRO's Data61, Eveleigh, Australia
dan@danmackinlay.name

Abstract. The Ensemble Kalman Filter (EnKF) is a widely used method for data assimilation in high-dimensional systems, with an ensemble update step equivalent to an empirical version of the Matheron update popular in Gaussian process regression—a connection that links half a century of data-assimilation engineering to modern path-wise GP sampling.

This paper provides a compact introduction to this simple but under-exploited connection, with necessary definitions accessible to all fields involved.

Source code is available at https://github.com/danmackinlay/paper_matheron_equals_enkf.

Keywords: Ensemble Kalman Filter · Matheron Update · Gaussian Process · Data Assimilation · Geostatistics

1 Introduction

The Ensemble Kalman Filter (EnKF) [7,8] is a cornerstone method that evolves ensembles of state vectors through model dynamics and updates them using observational data. The Matheron update provides a sample-based method for conditioning Gaussian random variables on observations [6,19,20], well-established in geostatistics but with little-known connections to ensemble data assimilation.

We establish that the ensemble update step in the EnKF is equivalent to an empirical Matheron update by putting them on a common probabilistic footing. This connection provides an alternative foundation for the EnKF and suggests improvements by leveraging computational optimizations from both communities.

1.1 Historical Context and Related Work

Although the affine residual update first appeared in geostatistics under the banner of *conditioning by kriging* [4], it was rediscovered many times:

M. Liu et al. (Eds.): AI 2025, LNAI 16371, pp. 56–68, 2026.
https://doi.org/10.1007/978-981-95-4972-6_5

- **Optimal interpolation (OI).** Early numerical-weather-prediction (NWP) systems wrote the analysis as $x_a = x_b + K(y - Hx_b)$ with a climatological Kalman gain; this is exactly the Matheron rule with fixed covariances [9].
- **Stochastic EnKF.** [7] replaced the static climatological covariances of OI with an empirical ensemble, creating the now-ubiquitous EnKF.
- **Fast conditional simulation.** [6] noticed that one can obtain conditional Gaussian draws *without* a Cholesky factorisation by adding a linear residual—again the same formula.
- **Pathwise GP sampling.** The machine-learning community re-invented the idea as *path-wise conditioning* for scalable Gaussian-process (GP) posterior draws [2,19,20].
- **Reservoir inversion and EnRML.** Iterative ensemble smoothers such as EnRML exploit the very same affine map in the context of PDE-constrained Bayesian inversion [3].
- **Theory of ensemble inversion.** Convergence proofs for ensemble inversion [17] and extensions via transport maps [18] all start from the linear Matheron/EnKF identity.

This equivalence forms the algebraic core of at least six research communities, turning a patchwork of field-specific heuristics into transferable technology.

1.2 Notation

We write random variates sans serif, x. Equality in distribution is $\stackrel{\mathrm{d}}{=}$. The law, or measure, of a random variate x is denoted μ_{x}, so that $(\mu_{\mathsf{x}} = \mu_{\mathsf{y}}) \Rightarrow \left(\mathsf{x} \stackrel{\mathrm{d}}{=} \mathsf{y}\right)$. Mnemonically, samples drawn from the μ_{x} are written with a serif $\boldsymbol{x} \sim \mu_{[x]}$. We use a hat to denote empirical estimates, e.g. $\widehat{\mu}_{\mathrm{X}}$ is the empirical law induced by the sample matrix X. When there is no ambiguity we suppress the sample matrix, writing simply $\widehat{\mu}$. We follow standard measure notation; Appendix A provides a density-based translation.

2 Matheron Update

The Matheron update is a technique for sampling from the conditional distribution of a Gaussian random variable given observations, without explicitly computing the posterior covariance [6,19,20].

Lemma 1 (Matheron Update). *Given a jointly Gaussian vector*

$$\begin{bmatrix}\mathsf{x}\\ \mathsf{y}\end{bmatrix} \sim \mathcal{N}\left(\begin{bmatrix}m_{\mathsf{x}}\\ m_{\mathsf{y}}\end{bmatrix}, \begin{bmatrix}\mathrm{C}_{\mathsf{xx}} & \mathrm{C}_{\mathsf{xy}}\\ \mathrm{C}_{\mathsf{yx}} & \mathrm{C}_{\mathsf{yy}}\end{bmatrix}\right), \tag{1}$$

the conditional $\mathsf{x}|\mathsf{y}{=}\boldsymbol{y}^*$ *is equal in distribution to*

$$(\mathsf{x} \mid \mathsf{y}{=}\boldsymbol{y}^*) \stackrel{\mathrm{d}}{=} \mathsf{x} + \mathrm{C}_{\mathsf{xy}}\mathrm{C}_{\mathsf{yy}}^{-1}(\boldsymbol{y}^* - \mathsf{y}). \tag{2}$$

Proof. A standard property of the Gaussian (e.g., [15]) is that the conditional distribution of a Gaussian variate x given $\mathsf{y} = \boldsymbol{y}^*$ defined as in (1) is again Gaussian

$$(\mathsf{x} \mid \mathsf{y}{=}\boldsymbol{y}^*) \sim \mathcal{N}(\boldsymbol{m}_{\mathsf{x}|\mathsf{y}}, \mathrm{C}_{\mathsf{x}|\mathsf{y}}) \tag{3}$$

with moments

$$\begin{aligned}
\boldsymbol{m}_{\mathsf{x}|\mathsf{y}} &= \mathbb{E}[\mathsf{x} \mid \mathsf{y}{=}\boldsymbol{y}^*] && (4)\\
&= \boldsymbol{m}_{\mathsf{x}} + \mathrm{C}_{\mathsf{xy}}\mathrm{C}_{\mathsf{yy}}^{-1}\left(\boldsymbol{y}^* - \boldsymbol{m}_{\mathsf{y}}\right), && (5)\\
\mathrm{C}_{\mathsf{x}|\mathsf{y}} &= \mathrm{Var}\left(\mathsf{x} \mid \mathsf{y}{=}\boldsymbol{y}^*\right) && (6)\\
&= \mathrm{C}_{\mathsf{xx}} - \mathrm{C}_{\mathsf{xy}}\mathrm{C}_{\mathsf{yy}}^{-1}\mathrm{C}_{\mathsf{yx}}. && (7)
\end{aligned}$$

Taking moments of the right hand side of (2)

$$\begin{aligned}
\mathbb{E}\left[\mathsf{x} + \mathrm{C}_{\mathsf{xy}}\mathrm{C}_{\mathsf{yy}}^{-1}(\boldsymbol{y}^* - \mathsf{y})\right] &= \boldsymbol{m}_{\mathsf{x}} + \mathrm{C}_{\mathsf{xy}}\mathrm{C}_{\mathsf{yy}}^{-1}(\boldsymbol{y}^* - \boldsymbol{m}_{\mathsf{y}}) && (8)\\
&= \boldsymbol{m}_{\mathsf{x}|\mathsf{y}} && (9)\\
\mathrm{Var}\left[\mathsf{x} + \mathrm{C}_{\mathsf{xy}}\mathrm{C}_{\mathsf{yy}}^{-1}(\boldsymbol{y}^* - \mathsf{y})\right] &= \mathrm{Var}[\mathsf{x}] + \mathrm{Var}[\mathrm{C}_{\mathsf{xy}}\mathrm{C}_{\mathsf{yy}}^{-1}(\boldsymbol{y}^* - \mathsf{y})] \\
&\quad + \mathrm{Var}(\mathsf{x}, \mathrm{C}_{\mathsf{xy}}\mathrm{C}_{\mathsf{yy}}^{-1}(\boldsymbol{y}^* - \mathsf{y})) \\
&\quad + \mathrm{Var}(\mathsf{x}, \mathrm{C}_{\mathsf{xy}}\mathrm{C}_{\mathsf{yy}}^{-1}(\boldsymbol{y}^* - \mathsf{y}))^\top && (10)\\
&= \mathrm{C}_{\mathsf{xx}} + \mathrm{C}_{\mathsf{xy}}\mathrm{C}_{\mathsf{yy}}^{-1}\mathrm{C}_{\mathsf{yy}}\mathrm{C}_{\mathsf{yy}}^{-1}\mathrm{C}_{\mathsf{yx}} - 2\mathrm{C}_{\mathsf{xy}}\mathrm{C}_{\mathsf{yy}}^{-1}\mathrm{C}_{\mathsf{yx}} && (11)\\
&= \mathrm{C}_{\mathsf{xx}} - \mathrm{C}_{\mathsf{xy}}\mathrm{C}_{\mathsf{yy}}^{-1}\mathrm{C}_{\mathsf{yx}} = \mathrm{C}_{\mathsf{x}|\mathsf{y}} && (12)
\end{aligned}$$

we see that both first and second moments match. Since $\mu\left\{\mathsf{x} \mid \mathsf{y}{=}\boldsymbol{y}^*\right\}$ and $\mathcal{N}(\boldsymbol{m}_{\mathsf{x}|\mathsf{y}}, \mathrm{C}_{\mathsf{x}|\mathsf{y}})$ are Gaussian distributions with the same moments, they define the same distribution.

This *pathwise* approach to conditioning Gaussian variates has gained currency in machine learning as a tool for sampling and inference in Gaussian processes [19,20], notably in generalising to challenging domains such as Riemannian manifolds [2].

3 Kalman Filter

We begin by recalling that in state filtering the goal is to update our estimate of the system state when a new observation becomes available. In a general filtering problem the objective is to form the posterior distribution $\mu_{\mathsf{x}_{t+1}|\mathsf{x}_t,\boldsymbol{y}^*}$ from the prior $\mu_{\mathsf{x}_{t+1}|\mathsf{x}_t}$ by incorporating the new information in the observation $\boldsymbol{y}^*$.

We assume a known observation operator $\mathcal{H}$ such that observations y are a priori related to state x by $\mathsf{y} = \mathcal{H}(\mathsf{x})$; Thus there exists a joint law for the prior random vector

$$\begin{bmatrix}\mathsf{x}\\ \mathsf{y}\end{bmatrix} = \begin{bmatrix}\mathsf{x}\\ \mathcal{H}(\mathsf{x})\end{bmatrix} \tag{13}$$

which is determined by the prior state distribution μ_{x} and the observation operator $\mathcal{H}$. The *analysis* step, in state filtering parlance, is the update at time t of μ_{x_t}, into the posterior $\mu_{\mathsf{x}_t|(\mathsf{y}_t=\boldsymbol{y}_t^*)}$ i.e. incorporating the likelihood of the observation $\boldsymbol{y}_t^* = \boldsymbol{y}_t$. Although recursive updating in t is the focus of the classic Kalman filter, in this work we are concerned only with the observational update step. Hereafter we suppress the time index t, and consider an individual analysis update.

Suppose that the state and observation noise are independent, and all variates are defined over a finite dimensional real vector space $\boldsymbol{x} \in \mathbb{R}^{D_{\mathsf{x}}}, \boldsymbol{y} \in \mathbb{R}^{D_{\mathsf{y}}}$. Suppose, moreover, that at the update step our prior belief about x is Gaussian mean $\boldsymbol{m}_{\mathsf{x}}$ and covariance C_{xx}, that the observation noise is centred Gaussian with covariance R, and the observation operator is linear with matrix H, so that the observation is related to the state via

$$\boldsymbol{y} = \mathrm{H}\boldsymbol{x} + \boldsymbol{\varepsilon}, \quad \boldsymbol{\varepsilon} \sim \mathcal{N}(0, \mathrm{R}).$$

Then the joint distribution of the prior state and observation is Gaussian and (13) implies that it is

$$\begin{bmatrix}\mathsf{x}\\ \mathsf{y}\end{bmatrix} \sim \mathcal{N}\left(\begin{bmatrix}\boldsymbol{m}_{\mathsf{x}}\\ \boldsymbol{m}_{\mathsf{y}}\end{bmatrix}, \begin{bmatrix}\mathrm{C}_{\mathsf{xx}} & \mathrm{C}_{\mathsf{xy}}\\ \mathrm{C}_{\mathsf{yx}} & \mathrm{C}_{\mathsf{yy}}\end{bmatrix}\right) \tag{14}$$

$$= \mathcal{N}\left(\begin{bmatrix}\boldsymbol{m}_{\mathsf{x}}\\ \mathrm{H}\boldsymbol{m}_{\mathsf{x}}\end{bmatrix}, \begin{bmatrix}\mathrm{C}_{\mathsf{xx}} & \mathrm{C}_{\mathsf{xx}}\mathrm{H}^\top\\ \mathrm{H}\,\mathrm{C}_{\mathsf{xx}} & \mathrm{H}\,\mathrm{C}_{\mathsf{xx}}\mathrm{H}^\top + \mathrm{R}\end{bmatrix}\right) \tag{15}$$

When an observation $\boldsymbol{y}^*$ is obtained, we apply the formulae (3), (5), and (7) to calculate

$$(\mathsf{x} \mid \mathsf{y}{=}\boldsymbol{y}^*) \sim \mathcal{N}(\boldsymbol{m}_{\mathsf{x}|\mathsf{y}}, \mathrm{C}_{\mathsf{x}|\mathsf{y}}) \tag{16}$$

$$\boldsymbol{m}_{\mathsf{x}|\boldsymbol{y}} = \boldsymbol{m}_{\mathsf{x}} + \mathrm{K}\left(\boldsymbol{y}^* - \boldsymbol{m}_{\mathsf{y}}\right), \tag{17}$$

$$= \boldsymbol{m}_{\mathsf{x}} + \mathrm{K}\left(\boldsymbol{y}^* - \mathrm{H}\,\boldsymbol{m}_{\mathsf{x}}\right), \tag{18}$$

$$\mathrm{C}_{\mathsf{x}|\mathsf{y}} = \mathrm{C}_{\mathsf{xx}} - \mathrm{K}\mathrm{C}_{\mathsf{yx}} \tag{19}$$

$$= \mathrm{C}_{\mathsf{xx}} - \mathrm{K}\,\mathrm{H}\,\mathrm{C}_{\mathsf{xx}}. \tag{20}$$

where

$$\mathrm{K} :- \mathrm{C}_{\mathsf{xy}}\,\mathrm{H}^\top\left(\mathrm{C}_{\mathsf{yy}}\right)^{-1}, \tag{21}$$

$$:- \mathrm{C}_{\mathsf{xx}}\,\mathrm{H}^\top\left(\mathrm{H}\,\mathrm{C}_{\mathsf{xx}}\,\mathrm{H}^\top + \mathrm{R}\right)^{-1}, \tag{22}$$

is the *Kalman gain*. Hereafter we consider a constant diagonal R for simplicity, so that $\mathrm{R} = \rho^2\mathrm{I}_{D_{\mathsf{y}}}$.

4 Ensemble Kalman Filter

In high-dimensional or nonlinear settings, directly computing these posterior updates is often intractable. The Ensemble Kalman Filter (EnKF) addresses this issue by representing the belief about the state empirically, via an ensemble of N state vectors sampled from the prior distribution,

$$\mathrm{X} = \left[\boldsymbol{x}^{(1)}\ \boldsymbol{x}^{(2)}\ \cdots\ \boldsymbol{x}^{(N)}\right], \tag{23}$$

and working with the empirical measure $\widehat{\mu}_{\mathrm{X}} \approx \mu$.

Gaussian measures are specified entirely by their first two moments, so we aim to construct empirical measures which match the desired target in terms of these moments. For convenience, we introduce notation for, respectively, matrix mean and deviations,

$$\bar{\mathrm{X}} :- \frac{1}{N}\sum_{i=1}^{N} \boldsymbol{x}^{(i)} \qquad \breve{\mathrm{X}} :- \frac{1}{\sqrt{N-1}}\left(\mathrm{X} - \bar{\mathrm{X}}\,\mathbf{1}^{\top}\right) \tag{24}$$

where $\mathbf{1}^{\top}$ is a row vector of N ones. The ensemble mean (25) and covariance (27) are computed from the empirical measure,

$$\widehat{\mathbb{E}}[\mathsf{x}] :- \mathbb{E}_{\mathsf{x}\sim\widehat{\mu}_{\mathrm{X}}}[\mathsf{x}] = \widehat{\boldsymbol{m}}_{\mathsf{x}} = \bar{\mathrm{X}} \tag{25}$$

$$\widehat{\mathrm{Var}}(\mathsf{x}) :- \mathrm{Var}_{\mathsf{x}\sim\widehat{\mu}_{\mathrm{X}'}}(\mathsf{x}) = \widehat{\mathrm{C}}_{\mathsf{xx}} :- \frac{1}{N-1}\sum_{i=1}^{N}\left(\boldsymbol{x}^{(i)} - \widehat{\boldsymbol{m}}_{\mathsf{x}}\right)\left(\boldsymbol{x}^{(i)} - \widehat{\boldsymbol{m}}_{\mathsf{x}}\right)^{\top} \tag{26}$$

$$= \breve{\mathrm{X}}\breve{\mathrm{X}}^{\top} + (\xi^2 \mathrm{I}_{D_{\mathsf{x}}}), \tag{27}$$

where ξ^2 is a scalar constant that introduces a regularisation to the empirical covariance to ensure it is invertible. Abusing notation, we associate a Gaussian distribution with the empirical measure

$$\widehat{\mu}_{\mathrm{X}} \approx \mathcal{N}(\widehat{\mathbb{E}}[\mathsf{x}], \widehat{\mathrm{Var}}(\mathsf{x})) = \mathcal{N}(\bar{\mathrm{X}}, \breve{\mathrm{X}}\breve{\mathrm{X}}^{\top} + \xi^2 \mathrm{I}_{D_{\mathsf{x}}}). \tag{28}$$

For posterior inference about x given $\mathsf{y} = \boldsymbol{y}^*$, we seek an ensemble X' such that[1]

$$\mathbb{E}_{\mathsf{x}\sim\widehat{\mu}_{\mathrm{X}'}}[\mathsf{x}] \approx \mathbb{E}_{\mathsf{x}\sim(\mu_{\mathsf{x}|\mathsf{y}=\boldsymbol{y}^*})}[\mathsf{x}] \tag{29}$$

$$\mathrm{Var}_{\mathsf{x}\sim\widehat{\mu}_{\mathrm{X}'}}(\mathsf{x}) \approx \mathrm{Var}_{\mathsf{x}\sim(\mu_{\mathsf{x}|\mathsf{y}=\boldsymbol{y}^*})}(\mathsf{x}) \tag{30}$$

We overload the observation operator to apply to ensemble matrices, writing

$$\mathrm{Y} :- \mathcal{H}\mathrm{X} :- \left[\mathcal{H}(\boldsymbol{x}^{(1)})\ \mathcal{H}(\boldsymbol{x}^{(2)}) \ldots \mathcal{H}(\boldsymbol{x}^{(N)})\right]. \tag{31}$$

This also induces an empirical joint

$$\widehat{\mu}_{\left[\begin{smallmatrix}\mathrm{X}\\ \mathrm{Y}\end{smallmatrix}\right]} :- \mathcal{N}\left(\begin{bmatrix}\bar{\mathrm{X}}\\ \bar{\mathrm{Y}}\end{bmatrix}, \begin{bmatrix}\breve{\mathrm{X}}\breve{\mathrm{X}}^{\top} + \xi^2\mathrm{I}_{D_{\mathsf{x}}} & \breve{\mathrm{X}}\breve{\mathrm{Y}}^{\top}\\ \breve{\mathrm{Y}}\breve{\mathrm{X}}^{\top} & \breve{\mathrm{Y}}\breve{\mathrm{Y}}^{\top} + \upsilon^2\mathrm{I}_{D_{\mathsf{y}}}\end{bmatrix}\right). \tag{32}$$

Here the regularisation term $\upsilon^2\mathrm{I}_{D_{\mathsf{y}}}$ is introduced to ensure the empirical covariance is invertible, given that the ensemble of observations is typically rank deficient when $D_{\mathsf{y}} > N$.

[1] see [5,10,12–14] for finite-ensemble convergence.

The Kalman gain in the ensemble setting is constructed by plugging in the empirical ensemble estimates (32) to (21) obtaining

$$\widehat{\mathrm{K}} = \widehat{\mathrm{C}}_{\mathsf{xx}} \mathcal{H}^\top \left(\mathcal{H} \widehat{\mathrm{C}}_{\mathsf{xx}} \mathcal{H}^\top + \upsilon^2 \mathrm{I}_{D_\mathsf{y}} + \rho^2 \mathrm{I}_{D_\mathsf{y}}\right)^{-1} \tag{33}$$

where $\rho^2 \mathrm{I}_{D_\mathsf{y}}$ is the observation error covariance matrix. The term $\upsilon^2 \mathrm{I}_{D_\mathsf{y}}$ comes from the regularization of the empirical observation covariance (as defined by the ensemble of Y), while $\rho^2 \mathrm{I}_{D_\mathsf{y}}$ represents the known covariance of the observation noise. Combining these two gives the effective covariance in the observation space.[2]

For compactness we define $\gamma^2 :- \upsilon^2 + \rho^2$ so that the combined covariance is $\mathrm{C}_{\mathsf{yy}} = \breve{\mathrm{Y}}\breve{\mathrm{Y}}^\top + \gamma^2 \mathrm{I}_{D_\mathsf{y}}$. We also define $\mathrm{Y}^* = \boldsymbol{y}^* \mathbf{1}^\top$, so that each column of Y^* equals the observation $\boldsymbol{y}^*$. Then, the analysis update for the ensemble is

$$\mathrm{X}' = \mathrm{X} + \widehat{\mathrm{K}} \left(\mathrm{Y}^* - \mathcal{H}\mathrm{X}\right). \tag{34}$$

i.e. each ensemble member is updated $\boldsymbol{x}^{(i)\prime} \leftarrow \boldsymbol{x}^{(i)} + \widehat{\mathrm{K}} \left(\boldsymbol{y}^* - \mathcal{H} \boldsymbol{x}^{(i)}\right)$. Equating moments, we see that $\widehat{\mu}_{\mathsf{x} \sim \mathrm{X}'} \approx \mu_{\mathsf{x}|\mathsf{y}=\boldsymbol{y}^*}$ as desired. That is, the EnKF analysis equations (33) and (34) can be justified in terms of an empirical approximation to the Gaussian posterior update.

5 Core Result

Proposition 1. *The Empirical Matheron Update is equivalent to Ensemble Kalman Update*

Proof. Under the substitution

$$\boldsymbol{m}_\mathsf{x} \to \bar{\mathrm{X}}, \qquad \mathrm{C}_{\mathsf{xx}} \to \breve{\mathrm{X}}\breve{\mathrm{X}}^\top + \xi^2 \mathrm{I}_{D_\mathsf{x}}, \tag{35}$$

$$\boldsymbol{m}_\mathsf{y} \to \bar{\mathrm{Y}}, \qquad \mathrm{C}_{\mathsf{yy}} \to \breve{\mathrm{Y}}\breve{\mathrm{Y}}^\top + \gamma^2 \mathrm{I}_{D_\mathsf{y}}, \tag{36}$$

$$\mathrm{C}_{\mathsf{xy}} \to \breve{\mathrm{X}}\breve{\mathrm{Y}}^\top, \qquad \mathrm{C}_{\mathsf{yx}} \to \breve{\mathrm{Y}}\breve{\mathrm{X}}^\top, \tag{37}$$

the Matheron update (2) becomes identical to the ensemble update (34).

6 Computational Complexity

The EnKF avoids computing full covariance matrices, instead using empirical ensemble statistics with cost $\mathcal{O}(D_\mathsf{y} N^2 + N^3 + D_\mathsf{x} D_\mathsf{y} N)$. Since ensemble size N is typically much smaller than dimensions D_x, D_y, this is significantly more efficient than the naive Kalman update's $\mathcal{O}(D_\mathsf{y}^3)$ cost.

[2] The EnKF extends to nonlinear observation operators, though Gaussian assumptions no longer hold exactly [8].

7 Capabilities Unlocked by the Equivalence

(i) Linear-time GP sample paths. EnKF algebra lets us compute the Matheron update with $\mathcal{O}(N^2 d)$ cost—or $\mathcal{O}(Nd)$ with localisation—rather than $\mathcal{O}(d^3)$ Cholesky factorizations, enabling multi-million-point GP draws [9,19].

(ii) Streaming and hyper-parameter learning. Joint-state augmentation in EnKF permits on-line estimation of GP length-scales and noise variances [11], giving real-time Bayesian regression as data arrive.

(iii) Differentiable data-assimilation layers. Because the affine map is a re-parameterisation, the complete update is compatible with auto-differentiation. One can embed an EnKF/Matheron block inside a neural network and back-propagate end-to-end, an approach already explored in differentiable simulators [17,18].

(iv) Sparse covariance structure via localisation. Local ensemble transform Kalman filters (LETKF) taper covariances in physical space, yielding block-banded precisions that transfer verbatim to spatial or graph-based GPs [9].

(v) Rigorous error bounds. Inverse-problem analyses prove contraction and bias properties of finite-ensemble EnKF [17]; those proofs now bound the error of path-wise GP samplers for free.

Take-away. Treating EnKF analysis as an empirical Matheron brings mature geophysical tool-kits—localisation, adaptive inflation, square-root variants—into scalable machine-learning while exporting automatic differentiation and sparse GP tricks back to data assimilation.

8 Numerical Illustration: 1D Kriging by Ensemble Kalman Filer

We revisit the classic 1D kriging problem to demonstrate both the direct equivalence of the EnKF update to GP regression and the power of extending this link. We compare three solvers side-by-side: a standard GP implementation, its equivalent EnKF formulation, and the Local Ensemble Transform Kalman Filter (LETKF) [1,9], a popular data assimilation variant that introduces covariance localization. This illustrates how mature geophysical tools can be immediately repurposed for scalable GP inference. Code is available in the supplement. We note that the Ensemble Kalman solvers used here are chosen for ease of implementation rather than efficiency of performance, and so we do not expect to see massive performance gains on these small problems.

Task. Recover a latent field $x \in \mathbb{R}^d$ on a unit grid from $m = d/5$ noisy point samples (scaling with dimension) $y = Hx + \eta$, $\eta \sim \mathcal{N}(0, \tau^2 I)$ with $\tau = 0.2$. The prior is $\mathcal{N}(0, K)$, $K_{ij} = \sigma^2 \exp(-\|r_i - r_j\|^2 / 2\ell^2)$ with $\sigma = 1$, $\ell = 0.2$. We test grid sizes $d \in \{200, 400, 600, 800\}$.

Methods.

- **GP Regression:** The exact posterior computed via `Gaussian ProcessRegressor` from SCIKIT-LEARN. This serves as our ground truth and scales as $\mathcal{O}(m^3)$ with observations.
- **EnKF:** An implementation of the standard Ensemble Kalman Filter using DAPPER [16]. It is the direct, empirical equivalent of the Matheron update and is expected to be computationally cheaper than the exact GP.
- **LETKF:** The Local Ensemble Transform Kalman Filter [1,9], also implemented in DAPPER. This method introduces localization, where each state variable is updated using only a subset of nearby observations. This is a common technique in data assimilation for improving accuracy in high-dimensional systems.

We report **median fit and predict times** obtained with our in-process timer over multiple runs to ensure statistical robustness.

Accuracy. Fig. 1 shows posterior draws from all three methods using squared-exponential basis functions matched to the generating process (the "true prior" setting). We see similar RMSE for the mean estimate between the methods, with a slightly overconfident posterior uncertainty in the ensemble methods.

Speed. Fig. 2 reports fit (training) and predict times as we sweep (top) the number of observations m at fixed state dimension d, and (bottom) the state dimension d at fixed m. The exact GP (scikit-learn) shows the expected $\mathcal{O}(m^3)$ fit scaling. Its predict time grows with d and m—mean-only scales as $\mathcal{O}(d\,m)$, while computing both mean and standard deviation (our default) scales as $\mathcal{O}(d\,m^2)$. In contrast, EnKF and LETKF fits scale roughly linearly in m for fixed ensemble size N (and linearly in d with localization), while prediction is linear in d to read out the analysis mean (and $\mathcal{O}(Nd)$ if we also materialize an ensemble). We do not vary N in these plots. Consequently, as d grows large at fixed m, the DA methods yield flatter predict-vs-d curves than the exact GP, and as m grows at fixed d, DA fits avoid the cubic blow-up of exact GP regression.

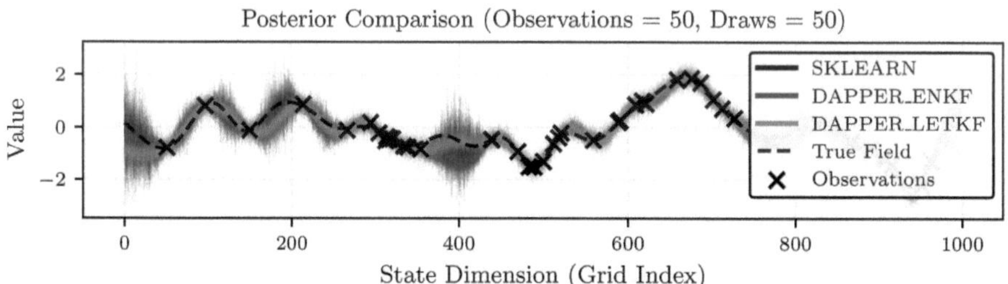

Fig. 1. Posterior draws from the exact GP (sklearn), EnKF, and LETKF.

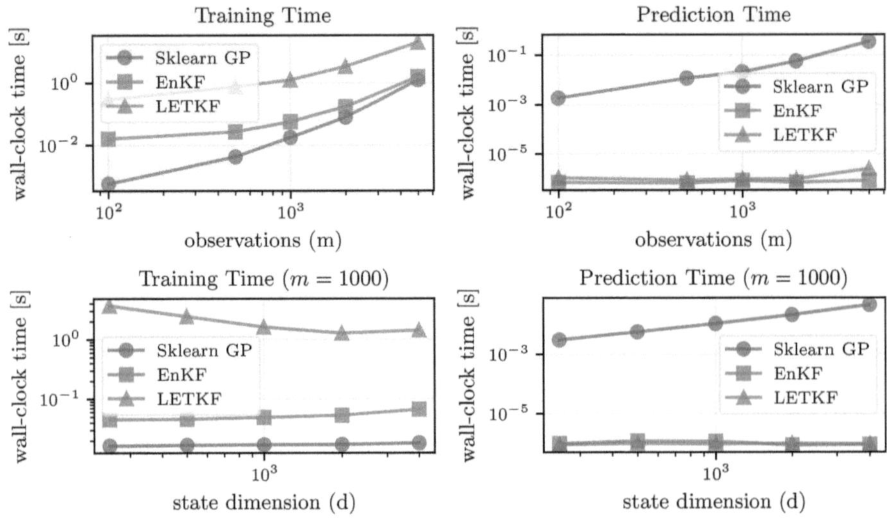

Fig. 2. Wall-clock timings for training and prediction with ensemble and classical GP regression.

Take-Away. The results confirm that a standard EnKF is a computationally equivalent method for path-wise GP sampling. More importantly, the equivalence acts as a bridge to a rich ecosystem of data assimilation techniques. The LETKF demonstrates this: by simply changing one line of code to invoke a localized filter, we gain access to a highly scalable inference method that parallels sparse or domain-decomposition approaches in the GP literature. This opens the door for the GP community to leverage decades of mature, practical DA engineering "for free."

9 Conclusion and Implications

The affine residual identity is a Rosetta stone linking kriging, optimal-interpolation, ensemble-Kalman, and path-wise GP sampling. This synthesis upgrades each field: geoscientists inherit differentiable, variational, and sparse-GP tools; machine-learning researchers gain decades of localisation, inflation, and convergence theory; and inverse-problem analysts get a unifying algebra that bridges finite-ensemble and infinite-dimensional limits.

The literature of the EnKF is vast, and we have only scratched the surface of the many variants and extensions that have been proposed. Many of the ideas developed for the EnKF are thus potentially applicable to Gaussian Process regression via the Matheron update. For example, it provides a clear probabilistic interpretation of the ensemble update that may inspire improved regularization and localization strategies.

By recasting the EnKF update as an empirical Matheron update, we've exposed a straightforward, unifying mechanism behind two seemingly different approaches. This connection cuts through the usual complexity and suggests that many of the practical tricks developed for the EnKF—like improved regularization—could be reinterpreted and possibly enhanced when applied to Gaussian process problems. It's a reminder that sometimes a simple change in perspective can open up entirely new avenues for improvement.

The author thanks CSIRO's Machine Learning and Artifical Intelligence Future Science Platform for supporting this research. The original of this content was drafted by human hands, with the exception of the experiments section and supporting code which was generated by LLMs.

Acknowledgments. An LLM was used to 1) translate an initial draft from markdown 2) propose copy edits to the document and 3) to generate the code scaffolding for the experiments.

Disclosure of Interests. The author has no competing interests to declare that are relevant to the content of this article.

A Measures as Densities

In many machine learning and data assimilation applications, probability measures are represented via probability density functions (pdfs) with respect to the Lebesgue measure. For a continuous random variable x taking values in $\mathbb{R}^{D_{\mathsf{x}}}$, we denote its density by $p_{\mathsf{x}}(\boldsymbol{x})$ so that for any measurable set $A \subset \mathbb{R}^{D_{\mathsf{x}}}$,

$$\mathbb{P}(\mathsf{x} \in A) = \int_A p_{\mathsf{x}}(\boldsymbol{x})\, \mathrm{d}\boldsymbol{x}. \tag{38}$$

The expectation of any measurable function $f : \mathbb{R}^{D_{\mathsf{x}}} \to \mathbb{R}$ is given by

$$\mathbb{E}[f(\mathsf{x})] = \int_{\mathbb{R}^{D_{\mathsf{x}}}} f(\boldsymbol{x})\, p_{\mathsf{x}}(\boldsymbol{x})\, \mathrm{d}\boldsymbol{x}. \tag{39}$$

We can write out the results of this paper in terms of densities, but since the primary object of our interest is the empirical measure and the moments of the theoretical and empirical measures, densities are not strictly necessary and result in a rather longer exposition. Instead, we hope to reassure machine-learners that the two are equivalent in this appendix.

Dirac Functionals and Empirical Measures

In many practical settings, we work with a finite sample $\{\boldsymbol{x}^{(i)}\}_{i=1}^N$ drawn from the true distribution $p_{\mathsf{x}}(\boldsymbol{x})$. The empirical measure associated with this sample is defined as

$$\widehat{\mu}_{\mathrm{X}}(A) = \frac{1}{N} \sum_{i=1}^{N} \mathbf{1}_A(\boldsymbol{x}^{(i)}), \tag{40}$$

where $\mathbf{1}_A$ is the indicator function for the set A. When we wish to express the empirical measure in density notation (with respect to the Lebesgue measure), we represent it as a sum of Dirac delta functions:

$$p_{\text{emp}}(\boldsymbol{x}) = \frac{1}{N}\sum_{i=1}^{N}\delta(\boldsymbol{x}-\boldsymbol{x}^{(i)}). \tag{41}$$

Note that the *Dirac delta* is not a function in the classical sense but a *distribution* (or generalized function). In functional analysis, a Dirac delta centered at $\boldsymbol{x}^{(i)}$ is defined as a linear functional $\delta_{\boldsymbol{x}^{(i)}}$ on a space of test functions (typically smooth functions with compact support) such that

$$\langle \delta_{\boldsymbol{x}^{(i)}}, f\rangle = f(\boldsymbol{x}^{(i)}), \tag{42}$$

for any test function f. Here, the angle brackets $\langle\cdot,\cdot\rangle$ denote the action of the distribution on f. This *sifting property* ensures that when integrating a test function against the empirical density, we recover the sample average:

$$\int_{\mathbb{R}^{D_{\mathsf{x}}}} f(\boldsymbol{x})\, p_{\text{emp}}(\boldsymbol{x})\,\mathrm{d}\boldsymbol{x} = \frac{1}{N}\sum_{i=1}^{N}\langle \delta_{\boldsymbol{x}^{(i)}}, f\rangle = \frac{1}{N}\sum_{i=1}^{N} f(\boldsymbol{x}^{(i)}). \tag{43}$$

In this sense, the empirical measure is exactly the sum of Dirac functionals centered at each sample point, and the "density" p_{emp} should be understood in the distributional sense. This formulation is particularly useful when comparing the empirical measure to the true measure p_{x}. Although the empirical measure is singular with respect to the Lebesgue measure (since it concentrates mass on a finite set of points), many operations (such as taking expectations of smooth functions) remain well-defined.

From Measures to Densities in Gaussian Conditioning

For example, suppose the true prior for x is given by a density $p_{\mathsf{x}}(\boldsymbol{x})$ (say, a Gaussian)

$$p_{\mathsf{x}}(\boldsymbol{x}) = \frac{1}{(2\pi)^{D_{\mathsf{x}}/2}\,|\mathrm{C}_{\mathsf{xx}}|^{1/2}}\exp\!\left(-\tfrac{1}{2}(\boldsymbol{x}-\boldsymbol{m}_{\mathsf{x}})^{\top}\mathrm{C}_{\mathsf{xx}}^{-1}(\boldsymbol{x}-\boldsymbol{m}_{\mathsf{x}})\right),$$

and the likelihood $p_{\mathsf{y}|\mathsf{x}}(\boldsymbol{y}|\boldsymbol{x})$ is similarly specified. Then the joint density is

$$p_{\mathsf{x},\mathsf{y}}(\boldsymbol{x},\boldsymbol{y}) = p_{\mathsf{x}}(\boldsymbol{x})\,p_{\mathsf{y}|\mathsf{x}}(\boldsymbol{y}|\boldsymbol{x}), \tag{44}$$

and the conditional density is given by

$$p_{\mathsf{x}|\mathsf{y}}(\boldsymbol{x}|\boldsymbol{y}^*) = \frac{p_{\mathsf{x}}(\boldsymbol{x})\,p_{\mathsf{y}|\mathsf{x}}(\boldsymbol{y}^*|\boldsymbol{x})}{\int_{\mathbb{R}^{D_{\mathsf{x}}}} p_{\mathsf{x}}(\boldsymbol{x}')\,p_{\mathsf{y}|\mathsf{x}}(\boldsymbol{y}^*|\boldsymbol{x}')\,\mathrm{d}\boldsymbol{x}'}. \tag{45}$$

In our work, while we initially denote measures by μ. (which emphasizes their abstract nature), one can equivalently work with the densities described above. Thus, updating the ensemble according to the empirical version of the Kalman or Matheron update can be seen as operating directly on these Dirac-based representations. In practical implementations, one does not manipulate the Dirac deltas directly; instead, one works with the sample values and their empirical moments, which—as shown earlier—are equivalent (in the limit of large N) to the full probabilistic update based on the density.

This density-based view, incorporating Dirac functionals to represent empirical measures, is particularly common in machine learning, where it provides an intuitive bridge between abstract measure-theoretic probability and the concrete computations performed with finite samples.

The need to throw around integrals, introduce and worry about density normalisation and so on, we find a little distracting and thus it is not used in the main text.

References

1. Bocquet, M., Farchi, A., Malartic, Q.: Online learning of both state and dynamics using ensemble Kalman filters. Found. Data Sci. **3**(3), 305–330 (2020). https://doi.org/10.3934/fods.2020015
2. Borovitskiy, V., Terenin, A., Mostowsky, P., Deisenroth, M.P.: Matérn Gaussian processes on Riemannian manifolds. In: Advances in Neural Information Processing Systems. arXiv (2023). https://doi.org/10.48550/arXiv.2006.10160
3. Chen, Y., Oliver, D.S.: Ensemble randomized maximum likelihood method as an iterative ensemble smoother. Math. Geosci. **44**(1), 1–26 (2012). https://doi.org/10.1007/s11004-011-9376-z
4. Chilès, J.-P., Desassis, N.: Fifty years of kriging. In: Daya Sagar, B.S., Cheng, Q., Agterberg, F. (eds.) Handbook of Mathematical Geosciences, pp. 589–612. Springer, Cham (2018). https://doi.org/10.1007/978-3-319-78999-6_29
5. Del Moral, P., Kurtzmann, A., Tugaut, J.: On the stability and the uniform propagation of chaos of a class of extended ensemble kalman-bucy filters. SIAM J. Control. Optim. **55**(1), 119–155 (2017). https://doi.org/10.1137/16M1087497
6. Doucet, A.: A Note on Efficient Conditional Simulation of Gaussian Distributions. Technical Report, University of British Columbia (2010)
7. Evensen, G.: The ensemble kalman filter: theoretical formulation and practical implementation. Ocean Dyn. **53**(4), 343–367 (2003). https://doi.org/10.1007/s10236-003-0036-9
8. Evensen, G.: Data Assimilation - The Ensemble Kalman Filter. Springer, Heidelberg (2009). https://doi.org/10.1007/978-3-642-03711-5
9. Hunt, B.R., Kostelich, E.J., Szunyogh, I.: Efficient data assimilation for spatiotemporal chaos: a local ensemble transform Kalman filter. Physica D **230**(1), 112–126 (2007). https://doi.org/10.1016/j.physd.2006.11.008
10. Kelly, D.T.B., Law, K.J.H., Stuart, A.M.: Well-posedness and accuracy of the ensemble Kalman filter in discrete and continuous time. Nonlinearity **27**(10), 2579 (2014). https://doi.org/10.1088/0951-7715/27/10/2579
11. Kuzin, D., Yang, L., Isupova, O., Mihaylova, L.: Ensemble kalman filtering for online gaussian process regression and learning. In: 2018 21st International Conference on Information Fusion (FUSION), pp. 39–46 (2018). https://doi.org/10.23919/ICIF.2018.8455785

12. Kwiatkowski, E., Mandel, J.: Convergence of the square root ensemble kalman filter in the large ensemble limit. SIAM/ASA J. Uncertainty Quant. **3**(1), 1–17 (2015). https://doi.org/10.1137/140965363
13. Le Gland, F., Monbet, V., Tran, V.D.: Large sample asymptotics for the ensemble Kalman filter. In: The Oxford Handbook of Nonlinear Filtering,, p. 598. Oxford University Press (2011)
14. Mandel, J., Cobb, L., Beezley, J.D.: On the convergence of the ensemble Kalman filter. Appl. Math. **56**(6), 533–541 (2011). https://doi.org/10.1007/s10492-011-0031-2
15. Petersen, K.B., Pedersen, M.S.: The Matrix Cookbook. Technical report (2012)
16. Raanes, P.N., Chen, Y., Grudzien, C., Tondeur, M., Dubois, R.: DAPPER: data assimilation with python: a package for experimental research (2024). https://doi.org/10.5281/zenodo.10710355
17. Schillings, C., Stuart, A.M.: Analysis of the ensemble kalman filter for inverse problems. SIAM J. Numer. Anal. **55**(3), 1264–1290 (2017). https://doi.org/10.1137/16M105959X
18. Spantini, A., Baptista, R., Marzouk, Y.: Coupling techniques for nonlinear ensemble filtering. SIAM Rev. **64**(4), 921–953 (2022). https://doi.org/10.1137/20M1312204
19. Wilson, J.T., Borovitskiy, V., Terenin, A., Mostowsky, P., Deisenroth, M.: Efficiently sampling functions from Gaussian process posteriors. In: Proceedings of the 37th International Conference on Machine Learning, pp. 10292–10302. PMLR (2020)
20. Wilson, J.T., Borovitskiy, V., Terenin, A., Mostowsky, P., Deisenroth, M.P.: Pathwise conditioning of gaussian processes. J. Mach. Learn. Res. **22**(105), 1–47 (2021)

Using Individual Problem Instances for Exploratory Black Box Optimisation Benchmarking: A Case Study Using XOR Neural Networks

Marcus Gallagher(✉)

School of Electrical Engineering and Computer Science, University of Queensland, Q. 4072, Brisbane, Australia
marcusg@uq.edu.au

Abstract. Considerable progress has been made recently in the development of methodologies, software libraries, problem sets and landscape analysis towards improving the benchmarking of black box optimisation algorithms. The majority of this work focuses on algorithm evaluation at a large/broad scale, by evaluating algorithms on large sets of different problem instances, across multiple dimensionalities. Broad evaluations are useful in providing a more complete picture of the behaviour of an algorithm. However they necessarily involve the summarization of experimental results, which can mean insights at the individual problem level are lost. This paper presents a case study of exploratory, per-instance algorithm evaluation, using real-world-representative problem instances from training small neural networks. The results illustrate that insights can be gained from looking at individual problem instances that would not be seen in a broad summative analysis. The work also highlights some potential problems that can exist in summative comparisons if care is not taken at the instance level.

Keywords: Black box optimisation · Algorithm benchmarking · Neural networks · XOR

1 Introduction and Background

Black box optimisation algorithms such as metaheuristics and evolutionary algorithms have seen a significant amount of research and have been applied to a wide variety of practical problems. They make minimal assumptions about the problem and are relatively easy to implement and deploy, however there are considerable challenges in carrying out experimental evaluation, comparison and benchmarking of these algorithms [1]. There has been significant interest in this topic in recent years, leading to advances in experimental methodologies and practice, as well as software libraries [4,5,16,23] and benchmark problem sets [10] that facilitate more rigorous evaluation and comparison of algorithms,

M. Liu et al. (Eds.): AI 2025, LNAI 16371, pp. 69–80, 2026.
https://doi.org/10.1007/978-981-95-4972-6_6

including better reproducibility of experimental studies [9,12]. In addition, landscape analysis techniques provide more insight into understanding the nature of optimisation problems, which is essential in order to understand what types of problems a given algorithm performs well/poorly on and why.

Much of the research in this area facilitates and carries out large scale experimental studies, where numerous algorithm instances are compared on numerous problem instances across different dimensionalities. This approach is valuable as it provides an overall picture of how algorithms compare on problems with different features and avoids the possibility that, e.g. if two algorithms (A and B) are compared on a small selected set of problems, we might see that A always outperforms B, but the opposite could be true for problems that are different to those in the small selected problem set. However, to make sense of a large set of experimental results, it is necessary to summarize and quantize the data, to extract information such as a ranking of algorithm performance across all problems tested. Doing so implies a loss of information, meaning that insights regarding algorithm performance on individual problems (or even individual trials) are lost. Ideally, the experimental evaluation of algorithms should include a mixture of both large/broad scale, as well as exploratory per-instance studies to provide maximum insight. Exploratory analysis can be useful as both a precursor to help frame/specify large/broad scale experiments, or afterwards to "zoom in" and examine anomalies or specific trends observed in the results data. It can be particularly useful for real-world problem instances where some properties of the problems are unknown or less clear than artificially designed benchmark problems.

Training neural networks is a class of real-world optimisation problems of enormous interest, given the successes and progress of deep learning models in recent years [17]. One of the fundamental questions in this area is understanding why variants of stochastic gradient descent are so successful in practice at optimising networks with millions - trillions of parameters, despite theoretically being a difficult, highly nonconvex problem (see, e.g. [18]). The exclusive-OR (XOR) problem was one of the first examples that demonstrated the potential power of multilayer perceptron/feedforward neural networks to learn nonlinearly separable classification problems [21], as opposed to earlier perception models [15,20]. Although the XOR problem is tiny compared to current deep neural networks, it is very well-studied (see [2] for a review of earlier work). Recent work has made progress in analysing neural network loss landscapes (including XOR) in terms of the mathematical properties of the critical points of the landscapes and convergence behaviour of gradient descent [6,24]. Neural network loss landscapes have also been studied from a (black box) fitness landscape perspective [3,14] and benchmark problems have been proposed using neural network training [13,19].

While considerable progress has been made, there are still some gaps to connect and determine how to translate insights from theory and landscape analysis into better understanding of the experimental behaviour of black box algorithms. This paper aims to show that one way to close this gap is to carry out exploratory, per-instance algorithm evaluation. The results illustrate that insights can be gained from looking at individual problem instances that would not be seen in a broad summative analysis. The work also highlights some potential problems that can exist in summative comparisons if care is not taken at the instance level.

The focus of this work is on developing better methodologies for the benchmarking of optimisation algorithms. From this point of view, using XOR and small neural networks is reasonable since this provides a challenging source of problem instances that are also fast to evaluate. However the techniques proposed can also be applied to any continuous optimisation problem, including instances derived from training large, deep neural networks such as those used in current applications of AI.

An outline of the rest of the paper is as follows. In Sect. 2, the XOR neural network problem instances are defined. Section 3 describes the algorithm instances used and presents experimental results and analysis. Section 4 discusses the limitations of the experiments, Sect. 5 comments on reproducibility of the work and Sect. 6 provides conclusions.

2 XOR and Neural Network Models

The binary XOR function is specified as a supervised learning problem with 2 inputs, 1 output and 4 data points: $(0, 0 \rightarrow 1), (0, 1 \rightarrow 1), (1, 0 \rightarrow 1), (1, 1 \rightarrow 0)$. Note that there is no test set: the goal is simply to train a neural network that can approximate these data points (input-output mapping) as closely as possible. Creating an optimisation problem instance for this task requires the specification of the network architecture and objective function. The output of a standard feedforward network for this problem is given by:

$$\hat{y} = h(h(\mathbf{x}\mathbf{w}_1 + \mathbf{w}_{b1})\mathbf{w}_2 + w_{b2}) \tag{1}$$

where $\mathbf{w}_i$ is a vector of weights in the i-th layer, $\mathbf{w}_{b1}$ the bias weights in the 1st layer, w_{b2} the bias weight in the 2nd layer and $h()$ is the logistic activation function:

$$h(z) = \frac{1}{1 + e^{-z}} \tag{2}$$

For the objective function we use mean-squared error (MSE):

$$f(\theta) = \frac{1}{4}\sum_{n=1}^{4}(y_n - \hat{y}_n(\theta)^2) \tag{3}$$

where θ is a vector of all of the weight and bias weight parameters in the network. Note that the true outputs are 0 or 1 and predicted outputs are bounded between 0 and 1. For predicted outputs to approach 0 or 1, one or more of the weights will be driven towards $\pm\infty$. For the experiments below, we choose a target threshold to be an objective function value of 0.01, since all outputs will be very close to their target values when this is achieved.

2.1 2-2-1 XOR Network

The most common network used for XOR is a 2-2-1 network (i.e. 2 inputs, 2 hidden units and 1 output unit). The model has 9 weight parameters (including bias weights in each layer).

The loss landscape (error surface) of this optimisation problem has been previously analysed [8,11]. In terms of critical points, it has been shown that there are no local optima[1], however there are several regions of the landscape that are saddle points as well as the global optima. The critical points can be described in terms of the 4 outputs $\{y_1, y_2, y_3, y_4\}$ produced by the network:

(a) (Saddle points): $\{\frac{1}{2}, \frac{1}{2}, \frac{1}{2}, \frac{1}{2}\} \rightarrow f(\theta) = 0.25$
(b) (Saddle points): $\{\frac{1}{2}, \frac{1}{2}, 1, 0\}$ or $\{0, 1, \frac{1}{2}, \frac{1}{2}\} \rightarrow f(\theta) = 0.125$
(c) (Saddle points): $\{\frac{2}{3}, \frac{2}{3}, \frac{2}{3}, 0\}$ or $\{0, \frac{2}{3}, \frac{2}{3}, \frac{2}{3}\} \rightarrow f(\theta) = 0.1\dot{6}$
(d) (Saddle points): $\{\frac{1}{3}, 1, \frac{1}{3}, \frac{1}{3}\}$ or $\{\frac{1}{3}, \frac{1}{3}, 1, \frac{1}{3}\} \rightarrow f(\theta) = 0.4\dot{9}$
(e) Global optima: $\{0, 1, 1, 0\} \rightarrow f(\theta) = 0$

This is a continuous, unconstrained optimisation problem. Saddle points (b)-(d), as well as the global optima, are asymptotic since they require one or more of the weights to be infinite. Such points can be attractors for optimisation algorithms and are approached numerically. It is also known that the landscape has a (permutation) symmetry due to the possibility of swapping the two hidden units. The loss landscape is composed of $2! = 2$ equivalent regions. This symmetry is evident in (b)-(d) above but also exists for (a) and (e).

This is a 9-dimensional optimisation problem. We use a soft boundary $[-10, 10]^9$ for the search space. This is used for initialisation of some algorithms but it is not a hard constraint - algorithms can search outside of this boundary (and would do so if attracted to some of the critical points).

2.2 2-3-1 XOR Network

Given that the 2-2-1 network is capable of solving the XOR problem, it follows that any network with 3 or more hidden units is also capable of solving the problem. These larger networks contain the 2-2-1 network as one possible solution (with other weights set to zero), but there may be other solutions in the weight space [22]. This means that benchmark optimisation problems can be created where the dimensionality of the optimisation problem grows. Here, we use a 2-3-1 network, which is a 13-dimensional problem. The symmetry is greater on this landscape: $3! = 6$ symmetric regions.

2.3 2-1-1 XOR Network

It is possible to solve the XOR problem using a network with only a single hidden unit, if "skip" weights/connections are added connecting the inputs directly to the output [11]. This creates a 7-dimensional problem instance. There is no permutation symmetry since we have only 1 hidden unit.

[1] As shown in [8], neural network loss landscapes require a careful definition of a local optimum, since they often have attractors at infinity.

2.4 Landscape Properties

The fitness/loss landscapes of XOR neural networks have some interesting properties that are different to many artificial benchmark functions. The landscapes have no isolated local optima and are dominated by saddle point regions (i.e. a high degree of neutrality) with attractors at infinity. There is also the symmetry property mentioned above.

3 Analysis

3.1 Algorithm Instances

The focus of this paper is *not* to apply a large set of "state-of-the-art" algorithms (and tune their hyperparameters) to try and achieve the best performance on the problem instances. Rather, the goal is to illustrate the kinds of questions that can be asked and insights to be gained on a per-instance exploratory study. We also make the distinction between an "algorithm" (which often means something generic) and an *algorithm instance*, which means that all hyperparameters and experimental settings are completely specified (such that the experiment can be precisely reproduced). To this end, we evaluate and compare three simple algorithm instances:

- Gradient Descent (GD): full-batch gradient descent via backpropagation, with a learning rate equal to 0.5. The search is initialized with each weight chosen randomly from $U[-1, 1]$[2].
- Uniform Random Search (URS): points sampled randomly from $[-10, 10]^d$.
- (1+1)-Evolution Strategy (ES): a simple evolution strategy with a population size of 1, which is effectively a stochastic hill-climber. Mutation is from a Gaussian with standard deviation 0.1.

Unless otherwise stated, algorithms were given a fixed budget of 10^5 function evaluations and were run for 100 trials on each problem. This was intended to be a relatively large budget as the focus is more on gaining insight into the convergence of the algorithms rather than performance in a limited budget setting.

3.2 2-2-1 Network Problem Instance

To establish a performance baseline, perhaps the simplest approach is to run URS on this problem. Figure 1 (left) shows that a budget of 10^5 function evaluations is sufficient to meet the performance threshold with high probability[3]. In the limit, uniform random search will become arbitrarily close to the global optimum. Nevertheless, in practice we see the expected slow convergence rate towards the optimal solution value.

---

[2] Note that this choice is a use of prior knowledge (non-black box), in addition to using the gradient.

[3] 10^7 evaluations were arranged as 100 trials of 10^5 evaluations for ease of comparison.

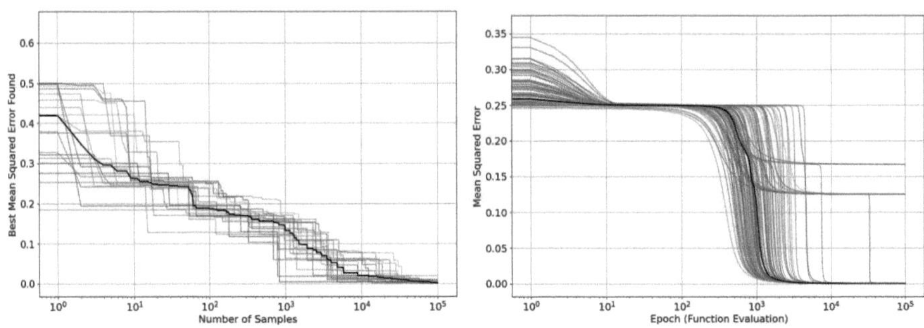

Fig. 1. Convergence curves (100 trials) for URS (left) and GD (right) on the 2-2-1 XOR problem. Black curve shows the median.

Figure 1 (right) shows the performance of GD on this problem. As expected, we see that gradient descent is typically able to solve the problem with a considerably smaller budget compared to URS. Note however that the algorithm can be attracted to one of the saddle point regions. Some runs eventually escape regions of very shallow gradient, while other runs may never reach a global optimum, no matter how large the budget.

For the specific GD instance tested, the results show that the progress of all of algorithm runs are slowed dramatically in the vicinity of saddle point region (a). However, all of the runs eventually escape this value. A small number of runs then become trapped by saddle point regions (b) or (c). They do not escape for the budget given. Further investigation could determine whether these runs are likely to escape, e.g. by looking at the change in weights, gradient and Hessian matrix over time.

A simple Evolution Strategy was also run on the problem. Figure 2 shows the empirical runtime distribution curves for all three algorithms to reach the 0.01 target objective value. We can make some interesting observations:

- For a small budget (less than $\sim 10^3$), the (1+1)-ES is the only algorithm with some chance of reaching the target, although this would happen at most around 20% of the time.
- For a budget between 10^3 and 5×10^4, GD has much better performance, with an 80% chance of finding a solution that meets the target.
- For a large budget (more than $\sim 5 \times 10^4$), URS becomes the algorithm with the highest success rate. In general this is probably a regime where the budget is too large for meaningful comparison of algorithms, but we see here the effect of some runs of GD being stuck in the vicinity of a saddle point.

The empirical runtime distribution results in Fig. 2 illustrate the advantage of the fixed target approach and the anytime assessment of the algorithms as opposed to a fixed budget experiment (where clearly a choice can be made that results in any one of the algorithms having the best performance) [9].

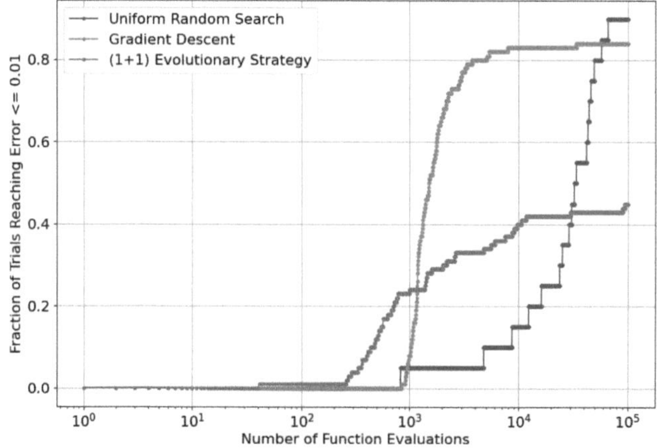

Fig. 2. Empirical runtime distribution curves for the URS, GD and (1+1)-ES algorithm instances to reach the target value on the 2-2-1 XOR network.

3.3 2-3-1 Network Problem Instance

Results for the algorithm instances on the 2-3-1 network problem are shown in Fig. 3 in terms of empirical runtime distribution curves to reach the target value. While the picture is somewhat similar to Fig. 2, the results also suggest that the problem is easier to solve. This is evident because all three curves rise sooner and more steeply. The (1+1)-ES curve has the largest difference. Convergence to non-optimal critical points has become less of a problem for GD; the URS curve only moves above GD at the limit of the budget. The URS curve is much smoother in Fig. 3 compared to Fig. 2, suggesting that the objective values that dominate in the 2-2-1 network are less so for the 2-3-1 network.

3.4 2-1-1 Network Problem Instance

Figure 4 shows the empirical runtime distribution curves for the algorithm instances on the 2-1-1 problem. The results show that GD performs very well on this problem, with the target approximately guaranteed after $\sim 10^4$ evaluations. Comparing with Fig. 2 and Fig. 3, both the 2-1-1 and 2-3-1 instances are easier for GD than the classic 2-2-1 instance. In contrast, the (1+1)-ES does not perform as well on the 2-1-1 problem compared to the 2-3-1 instance. Most surprising in Fig. 4 is the dramatic drop in performance of URS. The budget for this algorithm was extended to 5×10^5 to find any trials capable of meeting the target. It seems that a much larger budget would be required to increase the success rate. One might assume that this problem would be easier for URS given the lower dimensionality. However the results suggest that solutions that meet the target are much rarer on the landscape. Nevertheless, the gradient properties of the landscape make the problem easy for GD.

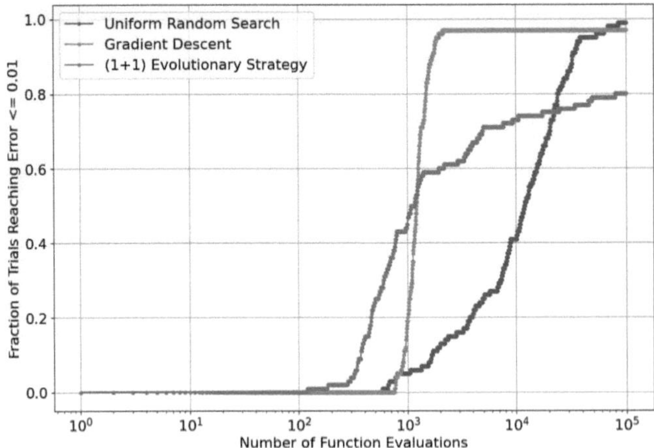

Fig. 3. Empirical runtime distribution curves for the URS, GD and (1+1)-ES algorithm instances to reach the target value on the 2-3-1 XOR network.

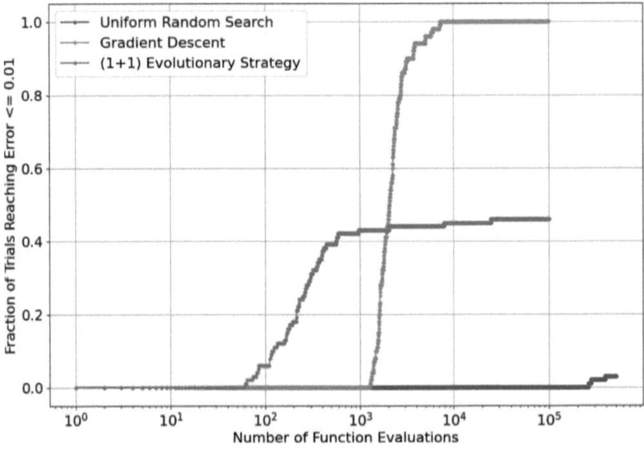

Fig. 4. Empirical runtime distribution curves for the URS, GD and (1+1)-ES algorithm instances to reach the target value on the 2-1-1 XOR network.

3.5 Inverting XOR to Learn XNOR

One of the general advantages of black box optimizers is that they make minimal assumptions about the problem instance. From a benchmarking point of view, this allows researchers to easily explore a wider variety of problem types which may help gain a better understanding of algorithm strengths and weaknesses. While training an XOR neural network is a much-studied problem, the critical points identified in Sect. 2.1 do not include *maxima*. If the network predicts the binary opposite value for each of the four data points, the (global maximum) MSE value will be 1. This network will actually implement the exclusive-NOR function, hence it is possible to invert an XOR problem instance and maximize (or minimize the negative) of the MSE objective function. This approach is not generally recommended, since the objective function is still based on the squared difference between predictions and outputs for the XOR data. Hence, we might expect that the topology of the problem landscape would not be very informative for learning XNOR. Figure 5 shows the results of running our GD, URS and (1+1)-ES instances on this "inverse XOR" problem, using a 2-3-1 network. Clearly, GD is never able to solve this problem (Fig. 5, left); the algorithm is always attracted to a saddle point region (d). In constrast, URS explores the landscape at higher MSE values and is approaching the optimal value (Fig. 5, centre). The performance of the (1+1)-ES is considerably different from both GD and URS (Fig. 5, right). The results show that more than half of the ES runs are stuck in saddle point region (d), but the other runs escape or explore larger values. Out of 100 runs, most get stuck at $f() = 0.75$, while 4 of the runs converge towards the global optimum. The maxima of these loss landscapes do not seem to have been previously studied, so it is interesting to consider if challenging black box optimisation benchmarks can be created in this way. Solutions with $f() = 0.75$ can be seen as ridge-like regions in 2D projections of the landscape in [24].

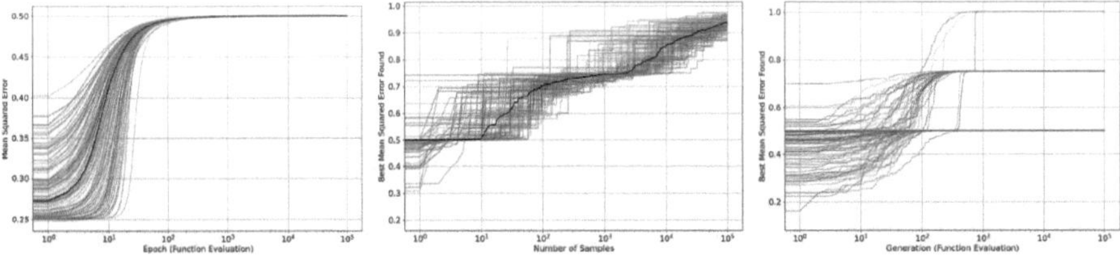

Fig. 5. Convergence curves for GD (left), URS (centre) and (1+1)-ES (right) on the inverse-XOR-2-3-1 problem. The median (over the 100 trials) is shown in black.

3.6 Runtime Distributions and Convergence Distributions

Empirical runtime distributions have been used above as they provide an anytime view of algorithm performance [9]. While limited budget scenarios are clearly of practical interest, it is also interesting to consider the behaviour of an algorithm when the budget can grow arbitrarily large. That is, what is the distribution of points that an algorithm will converge to given a large budget? URS (and many other stochastic algorithms) can be shown to converge to the global optimum, where this distribution would be a Dirac delta function over the global optimum value. This regime is not particularly interesting and for many problems is infeasible in experimentation. However, if an algorithm has a stable convergence distribution for a smaller (but still large) budget, it provides insights into the operators and dynamics of the algorithm. For example, Fig. 5 (right) suggests that the convergence distribution for the (1+1)-ES instance tested has three distinct values for the inverse XOR-2-3-1 problem. Consider running a different black box algorithm which converges to only 2 values. What mechanism in this new algorithm leads to it avoiding one of the convergence values? These types of questions are tied closely to both the algorithm and problem instance landscape.

4 Limitations

It is important to note that the experiments results presented in this paper are highly dependent on the configuration and hyperparameter settings of the algorithm instances used. While URS is essentially parameter-free, the performance of GD and the (1+1)-ES could be very different if different configuration/hyperparameters were used. For example:

- GD is known to be sensitive to the initialization. Changing this can make the performance dramatically better/worse. The value of the learning rate is similarly a key hyperparameter. Finally, gradient descent provides a good standard baseline but is not a black box algorithm so it may not be fair to make a direct comparison.
- The (1+1)-ES is likely to be limited by the fixed mutation variance. An adaptive approach (e.g. 1/5 success rule) would likely perform much better. Also, increasing the parent/offspring populations would likely improve performance (but tuning of these values is required).
- Changing aspects of the problem, such as the activation function used in the neural network, the objective function or the boundary constraint will create different problem instances. Previous research has made different choices for these factors, therefore some results may be incomparable or not generalize.

It is possible to make these claims since these algorithms are relatively simple and well-understood. This would be difficult for some complex algorithms, however.

5 Resources for Reproducibility

The problem instances identified in this paper should provide a small, useful benchmark set for further research (6 problems: XOR_2-2-1, XOR_2-3-1, XOR_2-1-1 and the inverse (maximization problem) for each), with a soft boundary constraint $[-10, 10]^d$ for each and a target objective value of 0.01 (or 0.99 for the inverse problems). Algorithm budgets can be guided by the results above - roughly speaking, $10^2 - 10^5$ should be appropriate. Code to implement these problem instances as well as reproduce the experiments is available [7]. Generative AI tools were not used in writing this paper but were used to co-develop some of the code used in the experiments.

6 Conclusion

This paper argues for the use of exploratory per-instance studies for the evaluation and benchmarking of black box optimisation algorithms, as a complementary method to more large/broad-scale experimental studies. A case study on XOR neural network training is presented to show some of the advantages of this approach. While three very simple algorithm instances were used here, the results still led to some unexpected and interesting insights regarding their relative performances.

The methodological point made here is that benchmarking on large sets of problem instances and focusing on end performance is insufficient to provide a deep understanding of the behaviour of black box optimisation algorithms. However an important next step is to actually demonstrate the utility of per instance evaluation, e.g. by carring out a detailed experimental evaluation of some more sophisticated, state-of-the-art black box optimizers on these problems. Integration into an existing framework such as the IOH suite [4] would help facilitate this work. It may also be possible to translate some of the knowledge of critical points in XOR problems to the understanding of black box algorithms.

Disclosure of Interests. The author has no competing interests to declare that are relevant to the content of this article.

References

1. Bartz-Beielstein, T., et al.: Benchmarking in optimization: best practice and open issues. arXiv preprint arXiv:2007.03488 (2020)
2. Bland, R.: Learning XOR: exploring the space of a classic problem. Department of Computing Science and Mathematics, University of Stirling Stirling (1998)
3. Bosman, A.S., Engelbrecht, A., Helbig, M.: Visualising basins of attraction for the cross-entropy and the squared error neural network loss functions. Neurocomputing **400**, 113–136 (2020)
4. Doerr, C., Wang, H., Ye, F., van Rijn, S., Bäck, T.: IOHprofiler: a benchmarking and profiling tool for iterative optimization heuristics. arXiv e-prints:1810.05281 (2018). https://arxiv.org/abs/1810.05281
5. Dreo, J., et al.: Paradiseo: from a modular framework for evolutionary computation to the automated design of metaheuristics. In: Proceedings of the Genetic and Evolutionary Computation Conference Companion, pp. 1522–1530 (2021)
6. Frye, C.G., Simon, J., Wadia, N.S., Ligeralde, A., DeWeese, M.R., Bouchard, K.E.: Critical point-finding methods reveal gradient-flat regions of deep network losses. Neural Comput. **33**(6), 1469–1497 (2021)
7. Gallagher, M.: Reproducibility and additional files. Zenodo (2025). https://doi.org/10.5281/zenodo.17172657
8. Hamey, L.G.: XOR has no local minima: a case study in neural network error surface analysis. Neural Netw. **11**(4), 669–681 (1998)
9. Hansen, N., Auger, A., Brockhoff, D., Tušar, T.: Anytime performance assessment in blackbox optimization benchmarking. IEEE Trans. Evol. Comput. **26**(6), 1293–1305 (2022)

10. Hansen, N., Auger, A., Ros, R., Mersmann, O., Tušar, T., Brockhoff, D.: Coco: a platform for comparing continuous optimizers in a black-box setting. Optim. Methods Softw. **36**(1), 114–144 (2021)
11. Lisboa, P., Perantonis, S.: Complete solution of the local minima in the XOR problem. Netw. Comput. Neural Syst. **2**(1), 119–124 (1991)
12. López-Ibáñez, M., Branke, J., Paquete, L.: Reproducibility in evolutionary computation. ACM Trans. Evolutionary Learn. Optim. **1**(4), 1–21 (2021)
13. Malan, K.M., Cleghorn, C.W.: A continuous optimisation benchmark suite from neural network regression. In: Rudolph, G., Kononova, A.V., Aguirre, H., Kerschke, P., Ochoa, G., Tušar, T. (eds.) Parallel Problem Solving from Nature – PPSN XVII. PPSN 2022. LNCS, vol. 13398, pp. 177–191. Springer, Cham (2022). https://doi.org/10.1007/978-3-031-14714-2_13
14. Mehta, D., Zhao, X., Bernal, E.A., Wales, D.J.: Loss surface of XOR artificial neural networks. Phys. Rev. E **97**(5), 052307 (2018)
15. Minsky, M., Papert, S.: Perceptrons: an introduction to computational geometry. Cambridge tiass., HIT **479**(480), 104 (1969)
16. de Nobel, J., Ye, F., Vermetten, D., Wang, H., Doerr, C., Bäck, T.: IOHexperimenter: benchmarking platform for iterative optimization heuristics. arXiv e-prints:2111.04077 (2021). https://arxiv.org/abs/2111.04077
17. Palagi, L.: Global optimization issues in deep network regression: an overview. J. Glob. Optim. **73**(2), 239–277 (2019)
18. Petzka, H., Sminchisescu, C.: Non-attracting regions of local minima in deep and wide neural networks. J. Mach. Learn. Res. **22**(143), 1–34 (2021)
19. Rapin, J., Gallagher, M., Kerschke, P., Preuss, M., Teytaud, O.: Exploring the MLDA benchmark on the nevergrad platform. In: Proceedings of the Genetic and Evolutionary Computation Conference Companion, pp. 1888–1896 (2019)
20. Rosenblatt, F.: The perceptron: a probabilistic model for information storage and organization in the brain. Psychol. Rev. **65**(6), 386 (1958)
21. Rumelhart, D.E., Hinton, G.E., Williams, R.J.: Learning representations by back-propagating errors. Nature **323**(6088), 533–536 (1986)
22. Sprinkhuizen-Kuyper, I.G., Boers, E.: A local minimum for the 2-3-1 XOR network. IEEE Trans. Neural Netw. **10**(4), 968–971 (1999)
23. Wang, H., Vermetten, D., Ye, F., Doerr, C., Bäck, T.: IOHanalyzer: detailed performance analyses for iterative optimization heuristics. ACM Trans. Evolutionary Learn. Optim. **2**(1) (2022)
24. Yang, X., Arora, K., Bachmann, M.: Dissecting a small artificial neural network. J. Phys. A: Math. Theor. **58**(2), 025001 (2024)

On Knowledge-Informed Deep Learning for Modeling Complex Spatiotemporal Systems

Minh Ngoc Vu[1,2(✉)] and Son N. Tran[1]

[1] Deakin University, Melbourne, VIC, Australia
{n.vu,son.tran}@deakin.edu.au
[2] The University of Melbourne, Melbourne, VIC, Australia

Abstract. Modeling complex spatiotemporal systems remains a challenging task due to nonlinear interactions, chaotic behavior, and the frequent lack of complete mechanistic knowledge. Traditional approaches rely on approximate mathematical representations, while purely data-driven models often demand large volumes of high-quality data and struggle to generalize to unseen regimes. In this work, we explore hybrid deep learning approaches that integrate imperfect prior knowledge with observational data to model complex dynamics. We focus on two representative systems—the Lorenz system and the KuramotoSivashinsky (KS) equation—to study how knowledge quality and data availability influence predictive performance. Two architectures are employed: Echo State Networks (ESNs) and Long Short-Term Memory (LSTM) networks, both adapted to incorporate knowledge-based predictions. Our empirical analysis shows that incorporating prior knowledge substantially improves the performance of ESNs, while LSTMs demonstrate strong baseline predictive capability with gains from knowledge integration and large data. The results highlight the importance of balancing knowledge and data in hybrid modeling and provide empirical evidence for designing robust deep learning frameworks for spatiotemporal chaotic systems.

Keywords: Deep learning · knowledge-informed · complex systems · spatiotemporal dynamics

1 Introduction

Complex systems comprise multiple interacting components that operate within strict requirements, making them essential to real-world applications across diverse domains. These systems are widely utilised in areas such as manufacturing, software engineering, healthcare, and science discovery, where their structured interactions are expected to enable efficient and reliable decision-making in intricate environments. Due to their intrinsic complexity, modeling complex systems and inferring interaction patterns are challenging.

M. Liu et al. (Eds.): AI 2025, LNAI 16371, pp. 81–92, 2026.
https://doi.org/10.1007/978-981-95-4972-6_7

Traditional approaches to modeling complex systems—especially those exhibiting spatiotemporal chaos—depend heavily on mechanistic knowledge. Researchers derive approximate mathematical representations of system dynamics using ordinary or partial differential equations, informed by the underlying physical or biological processes [3,8,10,24]. Knowledge-based modeling of complex systems faces several challenges. Key limitations include incomplete or inaccurate understanding of system interactions, difficulty in capturing nonlinear and dynamic behaviors, and reliance on oversimplified assumptions that may compromise model validity. Building and simulating such models often requires significant time, resources, and domain expertise, making them computationally expensive and challenging to scale. Additionally, these models struggle with emergent behaviors, adaptability to evolving systems, and integrating interdisciplinary knowledge.

With the recent emergence of advanced data-driven approaches, such as deep learning, scientists have discovered a scalable and effective method for modeling complex systems [7,11,13]. These approaches excel at capturing intricate nonlinear relationships and dynamic behaviors, effectively handling high-dimensional data and scaling to accommodate complex interactions. They enable rapid model development through automation and iterative refinement, making them adaptable to evolving systems. Data-driven approaches, while powerful, have several limitations in modeling complex systems. They heavily depend on the quality and quantity of data, with poor or biased data leading to inaccurate or unreliable models. These models often require significant computational resources and can struggle to generalize to new contexts or unseen conditions.

Although there have been initial attempts to combine incomplete knowledge with data for learning the behavior of complex systems—such as modeling spatiotemporal chaotic dynamics [7,18]—further investigation is still needed to understand the potential contributions of deep learning to this process. In particular, the research community is increasingly interested in exploring the trade-offs between the quality of prior knowledge and the quantity of available data, and how these factors jointly influence the performance of integrated modeling approaches. Addressing these questions is essential for assessing the relative value of knowledge-driven versus data-driven components in modeling complex systems, with a particular focus on spatiotemporal chaos. Such a study would provide valuable insights into how different levels of knowledge completeness and data availability impact predictive accuracy and robustness, thereby guiding the design of hybrid learning frameworks.

In this paper, we investigate the application of various deep learning models for modeling complex systems in the presence of noisy or incomplete prior knowledge. To this end, we develop a hybrid approach that integrates observational data with imperfect knowledge by adapting a Long Short-Term Memory (LSTM) architecture. In addition, we extend the use of reservoir computing, a well-established architecture for learning spatiotemporal dynamics [2,18], and employ it as a comparative baseline.

Our empirical study focuses on two widely used benchmark problems: the Lorenz63 system and the KuramotoSivashinsky (KS) equation, both of which exhibit rich nonlinear and chaotic behaviors. We perform extensive experiments to assess how two critical factors—the size of the training dataset and the quality of prior knowledge—affect model performance. This systematic analysis reveals two key findings: (i) Long Short-Term Memory (LSTM) networks show strong potential for modeling complex spatiotemporal dynamics, highlighting opportunities for exploring other advanced deep learning architectures; and (ii) incorporating prior knowledge, even when imperfect or noisy, significantly enhances predictive accuracy, demonstrating that data quantity alone cannot fully compensate for the absence of domain knowledge. We also discuss the practical challenges faced by deep learning architectures, such as LSTMs, when applied to complex dynamical systems. These challenges include handling long-range dependencies, coping with noise in knowledge sources, and ensuring model generalization in highly nonlinear regimes.

The contributions of this paper are twofold:

- We investigate deep learning models, particularly Long Short-Term Memory (LSTM) networks, within a hybrid framework that integrates data and prior knowledge to capture the complexity of spatiotemporal dynamics.
- We examine the trade-off between data quality and knowledge quality, analyzing how these factors influence model performance and generalization.

The remainder of this paper is structured as follows. Section 2 provides a review of the relevant literature, highlighting key developments and positioning our work in relation to existing approaches. Section 3 details the proposed methodology, including its theoretical foundations and practical considerations. In Sect. 4, we present the experimental design, describe the datasets used, and discuss the results obtained from our evaluation. Finally, Sect. 5 summarizes the main contributions of this work and outlines several promising directions for future research.

2 Related Work

Physics-informed machine learning incorporates knowledge of physical laws governing a dataset into the learning process [9]. These approaches rely on well-defined prior knowledge, such as logic rules, relational structures, or general physical laws. In contrast, our project addresses a more challenging scenario: the knowledge of complex systems is often only partially approximate. This is a critical consideration because, in many real-world applications, we lack precise or complete knowledge about the systems we aim to model. By tackling this challenge, our approach seeks to bridge the gap between incomplete knowledge and data-driven learning, creating more adaptable and robust solutions for modeling complex systems.

The past few years have seen rapid algorithmic advances in machine learning (ML) and data-driven modeling, leading to their increasing adoption for simulating and predicting nonlinear dynamical systems [6,11,23]. A key appeal of these data-driven methods is their ability to construct fast and accurate surrogate models trained on high-fidelity data, which can significantly accelerate or improve simulations of complex dynamical systems. In cases where processes are poorly understood but sufficient observational data are available—such as cloud microphysics—data-driven surrogates have even shown potential to outperform traditional physics-based approximations [21,23].

These methods have demonstrated promising results in atmospheric and oceanic modeling, with data-driven parameterizations of subgrid processes improving predictive skill and computational efficiency [1,4,17,20]. Within the turbulence and dynamical systems communities, similar successes have been reported, including accurate recovery of governing equations [19,22], learning closure models for turbulent flows [12], and forecasting chaotic systems such as Lorenz-96 and KuramotoSivashinsky (KS) using recurrent neural networks and reservoir computing [15,16,18,25,26].

Deep learning architectures have been central to these advances. Recurrent models, such as Long Short-Term Memory (LSTM) networks, have been widely used to capture temporal dependencies in nonlinear systems, while convolutional and graph-based networks have been applied to model spatial structures. Reservoir Computing (RC), and in particular Echo State Networks (ESNs), have been shown to efficiently predict high-dimensional chaotic spatiotemporal systems, often outperforming more complex neural architectures when training data are limited [2,5,18]. Hybrid approaches, which integrate incomplete physical knowledge with machine learning models, are also gaining traction, demonstrating improved accuracy and interpretability in modeling complex spatiotemporal dynamics [7,13].

Despite these advances, challenges remain. Many ML models require large amounts of high-quality data and may struggle with generalization to unseen dynamical regimes or parameter changes. Understanding the trade-off between prior knowledge and data availability, as well as developing robust hybrid methods, continues to be an active area of research. This paper addresses these challenges by focusing specifically on these aspects.

3 Methodology

We consider a dynamic system of measurable, state-dependent variables $\mathbf{x} = \{x_1, x_2, ..., x_M\}$. At time t, the state of the system is presented as an M-dimensional vector $\mathbf{u}(t) \in \mathbb{R}^M$. In real-life scenarios, an exact model of the system is often unavailable. However, we may have a theoretical model that, while related, does not fully capture the system's behavior. Such models are typically derived from human understanding of the natural world. We consider it as a knowledge-based approximator $\mathcal{K}$, which predicts the future state given the current one.

$$\mathcal{K}[\mathbf{u}(t)] \approx \mathbf{u}(t + \Delta t) \tag{1}$$

Besides the knowledge of the system, we also have observed data $\mathcal{D} = \{\mathbf{u}(t)|t = -T, ..., 0\}$ obtained from recordings of the systems. With the data, we can train a neural network to represent the system:

$$N[\mathbf{u}(t)] \approx \mathbf{u}(t + \Delta t) \tag{2}$$

Here, $\mathcal{K}$ is interpretable and presents the underlying mechanism of the complexity in the system, although its estimation performance depends on the matching between the knowledge and the system. Meanwhile, N is a black-box that can be effective as long as we have adequately large data for training.

The paper studies a methodology to integrate the knowledge-based estimator $\mathcal{K}$ with the neural network N for more accurate modeling of dynamic systems. The challenge here is that $\mathcal{K}$ does not approximate. This is a critical consideration because, in many real-world applications, we lack precise or complete knowledge about the systems we aim to model. However, approximate knowledge is available.

We begin by developing an integrated model (hybrid model) in which a neural network learns from data while being guided by insights from the knowledge estimator about the system's dynamics.

$$N[\mathbf{u}(t)|\mathcal{D}, \mathcal{K}] \approx \mathbf{u}(t + \Delta t) \tag{3}$$

The integration is first implemented via the input construction. During the training phase, at each time step t, the input to the hybrid model $\mathbf{x}(t) \in \mathbb{R}^{2M}$ is formed by concatenating the current state $\mathbf{u}(t)$ with the output of the knowledge-based model $\mathcal{K}[\mathbf{u}(t)]$.

$$\mathbf{x}(t) = \begin{bmatrix} \mathbf{u}(t) \\ \mathcal{K}[\mathbf{u}(t)] \end{bmatrix} \tag{4}$$

In the prediction phase, the model uses a closed-loop configuration, where the true state $\mathbf{u}(t)$ is replaced by the predicted output $\tilde{\mathbf{u}}(t)$ from the previous step $t - \Delta t$.

$$\mathbf{x}(t) = \begin{bmatrix} \tilde{\mathbf{u}}(t) \\ \mathcal{K}[\tilde{\mathbf{u}}(t)] \end{bmatrix} \tag{5}$$

We consider two hybrid architectures: hybrid ESN and hybrid LSTM. To the best of our knowledge, this is the first work to explore an LSTM-based hybrid approach for modeling spatiotemporal dynamics. For comparison, we also implement base versions of ESN and LSTM that rely solely on $\mathbf{u}(t)$ without any knowledge integration.

Hybrid ESN. ESN is a Recurrent Neural Network (RNN) with a fixed and large reservoir. The neurons in the reservoir are sparsely and randomly connected, forming a high-dimensional dynamical system. The reservoir state at time t is denoted by the vector $\mathbf{r}(t) \in \mathbb{R}^{D_r}$ and is updated according to

$$\mathbf{r}(t + \Delta t) = \tanh(\mathbf{A}\mathbf{r}(t) + \mathbf{W}_{\text{in}}\mathbf{x}(t)), \tag{6}$$

where $\mathbf{A} \in \mathbb{R}^{D_r \times D_r}$ is the adjacency matrix showing the connectivity among reservoir nodes, and $\mathbf{W}_{\text{in}} \in \mathbb{R}^{D_r \times 2M}$ is the input weight matrix.

The prediction is then given by

$$\tilde{\mathbf{u}}(t+\Delta t) = \mathbf{W}_{\text{out}} \begin{bmatrix} \mathcal{K}[\mathbf{u}(t)] \\ \mathbf{r}(t) \end{bmatrix}, \tag{7}$$

where $\mathbf{W}_{\text{out}} \in \mathbb{R}^{M\times(M+D_r)}$ is the output weight matrix.

ESNs train only the output weights, typically via Tikhonov regularized regression, while keeping the input and reservoir weights fixed, making them computationally more efficient than conventional RNNs.

Hybrid LSTM. In contrast to the Hybrid ESN model, which keeps the reservoir weights fixed and only trains the output weights, the Hybrid LSTM model optimizes all parameters, including the input weights, recurrent weights, and biases. At each time step t, the LSTM updates its hidden state $\mathbf{h}(t)$ and cell state $\mathbf{c}(t)$ based on the augmented input $\mathbf{x}(t)$ and previous hidden state $\mathbf{h}(t-\Delta t)$ as

$$\begin{aligned} \mathbf{i}_t &= \sigma\big(\mathbf{W}_{ii}\mathbf{x}(t) + \mathbf{b}_{ii} + \mathbf{W}_{hi}\mathbf{h}(t-\Delta t) + \mathbf{b}_{hi}\big), \\ \mathbf{f}_t &= \sigma\big(\mathbf{W}_{if}\mathbf{x}(t) + \mathbf{b}_{if} + \mathbf{W}_{hf}\mathbf{h}(t-\Delta t) + \mathbf{b}_{hf}\big), \\ \mathbf{g}_t &= \tanh\big(\mathbf{W}_{ig}\mathbf{x}(t) + \mathbf{b}_{ig} + \mathbf{W}_{hg}\mathbf{h}(t-\Delta t) + \mathbf{b}_{hg}\big), \\ \mathbf{o}_t &= \sigma\big(\mathbf{W}_{io}\mathbf{x}(t) + \mathbf{b}_{io} + \mathbf{W}_{ho}\mathbf{h}(t-\Delta t) + \mathbf{b}_{ho}\big), \\ \mathbf{c}(t) &= \mathbf{f}_t \odot \mathbf{c}(t-\Delta t) + \mathbf{i}_t \odot \mathbf{g}_t, \\ \mathbf{h}(t) &= \mathbf{o}_t \odot \tanh(\mathbf{c}(t)), \end{aligned} \tag{8}$$

where $\sigma(\cdot)$ is the sigmoid function and $\odot$ is the Hadamard product. The prediction for the next state is computed by applying a linear transformation to the concatenation of the hidden state and the output of the knowledge-based model:

$$\tilde{\mathbf{u}}(t+\Delta t) = \mathbf{W}_{\text{out}} \begin{bmatrix} \mathcal{K}[\mathbf{u}(t)] \\ \mathbf{h}(t) \end{bmatrix} + \mathbf{b}_{\text{out}} \tag{9}$$

Both the training and prediction phases use the sliding-window approach. Specifically, during training, the model takes the most recent L states, $\{\mathbf{u}(t-(L-1)\Delta t), \mathbf{u}(t-(L-2)\Delta t), \dots, \mathbf{u}(t)\}$, to predict subsequent state $\mathbf{u}(t+\Delta t)$. In the prediction phase, the window is updated in a rolling manner, where the oldest state is discarded and the newly predicted state is appended to the sequence, ensuring a fixed-length window during multi-step prediction.

4 Experiment

4.1 Dataset

Lorenz63 System. The Lorenz63 system [14] consists of three ordinary differential equations:

$$\begin{aligned}\frac{dx}{dt} &= \sigma(y - x),\\ \frac{dy}{dt} &= x(\rho - z) - y,\\ \frac{dz}{dt} &= xy - \beta z.\end{aligned} \tag{10}$$

For the experiment, we use $\sigma = 10, \rho = 28$, and $\beta = 2$ to generate ground truth trajectories. For the knowledge-based approximator, to model the imperfect situation in real-life scenarios, we introduce a small error by setting $\beta = 2.1$, while retaining ρ and σ.

Kuramoto-Sivashinsky Equation. Besides the Lorenz63 system, we also test the models on a high-dimensional spatiotemporal chaos system called the Kuramoto-Sivashinsky (KS) equation. The one-dimensional version of this equation for $y(x, t)$ is:

$$y_t + yy_x + y_{xx} + y_{xxxx} = 0 \tag{11}$$

For the experiment, we simulate the data from the KS equation with the domain size of $L = 35$ and the grid size of $N = 64$ grid points. For the knowledge-based model, we introduce a small error $\epsilon = 0.01$, resulting in a new equation:

$$y_t + yy_x + (1 + \epsilon)y_{xx} + y_{xxxx} = 0 \tag{12}$$

4.2 Setup

For the hybrid ESN, we use a reservoir of $D_r = 100$ units for the Lorenz63 system and $D_r = 500$ units for the KS equation. The non-zero entries of the adjacency matrix $\mathbf{A}$ were drawn from the uniform distribution over the interval $[-1, 1]$, and then the matrix was rescaled to have the spectral radius $\rho = 0.3$.

For the hybrid LSTM, we use a hidden size of 5 neurons for the Lorenz63 system and 10 neurons for the KS equation. The model is trained using the Adam optimizer with a learning rate of 10^{-4}, and a sliding-window approach with $L = 10$ is used to predict the subsequent state.

4.3 Experimental Results

Performance. To illustrate the performance of each model on each system, we simulate 1600 timesteps from each system with a step size of $\Delta t = 0.01$. For ESN models, we discard the first 100 time steps of the trajectory as transient steps to eliminate the influence of arbitrary initial conditions. The remaining 1500 time steps are split into 1000 for training and 500 for testing. For LSTM models, we train on the entire first 1100 time steps and test on the remaining data. The performance of the models is assessed using the Mean Squared Error (MSE) on the test set.

Table 1. MSE comparisons for the two systems

Model	MSE
Base ESN	1.0441
Hybrid ESN	**1.57e-07**
Base LSTM	0.3765
Hybrid LSTM	0.0617

(a) Lorenz63

Model	MSE
Base ESN	0.1614
Hybrid ESN	0.0508
Base LSTM	0.0756
Hybrid LSTM	**0.0449**

(b) KS

According to Table 1(a), the base ESN model has the highest MSE. This is expected since a random, fixed reservoir without additional information makes it difficult to learn the dynamics of chaotic systems. The knowledge-informed LSTM model outperforms the base version, achieving a lower MSE of 0.06 compared to 0.38. This indicates that integrating information from the knowledge-based model improves the performance of LSTMs. Surprisingly, the Hybrid ESN model achieves the best performance, accurately predicting the system with a significantly lower MSE compared to other models. This demonstrates that the hybrid approach remarkably improves the performance of the ESN model when predicting low-dimensional chaotic systems such as the Lorenz63 system.

For the KS equation, Table 1(b) shows that the base ESN model still has the highest MSE among all models. ESN with incorporated knowledge outperforms the base version, but the improvement in MSE is not as significant as in the Lorenz63 system. The hybrid LSTM model yields the lowest error, achieving an MSE of 0.04. This result illustrates the potential of complex deep learning models in forecasting high-dimensional chaotic systems.

Training Data. To investigate the effect of data quantity on model performance, we vary the length of the training dataset, denoted by $T \in [100, 5000]$.

As shown in Fig. 1(a), the MSE of the hybrid ESN model is substantially lower than that of the base model in the Lorenz63 system, regardless of the training sequence length. For the KS equation, Fig. 2(a) shows that the hybrid ESN consistently achieves lower error than the base ESN across different training sizes, although the improvement is less marked than in the Lorenz63 system.

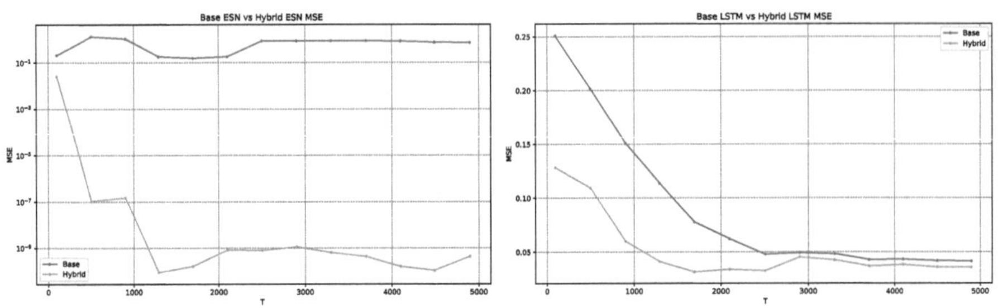

(a) Base ESN vs Hybrid ESN MSE (b) Base LSTM vs Hybrid LSTM MSE

Fig. 1. Performance of the models on different T for Lorenz63

These findings suggest that for the reservoir-based model, the contribution from the knowledge-based approximator is valuable and does not diminish, even if the training sequence length increases.

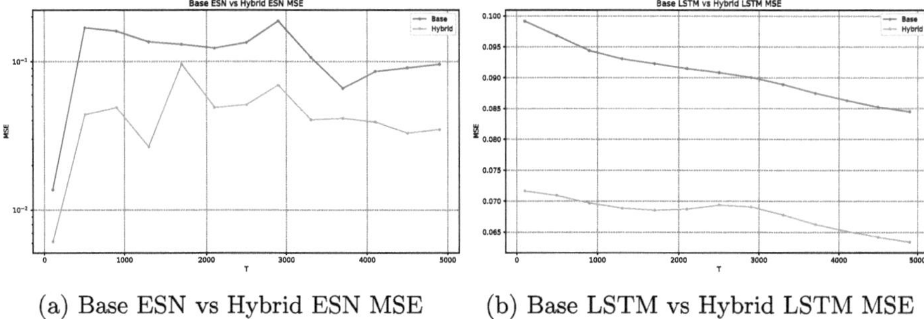

(a) Base ESN vs Hybrid ESN MSE (b) Base LSTM vs Hybrid LSTM MSE

Fig. 2. Performance of the models on different T for KS

For LSTM models, Fig. 1(b) shows that in the Lorenz63 system, the MSE of the hybrid model is initially lower than that of the base model, but the difference decreases as the training sequence length increases. For the KS equation, the hybrid model maintains a lower MSE than the base model, with a small difference of 0.03, as illustrated in Fig. 2(b). These results suggest that incorporating the knowledge-based model improves the performance of LSTMs, especially for small training datasets, while the improvement decreases for larger datasets.

Prediction Steps. After evaluating the impact of the training dataset size, we investigate the performance of models when predicting over longer horizons. Specifically, the training stage is set up similarly to what we did in the Performance section, and we vary the length of the test set within the interval [100, 4000].

Figure 3(a) and 4(a) show that for both dynamical systems, the hybrid ESN model consistently achieves lower MSE compared to the base model across all prediction horizons. This suggests that the incorporation of knowledge-based output improves the multi-step prediction accuracy of the ESN model.

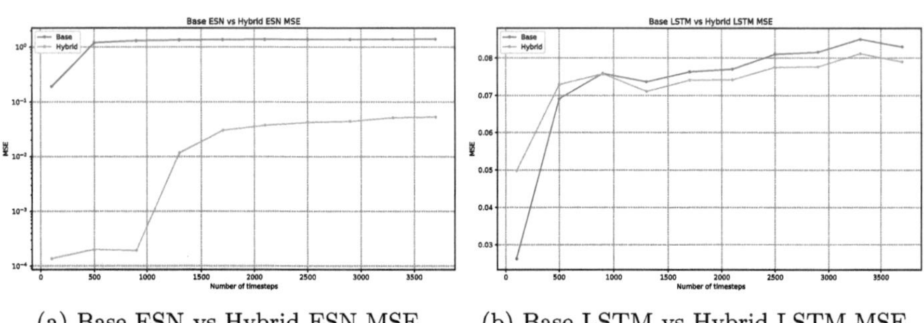

(a) Base ESN vs Hybrid ESN MSE (b) Base LSTM vs Hybrid LSTM MSE

Fig. 3. MSE over the prediction horizon for Lorenz63

For LSTMs, Fig. 3(a) and 4(b) illustrate that both hybrid and base models have relatively low MSEs for both the Lorenz63 system and the KS equation. This indicates that the pure LSTM model is already capable of predicting multiple steps, and the improvement of integrating a knowledge-based approximator is small. Moreover, LSTM models have a lower MSE plateau than the hybrid ESN model, showing that more sophisticated deep learning models provide more consistent performance than simpler approaches like ESNs over longer-term predictions.

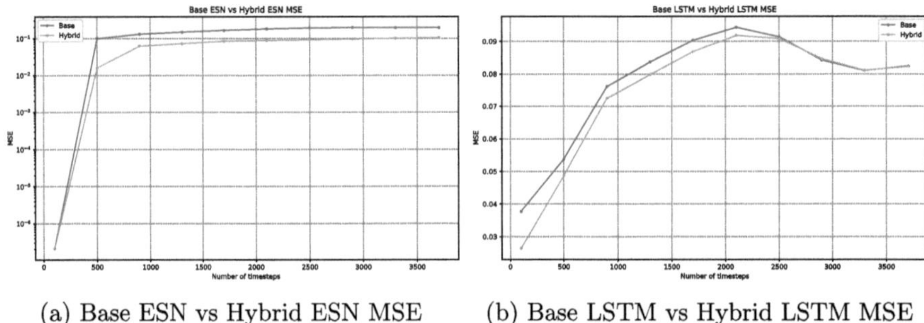

(a) Base ESN vs Hybrid ESN MSE (b) Base LSTM vs Hybrid LSTM MSE

Fig. 4. MSE over the prediction horizon for KS

5 Conclusion and Future Work

This paper investigated hybrid deep learning approaches for modeling complex spatiotemporal systems when only imperfect or noisy prior knowledge is available. We evaluated Echo State Networks (ESNs) and Long Short-Term Memory (LSTM) networks on two benchmark systems: the Lorenz system and the KuramotoSivashinsky (KS) equation. Our findings show that incorporating prior knowledge improves the performance of both ESNs and LSTMs, with particularly notable gains for ESNs in low-dimensional dynamics and meaningful improvements for LSTMs even in high-dimensional settings. Importantly, the benefits of incorporating knowledge persist even when large volumes of training data are available, highlighting its critical role in capturing the complexity of spatiotemporal dynamics.

These results underscore the value of hybrid approaches that combine domain knowledge with data-driven learning. Future work will extend this framework to more advanced architectures, such as Temporal Graph Neural Networks and Transformers, and explore adaptive mechanisms for integrating knowledge of varying quality to enhance the modeling of complex dynamical systems.

References

1. Brenowitz, N.D., Bretherton, C.S.: Prognostic validation of a neural network unified physics parameterization. Geophys. Res. Lett. **45**(12), 6289–6298 (2018). https://doi.org/10.1029/2018GL078510
2. Chattopadhyay, A., et al.: Data-driven predictions of a multiscale lorenz 96 chaotic system using machine-learning methods: reservoir computing, artificial neural network, and long short-term memory network. Nonlinear Process. Geophys. **27**(4), 373–389 (2020). https://doi.org/10.5194/npg-27-373-2020
3. Cross, M.C., Hohenberg, P.C.: Pattern formation outside of equilibrium. Rev. Mod. Phys. **65**(3), 851–1112 (1993). https://doi.org/10.1103/RevModPhys.65.851
4. Gagne, D.J., Christensen, H., Subramanian, A., Poulsen, C.J.: Machine learning for physics parameterization in climate models. J. Adv. Model. Earth Syst. **12**(9), e2020MS002231 (2020). https://doi.org/10.1029/2020MS002231
5. Gauthier, D.J., Bollt, E., Griffith, A., Barbosa, W.A.S.: Next generation reservoir computing. Nat. Commun. **12**(1) (2021). https://doi.org/10.1038/s41467-021-25801-2
6. Gentine, P., Pritchard, M., Rasp, S., Reinaudi, G., Yacalis, G.: Could machine learning break the convection parameterization deadlock? Geophys. Res. Lett. **45**(11), 5742–5751 (2018). https://doi.org/10.1029/2018GL078202
7. Ha, S., Jeong, H.: Unraveling hidden interactions in complex systems with deep learning. Sci. Rep. **11**(1), 12804 (2021). https://doi.org/10.1038/s41598-021-92156-5
8. Holmes, P., Lumley, J.L., Berkooz, G., Rowley, C.W.: Turbulence, Coherent Structures, Dynamical Systems and Symmetry, 2nd edn. Cambridge University Press, Cambridge (2012). https://doi.org/10.1017/CBO9780511919701
9. Karniadakis, G.E., Kevrekidis, I.G., Lu, L., Perdikaris, P., Wang, S., Yang, L.: Physics-informed machine learning. Nat. Rev. Phys. **3**(6), 422–440 (2021). https://doi.org/10.1038/s42254-021-00314-5
10. Kuramoto, Y.: Chemical Oscillations, Waves, and Turbulence. Chemistry Series, Dover Publications, p. 1984. Springer, New York, Heidelberg, Berlin (2003). https://doi.org/10.1007/978-3-642-69689-3. originally published
11. Kutz, J.N.: Deep learning in fluid dynamics. J. Fluid Mech. **814**, 1–4 (2017). https://doi.org/10.1017/jfm.2016.803
12. Ling, J., Kurzawski, A., Templeton, J.: Reynolds averaged turbulence modeling using deep neural networks with embedded invariance. J. Fluid Mech. **807**, 155–166 (2016). https://doi.org/10.1017/jfm.2016.615
13. Liu, C., Xu, F., Gao, C., Wang, Z., Li, Y., Gao, J.: Deep learning resilience inference for complex networked systems. Nat. Commun. **15**(1), 9203 (2024). https://doi.org/10.1038/s41467-024-53303-4
14. Lorenz, E.N.: Deterministic nonperiodic flow. J. Atmos. Sci. **20**(2), 130–141 (1963). https://doi.org/10.1175/1520-0469(1963)020⟨0130:DNF⟩2.0.CO;2
15. McDermott, P., Wikle, C.K.: Data-driven forecasting of dynamical systems. Environ. Ecol. Stat. **24**(2), 317–339 (2017). https://doi.org/10.1007/s10651-017-0376-6
16. McDermott, P., Wikle, C.K.: Modeling dynamical systems with recurrent neural networks. J. Time Ser. Anal. **40**(6), 892–912 (2019). https://doi.org/10.1111/jtsa.12444
17. O'Gorman, P.A., Dwyer, J.: The use of machine learning to improve parameterizations of clouds and convection. J. Adv. Model. Earth Syst. **10**, 2548–2563 (2018). https://doi.org/10.1029/2018MS001472

18. Pathak, J., Hunt, B.R., Girvan, M., Lu, Z., Ott, E.: Model-free prediction of large spatiotemporally chaotic systems from data: a reservoir computing approach. Phys. Rev. Lett. **120**(2), 024102 (2018). https://doi.org/10.1103/PhysRevLett.120.024102
19. Raissi, M., Perdikaris, P., Karniadakis, G.E.: Physics-informed neural networks: a deep learning framework for solving forward and inverse problems involving nonlinear partial differential equations. J. Comput. Phys. **378**, 686–707 (2019). https://doi.org/10.1016/j.jcp.2018.10.045
20. Rasp, S., Pritchard, M.S., Gentine, P.: Deep learning to represent subgrid processes in climate models. Proc. Natl. Acad. Sci. **115**(39), 9684–9689 (2018). https://doi.org/10.1073/pnas.1810286115
21. Reichstein, M., et al.: Deep learning and process understanding for data-driven earth system science. Nature **566**, 195–204 (2019). https://doi.org/10.1038/s41586-019-0912-1
22. Rudy, S.H., Brunton, S.L., Proctor, J.L., Kutz, J.N.: Data-driven discovery of partial differential equations. Sci. Adv. **4**(4), eaat2621 (2018). https://doi.org/10.1126/sciadv.aat2621
23. Schneider, T., Lan, S., Stuart, A., Teixeira, J.: Earth system modeling 2.0: a blueprint for models that learn from observations and targeted high-resolution simulations. Geophys. Res. Lett. **44**(24), 12,396–12,417 (2017). https://doi.org/10.1002/2017GL076101
24. Strogatz, S.H.: Nonlinear Dynamics and Chaos: With Applications to Physics, Biology, Chemistry and Engineering. Westview Press (2000)
25. Vlachas, P.R., Byeon, W., Wan, Z.Y., Sapsis, T.P., Koumoutsakos, P.: Data-driven forecasting of high-dimensional chaotic systems with long short-term memory networks. Proc. R. Soc. A **474**(2213), 20170844 (2018). https://doi.org/10.1098/rspa.2017.0844
26. Zhu, Y., Brunton, S.L., Kutz, J.N.: Deep learning for predicting spatiotemporal chaos. Phys. Rev. E **100**(3), 032208 (2019). https://doi.org/10.1103/PhysRevE.100.032208

Trajectory Segmentation and Self-Supervised Classification for Subtask Discovery

Runze Tang(✉) and Penny Sweetser

The Australian National University, Canberra, ACT, Australia
{runze.tang,penny.kyburz}@anu.edu.au

Abstract. Discovering and solving subtasks individually is a common approach to tackle long-horizon tasks. However, the existing approaches for subtask discovery primarily aim to accelerate downstream task training and often fail to identify semantically meaningful subtasks. We aim to discover semantically meaningful subtasks, in a robotic dataset where the demonstrations complete subtasks in varying orders, to enable sequential execution of subtask policies. We first segment the demonstrations into subtask working and transition phases in a semantically meaningful way. We then employ a self-supervised approach that is agnostic to the number of labels to classify the trajectory segments to obtain distinct subtasks. We trained a set of independent policies solely on the corresponding subtask demonstration segments. The policy learns to decide when to switch to the next subtask policy. Our experiment showed that executing these subtask policies sequentially achieved 98% of the total rewards of a baseline policy trained on complete demonstrations. This result demonstrates that our method identifies semantically meaningful subtasks with accurate boundaries, ensuring the seamless execution of multiple independent subtask policies.

Keywords: Robot Manipulation · Behaviour Cloning · Subtask Discovery · Self-supervised classification

1 Introduction

In hierarchical reinforcement learning, goal-conditioned policy [9,17] and the Options framework [1,28] are well-established approaches for addressing long-horizon tasks. Both methodologies tackle such tasks by defining sub-goals or solving subtasks. Behaviour cloning (BC) is one of the methods in imitation learning [30]. The BC policy learns from expert demonstrations in a supervised learning manner, enabling it to learn long-horizon tasks more efficiently than through exploration. We aim to identify subtasks from demonstrations that involve the completion of several subtasks in varying orders, enabling the sequential execution of subtask policies in different orders.

Most existing methods for subtask discovery focus on accelerating the downstream task training. One of the common approaches is maximising the likelihood of demonstrations using a latent variable model [11,13]. These methods

M. Liu et al. (Eds.): AI 2025, LNAI 16371, pp. 93–105, 2026.
https://doi.org/10.1007/978-981-95-4972-6_8

are effective in environments with discrete state spaces and short-horizon tasks. They often encounter difficulties in continuous environments with a large state space. Some works explored the discovery of subtasks for policy reuse, such as solving long-horizon tasks with varying subtask orders [29]. However, their discovered subtasks are more like short fragments of time without semantic meaning and are not aligned with the human-defined subtasks.

Our work identifies subtasks from offline long-horizon demonstrations in a continuous robot manipulation environment. We exploit the inherent structure of the dataset, wherein each demonstration consists of a sequence of subtasks in a specific order, to identify forking and merging states in the gripper's trajectory and segment each demonstration accordingly. We partition the trajectory segments into working and transition segments to comprehensively model the entire task. We then employ a self-supervised classification method to differentiate the working segments and obtain distinct subtasks. Our work makes three key contributions to the field of robotics subtask discovery. First, we address the demonstration alignment problem by employing a diffusion trajectory predictor and utilising dispersion as a criterion to identify forking and merging states as boundaries to segment the demonstration trajectory. Second, based on a self-supervised classification architecture similar to SwAV [2], we proposed a method to automatically merge similar clusters. This approach does not require specifying the number of clusters and yields a more nuanced and accurate classification compared to using a fixed number of clusters. Third, we proposed saturated-early-stop and voting-for-stopping strategies to mitigate the failures caused by minor deviations in the segmentation.

2 Related Works

Subtask discovery, akin to option discovery or skill discovery, identifies distinct primitives that can be combined to solve long-horizon tasks. Many works [5,17,24,29] discover and represent the subtasks as continuous sub-goals or dense latent representations, which guide the policy progressively at each state to solve long-horizon tasks. In contrast, our approach aims to discover subtasks in a discrete and explicit form.

Subtask discovery from demonstrations usually has two steps: (1) segment the demonstration trajectory and (2) classify the segments to obtain subtask trajectories. Srinivas et al. [26] and Zhu et al. [31] employ the clustering method to segment demonstration trajectories based on state features before identifying the subtasks within them. However, the objective of the clustering method does not align with the trajectory segmentation objective of obtaining semantically meaningful segments. Another commonly used method to discover options is through maximising the likelihood of the demonstration trajectory with a latent variable model [13,27], such as the variational auto-encoder (VAE) [12]. These methods identify the trajectory segment point and simultaneously discover the subtasks. However, the maximising likelihood method may yield an incorrect result due to an underspecified objective [11]. Jiang et al. [11] minimise the code length to discover semantically meaningful subtasks, but only in environments with discrete and small state and action sets. Our approach exploits the difference among multiple automatically aligned demonstration trajectories in continuous state space and is diagnostic of the action to find the segmentation points of the demonstration.

The landmark states are used as sub-goals in the goal-conditioned policy and initial states or termination conditions in the Options framework to lead the agent to reach the destination [9,10,19,26]. Instead, we discover the merging and forking states by the divergence of the preceding and succeeding states. The merging and forking states can be considered as the entrance and the exit of the bottleneck states, respectively. The bottleneck states in our work correspond to the working phase of a subtask, in which the trajectories are nearly identical.

Deep clustering learns representations that are clustering-friendly and applies clustering methods to the representation space to solve the self-supervised classification problem [20]. Contrastive learning [2,6,16] is a widely used method that learns representations from positive and negative sample pairs. Time is a suitable choice for defining positive and negative samples [22] in robot manipulation tasks. After learning the representation, classifying without requiring the number of clusters as prior knowledge is the next challenge. Ren et al. [21] exploits the characteristic of density-based clustering [14] to classify samples without knowing the number of clusters. Shah and Koltu [23] formulates the clustering as a continuous optimisation problem and merges the similar representatives of the cluster to obtain the final number of clusters. We use a contrastive architecture similar to SwAV [2,29], but we start with a large number of clusters and gradually merge similar ones to obtain distinct clusters without knowing the number of clusters at the beginning.

3 Approach

In this paper, we aim to discover subtasks from the demonstration dataset. We follow the basic setting of behaviour cloning, in which a policy is trained without interacting with an environment but by a dataset $\mathcal{D}$ containing demonstrations $\tau = ((s_0, a_0), (s_1, a_1), \ldots)$. Each demonstration completes a long-horizon task, consisting of several subtasks. We first align the trajectory via a sliding window to discover the forking and merging states in each demonstration. The forking and merging states are where the robot's 3D trajectory forks to or merges from different directions, caused by having different preceding and succeeding subtasks. Then, a self-supervised classifier is applied to the trajectory segments to obtain distinct subtask labels. Finally, the labelled subtask demonstration segments can be used to train policies for each subtask.

3.1 Forking and Merging State Discovery

We first introduce how we discover the forking state s_f. Intuitively, a forking state s_f at forking time step f of a set of trajectories is where the preceding

Algorithm 1 Discover forking states using dispersion

Input: A 3D point trajectory τ with length T, converged parameter θ of trajectory predictor p_θ, prediction window size n, prediction amount j and prediction mask h.
Output: A list of forking state time steps in the trajectory.

Initialise an empty dispersion-time curve C.
for $t \in [n, T-n)$ **do**
  $P = \emptyset$
  for j times **do**
    $\tilde{\tau} = p_\theta(\tau_{t-n:t})$ // Predict the future 3D trajectories.
    $P \leftarrow P \cup \tilde{\tau}_{-h:}$ // Record the last h predicted points.
  end for
  $P \leftarrow \text{Normalize}(P)$
  $X \leftarrow \text{ConvexHull}(P)$
  $\delta_t \leftarrow \sup_{x \in X} \{\min_{p \in P} \{||x, p||_2\}\}$ // Calculate dispersion.
  Add dispersion δ_t at time step t to curve C.
end for
Smooth curve C using Savitzky-Golay and moving window filters.
return The time step t of all peaks in curve C as forking state time steps.

trajectories $\tau_{f-n:f}$ are similar but the succeeding trajectories $\tau_{f:f+n}$ fork to different directions, where n is the aligning window size. The same forking state may occur at different time steps in different trajectories due to having different preceding subtasks of different lengths. How the similarity and dissimilarity metric should be defined to align the trajectory window and discover the forking states is case-sensitive and challenging because the trajectory diversity in each window may not be consistent.

We embed the alignment problem and the similarity metric into the training of a future trajectory predictor. The neural network predictor $p_\theta(\tau_{t:t+n}|\tau_{t-n:t})$ predicts the future trajectory $\tau_{t:t+n}$ given the past trajectory $\tau_{t-n:t}$ to align the trajectories window. Diffusion models [8] formulate the generation task as a denoising process, which provides strong multimodal fitting capability. We train a diffusion trajectory predictor to remember the distribution of the possible future trajectory given the past trajectory. To eliminate the interference of irrelevant information, such as the illumination and irrelevant objects in the observed images, we use the 3D position and rotation of the robot arm gripper as the trajectory for predictor training. The rotation information is not used for the following forking state discovery, but it helps the predictor distinguish states with the same position but different rotations.

After aligning the past trajectory, we use dispersion to evaluate the dissimilarity of the future trajectory. Dispersion is a metric that evaluates the degree of uniformity of a set of points. The dispersion of a finite sample set P in a metric space (X, ρ) is defined by

$$\delta(P) = \sup_{x \in X} \left\{ \min_{p \in P} \{\rho(x, p)\} \right\}.$$

Intuitively, if the metric space X is a 3D space with Euclidean distance as the metric ρ, the dispersion of a point set P is the radius of the largest ball at $x \in X$

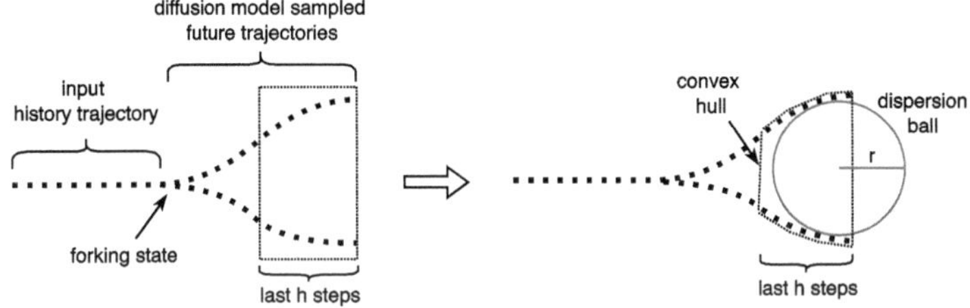

Fig. 1. An illustration of how the dispersion is calculated. Time progresses from left to right. The dots represent the position of the robot arm gripper. The dispersion is the radius of the largest ball that doesn't overlap with any trajectory points with centre inside the convex hull.

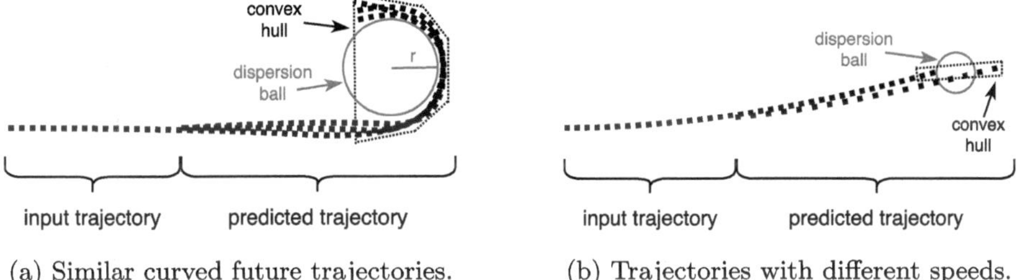

(a) Similar curved future trajectories. (b) Trajectories with different speeds.

Fig. 2. Example of predicted trajectories having unexpectedly large dispersion due to the use of too many (left) or too few points (right) for dispersion calculation. For similar future trajectories, the expected dispersion should be smaller than the distance between two consecutive trajectory points.

that can fit into the space without overlapping with any points p in the point set P. We use dispersion to represent the size of the cavity in the predicted future trajectories to evaluate the dissimilarity. Figure 1 illustrates how the dispersion is calculated. We sample the future trajectory within the aligned window by letting the diffusion predictor predict multiple times, given the same input past trajectory. For multiple predicted future trajectories $\tau_{t:t+n}$ with the same input past trajectory $\tau_{t-n:t}$, a larger dispersion means the trajectories forked to more separated directions rather than one direction with high variance. In practice, we use the last $h = 4$ positions $\tau_{t+n-h:t+n}$ of the predicted trajectory to calculate the dispersion. Setting $h = 4$ can prevent unexpectedly large dispersion caused by trajectories with curves or varying speeds, as Fig. 2 shows. We define the boundary of the space using the convex hull of the last h positions to limit the point $x \in X$. To calculate the approximate dispersion, we sample multiple points x inside the convex hull and then move them in the opposite direction of the nearest point until reaching the edge of the convex hull or a local minimum. Thus, the largest distance among all the sampled points to their nearest trajectory point is the approximate dispersion. After calculating the dispersion of the predicted trajectories at each time step, the peaks in the smoothed dispersion-

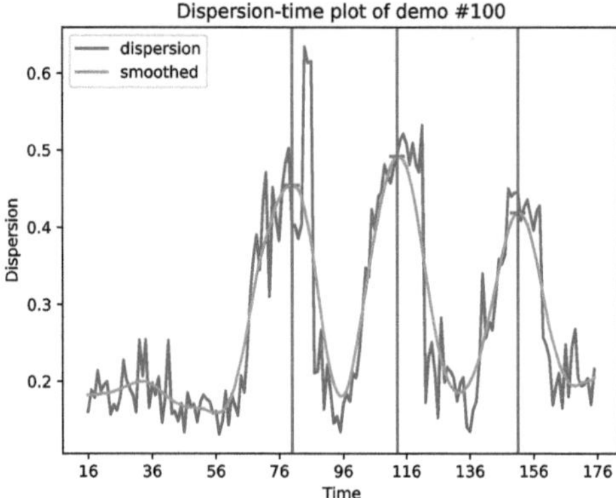

Fig. 3. An example dispersion-time curve. The blue curve is the raw dispersion-time curve. The orange curve is the smoothed curve. The peaks are marked by the magenta vertical lines. (Color figure online)

time curve are used as the forking states for the trajectory segmentation. The pseudo-code of discovering the forking states of a trajectory is shown in Algorithm 1. Figure 3 is an example dispersion-time curve.

The forking state s_f with forward time order is equivalent to the merging state s_m in a reversed time order. The forking states discovery algorithm can be applied after reversing the time of the trajectories to discover the merging states. After discovering all the forking and merging states in every trajectory of the dataset demonstration, we obtain the trajectory segments to be classified.

3.2 Self-Supervised Trajectory Segments Classification

We divide the trajectory segments into two categories, the working trajectory segment τ_u and the transition trajectory segment τ_v. In a working trajectory segment $\tau_{u_i} = \tau_{m_i:f_i}$, the robot arm is working on the i-th subtask in the episode. It initializes from the i-th merging state s_{m_i} and terminates at the i-th forking state s_{f_i}. The transition trajectory segment $\tau_{v_i} = \tau_{f_i:m_{i+1}}$, from the i-th forking state s_{f_i} to the $(i+1)$-th merging state $s_{m_{i+1}}$, represents the robot arm moving to the initial state set of the $(i+1)$-th subtask from the terminal state set of the i-th subtask. In self-supervised subtask classification, only the working trajectory segments τ_u are used. The transition trajectory segment τ_{v_i} will be given the same subtask label as its succeeding working trajectory segment $\tau_{u_{i+1}}$ so that the policy can learn to bridge the succeeding subtasks to the current one.

We use a self-supervised trajectory segment classification framework similar to SwAV [2,29], as Fig. 4 shows. Each working trajectory segment τ_u is first interpolated to the same length and augmented to extract the feature $\mathbf{z}$. Then the feature $\mathbf{z}$ is mapped to K normalized learnable prototypes $\mathbf{C} = [\mathbf{c}_1, \ldots, \mathbf{c}_K]$, which serve as centroids of clusters, to obtain the similarity $\mathbf{Q} = [\mathbf{q}_1, \ldots, \mathbf{q}_K]$ between feature $\mathbf{z}$ and each prototype $\mathbf{c}_k$. The number K is a hyperparameter. The model minimises a cross-entropy loss and an InfoNCE loss [18]. The

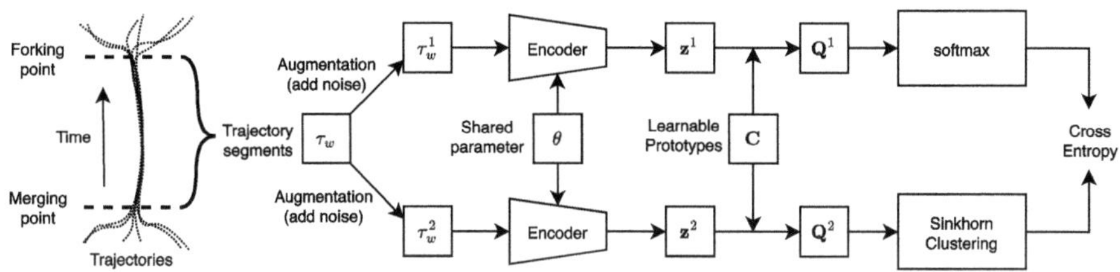

Fig. 4. The architecture of the self-supervised trajectory segment classifier.

cross-entropy loss is calculated between the softmax similarity $softmax(\mathbf{Q})$ of sample τ_u^1 and the result of Sinkhorn-Knopp clustering [2,4] on the similarity $\mathbf{Q}$ of sample τ_u^2. The positive sample pair τ_u^1 and τ_u^2 are obtained by perturbing the same trajectory segment τ_u. The noisy samples τ_u^1 and τ_u^2 have subtle differences but retain the general movement patterns. The InfoNCE loss serves as a contrastive loss between negative pairs. Since the agent rarely does the same subtask twice sequentially, we choose the adjacent working trajectory segments τ_{u_i} and $\tau_{u_{i+1}}$ as negative sample pairs. Finally, the index with the maximum value in the softmax similarity is chosen as the subtask label of trajectory segment τ_u. We establish the correspondence between the assigned labels and the subtask descriptions by visualising the averaged observation videos of the demonstration segments having the same label.

The trajectory segment feature encoder learns to separate negative pairs and cluster positive pairs in the feature space. A fixed number of randomly initialised prototypes may not be able to anchor the expected cluster features. It may split one cluster into two or assign two clusters to one prototype. Therefore, we set the initial number of prototypes K to be significantly larger than the actual number of labels, in order to cover a broader range of cluster features. Then, the similar prototypes are merged to obtain the final distinct prototypes. We use cosine similarity with threshold 0.5 to merge the similar normalised prototypes.

3.3 Subtask Policy Training

In the previous sections, we introduced how we discover the forking state s_f and merging state s_m to segment the demonstrations and how we classify the segments to obtain subtask labels. Here, we introduce how we train the BC policies using the demonstration segments with subtask labels. We trained (1) a subtask-conditioned diffusion policy (TCP) on the full task demonstration with a one-hot vector indicating which subtask to complete now, and (2) an Options framework including several independent diffusion policies where each policy is trained on corresponding subtask demonstration segments, consisting of the working phase and the preceding transition phase. The action space of the policy is a union of the environment action space and a STOP action [25]. The STOP action occurs at the end of each subtask, i.e., at the forking state s_f. Both the TCP and the Options policies are trained to give a STOP action when the current subtask is complete. In execution, the subtasks to be completed and their order are specified at the beginning of each episode. When the policy gives a STOP action, the subtask label or policy is changed to the next subtask. The policy should complete each of the subtasks in the given order, without being aware of the other subtasks.

The accuracy of the STOP action is crucial for policies to complete subtasks sequentially, as a BC policy will struggle to recover once it goes off track [15]. In the training of both TCP and Options policies, we used a saturated-early-stop strategy by marking all the 5 states before the actual forking state as having a STOP action. This strategy prevents the policy from stopping too late and reaching a state that is not in the initial state set of the succeeding subtask policy. To train the policies in the Options framework, we set the start of each subtask demonstration used for training 10 steps earlier to better bridge the gap between adjacent subtasks. In addition to the saturated-early-stop strategy, we applied a voting-for-stop (V4S) strategy during execution to alleviate the failure caused by stopping too early. The V4S strategy lets the diffusion policy infer whether to perform a STOP action multiple times and performs one if more than 40% of STOP actions are inferred. These two strategies work together to remove outlier STOP actions and stabilise the subtask handover point.

4 Experiment

We evaluate our method in the Franka Kitchen environment [7] with a demonstration dataset including 25 tasks and 7 subtasks. Each task has 4 subtasks in a unique order. We define the first state of a demonstration as a merging state and the last state as a forking state because the robot arm in the dataset begins working on a subtask immediately at the beginning and terminates right after completing the last subtask. We first evaluate the subtask discovery by visualising the trajectory segments. We then present the performance of policies trained on the discovered subtasks.

4.1 Subtask Discovery Evaluation

To train the diffusion trajectory predictor, we interpolate each demonstration trajectory to a length of 200. We choose the input trajectory window $n = 16$ and predict a trajectory of length 24. We use a larger prediction window than the input window to capture more diverse future trajectories, allowing the peaks in the dispersion-time curve to be more significant. This setting enables us to find the correct subtasks in almost all demonstration trajectories.

Our Method Can Discover Clear Subtask Boundaries. Figure 5 shows the classification result of subtask trajectory segments. The start and the end of each trajectory segment are close to each other, making the robot arm recognisable in averaged observation images. Besides, combining subtask demonstrations, including the transition phases from different preceding subtasks, enables

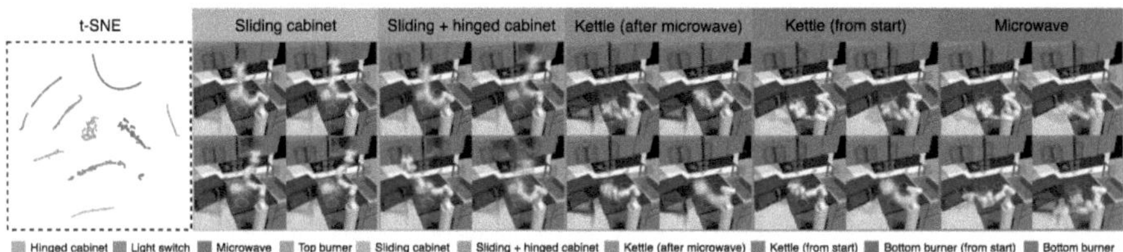

Fig. 5. The t-SNE plot of the softmax similarity vector $softmax(\mathbf{Q})$ on the left is coloured by the self-supervised labels, i.e., the argmax of the similarity vector. The 5 blocks on the right show the demonstration segments of 5 labels. The four images in each block are the averaged observations at 0%, 33%, 67%, and 100% of each demonstration segment. Subtasks are still distinguishable from the ghosting images that are averaged across all images with the same label, showing that our method can identify semantically meaningful subtasks with accurate boundaries.

Fig. 6. Trajectory sketch shows the multi-merging scenario (left) and the fork-state-missing scenario (right). The coloured and grey dotted lines represent the working and transition trajectory segments, respectively. (Color figure online)

the subtask policy to start at outlier initial states. In the policy execution experiments, we identified that a subtask policy starts by recovering from a deviation caused by the previous policy stopping late.

The Discovered Subtasks are Generalised with Irrelevant Observations. Since we use only the position and rotation of the robot arm gripper to discover the subtask, irrelevant visual information is ignored. A subtask with different visual information caused by different former subtasks is assigned the same label. Figure 5 shows that opening the sliding cabinet was assigned the same label regardless of whether the microwave door is opened, the kettle is moved, or the light is on.

Starting from a Large Number of Prototypes and Merging them Gradually yields a Correct and Detailed Classification. We set the initial number of prototypes to be $K = 32$. After merging similar ones during training, 10 prototypes remained, as shown in Fig. 5. The additional 3 classes are caused by the dataset characteristics at the beginning and the end, shown in Fig. 6. The subtask of moving the kettle is split into (1) after opening the microwave and (2) at the very beginning with the microwave closed. The merging points of these two conditions are too far away, causing the trajectory segments to have considerable differences. Forking-state-missing is another scenario that induces additional labels. Among all the dataset demonstrations, opening the hinged cabinet is the only possible succeeding subtask of opening the sliding cabinet, as shown in Fig. 6 (right). Therefore, there is no forking state to define the end

Table 1. The average order rewards and total rewards (in the bracket) for 25 tasks. The capital letters represent the subtasks: M for the microwave, K for the kettle, B for the bottom burner, T for the top burner, L for the light switch, S for the sliding cabinet and H for the hinged cabinet.

Task Number	1	2	3	4	5	6	7	8	9	10	11	12	13
Subtasks	KBSH	KBLS	KLSH	KBTS	KTLS	MBSH	MBLS	MKBH	MKBS	MKSH	MKLS	MKTH	MKTL
Oracle baseline	3.98	3.83	3.88	**3.98**	**3.95**	**4.00**	3.95	**4.00**	3.84	**4.00**	3.90	**4.00**	**4.00**
TCP	3.97	3.95	3.94	**3.98**	3.89	3.42	3.43	3.95	3.95	**4.00**	3.71	3.92	3.90
TCP(w/o V4S)	**4.00**	3.96	3.94	3.78	3.68	3.36	3.02	3.92	3.92	**4.00**	3.78	3.92	3.86
Options	3.99	**4.00**	**4.00**	3.95	2.58	**4.00**	**4.00**	3.99	**4.00**	**4.00**	**4.00**	2.86	2.93
Options(w/o V4S)	3.94	**4.00**	**4.00**	3.96	2.38	**4.00**	3.88	3.82	**4.00**	3.94	3.94	2.64	2.60
Task Number	14	15	16	17	18	19	20	21	22	23	24	25	Average
Subtasks	MBTH	MBTS	BTSH	BTLS	KBLH	KBTH	KBTL	MBLS	MKLH	MLSH	MBTL	MTLH	
Oracle baseline	3.98	**4.00**	**3.98**	3.88	**4.00**	**4.00**	**3.99**	3.98	**4.00**	**4.00**	3.83	**4.00**	**3.96** (**3.98**)
TCP	3.45	3.36	3.57	3.45	3.97	**4.00**	3.73	3.28	3.73	3.29	3.25	1.93	3.64 (3.81)
TCP(w/o V4S)	3.48	3.40	3.50	3.42	**4.00**	3.90	3.74	3.20	3.72	3.20	2.68	1.86	3.57 (3.78)
Options	**4.00**	3.99	3.86	**4.00**	3.98	3.87	3.93	**4.00**	**4.00**	3.98	**3.89**	2.29	3.76 (3.90)
Options(w/o V4S)	**4.00**	3.92	3.78	**4.00**	3.94	3.86	4.00	3.86	3.82	3.98	3.80	2.58	3.71 (3.87)

of opening the sliding cabinet to separate these two subtasks. Figure 5 shows the observations of subtasks caused by these two scenarios. To avoid these two scenarios, when creating the dataset, we recommend initialising the robot arm gripper at a random position and moving it to a random position after completing all the subtasks. In this case, we should define the beginning and the end of each demonstration as forking and merging states, respectively.

Locking the Number of Prototypes to be the Expected Final Number will Result in an incorrect Classification. It is not guaranteed that the three additional labels can be merged with the correct label if locking the number of prototypes to $K = 7$. Our experiment shows that (1) moving the kettle at the beginning and (2) moving the kettle after opening the microwave will always be classified into two classes if locking $K = 7$, resulting in the other two different subtasks being assigned the same label. Therefore, initialising more prototypes than the expected number of labels to cover more features and automatically merging the similar ones is necessary to obtain the correct classification.

4.2 Policy Evaluation

We compared the following policies trained using behaviour cloning to evaluate the accuracy of the discovered subtask boundaries and the correctness of the subtask classification.

- A task-conditioned diffusion policy [3] trained on the complete task demonstration as an Oracle baseline. It takes a one-hot encoded task label of length 25 to indicate which task to complete in this episode. The task label remains the same throughout the entire episode.
- A subtask-conditioned diffusion policy (TCP). It takes a one-hot encoded subtask label of length 7 to indicate which subtask to complete. The input subtask label remains the same until a STOP action is performed, at which point it changes to the next subtask label.

- The Options framework comprises a set of diffusion policies, each corresponding to a specific subtask. Each policy takes the state observation as input without any conditions. The policy is changed to the next one when the current policy gives a STOP action.

The Oracle baseline policy knows all the subtasks to complete, whereas the TCP and Options policies only know the current subtask to be completed. The order rewards mean how many subtasks it completes in the expected order before completing an unexpected subtask. The total rewards represent the number of expected subtasks the agent completes in total. As Table 1 shows, the TCP and the options policies reach close to the performance of the Oracle baseline policy. These results demonstrate that policies trained on the discovered subtask demonstrations can execute continuously across different preceding subtasks, indicating the potential to complete tasks having seen subtasks and transitions but in an unseen order. Besides, the V4S strategy increases the rewards.

The TCP mainly failed the subtask of opening the microwave. The STOP action was performed early, causing the robot arm to move to the next subtask before fully opening the microwave. These failure cases did not occur in the Options policies and the Oracle baseline policy. We believe the reason is that the TCP is trained on all the subtasks, hence it knows how to execute the next subtasks. Therefore, once the subtask indicator is changed to the next one, the TCP policy is eager to complete the next subtask and leaves the current one not fully completed. These failure cases did not occur in the Options policies because each policy was only trained on one subtask.

5 Conclusions

We proposed a method to discover semantically meaningful subtasks in a demonstration dataset that consists of several subtasks. We use dispersion on the gripper's 3D trajectory to segment demonstrations. This 3D trajectory is robust to irrelevant external information, such as illumination or background objects. We merge prototypes that anchor similar clusters to obtain a more accurate classification than using a fixed number of classes. We evaluate our method on a multi-subtask dataset in which the forking and merging points are not designed to be spatially and temporally aligned. The experiment shows that the discovered subtasks are semantically meaningful. The subtask policies achieve performance comparable to that of the baseline trained on the complete tasks. These results demonstrate that forking and merging points defined by dispersion are suitable segmentation points for obtaining semantically meaningful subtasks with accurate boundaries, thereby enabling sequential execution. The future work involves learning a latent space of states with a well-defined metric for dispersion calculation to expand the application to agents whose position is hard to obtain.

References

1. Bacon, P.L., Harb, J., Precup, D.: The option-critic architecture. In: Proceedings of the AAAI Conference on Artificial Intelligence, vol. 31 (2017)
2. Caron, M., Misra, I., Mairal, J., Goyal, P., Bojanowski, P., Joulin, A.: Unsupervised learning of visual features by contrasting cluster assignments. Adv. Neural Inf. Process. Syst. 1–23 (2020)
3. Chi, C., et al.: Diffusion policy: visuomotor policy learning via action diffusion. arXiv preprint arXiv:2303.04137 (2023)
4. Cuturi, M.: Sinkhorn distances: lightspeed computation of optimal transport. Adv. Neural Inf. Process. Syst. **26** (2013)
5. Eysenbach, B., Gupta, A., Ibarz, J., Levine, S.: Diversity is all you need: learning skills without a reward function. arXiv preprint arXiv:1802.06070 (2018)
6. Eysenbach, B., Zhang, T., Levine, S., Salakhutdinov, R.R.: Contrastive learning as goal-conditioned reinforcement learning. Adv. Neural Inf. Process. Syst. **35**, 35603–35620 (2022)
7. Gupta, A., Kumar, V., Lynch, C., Levine, S., Hausman, K.: Relay policy learning: solving long-horizon tasks via imitation and reinforcement learning. arXiv preprint arXiv:1910.11956 (2019)
8. Ho, J., Jain, A., Abbeel, P.: Denoising diffusion probabilistic models. Adv. Neural Inf. Process. Syst. **33**, 6840–6851 (2020)
9. Hoang, C., Sohn, S., Choi, J., Carvalho, W., Lee, H.: Successor feature landmarks for long-horizon goal-conditioned reinforcement learning. Adv. Neural Inf. Process. Syst. **34**, 26963–26975 (2021)
10. Huang, Z., Liu, F., Su, H.: Mapping state space using landmarks for universal goal reaching. Adv. Neural Inf. Process. Syst. **32** (2019)
11. Jiang, Y., Liu, E., Eysenbach, B., Kolter, J.Z., Finn, C.: Learning options via compression. Adv. Neural Inf. Process. Syst. **35**, 21184–21199 (2022)
12. Kingma, D.P., Welling, M.: Auto-encoding variational bayes. arXiv preprint arXiv:1312.6114 (2013)
13. Kipf, T., et al.: Compile: compositional imitation learning and execution. In: International Conference on Machine Learning, pp. 3418–3428. PMLR (2019)
14. Kriegel, H.P., Kröger, P., Sander, J., Zimek, A.: Density-based clustering. Wiley Interdiscip. Rev. Data Min. Knowl. Discov. **1**(3), 231–240 (2011)
15. Laskey, M., Lee, J., Fox, R., Dragan, A., Goldberg, K.: Dart: noise injection for robust imitation learning. In: Conference on Robot Learning, pp. 143–156. PMLR (2017)
16. Laskin, M., Srinivas, A., Abbeel, P.: Curl: contrastive unsupervised representations for reinforcement learning. In: International Conference on Machine Learning, pp. 5639–5650. PMLR (2020)
17. Nachum, O., Gu, S., Lee, H., Levine, S.: Near-optimal representation learning for hierarchical reinforcement learning. arXiv preprint arXiv:1810.01257 (2018)
18. Oord, A.V.D., Li, Y., Vinyals, O.: Representation learning with contrastive predictive coding. arXiv preprint arXiv:1807.03748 (2018)
19. Ramesh, R., Tomar, M., Ravindran, B.: Successor options: an option discovery framework for reinforcement learning. arXiv preprint arXiv:1905.05731 (2019)
20. Ren, Y., et al.: Deep clustering: a comprehensive survey. IEEE Trans. Neural Netw. Learn. Syst. (2024)
21. Ren, Y., Wang, N., Li, M., Xu, Z.: Deep density-based image clustering. Knowl.-Based Syst. **197**, 105841 (2020)

22. Sermanet, P., et al.: Time-contrastive networks: self-supervised learning from video. In: 2018 IEEE International Conference on Robotics and Automation (ICRA), pp. 1134–1141. IEEE (2018)
23. Shah, S.A., Koltun, V.: Deep continuous clustering. arXiv preprint arXiv:1803.01449 (2018)
24. Sharma, A., Gu, S., Levine, S., Kumar, V., Hausman, K.: Dynamics-aware unsupervised discovery of skills. arXiv preprint arXiv:1907.01657 (2019)
25. Shiarlis, K., Wulfmeier, M., Salter, S., Whiteson, S., Posner, I.: Taco: learning task decomposition via temporal alignment for control. In: International Conference on Machine Learning, pp. 4654–4663. PMLR (2018)
26. Srinivas, A., Krishnamurthy, R., Kumar, P., Ravindran, B.: Option discovery in hierarchical reinforcement learning using spatio-temporal clustering. arXiv preprint arXiv:1605.05359 (2016)
27. Tanneberg, D., Ploeger, K., Rueckert, E., Peters, J.: Skid raw: skill discovery from raw trajectories. IEEE Robot. Autom. Lett. **6**(3), 4696–4703 (2021)
28. Tessler, C., Givony, S., Zahavy, T., Mankowitz, D., Mannor, S.: A deep hierarchical approach to lifelong learning in minecraft. In: Proceedings of the AAAI Conference on Artificial Intelligence, vol. 31 (2017)
29. Xu, M., Xu, Z., Chi, C., Veloso, M., Song, S.: XSkill: cross embodiment skill discovery. Proc. Mach. Learn. Res. **229**(CoRL) (2023)
30. Zare, M., Kebria, P.M., Khosravi, A., Nahavandi, S.: A survey of imitation learning: algorithms, recent developments, and challenges. IEEE Trans. Cybern. (2024)
31. Zhu, Y., Stone, P., Zhu, Y.: Bottom-up skill discovery from unsegmented demonstrations for long-horizon robot manipulation. IEEE Robot. Autom. Lett. **7**(2), 4126–4133 (2022)

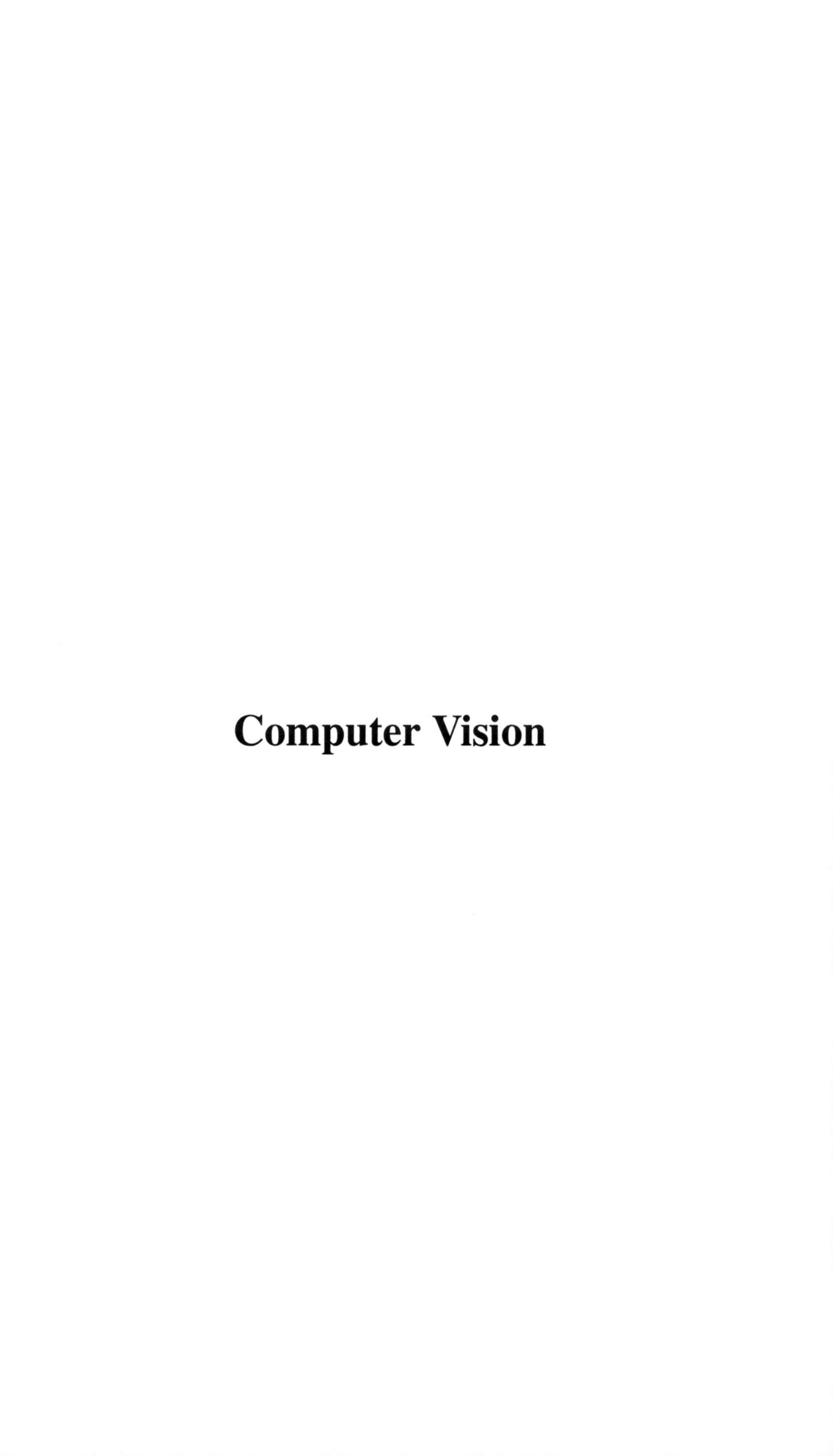

Computer Vision

NoPo3DFusion: Scaling Two Images to Long Videos via 3D-Aware Iterative Diffusion

Xiaolong Shen(✉), Cheng Zhang, Donny Y. Chen, Tien-Tsin Wong, and Jianfei Cai

Faculty of Information Technology, Monash University, Melbourne, Australia
shenxl2000@outlook.com, {cheng.zhang,tt.wong,jianfei.cai}@monash.edu, donny.chen@outlook.sg

Abstract. We present NoPo3DFusion, a scalable novel view synthesis (NVS) framework that generates long, multi-view consistent video sequences from just two unposed RGB images with controllable camera trajectories. NoPo3DFusion combines NoPoSplat, a pose-free feed-forward 3D Gaussian Splatting (3DGS) based method designed for paired inputs, with CogVideo-X, a transformer-based video diffusion model. To enhance spatial consistency and guide the generative process, we introduce an observable mask as an additional condition for the diffusion model. We further propose an iterative inference strategy that enables efficient generation of long sequences while maintaining high visual fidelity and multi-view consistency. Experiments on RealEstate-10K and ACID show that NoPo3DFusion sets a new state of the art for feed-forward NVS from sparse and unposed images, highlighting the effectiveness of combining pose-free 3D reconstruction with video diffusion model. Code and models are available at https://github.com/maxShen701/NoPo3DFusion.

Keywords: Novel View Synthesis · Video Diffusion Model · 3DGS

1 Introduction

Novel view synthesis (NVS) is the task of generating 3D-consistent images from novel viewpoints, and has been a fundamental challenge in computer vision and graphics for decades. NVS supports a wide range of real-world applications. For example, it enables dynamic camera movement from limited input views for content creation, enhances immersion in virtual environments, allows simulation of unseen viewpoints to improve the robustness of autonomous driving systems.

With the emergence of NVS applications, increasing efforts have been devoted to making models more effective in real-world scenarios. A promising development in this direction is feed-forward NVS [24,39], utilizing NeRF [14] or

This work was conducted while X. Shen was a Master student at Monash University. D. Y. Chen's contributions were made while he was a PhD student at Monash University.

M. Liu et al. (Eds.): AI 2025, LNAI 16371, pp. 109–120, 2026.
https://doi.org/10.1007/978-981-95-4972-6_9

3DGS [9] representations. Unlike traditional NeRF or 3DGS, which require dense input views and lengthy per-scene optimization, feed-forward approaches [2,3, 5,26,33] integrate an encoder trained on large-scale datasets, enabling inference on unseen scenes using only a few input images in a single forward pass. Despite significant progress in improving encoders to capture richer input information, feed-forward NVS models remain largely deterministic. This limitation hinders their ability to model complex textures in real-world scenes and to generate *unobserved regions*, often resulting in severe artifacts such as floaters and discontinuities (see "NoPoSplat" in Fig. 3).

Another recent advancement in real-world NVS leverages video diffusion models [1,36]. Trained on large-scale video datasets, video diffusion models learn strong priors for modeling the 3D world. By conditioning on camera pose information, these models can be adapted for NVS tasks. Although diffusion models excel at generating unobserved regions, they are fundamentally video-based and do not explicitly model geometric structure. As a result, existing NVS approaches based on video diffusion models [12,15,17,40] struggle to maintain multi-view consistency and fine-grained pose control (see "ViewCrafter" in Fig. 2), both of which require accurate 3D geometry modeling.

To make NVS more effective in real-world scenes, we introduce a novel framework, NoPo3DFusion. Our approach bridges the geometric accuracy of feed-forward 3DGS models with the generative capabilities of video diffusion models, enabling photorealistic and geometrically consistent video generation from only two RGB input images, without requiring any camera pose information. Unlike typical feed-forward 3DGS models that primarily interpolate between input views, our NoPo3DFusion supports extrapolation far beyond the inputs. The most similar approach to ours is MVSplat360 [6], however, it requires accurate camera poses for input views, making it less suitable for real-world scenarios.

In particular, our NoPo3DFusion builds upon two recent advances: NoPoSplat [37] and CogVideo-X [36]. The core idea is to refine the noisy outputs of NoPoSplat using the generative capabilities of CogVideo-X. However, naively combining these two models is suboptimal because **1)** CogVideo-X is unaware of which regions to refine and which to preserve, and **2)** CogVideo-X is trained on a fixed number of video frames and struggles to generalize to longer sequences. To address these challenges, we propose a novel conditioning method by introducing an "observable mask" extracted from the rendered depth maps to guide the diffusion model in refining the target regions. Furthermore, we introduce a training-free iterative inference strategy to enable CogVideo-X to generate an arbitrary number of video frames.

We conducted extensive ablation studies to validate each component, highlighting the effectiveness of our improved diffusion conditioning and iterative inference strategy. By evaluating on two widely used large-scale benchmarks, RealEstate-10K [43] and ACID [11], our NoPo3DFusion outperforms several state-of-the-art models both quantitatively and qualitatively. Our main contributions are summarized as follows.

- We introduce a novel framework NoPo3DFusion, which leverages the geometric accuracy of feed-forward 3DGS models and the generative capability

of video diffusion models, enabling multi-view consistent and long-sequence (150+ frames) novel view synthesis from only two unposed RGB inputs.
- To enhance the effectiveness of video diffusion models for conditional generation and arbitrary-length rendering, we propose a novel observable mask guidance mechanism and an iterative inference strategy.
- Our NoPo3DFusion outperforms related state-of-the-art approaches on two real-world photorealistic benchmarks RealEstate-10K and ACID.

2 Related Works

2.1 Sparse-View Scene Reconstruction and Rendering

Reconstructing and rendering 3D scenes from sparse views (*e.g.*, 1–6) remains a significant challenge. It has recently attracted increasing attention, driven by advances in neural 3D representations [9,14]. Early work addressed this task via generalizable Neural Radiance Fields (NeRF) [4,14,34]. More recent work uses feed-forward 3D Gaussian Splatting (3DGS) [9], where models predict 3DGS parameters in one forward pass. Pioneer works [23,24] infer 3DGS from a single image. Follow-ups extend to take two or more images, using epipolar attention [2] or cost volumes [5]. Further improvements include stacked volumes [30], voxel-aligned representation [29], efficient architecture [28], using monocular depth [35] or pointmap priors [18], and 4K panorama support [41]. However, most methods require known camera poses, which hinders real-world use. Pose-free solutions [19,37] address this using geometric priors [10,27]. Still, existing approaches are deterministic and struggle with unobserved regions. Our NoPo3DFusion combines 3DGS with diffusion models in a pose-free setting, pushing sparse-view NVS toward more general, in-the-wild scenarios.

2.2 Novel View Synthesis via Diffusion Models

Diffusion models have enabled novel view synthesis (NVS) from sparse inputs by generating unseen views. *(a) Implicit 3D reasoning via pose-conditioned diffusion:* These methods learn view synthesis without explicit geometry, relying on pose-conditioned training. 3DiM [32] uses stochastic denoising for 3D consistency, Zero-1-to-3 [12] trains on synthetic renderings to generalize to real images, and SV3D [25] adds CLIP-based conditioning for single-image NVS. Consistent-1-to-3 [38] employs transformer encoders to handle large viewpoint changes. However, lacking explicit geometry often limits spatial consistency and detail. *(b) Hybrid methods with explicit 3D priors:* These integrate geometric cues like depth, point clouds, or 3DGS to enhance realism. ZeroNVS [17] adds camera-aware conditioning, MultiDiff [15] uses monocular depth, and ViewCrafter [40] refines point clouds with video diffusion. LatentSplat [33] applies variational splatting for real-time NVS. CAT3D [7] generates pseudo views before full 3D reconstruction. While effective, these methods still struggle with consistency across long trajectories. Our NoPo3DFusion introduces an observable mask as a better diffusion condition and iterative inference for improved long-range generation.

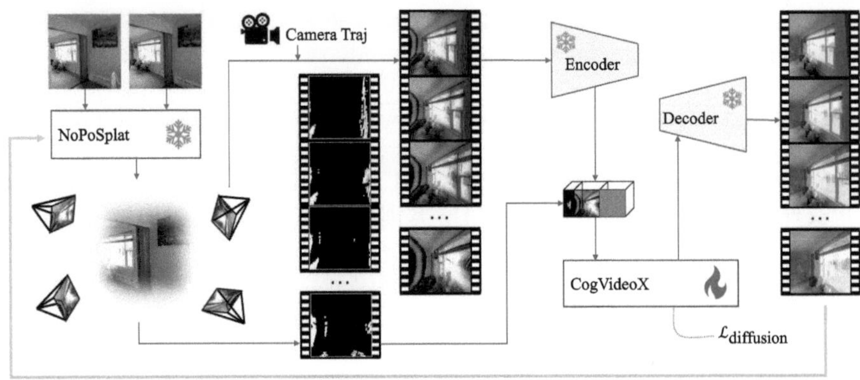

Fig. 1. Overview of NoPo3DFusion. Given two unposed input images, NoPo3DFusion first predicts a 3DGS representation, then renders novel views and corresponding masks along a predefined camera trajectory. These rendered views and masks are concatenated with randomly sampled noise as conditioning inputs for video generation.

3 Method

3.1 Preliminary

Diffusion model [8,20,22] is a generative model that learns to reverse a forward noising process to generate samples from a data distribution. Latent Diffusion Model (LDM) [21] extends diffusion to latent space, enabling efficient training and inference by operating on compressed representations. Formally, a LDM consists of three main components:

- Encoder $\mathcal{E}$, which maps input data to a latent space (used in latent diffusion frameworks);
- Denoising network p_θ (e.g., U-Net, Transformer), parameterized by θ, which learns to reverse the forward noising process by predicting either the clean data, noise, or intermediate states;
- Decoder $\mathcal{D}$, which maps latent samples back to image or video space.

Given a clean input sample $\mathbf{x}\_0 \sim q(\mathbf{x})$, the forward process gradually adds Gaussian noise over T steps:

$$q(\mathbf{x}_t \mid \mathbf{x}_0) = \mathcal{N}(\mathbf{x}_t; \sqrt{\bar{\alpha}_t}\mathbf{x}_0, (1-\bar{\alpha}_t)\mathbf{I}),$$

where $\bar{\alpha}_t = \prod_{s=1}^{t} \alpha_s$ and $\alpha_t = 1 - \beta_t$ is defined by a noise schedule, typically following a cosine-based schedule [16]. The goal is to learn the reverse process:

$$p_\theta(\mathbf{x}_{t-1} \mid \mathbf{x}_t, \mathbf{c}) = \mathcal{N}(\mathbf{x}_{t-1}; \mu_\theta(\mathbf{x}_t, t, \mathbf{c}), \sigma_t^2\mathbf{I}),$$

conditioned on auxiliary input $\mathbf{c}$ (e.g., text). The training objective typically minimises the diffusion loss, often simplified to a denoising score-matching loss:

$$\mathcal{L} = \mathbb{E}_{\mathbf{x}_0,\epsilon,t}\left[\|\epsilon_\theta(\mathbf{x}_t, t, \mathbf{c}) - \epsilon\|^2\right],$$

where $\epsilon \sim \mathcal{N}(0, \mathbf{I})$ is the noise added at timestep t, and ϵ_θ is the prediction.

3.2 Model Architecture

As shown in Fig. 1, the core of our pipeline is a diffusion model. Given sparse unposed RGB images $\mathcal{I} = \{I_v\}_{v=1}^{V}$ with their corresponding camera intrinsic $\mathcal{K} = \{K_v\}_{v=1}^{V}$ as input, the goal is to achieve scene-level novel view synthesis (NVS) at arbitrary camera trajectories $\mathcal{P}_{\text{tgt}} = \{P_f\}_{f=1}^{F}$, where F is the number of target views to be generated.

3D Gaussians Reconstruction. Inspired by MVSplat360 [6], we first reconstruct a 3D Gaussian representation $\mathcal{G}$ to provide 3D geometry awareness to the diffusion model. However, unlike MVSplat360 which require camera poses $\mathcal{P}_{\text{ctx}}$ of input images, our pipeline removes this dependency by leveraging a pose-free 3D Gaussian network NoPoSplat [37]. Formally, f_θ is a feed-forward function that maps the input images $\mathcal{I}$ and camera intrinsic parameters $\mathcal{K}$ to a set of 3D Gaussian parameters:

$$f_\theta : (\mathcal{I}, \mathcal{K}) \longrightarrow \mathcal{G} : \{\cup(\boldsymbol{\mu}_j^v, \alpha_j^v, \boldsymbol{r}_j^v, \boldsymbol{s}_j^v, \mathbf{c}_j^v)\}_{j=1,\cdots,H\times W}^{v=1,\cdots,V}$$

where θ are the learned parameters and the resulting 3D Gaussian parameters $\mathcal{G}$ are defined as the centre of 3D Gaussian ellipsoids $\boldsymbol{\mu} \in \mathbb{R}^3$, opacity $\alpha \in \mathbb{R}$, rotation in quaternion $\boldsymbol{r} \in \mathbb{R}^4$, scale $\boldsymbol{s} \in \mathbb{R}^3$, and spherical harmonics (SH) $\boldsymbol{c} \in \mathbb{R}^k$ with k being the number of SH coefficients.

Diffusion Model Conditioning. Given camera trajectories with F target views $\mathcal{P}_{\text{tgt}} = \{P_f\}_{f=1}^{F}$, we then render an image sequence $\mathcal{I}_{in} \in \mathbb{R}^{F\times3\times H\times W}$ and corresponding depth map $\mathcal{D} \in \mathbb{R}^{F\times1\times H\times W}$ from the 3D Gaussian representation $\mathcal{G}$ using the standard rasterization process [37]. From the depth map, we compute a binary observable mask to indicate occluded regions, which is used to guide the diffusion model during training and inference. Specifically, a threshold τ is used to determine valid pixels, where pixels with depth values below τ are marked as invalid (occluded). Then we use morphological dilation to expand occluded regions to improve robustness, resulting a binary mask $\mathcal{D} \overset{\leq\tau,\text{dilate}}{\longrightarrow} \mathcal{M} \in \mathbb{R}^{F\times1\times H\times W}$.

Diffusion-based Novel View Refinement. The incomplete video $\mathcal{I}_{\text{in}}$ is refined using a modified CogVideoX [36] that incorporates the occlusion mask $\mathcal{M}$. The mask is downsampled by a single convolution layer and temporal frame skipping (4×, excluding the first), producing a feature aligned with the latent $z = \mathcal{E}(\mathcal{I}_{in})$. This mask feature is concatenated with z and noisy latent z_{noise} along the channel axis, then projected back to the original channel dimension.

3.3 Iteratively Long Sequence Generation

Generating long sequences remains a challenge for video diffusion models, which are typically designed for short clips (e.g., 49 frames in CogVideoX [36]). To address this without re-training, we introduce a training-free iterative strategy that divides a camera trajectory into overlapping 49-frame clips with 13-frame overlap. Rather than using a naive method, which processes all clips together in a single diffusion pass but suffers from texture inconsistency, we adopt a sequential generation approach. By treating overlapping frames as temporal bridges, our method preserves visual continuity across segments. The first clip is conditioned on two known views, while each subsequent clip uses image pairs from the previous segment as $\mathcal{I}$ for generation. We compare this iterative strategy against the joint baseline in our experiments.

3.4 Training Objectives

We fine-tune from CogVideoX [36] pretrained weights. Given ground-truth video frames $\mathcal{I}_{gt}$, their latent representation $\mathbf{x}_0 = \mathcal{E}(\mathcal{I}_{gt})$ is extracted by the encoder. The training objective minimizes the diffusion loss:

$$\mathcal{L} = \mathbb{E}_{\mathbf{x}_0,\epsilon,t}\left[||\epsilon_\theta(\mathbf{x}_t, t, \mathcal{I}_{in}, \mathcal{M}) - \epsilon||^2\right],$$

where $\mathbf{x}_t$ is a noisy version of $\mathbf{x}_0$ in step t, obtained by adding sampled random noise under DDIM scheduling.

4 Experiment

4.1 Settings

Datasets. We perform model fine-tuning on the RealEstate10K dataset [43], and perform evaluations on both RealEstate10K and ACID datasets [11]. Specifically, the RealEstate10K dataset comprises indoor real estate videos, with its training subset containing 66,033 scenes. For evaluation, we randomly select 15 scenes (each with 157 frames) from the test split. In contrast, the ACID dataset captures natural scenes via aerial drones, from which 16 scenes (each with 157 frames) are randomly sampled from the test split for evaluation. Both datasets provide estimated camera intrinsic and extrinsic parameters for each frame.

Evaluation Metrics. For quantitative results, we report standard image quality metrics, including pixel-level PSNR, patch-level SSIM [31], and feature-level LPIPS [42]. For metrics computation, all computations are performed at 256×256 resolution following prior works [2,37].

Implementation Details. Our method is implemented with PyTorch based on the standard NoPoSplat [37], and CogVideoX-5B-I2V [36], an additional projection layer is introduced at the beginning of its transformer block to map the mask, which is a 2D convolution with 33 input channels, 32 output channels, kernel size 1, stride 1, and no padding. The τ was set to 10^{-5} for generate mask. During fine-tuning, we update the feedforward, query, and value layers of all 42 transformer blocks in the original diffusion model. Additionally, we fine-tune the projection layer in the patch embedding module and the final projection layer of transformer block, resulting in approximately resulting in approximately 71 % of the original model's parameters. We train our model for 5000 steps on RealEstate10K using 4 NVIDIA A100 GPUs with mixed precision. We employ

Table 1. Quantitative comparison on RealEstate10K [43], showing that our method outperforms prior models in both pixel-level and perceptual metrics. Results on ACID [11] showing the superiority of our method under challenging out-door scenarios.

Method	w/o $\mathcal{P}_{\text{ctx}}$	RealEstate10K [43]			ACID [11]		
		PSNR ↑	SSIM ↑	LPIPS ↓	PSNR ↑	SSIM ↑	LPIPS ↓
NoPo3DFusion	✓	**17.22**	**0.583**	**0.281**	**16.45**	**0.390**	0.407
NoPoSplat [37]		15.75	0.525	0.290	14.40	0.346	**0.359**
ViewCrafter [40]		13.57	0.378	0.432	10.33	0.149	0.699
MVSplat360 [6]	✗	16.10	0.508	0.316	16.72	0.344	0.395

Fig. 2. Qualitative comparison on RealEstate10K and ACID. Each row shows one scene; columns include input views, predictions from ViewCrafter, NoPoSplat, NoPo3DFusion, and GT. NoPo3DFusion produces consistent, high-fidelity results across domains despite training on indoor data only.

distributed training with a per-GPU batch size of 1 and 8-step gradient accumulation, resulting in an effective batch size of 32. The optimiser is AdamW [13] with a learning rate of 10^{-5}, $\beta_1 = 0.9$, $\beta_2 = 0.999$, and weight decay of 10^{-2}.

4.2 Main Results

We compare our method with ViewCrafter [40], MVSplat360 [6], and the original NoPoSplat [37]. The input frames $\mathcal{I}_{in}$ are the 1st and 13th frames. Due to the 49-frame limit of CogVideoX [36], each sequence is divided into four overlapping sub-clips (13-frame overlap). We adopt the iterative generation strategy described in Sect. 3.3, and include a naive baseline in our ablation. For reference, NoPoSplat uses a RealEstate10K-trained checkpoint. ViewCrafter is evaluated in its sparse-view setting without iterative refinement.

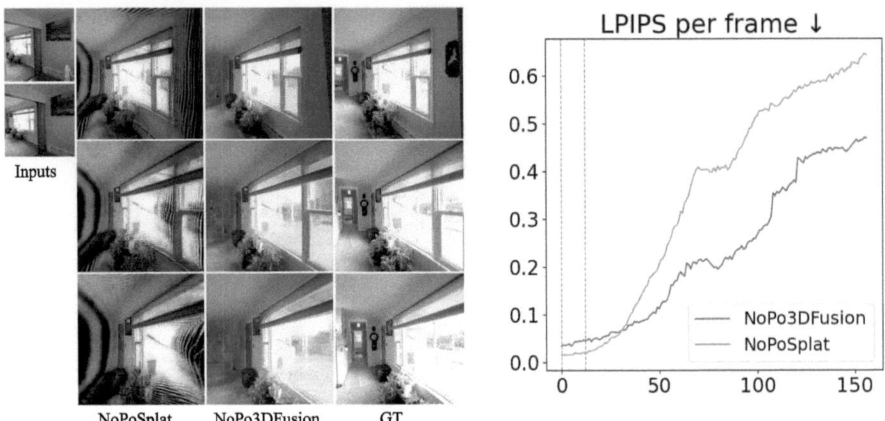

Fig. 3. Left: NoPo3DFusion consistently improve the visual quality compared to previous SoTA model. Right: LPIPS over 157 frames shows better perceptual quality.

Assessing Image Quality. We present a visual comparison in Fig. 2. NoPoSplat, lacking generative capabilities, fails to render regions under large camera motion, leading to black holes in the output. ViewCrafter, while leveraging a generative model, produces more complete results with fewer missing areas. However, without iterative point cloud optimization, its use of generation alone is insufficient to maintain high-quality outputs.

In Table 1, we report the quantitative metrics on the RealEstate10K and ACID dataset. For each sequence, we compute pairwise evaluation metrics between all generated frames and report the average. Notably, MVSplat360 requires explicit camera poses for the context views, while our method, along with ViewCrafter and NoPoSplat, operates in a pose-free setting. Despite this, our method achieves superior scores, validating its effectiveness for pose-free novel view synthesis under sparse input conditions. In Fig. 3 (right), we analyze a full 157-frame sequence where the two input views $\mathcal{I}_{in}$ have significant overlap, limiting spatial diversity. NoPoSplat performs well for the first 30–40 frames but degrades beyond the input region due to the lack of generative capability. By leveraging geometric priors and integrating a diffusion-based generative model, our method significantly improves synthesis quality—especially beyond frame 40—achieving high-fidelity reconstructions in occluded or ambiguous regions. This explains the superior performance of our method over NoPoSplat in Table 1, with side-by-side comparisons illustrated in Fig. 3.

Assessing Cross-dataset Generalisation. We test our model on the ACID dataset to assess generalisation beyond the indoor RealEstate10K domain. As shown in Fig. 2, our model produces temporally coherent and geometrically consistent videos. Quantitative results are reported in Table 1, demonstrating that our p_θ maintains strong performance even under significant domain shift.

Table 2. Ablation results showing the impact of removing Mask, the CogVideoX module on visual quality.

Method	PSNR ↑	SSIM ↑	LPIPS ↓
Full (Iter)	**17.33**	**0.581**	**0.283**
w/o $\mathcal{M}$	16.90	0.592	0.221
w/o CogVideoX (NoPoSplat)	14.40	0.346	0.359

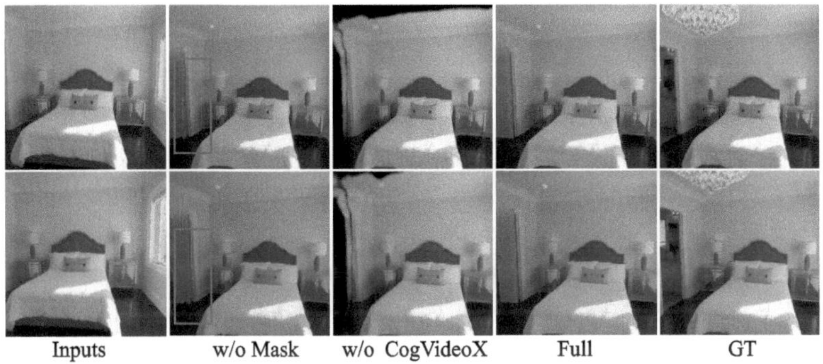

Fig. 4. Effectiveness of the mask feature $\mathcal{M}$ in in-painting. Without $\mathcal{M}$, artifacts from 3DGS (highlighted in the red box) are not properly removed.

4.3 Ablation

Assessing Mask Condition and Diffusion Model Refinement. In Fig. 4, we present a visual comparison between models trained with and without mask feature input. Without the mask feature, the output video shows visible artifacts from 3DGS rendering. The Table 2 further highlights the effectiveness of each component in improving video quality. This demonstrates that even a simple mask-aware pre-processing can significantly enhance the rendering fidelity. We also show the result without Diffuion model in ablation table, which is the NoPoSplat's performance.

Assessing Iterative Long-sequence Generation Strategy. The Fig. 5 compares our iterative generation strategy with a naive baseline. In the iterative method, the transition frames between clips are fixed, forming three overlapping regions across four clips for a 157-frame video (same as naive method). We compute the PSNR in these overlaps across 15 scenes and report the average. Results show that the iterative method maintains consistent image quality (PSNR $\geq$ 25), preserving texture across adjacent clips. In contrast, the naive method exhibits a PSNR decline over time, indicating texture degradation. A qualitative comparison (Fig. 5, left) further highlights inconsistent details in the naive results (marked in red).

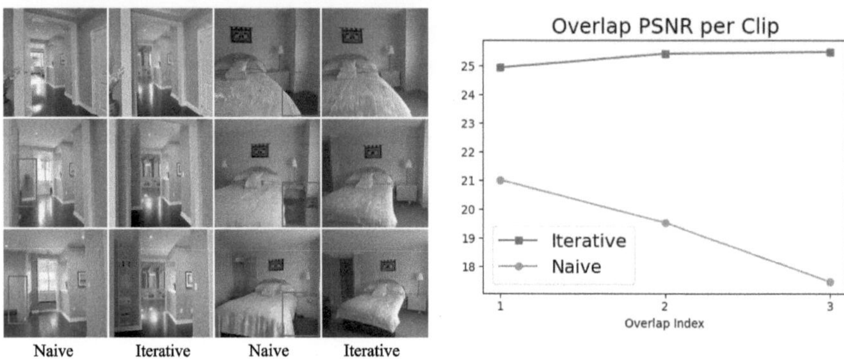

Fig. 5. Comparison of consistency between the naive and iterative methods. Naively forwarding noisy inputs into diffusion leads to inconsistent refinement, while our iterative strategy ensures temporally consistent refinement for unobserved regions (left portion). The improvement becomes larger as the rendered viewpoint moves further away from the input context (right portion).

5 Conclusion

We introduced NoPo3DFusion, a novel framework that advances pose-free novel view synthesis by combining feed-forward 3D Gaussian Splatting with video diffusion models. Our model requires only two unposed RGB images and generates a video sequence following a provided camera trajectory. By integrating observable mask conditioning and an iterative inference strategy, NoPo3DFusion achieves superior spatial consistency, long-range view extrapolation, and fine-grained camera control. Extensive ablation studies validate the effectiveness of each component, and experiments on two large-scale benchmarks demonstrate that NoPo3DFusion significantly outperforms related state-of-the-art approaches. In future work, we plan to extend our model to handle arbitrary numbers of input views by incorporating multi-view feed-forward 3DGS models, and to address more diverse real-world scenarios by leveraging more advanced video diffusion models.

References

1. Blattmann, A., et al.: Stable video diffusion: scaling latent video diffusion models to large datasets. arXiv preprint arXiv:2311.15127 (2023)
2. Charatan, D., Li, S.L., Tagliasacchi, A., Sitzmann, V.: Pixelsplat: 3d gaussian splats from image pairs for scalable generalizable 3d reconstruction. In: CVPR, pp. 19457–19467 (2024)
3. Chen, A., et al.: Mvsnerf: fast generalizable radiance field reconstruction from multi-view stereo. In: ICCV, pp. 14124–14133 (2021)
4. Chen, Y., Xu, H., Wu, Q., Zheng, C., Cham, T.J., Cai, J.: Explicit correspondence matching for generalizable neural radiance fields. TPAMI (2025)

5. Chen, Y., et al.: MVSplat: efficient 3D gaussian splatting from sparse multi-view images. In: Leonardis, A., Ricci, E., Roth, S., Russakovsky, O., Sattler, T., Varol, G. (eds.) Computer Vision – ECCV 2024. ECCV 2024. LNCS, vol. 15079, pp. 370–386. Springer, Cham (2025). https://doi.org/10.1007/978-3-031-72664-4_21
6. Chen, Y., et al.: Mvsplat360: feed-forward 360 scene synthesis from sparse views. NeurIPS **37**, 107064–107086 (2024)
7. Gao, R., et al.: Cat3d: create anything in 3d with multi-view diffusion models. NeurIPS **37**, 75468–75494 (2024)
8. Ho, J., Jain, A., Abbeel, P.: Denoising diffusion probabilistic models. NeurIPS **33**, 6840–6851 (2020)
9. Kerbl, B., Kopanas, G., Leimkühler, T., Drettakis, G.: 3D gaussian splatting for real-time radiance field rendering. ACM Trans. Graph. **42**(4), 139–1 (2023)
10. Leroy, V., Cabon, Y., Revaud, J.: Grounding image matching in 3D with MASt3R. In: Leonardis, A., Ricci, E., Roth, S., Russakovsky, O., Sattler, T., Varol, G. (eds.) Computer Vision – ECCV 2024. ECCV 2024. LNCS, vol. 15130, pp. 71–91. Springer, Cham (2025). https://doi.org/10.1007/978-3-031-73220-1_5
11. Liu, A., Tucker, R., Jampani, V., Makadia, A., Snavely, N., Kanazawa, A.: Infinite nature: perpetual view generation of natural scenes from a single image. In: ICCV, pp. 14458–14467 (2021)
12. Liu, R., Wu, R., Van Hoorick, B., Tokmakov, P., Zakharov, S., Vondrick, C.: Zero-1-to-3: zero-shot one image to 3d object. In: ICCV, pp. 9298–9309 (2023)
13. Loshchilov, I., Hutter, F.: Decoupled weight decay regularization. In: ICLR (2019)
14. Mildenhall, B., Srinivasan, P.P., Tancik, M., Barron, J.T., Ramamoorthi, R., Ng, R.: Nerf: representing scenes as neural radiance fields for view synthesis. In: ECCV, pp. 405–421 (2020)
15. Müller, N., et al.: Multidiff: consistent novel view synthesis from a single image. In: CVPR, pp. 10258–10268 (2024)
16. Nichol, A.Q., Dhariwal, P.: Improved denoising diffusion probabilistic models. In: ICML, pp. 8162–8171. PMLR (2021)
17. Sargent, K., et al.: Zeronvs: zero-shot 360-degree view synthesis from a single image. In: CVPR, pp. 9420–9429 (2024)
18. Shi, D., et al.: Revisiting depth representations for feed-forward 3d gaussian splatting. arXiv preprint arXiv:2506.05327 (2025)
19. Smart, B., Zheng, C., Laina, I., Prisacariu, V.A.: Splatt3r: zero-shot gaussian splatting from uncalibrated image pairs. arXiv preprint arXiv:2408.13912 (2024)
20. Sohl-Dickstein, J., Weiss, E., Maheswaranathan, N., Ganguli, S.: Deep unsupervised learning using nonequilibrium thermodynamics. In: ICML, pp. 2256–2265. pmlr (2015)
21. Song, J., Meng, C., Ermon, S.: Denoising diffusion implicit models. In: ICLR (2021)
22. Song, Y., Ermon, S.: Generative modeling by estimating gradients of the data distribution. NeurIPS **32** (2019)
23. Szymanowicz, S., et al.: Flash3d: feed-forward generalisable 3d scene reconstruction from a single image. In: 3DV (2025)
24. Szymanowicz, S., Rupprecht, C., Vedaldi, A.: Splatter image: ultra-fast single-view 3d reconstruction. In: CVPR, pp. 10208–10217 (2024)
25. Voleti, V., et al.: SV3D: novel multi-view synthesis and 3d generation from a single image using latent video diffusion. In: Leonardis, A., Ricci, E., Roth, S., Russakovsky, O., Sattler, T., Varol, G. (eds.) Computer Vision – ECCV 2024. ECCV 2024. LNCS, vol. 15059, pp. 439–457. Springer, Cham (2025). https://doi.org/10.1007/978-3-031-73232-4_25

26. Wang, Q., et al.: Ibrnet: learning multi-view image-based rendering. In: CVPR, pp. 4690–4699 (2021)
27. Wang, S., Leroy, V., Cabon, Y., Chidlovskii, B., Revaud, J.: Dust3r: geometric 3d vision made easy. In: CVPR, pp. 20697–20709 (2024)
28. Wang, W., Chen, D.Y., Zhang, Z., Shi, D., Liu, A., Zhuang, B.: Zpressor: bottleneck-aware compression for scalable feed-forward 3dgs. In: NeurIPS (2025)
29. Wang, W., et al.: Volsplat: rethinking feed-forward 3d gaussian splatting with voxel-aligned prediction. arXiv preprint arXiv:2509.19297 (2025)
30. Wang, Y., Huang, T., Chen, H., Lee, G.H.: Freesplat: generalizable 3d gaussian splatting towards free view synthesis of indoor scenes. NeurIPS **37**, 107326–107349 (2024)
31. Wang, Z., Bovik, A.C., Sheikh, H.R., Simoncelli, E.P.: Image quality assessment: from error visibility to structural similarity. TIP **13**(4), 600–612 (2004)
32. Watson, D., Chan, W., Martin-Brualla, R., Ho, J., Tagliasacchi, A., Norouzi, M.: Novel view synthesis with diffusion models. In: ICLR (2023)
33. Wewer, C., Raj, K., Ilg, E., Schiele, B., Lenssen, J.E.: LatentSplat: autoencoding variational gaussians for fast generalizable 3D reconstruction. In: Leonardis, A., Ricci, E., Roth, S., Russakovsky, O., Sattler, T., Varol, G. (eds.) Computer Vision – ECCV 2024. ECCV 2024. LNCS, vol. 15145, pp. 456–473. Springer, Cham (2025). https://doi.org/10.1007/978-3-031-73021-4_27
34. Xu, H., et al.: Murf: multi-baseline radiance fields. In: CVPR, pp. 20041–20050 (2024)
35. Xu, H., et al.: Depthsplat: connecting gaussian splatting and depth. In: CVPR, pp. 16453–16463 (2025)
36. Yang, Z., et al.: Cogvideox: text-to-video diffusion models with an expert transformer. In: ICLR (2025)
37. Ye, B., et al.: No pose, no problem: surprisingly simple 3d gaussian splats from sparse unposed images. In: ICLR (2025)
38. Ye, J., Wang, P., Li, K., Shi, Y., Wang, H.: Consistent-1-to-3: consistent image to 3d view synthesis via geometry-aware diffusion models. In: 3DV, pp. 664–674. IEEE (2024)
39. Yu, A., Ye, V., Tancik, M., Kanazawa, A.: Pixelnerf: neural radiance fields from one or few images. In: CVPR, pp. 4578–4587 (2021)
40. Yu, W., et al.: Viewcrafter: taming video diffusion models for high-fidelity novel view synthesis. arXiv preprint arXiv:2409.02048 (2024)
41. Zhang, C., Xu, H., Wu, Q., Gambardella, C.C., Phung, D., Cai, J.: Pansplat: 4k panorama synthesis with feed-forward gaussian splatting. In: CVPR, pp. 11437–11447 (2025)
42. Zhang, R., Isola, P., Efros, A.A., Shechtman, E., Wang, O.: The unreasonable effectiveness of deep features as a perceptual metric. In: CVPR, pp. 586–595 (2018)
43. Zhou, T., Tucker, R., Flynn, J., Fyffe, G., Snavely, N.: Stereo magnification: learning view synthesis using multiplane images. ACM Trans. Graph (2018)

Gaussian Alignment for Relative Camera Pose Estimation via Single-View Reconstruction

Yumin Li(✉) and Dylan Campbell

Australian National University, Canberra, Australia
{yumin.li,dylan.campbell}@anu.edu.au

Abstract. Estimating metric relative camera pose from a pair of images is of great importance for 3D reconstruction and localisation. However, conventional two-view pose estimation methods are not metric, with camera translation known only up to a scale, and struggle with wide baselines and textureless or reflective surfaces. This paper introduces GARPS, a training-free framework that casts this problem as the direct alignment of two independently reconstructed 3D scenes. GARPS leverages a metric monocular depth estimator and a Gaussian scene reconstructor to obtain a metric 3D Gaussian Mixture Model (GMM) for each image. It then refines an initial pose from a feed-forward two-view pose estimator by optimising a differentiable GMM alignment objective. This objective jointly considers geometric structure, view-independent colour, anisotropic covariance, and semantic feature consistency, and is robust to occlusions and texture-poor regions without requiring explicit 2D correspondences. Extensive experiments on the RealEstate10K dataset demonstrate that GARPS outperforms both classical and state-of-the-art learning-based methods, including MASt3R. These results highlight the potential of bridging single-view perception with multi-view geometry to achieve robust and metric relative pose estimation.

Keywords: Relative Pose Estimation · Gaussian Mixture Alignment · Single-View Reconstruction

1 Introduction

Estimating the relative pose between two camera views—comprising 3D rotation and translation—is a fundamental problem in computer vision, underpinning applications such as robot navigation, augmented reality, and 3D reconstruction [6,22]. In challenging indoor environments, textureless walls, repeated structures, and varying illumination often lead to ambiguous correspondences, making reliable pose estimation particularly difficult.

Classical two-view methods estimate pose by matching local features under epipolar constraints [13], typically by computing the essential matrix using algorithms such as the 5-point method [16], which requires known camera intrinsics. However, these methods may fail in textureless or low-feature environments.

M. Liu et al. (Eds.): AI 2025, LNAI 16371, pp. 121–132, 2026.
https://doi.org/10.1007/978-981-95-4972-6_10

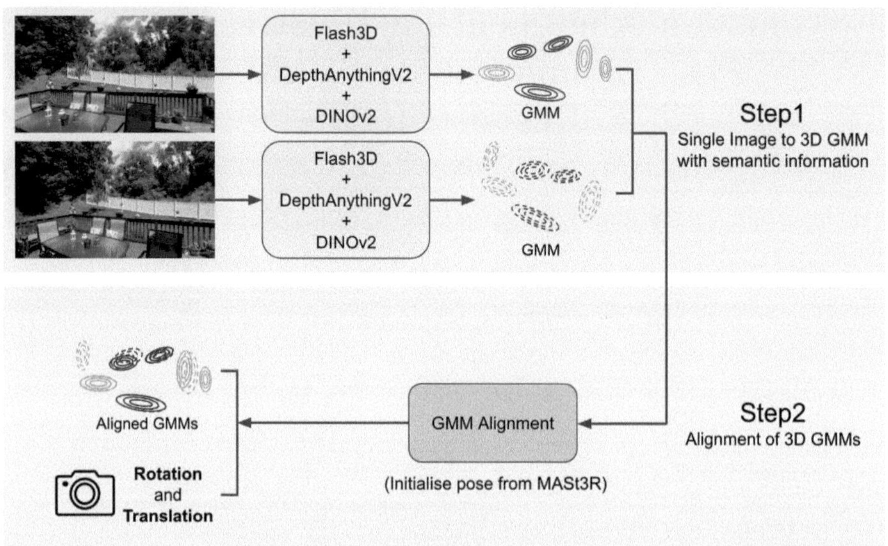

Fig. 1. Overview of the proposed GARPS pipeline. Given two RGB images, we predict a 3D Gaussian mixture model (GMM) for each. The relative pose is then estimated by directly aligning these GMMs in 3D space using our novel optimisation objective.

While multi-view Structure-from-Motion (SfM) pipelines like COLMAP [20,21] provide more accurate reconstructions by jointly optimising multiple views, they require multiple images and significant computation, limiting their use in lightweight, pairwise scenarios. Both approaches suffer from scale ambiguity.

The recent advent of single-view 3D reconstruction has created a new paradigm. Methods like Flash3D [23] can now generate a dense, explicit 3D scene representation from a single RGB image by leveraging powerful monocular depth estimators. The speed and accuracy of single-view scene reconstructors makes an alternative approach more feasible: instead of matching ambiguous 2D pixels, we can directly align the reconstructed 3D models.

Motivated by this, we introduce GARPS—Gaussian Alignment for Relative Pose via Single-view reconstruction—a novel method that casts relative pose estimation as an alignment problem between two 3D Gaussian Mixture Models (GMMs). As shown in Fig. 1, we first reconstruct each image into a metrically-scaled GMM. We then directly optimise for the relative pose by maximising a differentiable objective function that jointly considers geometric structure, anisotropic covariance, view-independent colour, and semantic features. Our approach bypasses the need for explicit keypoint matching and is training-free.

GARPS is inherently robust to outliers and local ambiguities compared to classical ICP [2,18] and, unlike learning-based methods such as DUSt3R [26], it is interpretable and applicable to unseen domains without fine-tuning. We make the following contributions:

1. GARPS, a novel method for relative pose estimation based on 3D GMM alignment from monocular reconstructions with metric scale; and
2. an alignment objective that combines 3D geometry, view-independent colour, anisotropic covariance and semantic features for robust pose optimisation.

We evaluate our method on the RealEstate10K dataset [33], achieving state-of-the-art performance compared to both classical and learning-based baselines.

2 Related Work

Our work builds on recent progress in estimating relative camera pose and reconstructing 3D scenes from a single image. In this section, we review related work in these two areas and explain how our method fits into this context.

2.1 Two-View Relative Pose Estimation

Estimating relative pose from two views is a long-standing problem in computer vision. We categorise prior work into three main streams.

Correspondence-based methods match 2D features, from classical descriptors [13] with robust estimators [5], to modern learned matchers [4,11,19]. These correspondences form the foundation for full Structure-from-Motion (SfM) pipelines like COLMAP [20,21], which leverage multi-view bundle adjustment to achieve globally consistent reconstructions. However, these approaches require robust image features for matching and perform less well in areas that are texture-poor, have view-dependent appearance, or have local or global symmetries.

3D registration methods, in contrast, operate on point clouds. The Iterative Closest Point (ICP) algorithm [2] and its variants [18] are widely used but are prone to local minima. Probabilistic approaches [7,15] and globally-optimal solvers [3,27] offer more robustness. However, they can be computationally expensive or rely on simplifying assumptions, such as GOGMA's [3] use of isotropic covariances.

Learning-based methods directly regress pose or leverage learned scene representations. Some infer dense 3D pointmaps from which pose is recovered [10,26]. Others use structures like 3D Gaussians for pose estimation or refinement [12,30,34]. Beyond these categories, some approaches jointly predict camera poses and scene geometry in an end-to-end framework, rather than first predicting the geometry [25]. Despite their success, the performance of these methods is tied to large-scale supervision, and recovering metric scale often requires specific training datasets.

Our work, GARPS, contributes a new perspective to this field. By framing pose estimation as the alignment of two independently generated 3D models, it combines the robustness of probabilistic 3D registration with the benefits of modern single-view reconstruction, avoiding direct 2D correspondence matching and end-to-end learning.

2.2 Single-View 3D Reconstruction

Our approach is enabled by recent progress in single-view 3D reconstruction, which itself builds upon monocular depth estimation.

Monocular depth estimation aims to infer pixel-wise depth from a single image. Deep learning methods have largely superseded classical geometric approaches [1,29]. State-of-the-art models like Depth Anything V2 [28] leverage large-scale training with teacher–student frameworks to produce highly accurate and robust metric depth maps, showing strong generalisation capabilities.

These 2.5D depth maps can be upgraded to full 3D models by using them as a high-quality prior for 3D reconstruction. While optimisation-based reconstruction methods like NeRF [14] and 3DGS [8] require multiple views, feed-forward methods like Splatter Image [24] and Flash3D [23] can generate an explicit 3D Gaussian-based scene representation from a single image and its predicted depth map. Unlike 2.5D depth maps, which only describe the parts of a scene visible from a single viewpoint, these 3D representations can also infer hidden or occluded areas—for example, Splatter Image can predict the back side of an object that is not visible in the input image.

In this work, we leverage this pipeline as the foundation for our method. We enhance the Flash3D framework by integrating it with the metric Depth Anything V2 backbone. This provides us with a high-quality, metrically-scaled Gaussian mixture model for each input image, which serves as the input to our alignment-based pose estimation algorithm.

3 Single-View Reconstuction for Relative Pose Estimation

In this section, we propose GARPS—Gaussian Alignment for Relative Pose via Single-view Reconstruction—a method that estimates the metric relative pose between two RGB images by aligning their single-view 3D reconstructions. Each image is independently processed using Flash3D [23] to obtain a coloured 3D Gaussian Mixture Model (GMM), capturing both geometry, appearance and semantic information. We estimate the relative pose by aligning these two GMMs via a differentiable L_2 distance-based objective function.

3.1 Objective Formulation

Given two images, we first reconstruct them into two sets of 3D Gaussian components, denoted as source GMM $\mathcal{G}_x$ and target GMM $\mathcal{G}_y$. Each GMM is defined as $\mathcal{G} = \{(\boldsymbol{\mu}_i, \boldsymbol{\Sigma}_i, \boldsymbol{c}_i, \phi_i)\}_{i=1}^{K}$, where $\boldsymbol{\mu}_i$ is the mean, $\boldsymbol{\Sigma}_i$ the spatial covariance, $\boldsymbol{c}_i$ the RGB colour, and ϕ_i the mixture weight. Our goal is to find the relative pose $(\mathbf{R}^\star, \mathbf{t}^\star)$ with rotation $\mathbf{R}^\star \in \mathrm{SO}(3)$ and translation $\mathbf{t}^\star \in \mathbb{R}^3$ that best aligns mixtures $\mathcal{G}_x$ and $\mathcal{G}_y$.

We formulate this as minimising the L_2 distance between the two GMM probability density functions. Let $T(\mathcal{G}_x, \mathbf{R}, \mathbf{t})$ denote the transformation of the source GMM, where each component is transformed as $\boldsymbol{\mu}'_i = \mathbf{R}\boldsymbol{\mu}_i + \mathbf{t}$ and $\boldsymbol{\Sigma}'_i = \mathbf{R}\boldsymbol{\Sigma}_i\mathbf{R}^\top$. The squared L_2 distance is given by

$$\begin{aligned} d(\mathbf{R}, \mathbf{t})^2 &= \int_{\mathbb{R}^3} \left(p(\boldsymbol{p}|T(\mathcal{G}_x, \mathbf{R}, \mathbf{t})) - p(\boldsymbol{p}|\mathcal{G}_y)\right)^2 d\boldsymbol{p} \\ &= C - 2\int_{\mathbb{R}^3} p(\boldsymbol{p}|T(\mathcal{G}_x))\, p(\boldsymbol{p}|\mathcal{G}_y)\, d\boldsymbol{p}, \end{aligned} \tag{1}$$

where C is a constant independent of the transformation. Maximising the cross-term is equivalent to minimising the L_2 distance. This cross-term has a convenient closed-form solution for GMMs, given by

$$\int p(\boldsymbol{p}|T(\mathcal{G}_x))\, p(\boldsymbol{p}|\mathcal{G}_y)\, d\boldsymbol{p} = \sum_{i=1}^{K}\sum_{j=1}^{K} \phi_i\phi_j\varphi(\mathbf{0} \mid \boldsymbol{\mu}'_i - \boldsymbol{\mu}_j, \boldsymbol{\Sigma}'_i + \boldsymbol{\Sigma}_j), \tag{2}$$

where $\varphi(\mathbf{0} \mid \boldsymbol{\mu}, \boldsymbol{\Sigma})$ denotes the multivariate Gaussian probability density function with mean $\boldsymbol{\mu}$ and covariance $\boldsymbol{\Sigma}$, evaluated at $\mathbf{0}$.

To incorporate appearance, we augment each spatial Gaussian with a colour distribution, also modelled as a Gaussian. The joint probability density, shown in Eq. (4), allows us to extend the cross-term to include colour similarity in the same way as geometric similarity. Building on this, we also incorporate semantic information by defining a semantic consistency weight ω_{ij}^{sem} for each Gaussian pair (i, j), such that

$$\omega_{ij}^{\text{sem}} = \begin{cases} 1 & \text{if } \ell_i = \ell_j \\ \lambda_{\text{sem}} & \text{otherwise,} \end{cases} \tag{3}$$

where ℓ denotes the semantic label, and $\lambda_{\text{sem}} \in [0, 1)$ is a discount factor applied to pairs with mismatched semantics.

The final cross-term in the objective function is weighted by both colour similarity and semantic consistency, promoting alignment between structures that are geometrically, visually, and semantically consistent. It is given by

$$\sum_{i=1}^{K}\sum_{j=1}^{K} \omega_{ij}^{\text{sem}}\phi_i\phi_j \underbrace{\varphi(\mathbf{0} \mid \boldsymbol{\mu}'_i - \boldsymbol{\mu}_j, \boldsymbol{\Sigma}'_i + \boldsymbol{\Sigma}_j)}_{\text{Geometric Term}} \underbrace{\varphi(\mathbf{0} \mid \boldsymbol{c}_i - \boldsymbol{c}_j, \boldsymbol{\Sigma}^c_i + \boldsymbol{\Sigma}^c_j)}_{\text{Colour Term}}. \tag{4}$$

By expanding the Gaussian expressions and removing constant factors, we derive our final objective function to be maximised:

$$f(\mathbf{R}, \mathbf{t}) = \sum_{i=1}^{K}\sum_{j=1}^{K} \frac{\omega_{ij}^{\text{sem}}\phi_i\phi_j}{|\boldsymbol{\Sigma}'_i + \boldsymbol{\Sigma}_j|^{1/2}} \exp\left(-\frac{1}{2}D_{ij}^{\text{geo}} - \frac{1}{2}D_{ij}^{\text{col}}\right), \tag{5}$$

where $D_{ij}^{\text{geo}} = (\boldsymbol{\mu}'_i - \boldsymbol{\mu}_j)^\top(\boldsymbol{\Sigma}'_i + \boldsymbol{\Sigma}_j)^{-1}(\boldsymbol{\mu}'_i - \boldsymbol{\mu}_j)$ is the squared Mahalanobis distance in 3D space, and $D_{ij}^{\text{col}} = (\boldsymbol{c}'_i - \boldsymbol{c}_j)^\top(\boldsymbol{\Sigma}^c_i + \boldsymbol{\Sigma}^c_j)^{-1}(\boldsymbol{c}'_i - \boldsymbol{c}_j)$ is the corresponding distance in RGB space.

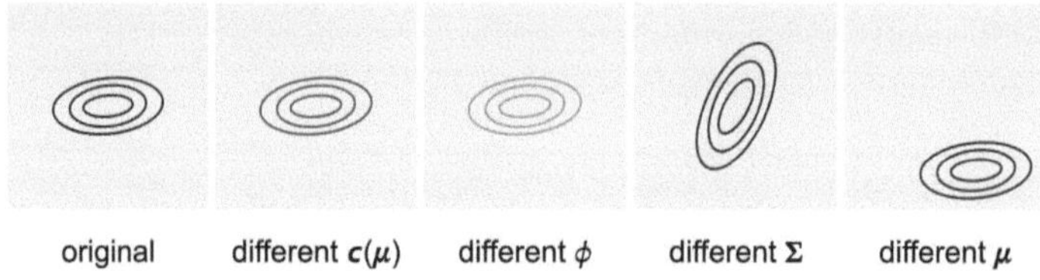

Fig. 2. Visual effect of modifying Gaussian parameters: mean $\boldsymbol{\mu}$, covariance $\boldsymbol{\Sigma}$, opacity ϕ, and colour $\boldsymbol{c}$.

Component Formulation. The parameters for each Gaussian component are derived from a pre-trained Flash3D model: the Gaussian means $\boldsymbol{\mu}_i$ and anisotropic spatial covariances $\boldsymbol{\Sigma}_i$ are copied directly from the Flash3D output, with the latter multiplied by a global scaling factor λ_{cov} to modulate the sharpness of the geometric alignment term; the colours $\boldsymbol{c}_i$ are taken from the view-independent zeroth-order spherical harmonics of Flash3D, lending robustness to reflective surfaces; the colour covariances $\boldsymbol{\Sigma}^c$ are fixed diagonal hyperparameter matrices that control sensitivity to colour deviations; the mixture weights ϕ_i are derived from the Flash3D opacities after L_1 normalisation; and the semantic labels ℓ are computed from a pre-trained DINOv2-based semantic segmentation network [17] with 150 classes corresponding to the categories from the ADE20K dataset [31,32]. Specifically, we use the pretrained model (`dinov2_vitg14_ade20k_m2f.pth`) released by the authors. Note that Flash3D predicts two Gaussians per pixel, allowing it to model occluded surfaces. We assign the same semantic label to both Gaussians, since usually these correspond to the front and back of the same object. The components are visualised in Fig. 2.

3.2 Optimisation Strategy

To ensure computational efficiency, we operate on a sparse and salient subset of the original Gaussians. This subset is formed by first discarding all components with an opacity value below a threshold τ_{opacity}. The remaining salient Gaussians are then uniformly subsampled with a stride of s_{sub}, which significantly reduces the number of components K for optimisation.

Furthermore, directly evaluating the double summation in Eq. (5) is computationally prohibitive due to its $O(K^2)$ complexity. To accelerate the process, we approximate the objective by considering only a sparse set of potential correspondences. Specifically, for each source Gaussian, we find its k_{nn} nearest neighbours in the target GMM based on the Euclidean distance between their means. These candidate pairs are further pruned by discarding any for which this distance exceeds a threshold τ_{dist}.

We represent rotation with a unit quaternion $\mathbf{q}$ and translation with a 3D vector $\mathbf{t} \in \mathbb{R}^3$. The pose is initialised using the estimate provided by MASt3R [10], which serves as a strong prior for both rotation and translation. The objective

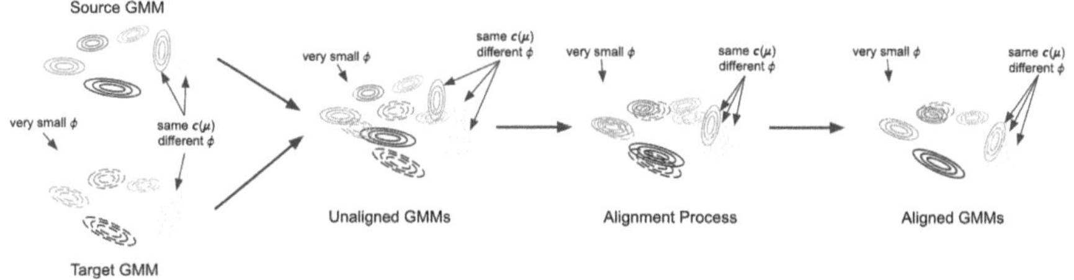

Fig. 3. Illustration of the alignment process. The value of the objective decreases as Gaussians align in both position and colour.

in Eq. (5) is maximised using the Adam optimiser [9], with separate learning rates η_{rot} and η_{trans} for optimising $\mathbf{q}$ and $\mathbf{t}$ respectively, to account for their differing sensitivities and magnitudes. All key hyperparameters—including η_{rot}, η_{trans}, colour covariance $\mathbf{\Sigma}^c$, spatial covariance scaling factor λ_{cov} and others—are determined via Bayesian optimisation on a validation set. The optimisation procedure is illustrated in Fig. 3, where it is shown that Gaussians with higher opacity contribute more strongly to the objective.

4 Experiments

We validate our method, GARPS, through extensive experiments, comparing it against classical and state-of-the-art methods on a challenging dataset.

Dataset. We use the RealEstate10K dataset [33], which covers a wide variety of real estate videos, including mainly indoor environments and a portion of outdoor views, all with natural camera motion. As our method is not a learning-based approach, no training set is used. We randomly sample 200 clips from the test split: 100 with small viewpoint changes (translation $\leq 0.3\,\text{m}$ and rotation $\leq 15^\circ$) and 100 with large viewpoint changes (translation $> 0.3\,\text{m}$ or rotation $> 15^\circ$), based on camera poses between the 1st and 30th frames. One image pair is generated per clip for evaluation. Similarly, we randomly sample 20 clips (10 from each category) from the dataset's training split for hyperparameter tuning.

Metrics. We evaluate pose accuracy using standard metrics: rotation error (E_{rot} in degrees), translation error (E_{trans} in metres), and a 3D alignment error (E_{3D} in metres), which measures the average displacement of transformed Gaussian centres. The latter is defined as the average Euclidean distance between the Gaussian centres of the first (source) GMM transformed by the estimated pose, and those transformed by the ground-truth pose, and is given by

$$E_{\text{3D}} = \frac{1}{N}\sum_{i=1}^{N}\left\|\hat{\boldsymbol{R}}\boldsymbol{\mu}_i + \hat{\boldsymbol{t}} - (\boldsymbol{R}\boldsymbol{\mu}_i + \boldsymbol{t})\right\|, \tag{6}$$

where $\{\boldsymbol{\mu}_i\}_{i=1}^N$ is a set of N source Gaussian means, $(\hat{\boldsymbol{R}}, \hat{\boldsymbol{t}})$ is the estimated pose, and $(\boldsymbol{R}, \boldsymbol{t})$ is the ground-truth pose. This error measures the discrepancy between the estimated and ground-truth transformations by applying both to the same source model and computing the average Euclidean distance between the resulting transformed points. Since ground-truth point clouds and Gaussian mixtures are unavailable, we use the output of Flash3D as pseudo ground truth geometry for this error measure.

Although Mahalanobis distance may better reflect the GMM structure, we adopt Euclidean distance for $E_{3\text{D}}$ as it offers a clearer geometric interpretation and avoids further reliance on the predicted covariances from Flash3D. This choice reflects a trade-off: Mahalanobis distance would fit better with the GMM formulation, but would make the evaluation even more dependent on the quality of the pseudo ground truth.

Implementation Details. Our method is implemented in PyTorch and runs on a single NVIDIA GTX 2080Ti GPU. We used the following key hyperparameters, tuned via Bayesian optimisation on the validation set: $\lambda_{\text{sem}} = 0.92$, $\lambda_{\text{cov}} = 90$, $\boldsymbol{\Sigma}^c = \text{diag}(0.4, 0.4, 0.4)$, $\tau_{\text{opacity}} = 0.5$, $\eta_{\text{trans}} = 9 \times 10^{-5}$, $\eta_{\text{rot}} = 3 \times 10^{-6}$, $s_{\text{sub}} = 32$, $k_{\text{nn}} = 30$, and $\tau_{\text{dist}} = 2.5$.

Comparison Methods. We compare GARPS against classical baselines and recent state-of-the-art methods. Baselines include the SfM pipeline COLMAP [20], geometry-based ICP [2] and Colored ICP [18], and the probabilistic CPD [15] method. SOTA methods include DUSt3R [26] and MASt3R [10], which regress aligned 3D pointmaps, and VGGT [25], which jointly predicts camera poses and scene geometry in an end-to-end manner. For point-cloud-based methods, we provide them with the point clouds from Flash3D with DepthAnythingV2 and a pose initialisation using the estimate from MASt3R.

Results. We define a successful pair as an image pair for which a method produces a valid relative pose. All reported statistics are computed only on the successful pairs for each method. COLMAP achieves a relatively low success rate of 22.5%, while all other methods attain a 100% success rate.

We report both mean ± standard deviation and median ± 0.5 × interquartile range (IQR) of pose estimation errors on 200 samples from the RealEstate10K test set. Tables 1 and 2 show that GARPS consistently outperforms both classical and learning-based baselines in most metrics. While VGGT [25] achieves the best rotation error, GARPS demonstrates robust overall performance and surpasses MASt3R in all metrics. Unlike the learning-based models, GARPS does not require task-specific training, yet delivers state-of-the-art performance through a purely optimisation-based pipeline.

In terms of runtime, the reported time for GARPS (17.69 s) includes the initial pose estimation by MASt3R. While GARPS is slower than lightweight geometric methods such as ICP, it is still considerably faster than classical non-rigid methods like CPD.

Table 1. Mean pose errors (± standard deviation) and average inference time per pair on the test set.

Method	$E_{\text{trans}} \downarrow$ (m)	$E_{\text{rot}} \downarrow$ (°)	$E_{\text{3D}} \downarrow$ (m)	Time (s)
COLMAP [20]	0.251 ± 0.274	9.422 ± 11.037	0.492 ± 0.451	3.18
ICP [2]	0.482 ± 0.628	6.246 ± 14.470	0.211 ± 0.257	**0.04**
Colored ICP [18]	0.427 ± 0.506	5.693 ± 12.869	0.189 ± 0.205	0.06
CPD [15]	0.492 ± 0.616	9.981 ± 10.328	0.506 ± 0.471	266
DUSt3R [26]	0.345 ± 0.278	1.058 ± 1.628	0.171 ± 0.133	13.07
MASt3R [10]	0.170 ± 0.200	0.995 ± 1.549	0.097 ± 0.110	9.07
VGGT [25]	0.286 ± 0.245	**0.902 ± 1.491**	0.139 ± 0.121	0.68
GARPS	**0.167 ± 0.200**	0.994 ± 1.545	**0.094 ± 0.109**	17.69

Table 2. Median pose errors (± 0.5 IQR) on the test set.

Method	$E_{\text{trans}} \downarrow$ (m)	$E_{\text{rot}} \downarrow$ (°)	$E_{\text{3D}} \downarrow$ (m)
COLMAP [20]	0.166 ± 0.104	5.207 ± 7.510	0.386 ± 0.251
ICP [2]	0.282 ± 0.202	2.607 ± 1.994	0.127 ± 0.086
Colored ICP [18]	0.275 ± 0.188	2.434 ± 1.822	0.125 ± 0.083
CPD [15]	0.316 ± 0.180	7.406 ± 5.712	0.397 ± 0.255
DUSt3R [26]	0.262 ± 0.152	0.522 ± 0.371	0.140 ± 0.072
MASt3R [10]	0.114 ± 0.066	0.570 ± 0.421	0.062 ± 0.031
VGGT [25]	0.212 ± 0.128	**0.439 ± 0.347**	0.110 ± 0.057
GARPS	**0.109 ± 0.065**	0.561 ± 0.400	**0.058 ± 0.030**

Table 3. Ablation study on the test set (mean pose errors). All values indicate relative change (%) with respect to the full method.

Variant	$E_{\text{trans}} \downarrow$	$E_{\text{rot}} \downarrow$	$E_{\text{3D}} \downarrow$
w/o image segmentation	+0.05%	+0.10%	+0.02%
w/o anisotropic covariances	+1.01%	+0.71%	+2.41%
w/o colour features	+0.45%	+0.58%	+1.11%

The qualitative results, shown in Fig. 4, support the quantitative findings. GARPS produces sharper and more consistent geometric alignment—for example, in challenging regions such as trees and trunks—while other methods exhibit visible translational or rotational drift.

Ablation Study. The ablation study in Table 3 highlights the contribution of each major component in our framework. Each variant removes one key element from the full method to isolate its impact on the performance, measured by the relative change (%) in mean pose errors: translation error (E_{trans}), rotation

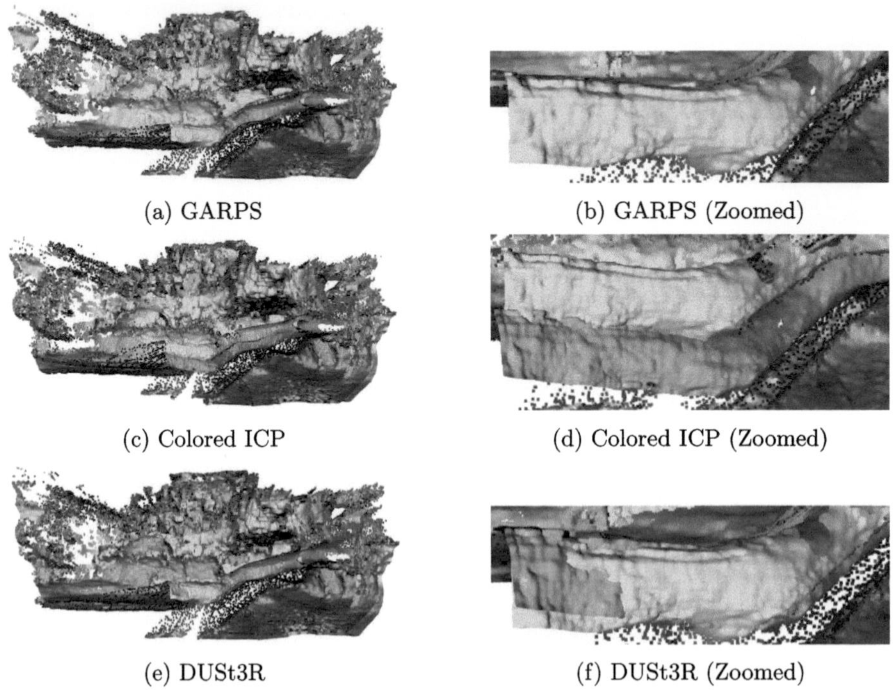

(a) GARPS
(b) GARPS (Zoomed)
(c) Colored ICP
(d) Colored ICP (Zoomed)
(e) DUSt3R
(f) DUSt3R (Zoomed)

Fig. 4. Qualitative comparison on a test scene (full/zoomed views). GARPS produces better geometric alignment compared to baselines and SOTA methods. The first point cloud is yellow, the second is blue.

error (E_{rot}), and 3D error (E_{3D}). The omission of the image segmentation leads to a minor degradation, indicating that it complements other modules without being the sole performance driver. Replacing the original anisotropic covariance in Flash3D with an isotropic covariance (i.e., an identity matrix) results in a significant performance drop, suggesting that modelling uncertainty plays a critical role in accurate pose estimation. Similarly, omitting colour information causes noticeable degradation across all metrics, highlighting its complementary role in enhancing geometric understanding. These results validate the effectiveness and necessity of each proposed component.

5 Conclusion

We present GARPS, a novel method for estimating metrically-scaled relative pose from two RGB images by aligning single-view 3D Gaussian mixtures. Our method leverages monocular 3D reconstructions from Flash3D [23], enhanced with Depth Anything V2 [28], semantic labels from DINOv2 [17], and prior pose estimates from MASt3R [10]. It aligns two independently generated coloured 3D GMMs using a robust L_2 objective that jointly considers geometry, colour, semantics, and anisotropic spatial uncertainties. GARPS sets a new state-of-the-art in relative pose estimation from two-view RGB images, outperforming both classical and learning-based methods across all evaluated metrics. Future work includes exploring monocular reconstruction methods beyond Flash3D that yield 3D GMMs, simplifying covariance computations to reduce runtime, and extending the method to dynamic scenes via object tracking, as well as testing improved initial pose estimation strategies and evaluating cross-dataset generalisation.

References

1. Bajcsy, R., Lieberman, L.: Texture gradient as a depth cue. Comput. Graphi. Image Process. **5**(1), 52–67 (1976)
2. Besl, P.J., McKay, N.D.: A method for registration of 3-D shapes. IEEE Trans. Pattern Anal. Mach. Intell. **14**(2), 239–256 (1992)
3. Campbell, D., Petersson, L.: GOGMA: globally-optimal gaussian mixture alignment. In: 2016 IEEE Conference on Computer Vision and Pattern Recognition (CVPR), pp. 5685–5694 (2016). https://doi.org/10.1109/CVPR.2016.613
4. DeTone, D., Malisiewicz, T., Rabinovich, A.: Superpoint: self-supervised interest point detection and description. In: Proceedings of the IEEE Conference on Computer Vision and Pattern Recognition Workshops, pp. 224–236 (2018)
5. Fischler, M.A., Bolles, R.C.: Random sample consensus: A paradigm for model fitting with applications to image analysis and automated cartography. In: Fischler, M.A., Firschein, O. (eds.) Readings in Computer Vision, pp. 726–740. Morgan Kaufmann, San Francisco (CA) (1987)
6. Hartley, R.: Multiple View Geometry in Computer Vision, vol. 665. Cambridge university press, Cambridge (2003)
7. Jian, B., Vemuri, B.C.: Robust point set registration using gaussian mixture models. IEEE Trans. Pattern Anal. Mach. Intell. **33**(8), 1633–1645 (2010)
8. Kerbl, B., Kopanas, G., Leimkühler, T., Drettakis, G.: 3D gaussian splatting for real-time radiance field rendering. ACM Trans. Graph. **42**(4), 139–1 (2023)
9. Kingma, D.P., Ba, J.: Adam: a method for stochastic optimization (2017). https://arxiv.org/abs/1412.6980
10. Leroy, V., Cabon, Y., Revaud, J.: Grounding image matching in 3D with MASt3R. In: ECCV (2024)
11. Lindenberger, P., Sarlin, P.E., Pollefeys, M.: LightGlue: local feature matching at light speed. In: ICCV (2023)
12. Liu, C., et al.: GS-CPR: efficient camera pose refinement via 3D gaussian splatting. In: The Thirteenth International Conference on Learning Representations (2025). https://openreview.net/forum?id=mP7uV59iJM
13. Lowe, D.G.: Distinctive image features from scale-invariant keypoints. Int. J. Comput. Vis. **60**, 91–110 (2004)
14. Mildenhall, B., Srinivasan, P.P., Tancik, M., Barron, J.T., Ramamoorthi, R., Ng, R.: NeRF: representing scenes as neural radiance fields for view synthesis. In: ECCV (2020)
15. Myronenko, A., Song, X., Carreira-Perpinan, M.: Non-rigid point set registration: coherent point drift. Adv. Neural Inf. Process. Syst. **19** (2006)
16. Nister, D.: An efficient solution to the five-point relative pose problem. IEEE Trans. Pattern Anal. Mach. Intell. **26**(6), 756–770 (2004)

17. Oquab, M., et al.: DINOv2: learning robust visual features without supervision (2023)
18. Park, J., Zhou, Q.Y., Koltun, V.: Colored point cloud registration revisited. In: 2017 IEEE International Conference on Computer Vision (ICCV). pp. 143–152 (2017). https://doi.org/10.1109/ICCV.2017.25
19. Sarlin, P.E., DeTone, D., Malisiewicz, T., Rabinovich, A.: Superglue: learning feature matching with graph neural networks. In: Proceedings of the IEEE/CVF Conference on Computer Vision and Pattern Recognition, pp. 4938–4947 (2020)
20. Schönberger, J.L., Frahm, J.M.: Structure-from-motion revisited. In: Conference on Computer Vision and Pattern Recognition (CVPR) (2016)
21. Schönberger, J.L., Zheng, E., Pollefeys, M., Frahm, J.M.: Pixelwise view selection for unstructured multi-view stereo. In: European Conference on Computer Vision (ECCV) (2016)
22. Szeliski, R.: Computer Vision: Algorithms and Applications. Springer, Cham (2022). https://doi.org/10.1007/978-3-030-34372-9
23. Szymanowicz, S., et al.: Flash3D: feed-forward generalisable 3D scene reconstruction from a single image (2024). https://arxiv.org/abs/2406.04343
24. Szymanowicz, S., Rupprecht, C., Vedaldi, A.: Splatter image: ultra-fast single-view 3D reconstruction. In: The IEEE/CVF Conference on Computer Vision and Pattern Recognition (CVPR) (2024)
25. Wang, J., Chen, M., Karaev, N., Vedaldi, A., Rupprecht, C., Novotny, D.: VGGT: visual geometry grounded transformer. In: Proceedings of the IEEE/CVF Conference on Computer Vision and Pattern Recognition (2025)
26. Wang, S., Leroy, V., Cabon, Y., Chidlovskii, B., Revaud, J.: DUSt3R: geometric 3D vision made easy. In: IEEE Conference on Computer Vision and Pattern Recognition, pp. 20697–20709. IEEE (2024)
27. Yang, J., Li, H., Campbell, D., Jia, Y.: Go-ICP: a globally optimal solution to 3D ICP point-set registration. IEEE Trans. Pattern Anal. Mach. Intell. **38**(11), 2241–2254 (2016). https://doi.org/10.1109/TPAMI.2015.2513405
28. Yang, L., et al.: Depth Anything V2 (2024). https://arxiv.org/abs/2406.09414
29. Zhang, R., Tsai, P.S., Cryer, J., Shah, M.: Shape-from-shading: a survey. IEEE Trans. Pattern Anal. Mach. Intell. **21**(8), 690–706 (1999)
30. Zhang, S., et al.: Flare: feed-forward geometry, appearance and camera estimation from uncalibrated sparse views. arXiv preprint arXiv:2502.12138 (2025)
31. Zhou, B., Zhao, H., Puig, X., Fidler, S., Barriuso, A., Torralba, A.: Scene parsing through ADE20K dataset. In: Proceedings of the IEEE Conference on Computer Vision and Pattern Recognition (2017)
32. Zhou, B., et al.: Semantic understanding of scenes through the ADE20K dataset. Int. J. Comput. Vis. **127**(3), 302–321 (2019)
33. Zhou, T., Tucker, R., Flynn, J., Fyffe, G., Snavely, N.: Stereo magnification: learning view synthesis using multiplane images. In: SIGGRAPH (2018)
34. Zhu, L., Li, Y., Sandström, E., Huang, S., Schindler, K., Armeni, I.: LoopSplat: loop closure by registering 3D gaussian splats. In: International Conference on 3D Vision (3DV) (2025)

Street Depth-Aware Gaussian for Modeling Dynamic Urban Scenes

Ke Xie, Runkai Zhao, Dingxin Zhang, and Weidong Cai(✉)

School of Computer Science, Faculty of Engineering, The University of Sydney, Sydney, NSW, Australia
{kxie0655,rzha9419,dzha2344,tom.cai}@sydney.edu.au

Abstract. Modeling dynamic urban streets is a critical challenge in the field of autonomous driving. With the evolution of 3D Gaussian Splatting (3DGS) technique, the efficient photo-realistic view synthesis can be achieved with high quality. However, existing dynamic scene modeling methods reconstruct the complex background of driving scenarios with unexpected blurries and color noises. To address this gap, we propose Street Depth-Aware Gaussian, dubbed as ***SeeDepthGaussian***, which introduces a depth regularization method by utilizing monocular depth into open autonomous driving scenarios. This enables the recovery of more accurate scene details under depth supervision. Furthermore, the incorporation of Global-Local Depth Normalization further refines depth regularization by better capturing depth variations at different scales. Extensive experiments on the Waymo dataset indicate that SeeDepthGaussian achieves superior rendering quality compared to mainstream advanced 3DGS-based methods, and requires fewer computational resources as well as less training time than NeRF-based methods.

Keywords: Novel View Synthesis · Autonomous Driving · 3D Gaussian Splatting · Generative AI · 3D Scene Reconstruction

1 Introduction

Urban street scene reconstruction is a key topic in autonomous driving tasks, benefiting applications such as simulation environments and data augmentation [3].

The fundamental distinction between urban street scene reconstruction and traditional 3D reconstruction or 2D novel-view synthesis tasks lies in the dynamic nature of the target scenes, requiring explicit modeling of temporal changes and moving objects.

StreetGaussian [20] introduced an influential approach for dynamic scene reconstruction in 3D Gaussian Splatting (3DGS), inspiring subsequent works [4] to adopt similar rendering pipelines. While these methods effectively handle dynamic elements, their use of category-specific rendering strategies can introduce certain challenges. In specific cases, particularly in distant regions, the method generates blurred or distorted imagery and occasional color noise, as illustrated in Fig. 1.

M. Liu et al. (Eds.): AI 2025, LNAI 16371, pp. 133–145, 2026.
https://doi.org/10.1007/978-981-95-4972-6_11

Fig. 1. Comparison of the state-of-the-art StreetGaussian [20] with our SeeDepthGaussian. Most current works focus on dynamic foregrounds, overlooking background blur and color noise from rapid changing background. With both global and local depth information, our framework leverages Hard-and-Soft Depth Regularization to optimize distant scene rendering, mitigating the above phenomena.

In our view, this phenomenon arises from two key factors: (i) the background modeling inherits the limitations of standard 3DGS, which struggles with consistency in large-scale scenes, and (ii) distant regions often occupy small areas in images with insufficient information, limiting the pipeline's ability to capture fine details necessary for robust optimization.

Recent depth-based 3DGS methods [1,7,8,19,23] have demonstrated that incorporating depth information as a regularization prior can effectively constrain the spatial distribution of 3D Gaussians and enhance rendering quality in static scenes. Given that monocular depth estimation has become a mature and readily available technology in autonomous driving systems, we propose SeeDepthGaussian, a framework that leverages monocular depth priors to address the aforementioned blurring, distortion, and color noise issues in dynamic urban scene modeling.

SeeDepthGaussian incorporates monocular depth estimation as an additional input, enabling the extraction of more enriched foreground-background features and local-global representations within the latent space. This is achieved through two key components integrated into the Depth Module: Hard-and-Soft Depth Regularization and Global-Local Depth Normalization. By normalizing depth at both global and local levels for regularization, we improve the spatial distribution of Gaussians and enhance rendering performance. Furthermore, as the background constitutes a crucial component of the synthesized image, we employ a neural renderer for background rendering. Compared to conventional spherical harmonics, this approach effectively avoids overfitting during training and improves rendering quality. Our experiments demonstrate that SeeDepthGaussian effectively alleviates the color noise and background blurring issues present in previous works. Our main contributions are as follows:

- We incorporate monocular-depth priors into the 3DGS pipeline for autonomous driving, boosting performance without extra training or inference time.
- We introduce a depth normalization and regularization method that is utilized in the loss function for the purpose of optimizing the spatial geometric distribution of 3D Gaussians, enabling higher-quality background modeling and rendering.
- We design a novel neural renderer to effectively model fine-grained background elements of the whole driving scene, rather than the conventional spherical harmonics, which could also avoid overfiting issue during training phase.
- We tested our approach on multiple scenarios from the Waymo Open Dataset, achieving a 0.21 improvement in PSNR when compared to state-of-the-art results. Extensive experiments further confirm its effectiveness.

2 Related Work

2.1 3D Scene Reconstruction

3D scene reconstruction is a classic research direction in computer vision. Since the emergence of NeRF [10], 3D scene reconstruction approaches based on deep learning have received increasing attention. By representing a continuous volumetric field with a multi-layer perceptron (MLP) [14], NeRF has achieved remarkable rendering results. Subsequently, as urban scene reconstruction holds broad application prospects, a series of works have extended NeRF to city-scale scenarios. In the domain of urban scene reconstruction, DNMP [9] leverages a set of deformable mesh vertices and associated feature vectors to parametrize local geometric and radiance information. Meanwhile, it decodes the mesh shape in a low dimensional latent space and employs rasterization for efficient feature interpolation, thereby reducing computational resource consumption. Nevertheless, the model remains primarily applicable to static scenes, and its initialization as well as optimization processes are notably complex.

2.2 3D Gaussian Splatting Methods

3DGS [5] defines a set of anisotropic Gaussian models in space. By controlling and optimizing each Gaussian's position, opacity, orientation, and scale, it constructs a highly realistic 3D scene. Then, by employing spherical harmonics to decode the color information obtained through an SfM-based initialization [15], it synthesizes photo realistic 2D views.

Currently, research on various derivative methods of 3DGS focuses on several major directions: i) enhancing the quality of generated results [1,7]; ii) optimizing computational resources [6]; and iii) expanding application domains and functionalities [13,18,24]. These research branches have achieved substantial progress, allowing 3DGS to gain rapid adoption in multiple fields. In addition, recent methods that incorporate monocular depth as a prior further refine the spatial distribution of 3D Gaussians, yielding advances in rendering equality [1,7], model generalization [8], and unsupervised training [1,19]. One noteworthy point is that most 3DGS-based methods are tailored to static scenes, and their performance in reconstructing dynamic scenes remains suboptimal.

2.3 Dynamic Scene Modeling

Dynamic scenes generally refer to scenes containing elements that change over time, where attributes such as position and light reflection vary as time progresses. For example, vehicles in autonomous driving scenarios are considered dynamic elements. In the context of dynamic scene modeling tasks, a common approach is to incorporate temporal dimension information as an input and encode it to construct a 4D neural field representation based on single-object scenarios. However, scene representations built for the entire environment are not instance-aware, which restricts their applicability in certain open-world settings, such as autonomous driving. Consequently, several methods have emerged that employ neural fields to separately render foreground objects and background [21], though these approaches still involve substantial memory and training time overhead.

As 3DGS continues to gain traction, a number of 3DGS-based frameworks for dynamic scene reconstruction have been introduced by leveraging additional prior information extracted from data and matching it with suitable methods. StreetGaussian (SG) [20] separates background and objects, and adds timestamps for dynamic objects (such as vehicles) to enable 3DGS-based urban dynamic scene modeling, achieving outstanding results. Subsequently, improved versions based on SG have also been proposed, such as S3Gaussian [4], which adopts self-supervised learning. Nevertheless, SG still faces challenges in reconstructing distant background and small objects, which can result in blurring and distortion.

3 Methods

As illustrated in Fig. 2, our proposed SeeDepthGaussian framework combines time-stamped sequential images and aggregated LiDAR point cloud data from the ego vehicle to initialize a Gaussian field. Drawing inspiration from StreetGaussian's framework, we employ a hierarchical scene representation that separates sky, objects, and background into distinct Gaussian sets for independent rendering. For the background, a trainable neural renderer replaces spherical harmonics to generate color information. All rendered results are then merged into a complete 2D image. In the depth regularization stage, we combine the Depth Estimator's output with SeeDepthGaussian's rendered depth, producing Hard Depth and Soft Depth. Their corresponding losses are computed via Global-Local Depth Normalization and integrated with other loss components to form a comprehensive loss function.

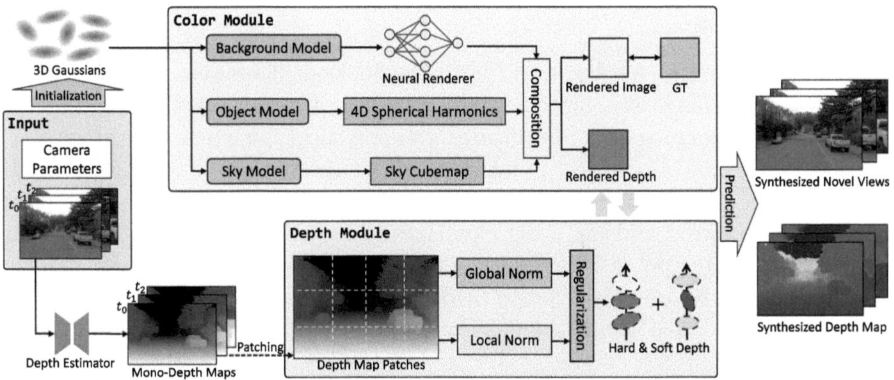

Fig. 2. SeeDepthGaussian Framework. Our framework first initializes a set of 3D Gaussians while a Depth Estimator predicts scene depths from the captured images. The 3D Gaussians are subsequently rendered with category-specific strategies and composited into a complete scene via the Composition module. The predicted depths are processed in the Depth Module through GlobalLocal Depth Normalization and HardSoft Depth Regularization, providing training constraints that optimize the Gaussian-field distribution during training, thereby improving overall rendering quality.

3.1 Preliminaries

3D Representation. 3D Gaussian Splatting [5] uses the distribution of a set of 3D Gaussians in a three-dimensional space to represent the 3D scene. Specifically, it employs a collection of Gaussian primitives $\theta_i = \{\mu_i, s_i, q_i, \alpha_i, f_i\}$, where i indicates the i-th Gaussian primitive. The parameters μ, s, q, α, and f, respectively, denote the 3D center coordinate, scaling factor, rotation quaternion, opacity coefficient, and color feature representation of each Gaussian primitive. Consequently, the basis function the i-th primitive G_i is given by:

$$G_i(x) = e^{-\frac{1}{2}(x-\mu_i)^\top \Sigma_i^{-1}(x-\mu_i)}. \tag{1}$$

Here, the covariance matrix Σ is used to describe the shape, scale, and orientation of the Gaussian kernel. The computation method for each Gaussian maintains universality, as it is derived from the product of the scaling matrix $\mathbf{S}$, which is constructed using the scaling factor s, and the rotation matrix $\mathbf{R}$, which is obtained from the corresponding rotation quaternion q:

$$\Sigma = \mathbf{R}\mathbf{S}\mathbf{S}^\top \mathbf{R}^\top. \tag{2}$$

Scene Decomposition. To enable high-quality rendering of dynamic scenes, we separate the background, objects, and sky, each adopting a distinct rendering strategy. For background Gaussians G_i^b, geometry is parameterized by $\Sigma_i^b = \mathbf{R}_i^b \mathbf{S}_i^b (\mathbf{S}_i^b)^\top (\mathbf{R}_i^b)^\top$ with scaling $\mathbf{S}_i^b$ and rotation $\mathbf{R}_i^b$ [20]. Object Gaussians use an SH-based appearance head with opacity to model view-dependent color, whereas the sky is represented by a high-resolution cubemap and blended with the Gaussian composite (no per-primitive SH for sky).

3.2 Rendering

Object Rendering. As shown in Fig. 2, the SeeDepthGaussian framework produces the final image by combining the outputs of each model and projecting them into 2D image space. Specifically, the spherical harmonic coefficients are first calculated, and the object point clouds are transformed into the world coordinate system [20]. These transformed object point clouds are then concatenated with the background and sky point clouds, forming a new, complete point cloud. Since the ultimate goal is 2D rendering, the camera extrinsic matrix $\mathbf{W}$ and the intrinsic matrix $\mathbf{K}$ are employed to project this unified 3D point cloud into 2D image space. For each point, the 2D Gaussian is computed as:

$$\mu' = \mathbf{K}\mathbf{W}\mu, \Sigma' = \mathbf{J}\mathbf{W}\Sigma\mathbf{W}^\top\mathbf{J}^\top, \tag{3}$$

where $\mathbf{K}$ denotes the camera intrinsics and $\mathbf{J}$ is the local Jacobian of the perspective projection (evaluated at μ); μ' and Σ' are the image-plane mean and covariance, respectively. The color $\mathbf{C}$ for each pixel is then calculated by:

$$\mathbf{C} = \sum_{i \in N} \mathbf{c}_i \tilde{\alpha}_i \prod_{j=1}^{i-1} (1 - \tilde{\alpha}_j), \tilde{\alpha}_i = \alpha_i G_i^{proj}(x_p). \tag{4}$$

Here, color c_i is derived from the spherical harmonic coefficients z and the view direction p, while $\tilde{\alpha}_i = \alpha_i G_i^{\text{proj}}(x_p)$, where $G_i^{\text{proj}}(x_p)$ denotes the projected 2D Gaussian density of the i-th 3D Gaussian parameterized by (μ_i', Σ_i') in Eq. 3.

Sky Cubemap Rendering. In open urban scenes, because 3D Gaussians are defined in Euclidean space, they are not well-suited for representing the distant sky. Consequently, employing a high-resolution cube map to map the view direction to the sky color is a more appropriate solution. The overall color of the scene is composed by blending $\mathbf{C}$ from Eq. 4, $\mathbf{C}_{neural}$, and the sky color $\mathbf{C}_{sky}$.

Background Neural Rendering. During training, spherical harmonics have relatively few parameters, which prevents overfitting in static scenes with smaller datasets. However, in urban scenes with larger and more complex datasets, the increased input complexity may lead to overfitting. This overfitting can degrade generalization ability, indirectly affecting rendering quality. Our model is designed for open scenes, particularly for backgrounds, which often feature complex textures and lighting conditions. Therefore, we believe that introducing a multi-layer perceptron (MLP) with a larger number of parameters is a better choice. Specifically, this neural renderer comprises an MLP and a grid encoder. The MLP consists of 5 layers, each with a hidden dimension of 64, while the hash encoder in the neural renderer includes 16 resolution levels, ranging from 16 to 512 [11].

3.3 Depth Module

In the context of autonomous driving, it is crucial to precisely capture and represent the geometric structure of a scene. By incorporating depth information, the relative positional relationships between different targets become clearer. Global-Local Depth Normalization and Hard-and-Soft Depth Regularization are designed as part of the loss function. During training, they help to adjust the spatial distribution of 3D Gaussians, optimizing rendering quality and alleviating issues such as background blurring and color noise.

Hard Depth Regularization. Hard Depth Regularization enhances the quality of scene reconstruction by enforcing strict constraints on the nearest Gaussian points, ensuring precise geometric distribution and surface completeness. The Hard Depth is calculated as follows:

$$D_{\text{hard}}(x_p) = \sum_{i \in N} \tau \left(1 - \tau\right)^{i-1} G_i^{\text{proj}}(x_p) \left\|\mu_i - o\right\|_2, \tag{5}$$

where the geometric weights are $\tau(1-\tau)^{i-1}$ that concentrate mass on the first few splats along the ray ($\tau = 0.95$ in all experiments, approximating a hard nearest-surface selection while remaining differentiable), o denotes the camera center, and x_p the pixel location. With $\tau \in (0,1)$, the sequence $\tau(1-\tau)^{i-1}$ is monotonically decreasing, acting as a near-biased prior over the depth ordering. This is consistent with transmittance decay in volumetric rendering, ensuring that closer Gaussians dominate D_{hard}.

To correct Gaussians that are erroneously positioned in this process, it is necessary to impose regularization via a L2 similarity loss over the target image region P, encouraging the hard depth D_{hard} to approximate the monocular depth $\tilde{\mathcal{D}}$ which is estimated from our depth estimator with DepthAnything V2 kernel [22]:

$$\mathcal{R}_{\text{hard}}(P) = \mathcal{L}_{\text{similar}}\left(\mathcal{D}_{\text{hard}}(P), \tilde{\mathcal{D}}(P)\right). \tag{6}$$

Soft Depth Regularization. To avoid rendering errors caused by semi-transparent surfaces, a soft depth regularization is introduced (intermittently enabled during training). This regularization complements the hard depth regularization, refining opacity. The specific formulation is as follows:

$$\begin{aligned} \mathcal{D}_{\text{soft}}(x_p) &= \sum_{i \in N} \|\tilde{\mu}_i - o\|_2 \times \tilde{\alpha}_i \prod_{j=1}^{i-1} (1 - \tilde{\alpha}_j), \\ \mathcal{R}_{\text{soft}}(P) &= \mathcal{L}_{\text{similar}}(\mathcal{D}_{\text{soft}}(P), \tilde{\mathcal{D}}(P)). \end{aligned} \tag{7}$$

The soft regularization is intermittently enabled during training to stabilize opacity in semi-transparent or sparse regions, acting as a complement to the hard term. By applying both Hard and Soft Depth Regularization, it becomes possible to constrain the nearest Gaussians so that they maintain sufficiently high opacity values while remaining at the correct positions. Consequently, a more complete and detailed surface representation can be formed.

Global-Local Depth Normalization. To mitigate noise in complex-textured regions when correcting minor depth errors, the loss function must prioritize smaller discrepancies. Accordingly, we segment the depth image into small patches and normalize both the predicted and monocular depth values within each patch:

$$\mathcal{D}^{\mathrm{LN}}(x) = \frac{\mathcal{D}(x) - \mathrm{mean}\big(\mathcal{D}(P)\big)}{\mathrm{std}\big(\mathcal{D}(P)\big) + \epsilon}, \quad \text{s.t. } x \in P, \tag{8}$$

where ϵ is a small value used for numerical stability. To capture global shape consistency, a global depth learning component is introduced alongside this local emphasis, where local normalization emphasizes fine-detail alignment to suppress texture noise, and global normalization uses an image-level standard deviation to maintain shape consistency. While it also employs a block-based normalization strategy akin to Local Depth Normalization, it differs by using the global standard deviation of the entire image depth $\mathcal{D}_I$, rather than the patch-level standard deviation:

$$\mathcal{D}^{\mathrm{GN}}(x) = \frac{\mathcal{D}(x) - \mathrm{mean}\big(\mathcal{D}(P)\big)}{\mathrm{std}\big(\mathcal{D}_I\big)}, \quad \text{s.t. } x \in P,\ P \subseteq I. \tag{9}$$

3.4 Learning Objectives

Depth Regularization, including R_{hard} and R_{soft}, incorporates both local and global normalization terms, which are derived from the aforementioned normalization computations. Let T represent either the hard or soft variant. Using L_2 loss to measure similarity, the regularization term is given by:

$$\mathcal{R}_T = \mathcal{L}_2\big(\mathcal{D}_T^{\mathrm{GN}}, \tilde{\mathcal{D}}^{\mathrm{GN}}\big) + \gamma\, \mathcal{L}_2\big(\mathcal{D}_T^{\mathrm{LN}}, \tilde{\mathcal{D}}^{\mathrm{LN}}\big). \tag{10}$$

It is then integrated into the following loss function:

$$\mathcal{L} = \mathcal{L}_{\mathrm{color}} + \lambda_1 \mathcal{L}_{\mathrm{lidar}} + \lambda_2 \mathcal{L}_{\mathrm{sky}} + \lambda_3 \mathcal{L}_{\mathrm{reg}} + \lambda_4 \mathcal{R}_{\mathrm{hard}} + \lambda_5 \mathcal{R}_{\mathrm{soft}}. \tag{11}$$

The depth-regularization terms ($\mathcal{R}_{\mathrm{hard}}$ and $\mathcal{R}_{\mathrm{soft}}$) are computed only during training, introducing zero inference overhead at test time.

Among these terms, $\mathcal{L}_{\mathrm{color}}$ is a reconstruction loss measuring the discrepancy between the rendered image and the observed image, which follows the original 3DGS [5]. $\mathcal{L}_{\mathrm{lidar}}$ is an L_1 loss between the rendered depth and the depth obtained by projecting sparse LiDAR points onto the camera plane. $\mathcal{L}_{\mathrm{sky}}$ is derived from a binary cross-entropy loss. $\mathcal{L}_{\mathrm{reg}}$ is an additional regularization term used to eliminate floaters caused by the object model and to reinforce decomposition effects.

4 Experiments and Results

4.1 Experimental Setup

Implementation Details. We adopt the Adam optimizer to train SeeDepth-Gaussian, setting the total number of iterations to 30,000 in accordance with

Fig. 3. Qualitative rendered image comparison: SeeDepthGaussian vs. StreetGaussian.

the 3DGS configuration [5]. The initial learning rates for the translation transform ΔT_t and the rotation transform ΔR_t are set to 5×10^{-3} and 1×10^{-3}, respectively, and they are exponentially decayed to 5×10^{-5} and 1×10^{-4}. The resolution of the Sky Cubemap is set to 1024, with its learning rate decaying from 1×10^{-2} to 1×10^{-4}. The predicted depth maps by the depth estimator [22] are normalized within the range of 0 to 65535. Loss function weights are adaptively configured based on specific scenes. The VIT-L [2] model is pre-loaded as the encoder for the depth estimator.

During the experiments, we conducted all training on a single RTX 4090 GPU. A typical full-scale monocular scene experiment required approximately 45 min and at most 12GB of GPU memory, depending on the specific scene and parameter settings.

Dataset. Experiments are conducted on the Waymo Open Dataset [16]. Waymo delivers higher-quality data, broader coverage, and most importantly, a richer variety of complex scenarios compared to other datasets. We selected multiple complex scenarios for experimental comparison, including elements such as dynamically moving vehicles, complicated road conditions, and extreme weather, in order to enhance the generality of the experimental results.

Following StreetGaussian [20], the same eight scene sequences are chosen, each containing a large number of moving objects, significant ego-vehicle motion, and complex lighting conditions. Approximately 100 consecutive frames from each scene are selected as the actual data; every fourth frame is used as the test set, while the remaining frames serve as the training set.

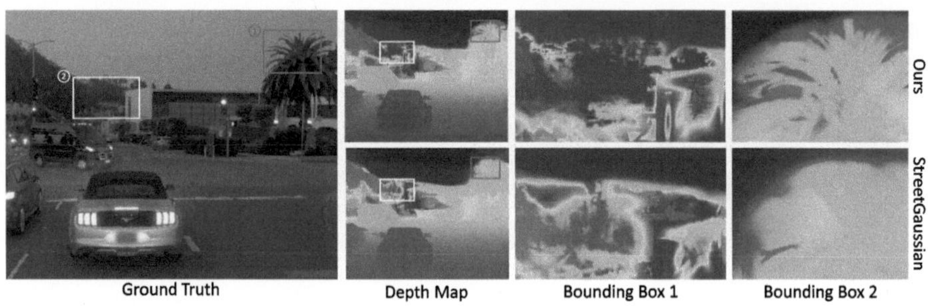

Fig. 4. Qualitative depth map comparison: SeeDepthGaussian vs. StreetGaussian.

Table 1. Quantitative results on the Waymo dataset. The rendering image resolution is 1066 × 1600.

	3DGS [5]	NSG [12]	MARS [17]	EmerNeRF [21]	StreetGaussian [20]	Ours
PSNR↑	29.64	28.31	29.75	30.87	34.91	**35.12**
SSIM↑	0.918	0.862	0.886	0.905	0.931	**0.936**
LPIPS↓	0.117	0.346	0.264	0.133	0.075	**0.074**
FPS↑	**205**	0.47	0.68	0.21	135	57

4.2 Comparisons with the State-of-the-Art Methods

In tests on eight Waymo dataset scenes, we keep the main parameters consistent with StreetGaussian [20], enabling direct comparisons to numerous prior works. We directly adopted certain experimental data of previously published models to save time and computational resources, because the metrics derived from the StreetGaussian dataset are fully reproducible. As shown in Table 1, our approach achieves the best overall performance, offering slight but consistent improvements over the state-of-the-art StreetGaussian across all metrics. Improvements concentrate on distant textures and occlusion boundaries (e.g., foliage/skyline interfaces), reducing color bleeding and blur. Because the background is modeled with a *neural renderer*, the rendering FPS is lower than StreetGaussian but remains substantially higher than NeRF-based baselines. Figure 3 presents selected visual comparisons. Figure 4 shows depth map rendering in complex urban scenes, where SeeDepthGaussian better reconstructs background structures (e.g., foliage and buildings in Bounding Boxes 1 & 2). The improved depth accuracy enables more precise Gaussian placement, enhancing occlusion handling and reducing color bleeding, thereby improving RGB rendering quality.

Additionally, we conducted experiments on multi-view scene reconstruction within the same environment. For each scene, data from three viewpoints were supplied as input, yielding the results reported in Table 2. This experiment involves three times the data volume of the standard setting and introduces more complex multi-view information, thereby increasing reconstruction complexity; the corresponding computational resource consumption is detailed in Table 3. These results further validate the feasibility and robustness of our design philosophy.

Table 2. Multi-view experiments on the Waymo dataset.

Configuration	PSNR↑	SSIM↑	LPIPS↓
StreetGaussian	31.06	0.901	0.182
Ours	**31.24**	**0.904**	**0.181**

4.3 Ablations and Analysis

To rigorously validate our approach and meticulously quantify the individual contributions of our proposed components, we conducted a comprehensive suite of ablation studies on the challenging Waymo dataset. As detailed in Table 3, the results unequivocally demonstrate that the introduction of our Depth Module yields the most substantial performance gain, underscoring its critical role in enhancing geometric accuracy. When integrated independently with a standard Spherical Harmonics (SH) background for rendering (i.e., without the Neural Renderer), the Depth Module achieves a frame rate comparable to the StreetGaussian baseline while delivering marked improvements in PSNR and SSIM metrics. Our full model, which combines all components, elevates the reconstruction quality even further, albeit at a predictable trade-off in rendering speed (e.g., a decrease from 134 to 57 FPS).

Crucially, because the Depth Module is utilized almost exclusively during the training phase to guide the optimization of Gaussian primitives, it imposes no additional computational overhead at inference time. The Depth Module introduces no computational overhead at inference time. This zero-overhead property works in tandem with the system's modular architecture, enabling the module to serve as a self-contained, "plug-and-play" component. Consequently, it can be seamlessly transplanted into other 3DGS-based pipelines, offering a straightforward path to consistently improve reconstruction fidelity across a diverse spectrum of related methods.

Further refining the output, the incorporation of the Neural Renderer also enhances the final results, though its impact is more modest compared to the foundational improvements from the Depth Module. Nevertheless, the evidence collectively suggests that, once fully matured, our Neural Renderer holds the potential to supplant the traditional spherical harmonics representation for color rendering, promising to capture more complex, view-dependent effects while preserving the high efficiency essential for real-time applications.

Table 3. Ablation studies on the Waymo dataset.

Metric	Neural Renderer only	Depth Module only	Full Model
PSNR↑	34.94	35.09	**35.12**
SSIM↑	0.932	0.935	**0.936**
LPIPS↓	0.076	**0.074**	**0.074**
FPS	57	**134**	57
Training Time (Mono)	43 min	30 min	45 min
Training Time (Multi)	129 min	88 min	132 min

5 Conclusion

This paper introduces SeeDepthGaussian, a novel framework for modeling dynamic urban street scenes. It features tailored rendering strategies specifically optimized for Gaussian models to enhance the final reconstruction. Extensive experiments across various scenarios validate the effectiveness of our Hard-and-Soft Depth Regularization, Global-Local Depth Normalization, and a neural renderer, which collectively improve image quality in these dynamic settings.

Nevertheless, we identify clear avenues for improvement. Empirically, while gains are quite apparent in the rendered depth maps, the final quality remains sensitive to the accuracy of the monocular depth estimator and the texture density in distant regions. Valuable directions for future work include incorporating sparse yet accurate LiDAR depth to calibrate this monocular prior, developing uncertainty modeling, and establishing a more principled data-preprocessing and depth-regularization pipeline.

References

1. Chung, J., Oh, J., Lee, K.M.: Depth-regularized optimization for 3D gaussian splatting in few-shot images. arXiv preprint: arXiv:2311.13398 (2023)
2. Dosovitskiy, A.: An image is worth 16x16 words: transformers for image recognition at scale. arXiv preprint: arXiv:2010.11929 (2020)
3. He, L., et al.: Neural radiance field in autonomous driving: a survey. arXiv preprint: arXiv:2404.13816 (2024)
4. Huang, N., et al.: S3Gaussian: self-supervised street gaussians for autonomous driving. arXiv preprint: arXiv:2405.20323 (2024)
5. Kerbl, B., Kopanas, G., Leimkühler, T., Drettakis, G.: 3D gaussian splatting for real-time radiance field rendering. ACM Trans. Graph. **42**(4), 1–14 (2023). https://doi.org/10.1145/3592433
6. Lee, J.C., Rho, D., Sun, X., Ko, J.H., Park, E.: Compact 3D Gaussian representation for radiance field. In: CVPR, pp. 21719–21728 (2024)
7. Li, J., et al.: DNGaussian: optimizing sparse-view 3D gaussian radiance fields with global-local depth normalization. arXiv preprint: arXiv:2403.06912 (2024)
8. Liu, Y., Fan, K., Yu, W., Li, C., Lu, H., Yuan, Y.: MonoSplat: generalizable 3D gaussian splatting from monocular depth foundation models. arXiv preprint: arXiv:2505.15185 (2025)

9. Lu, F., Xu, Y., Chen, G., Li, H., Lin, K.Y., Jiang, C.: Urban radiance field representation with deformable neural mesh primitives. In: ICCV, pp. 465–476 (2023)
10. Mildenhall, B., Srinivasan, P.P., Tancik, M., Barron, J.T., Ramamoorthi, R., Ng, R.: NeRF: representing scenes as neural radiance fields for view synthesis. Commun. ACM **65**(1), 99–106 (2022). https://doi.org/10.1145/3503250
11. Müller, T., Evans, A., Schied, C., Keller, A.: Instant neural graphics primitives with a multiresolution hash encoding. ACM Trans. Graph. **41**(4), 1–15 (2022). https://doi.org/10.1145/3528223.3530127
12. Ost, J., Mannan, F., Thuerey, N., Knodt, J., Heide, F.: Neural scene graphs for dynamic scenes. In: CVPR, pp. 2856–2865 (2021)
13. Qin, M., Li, W., Zhou, J., Wang, H., Pfister, H.: LangSplat: 3D language gaussian splatting. In: CVPR, pp. 20051–20060 (2024)
14. Rumelhart, D.E., Hinton, G.E., Williams, R.J.: Learning representations by back-propagating errors. Nature **323**(6088), 533–536 (1986)
15. Schönberger, J.L., Frahm, J.M.: Structure-from-motion revisited. In: CVPR, pp. 4104–4113 (2016). https://doi.org/10.1109/CVPR.2016.445
16. Sun, P., et al.: Scalability in perception for autonomous driving: Waymo open dataset. In: CVPR, pp. 2446–2454 (2020)
17. Wu, Z., et al.: Mars: an instance-aware, modular and realistic simulator for autonomous driving. CICAI (2023)
18. Xie, W., et al.: SurgicalGaussian: deformable 3D gaussians for high-fidelity surgical scene reconstruction. arXiv preprint: arXiv:2407.05023 (2024)
19. Xu, H., et al.: DepthSplat: connecting gaussian splatting and depth. In: CVPR (2025)
20. Yan, Y., et al.: Street Gaussians: modeling dynamic urban scenes with Gaussian splatting. In: ECCV (2024)
21. Yang, J., et al.: EmerNeRF: emergent spatial-temporal scene decomposition via self-supervision. arXiv preprint: arXiv:2311.02077 (2023)
22. Yang, L., et al.: Depth anything V2. arXiv preprint: arXiv:2406.09414 (2024)
23. Zhang, Q., Wysocki, O., Urban, S., Jutzi, B.: CDGS: confidence-aware depth regularization for 3D gaussian splatting. arXiv preprint: arXiv:2502.14684 (2025)
24. Zhou, X., Lin, Z., Shan, X., Wang, Y., Sun, D., Yang, M.H.: DrivingGaussian: composite gaussian splatting for surrounding dynamic autonomous driving scenes. In: CVPR, pp. 21634–21643 (2024)

Paired Hierarchical VAEs for Image-to-Image Translation and Cross Reconstruction

Daniel Hess[1(✉)], Tom Horrocks[1], Daniel Wedge[1], Eun-Jung Holden[2], and Ross P. Williams[3]

[1] The University of Western Australia, Crawley, WA 6009, Australia
daniel.hess@research.uwa.edu.au
[2] The University of Melbourne, Parkville, VIC 3010, Australia
[3] Rio Tinto, Perth, WA 6000, Australia

Abstract. This paper presents the use of Paired Hierarchical Variational Autoencoders (PHVAEs) for image-to-image translation and cross reconstruction. We explore a novel method for constructing and applying HVAEs for these types of tasks, and compare this model against existing GAN, transformer, and diffusion based methods, including pix2pix, StegoGAN, ResViT, and BBDM. The method utilizes deep, hierarchical VAEs with paired optimization to take advantage of the bidirectional latent space. The proposed model outperforms all alternative models in PSNR on all tested tasks, and attains consistently competitive SSIM, demonstrating its effectiveness and benefit for prediction-focused translation tasks such as cross reconstruction, while remaining parameter efficient.

Keywords: image-to-image translation · hierarchical variational autoencoders · cross-modality prediction · neural networks · convolutional neural networks

1 Introduction

Image-to-image translation represents a broad area within computer vision, encompassing tasks like image-infilling [5,19], colorization [8,19], style-transfer [8,11,31], and cross reconstruction [4,12,29]. These tasks aim to generate images in one domain based on conditions from another image from a different domain. Each task has distinct objectives that influence outcomes; for instance, a colorization model may prioritize color accuracy and structure over output diversity, unlike a style-transfer model.

The complexity of these tasks stems from the potential for various valid outputs for any given problem. This necessitates careful consideration of which features should be preserved from the input domain, and what can be generated depending on the task. Thus, these tasks range from generating images completely independent from the source domain at one extreme, to making completely deterministic predictions based on the input image at the other.

M. Liu et al. (Eds.): AI 2025, LNAI 16371, pp. 146–160, 2026.
https://doi.org/10.1007/978-981-95-4972-6_12

Evaluating such models is challenging: per-pixel error metrics like MSE or PSNR overlook structural nuances, while visual quality metrics like FID [7] lack information on "correctness". While diversity is important for a generative model, it also introduces uncertainty in the model's output. Consequently, evaluation criteria must align with specific task requirements.

Current methods for image-to-image translation include conditional GANs (cGANs) like pix2pix [8], which leverage random noise vectors and adversarial loss to enhance diversity and coherence. ResViT [4] adapts Vision Transformers for predicting medical images, while diffusion-based models such as Palette [19] and BBDM [11] utilize iterative denoising processes conditioned on inputs, significantly improving visual quality. Other methods based on Variational Autoencoders [9] are also used, however, often fall short in visual performance [18]. While variations such as Conditional VAEs [22] and Paired/Joint VAEs [25] extend the model's capabilities to perform image-to-image translation, they do not address issues of visual quality comprehensively. However, variants of Hierarchical VAEs [17] (HVAEs) have demonstrated notable improvements in output visuals in unimodal image generation [2,23,26], by increasing the stochastic depth and potential image complexity.

When compared to adversarial methods, the probabilistic nature of VAEs makes handling of uncertainty and variability more controllable. And while diffusion models excel at capturing fine detail and improving diversity, they are particularly prone to hallucinations [1]. In a prediction focussed setting, VAE-based methods may prove to be advantageous.

In this study, we introduce Paired Hierarchical Variational Autoencoders (PHVAEs) for image-to-image translation, and evaluate how effectively these models can perform cross reconstruction. For the cross reconstruction case, we assume that each image domain represents a different image *modality*, where our goal is to generate one modality in its entirety, from the other. To our knowledge, no prior study has explored the potential of PHVAEs for image-to-image translation using joint/paired optimization techniques. In this paper we discuss the architecture and training methodologies of PHVAEs and assess their performance on prediction-focused datasets, including multi-modal satellite imagery and paired medical CT-scans and MRIs [13].

2 Background

2.1 Variational Autoencoders

VAEs were first introduced as a general way to learn a generative probabilistic model [9]. The underlying assumption of VAEs is that for every sample of a dataset y, that sample is generated by some random process from some underlying random variable z. By sampling over $p_\theta(z)$ (i.e., some distribution parameterized by θ), and subsequently $p_\theta(y|z)$, we can approximate sampling directly across $p_\theta(y)$. The variational lower bound on the marginal log-likelihood $(\log p_\theta(y))$ is:

$$\mathcal{L}(\theta, \phi, y) = \mathbb{E}_{q_\phi(z|y)} [\log p_\theta(y|z)] - D_{KL} (q_\phi(z|y) || p_\theta(z)),$$

with D_{KL} representing the KL-divergence. The parameters of these distriibutions (i.e., θ and ϕ) are embedded in neural networks and optimized over the variational lower bound, effectively learning both $q_\phi(z|y)$ (i.e., the *approximation* of the posterior, $p_\theta(z|y)$), and $p_\theta(y|z)$ for a given prior $p_\theta(z)$.

There are a few variations which extend the VAE design for image-to-image translation, for tasks such as sketch-to-photo [31], style-transfer [25,31], and cross-reconstruction [12,29]. Conditional VAEs (CVAEs) [22] allow for an additional input variable x, where the prior $p_\theta(z|x)$, and posterior $p_\theta(y|x, z)$, are both conditional on the input variable. CVAEs are optimized over the conditional log-likelihood ($\log p_\theta(y|x)$).

Joint VAEs (JVAEs) [25] instead consider that the latent variables z generate two (or more) variables. JVAEs optimize over the joint log-likelihood (i.e., $\log p_\theta(x, y)$). The benefit of JVAEs over CVAEs is that they can be used as bidirectional translation models.

More generally, Paired VAEs (PVAEs) [12,29] model two distinct VAEs (one for each domain), and assume a shared latent space between paired samples of each domain. While JVAEs can be constructed in this way, we distinguish PVAEs in that they are not required to be optimized over the joint distribution [25].

2.2 Hierarchical VAEs

The issues regarding visual-quality in image-processing tasks in VAEs can partly be attributed to the limited complexity of the latent distribution [17,23], which does not affect other types of models to the same degree, such as autoregressive models [20] or image transformers [14]. This issue is not limited to the image-to-image translation case, but also apparent in unconditional image generation [18].

One method of increasing the visual expressivity is by introducing a top-down hierarchy of stochastic variable dependence [17]. Layers of distributions can become conditionally dependent on each other in more complex ways than permitted by the standard VAE. In the context of images, features can be learned which represent different resolutions, where the distribution of lower resolution variables (or *global* variables) strongly influences higher resolution variables (or *local* variables).

Ladder VAEs [23] are an early implementation of HVAEs which show improved performance in unconditional generation tasks over traditional VAEs. Two notable examples of HVAEs for unconditional generation in images are Nouveau VAEs [26], which model a residual parameterisation of the Normal distribution—and Very Deep VAEs [2], which explore significantly more hierarchical layers. These modifications have been shown to not only improve the visual quality of images, but also improve negative log-likelihood generative performance [2,23,26].

While hierarchical CVAEs are used in some contexts [15,24], to the best of our knowledge, hierarchical variants of VAEs have not been adapted for image-to-image translation. Additionally, many variants which are used for image-to-image translation rely on additional GAN components to aid the translation and alleviate the issues with image quality in VAEs [10,30,31]. However, translation is not strictly a generative task, but can also incorporate some degree of prediction. In certain contexts which rely on the predictive quality, the issues of mode collapse in GAN-based methods [21], or hallucination in diffusion-based methods [1], may be barriers towards adoption. As such, improvements in likelihood-based methods (such as VAEs), may be particularly useful.

3 Architecture

The PVAE method for generating image-to-image translation models constructs a distinct VAE for each image domain x and y. The neural networks in this case then learn two approximate posteriors ($q_\phi(\boldsymbol{z}_x|x)$ and $q_\phi(\boldsymbol{z}_y|y)$) and two generators ($p_\theta(x|\boldsymbol{z}_x)$ and $p_\theta(y|\boldsymbol{z}_y)$). Therefore, if there is some way of reconciling the latent spaces $\boldsymbol{z}_x$ and $\boldsymbol{z}_y$, we can use these components to perform translation. This is generally achieved by assuming a shared latent space (either by enforcing similarity in paired samples, or through joint optimization such as with JVAEs) which allows the model to assume $\boldsymbol{z}_x = \boldsymbol{z}_y$, and to perform translation through sampling the latent space from the source domain's posterior, and finally sampling from the target domain through the generator.

3.1 Paired Hierarchical VAEs

The hierarchical model is based primarily on the implementation of VDVAE [2]. For the domain x, we model the prior and approximate posterior with N hierarchical latent variables as:

$$p_\theta(\boldsymbol{z}_x) = p_\theta(z_{x,0})p_\theta(z_{x,1}|z_{x,0}) \dots p_\theta(z_{x,N}|z_{x,<N})$$

$$q_\phi(\boldsymbol{z}_x|x) = q_\phi(z_{x,0}|x)q_\phi(z_{x,1}|x, z_{x,0}) \dots q_\phi(z_{x,N}|x, z_{x,<N}),$$

and similarly for the domain y.

The neural network consists of two primary components for each modality: an encoder (E_x and E_y) and decoder (D_x and D_y). Each encoder-decoder pair is split into N layers corresponding to different resolution scales. The encoder path consists of a series of encoder-layers separated by pooling layers. Similarly the decoder path consists of a series of decoder-layers separated by unpooling layers.

3.2 Implementation Details

As VDVAE [2] is only designed for single-modality image generation, our contribution primarily focuses on the methodology which extends this architecture for image-to-image translation. This is accomplished architecturally by utilizing a VDVAE model for each modality with modifications that improve convergence and stability on models with small batch-sizes [6]. The convolution-layers are Swish-activated [16]. Unpooling layers are a convolution layer followed by a nearest-neighbour interpolation. Pooling layers are simply strided convolutions.

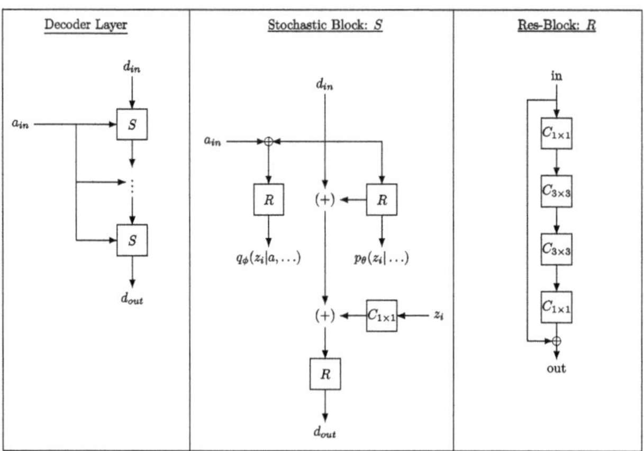

Fig. 1. Diagram of the HVAE structure in the decoder. A single layer in the decoder takes two inputs, one from the encoder (a_{in}), and one from the previous decoder-layer (d_{in}). Each decoder layer consists of a number of stochastic-blocks (labelled S). Convolution layers with kernel-size k are labelled C_k, and *Res-blocks* are labelled R.

The decoder-layers are constructed from of a number of stochastic-blocks as shown in Fig. 1, which allows the stochastic/hierarchical depth to be larger than N [2]. Additionally, each encoder-layers consists of a sequence of *res-blocks* (see Fig. 1). Finally, we model both prior and posteriors for each latent variable as Gaussian distributions.

3.3 Training Method

Although we are modelling $\boldsymbol{z}_x$ and $\boldsymbol{z}_y$ as being distinct latent spaces, we can infer a shared latent space as $\boldsymbol{z} = (\boldsymbol{z}_x, \boldsymbol{z}_y)$. This is permitted because the input to the decoder-layers from the encoder-layers is domain agnostic. This allows the model to be optimized over the log-likelihood $\log p_\theta(x, y)$. This involves estimating $p_\theta(\boldsymbol{z}|x, y)$, and so creates an issue during inference when the target domain is missing. In models such as JMVAE [25], the joint encoder $q_\phi(\boldsymbol{z}|x, y)$ is used, however, additional constraints are placed on the model to effectively learn both $q(\boldsymbol{z}|x)$ and $q(\boldsymbol{z}|y)$. The variational lower bound for joint optimization ($\mathcal{L}_J$) is:

$$\mathcal{L}_J = \mathbb{E}_{q_\phi(\boldsymbol{z}|x,y)}(\log p_\theta(x, y|\boldsymbol{z})) - D_{KL}(q_\phi(\boldsymbol{z}|x, y)||p_\theta(\boldsymbol{z}))$$

In our case, we use both $q_\phi(\boldsymbol{z}|x)$, and $q_\phi(\boldsymbol{z}|y)$ as estimates for $p_\theta(\boldsymbol{z}|x, y)$. In turn, this gives us two solutions to the variational lower bound:

$$\mathcal{L}_x = \mathbb{E}_{q_\phi(\boldsymbol{z}|x)}(\log p_\theta(x|\boldsymbol{z})) + \mathbb{E}_{q_\phi(\boldsymbol{z}|x)}(\log p_\theta(y|\boldsymbol{z})) - D_{KL}(q_\phi(\boldsymbol{z}|x)||p_\theta(\boldsymbol{z})),$$

$$\mathcal{L}_y = \mathbb{E}_{q_\phi(\boldsymbol{z}|y)}(\log p_\theta(x|\boldsymbol{z})) + \mathbb{E}_{q_\phi(\boldsymbol{z}|y)}(\log p_\theta(y|\boldsymbol{z})) - D_{KL}(q_\phi(\boldsymbol{z}|y)||p_\theta(\boldsymbol{z})).$$

As neither solution incorporates all of the network, we define the lower bound for paired, joint optimization ($\mathcal{L}_{P,J}$) with:

$$\mathcal{L}_{P,J} = \frac{\mathcal{L}_x + \mathcal{L}_y}{2}.$$

We attribute the discrepencies in these losses to the paired-VAE design of the model.

One-Way Vs Two-Way Translation. Part of the benefit of modelling the network as two VAEs is the bidirectional translation. However, when only one translation direction is required, only half the network needs to be optimized. Under the assumption that only translation from x to y is required, we can simplify the variational lower bound for paired, conditional optimization ($\mathcal{L}_{P,C}$) to:

$$\mathcal{L}_{P,C} = \mathbb{E}_{q_\phi(\boldsymbol{z}|x)}(\log\ p_\theta(y|\boldsymbol{z})) - D_{KL}(q_\phi(\boldsymbol{z}|x)||p_\theta(\boldsymbol{z})).$$

It is possible that optimization over $\mathcal{L}_{P,C}$ is more permissive to the latent space than $\mathcal{L}_{P,J}$ and could result in more effective parameter use. This, however, comes at the cost of bidirectional translation, and does not strictly learn a shared latent space. We distinguish these optimization methods by:

1. **One-way translation:** optimization over $\mathcal{L}_{P,C}$
2. **Two-way translation:** optimization over $\mathcal{L}_{P,J}$

One-way translation only optimizes one encoder, and one decoder of the network. Thus, training the model this way is effectively equivalent to conditional optimization of the original VDVAE model, however, we explore significantly smaller models in terms of stochastic depth than Child [2] due to hardware constraints.

Paired Latent Consistency Loss. As part of the paired-VAE design, we use both $q_\phi(\boldsymbol{z}|x)$, and $q_\phi(\boldsymbol{z}|y)$ as estimates for $p_\theta(\boldsymbol{z}|x,y)$. While ideal, it is not enforced that $q_\phi(\boldsymbol{z}|x) = q_\phi(\boldsymbol{z}|y)$ for a given pair x, y. Therefore, in the two-way translation case we can optionally include a term to encourage this relationship: $D_{KL}(q_\phi(\boldsymbol{z}|x)||q_\phi(\boldsymbol{z}|y)) + D_{KL}(q_\phi(\boldsymbol{z}|y)||q_\phi(\boldsymbol{z}|x))$.

4 Experiments

The types of evaluation for image-to-image translation tasks are broad, however, we evaluate in a prediction focussed context. Cross reconstruction can be considered a translation task between modalities of images. In this context, we use the term modality to refer to distinct domains of images where the context, value-types, or acquisition details differ. While we still expect a single input to have multiple valid outputs, preserving both structure and content of the target output is crucial. This makes cross reconstruction tasks well suited for this context.

4.1 Data Set 1: RGB/NDVI—Sentinel 2 Satellite Imagery

We have sourced multimodal satellite data, which contains paired true-colour/RGB images, and NDVI images, for various regions around Australia. The data set contains modified Copernicus Sentinel data (accessed 2023) processed by Sentinel Hub. NDVI (or Normalised Difference Vegetation Index) is an index used to assess the presence or abundance of vegetation in a region; and is useful for agriculture, tracking deforestation and many other land-use tasks. NDVI is calculated from a red spectral component (R), and an near-infrared component (NIR):

$$\mathrm{NDVI} = \frac{\mathrm{NIR} - \mathrm{R}}{\mathrm{NIR} + \mathrm{R}}.$$

The data set in total contains 10,740 pairs of images at a resolution of 256×256.

4.2 Data Set 2: CT/MRI—Male Pelvic Scans

CT scans and MRIs in health data acquire similar cross-sectional images of the human body, however, differ in terms of granularity and contrast to hard and soft tissues. Paired CT scans and MRIs have been made available as part of the Gold Atlas project [13], and contain registered and aligned pairs of each modality taken from the pelvic region of a number of different men. MRIs generally offer higher granularity, however, due to cost and availability, are not always feasible or appropriate for all diagnostic tasks. The data set contains 1,580 pairs of images at a resolution of 256×256.

4.3 Evaluation

As we are primarily interested in the performance with respect to structure and style, we use both PSNR and SSIM. Further, we evaluate the image-quality using FID.

The Structural Similarity Index Measure (SSIM) [27] is a sliding window-based metric to measure structural information via distributions around pixels, as opposed to an image-wide measure such as PSNR which uses mean squared-error. Fréchet Inception Distance (FID) [7] more substantially evaluates the visual similarity between sets of images, and ultimately describes how well output images "look" like they were drawn from the same distribution.

For comparison purposes, we evaluated a range of existing models on the same tasks. The first model, pix2pix [8], is a convolutional model which uses conditional GANs for training. It is a general-purpose model for image-to-image translation and has been evaluated on a number of tasks. While it is often outperformed by other task-specific models, it is frequently used as a comparison point because of its adequate all-round performance.

We model an additional GAN based method, StegoGAN [28] a model which uses steganography to improve non-bijective translations. StegoGAN shows significant improvements to other GAN based methods, especially on assymetric translation tasks.

ResViT [4] is a transformer based autoregressive model which has been evaluated on a few translation tasks, and has previously demonstrated state-of-the-art performance on the MRI/CT data set used in our experements. For MRI→CT translation (as well as other tasks), it was shown to outperform a number of other models in PSNR and SSIM.

Finally, BBDM [11] is a diffusion based method for image-to-image translation. It has been shown to outperform other GAN and diffusion models in terms of visual acuity on a number of tasks.

All models are trained and evaluated on the same machine for 100 epochs.

4.4 Results

Sample outputs from the various tasks are presented in Fig. 2, with each model showing unique characteristics. For the PHVAE samples, while the outputs appear to have greater regularity, this has also resulted in images which look smoother, losing many high frequency features from the target image. In contrast, the BBDM samples are generally structurally the least similar. There also

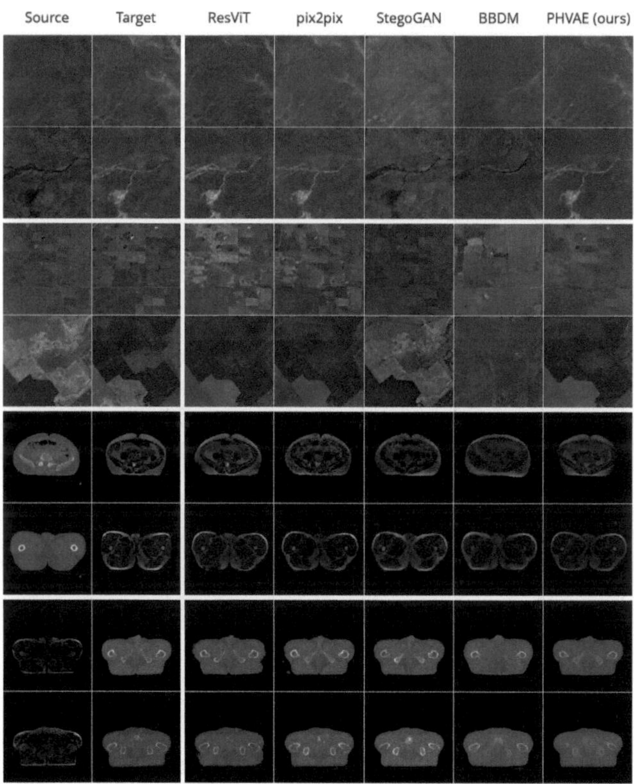

Fig. 2. Output samples. Each tested model generated 2 samples from four translation tasks: RGB→NDVI, NDVI→RGB, CT→MRI, and MRI→CT.

appears to be a significant colour-shift in the StegoGAN models in NDVI→RGB tasks.

As shown in Tables 1 and 2, the PHVAE model achieves the best PSNR across each of the four translation tasks. It additionally achieves the best SSIM on both tasks in data set 2, and remains competitive in data set 1. The parameter usage is also particularly low, meaning that relative to the performance, the model is particularly parameter efficient. However, the FID is the worst out of all the tested models, and explains the visual quality of the model samples. It is clear that the colour-shift in NDVI/RGB samples from StegoGAN model is picked up by both PSNR and SSIM, however, despite these issue, the FID remains especially good.

These results also explain the perceived lower structurally performance of the BBDM samples achieving consistently poor SSIM across the tasks. As diversity and visual quality are generally where diffusion models excel, the poorer PSNR and SSIM from BBDM is perhaps expected. However, even in terms of FID, the model is outperformed by both ResViT and pix2pix. This is perhaps due to the much higher predictive performance of these model, leading to much stronger FIDs as a consequence.

It is clear from these results that there is no single best metric for translation tasks. While PSNR evaluates numeric error, it is particularly biased towards low-frequency features. In contrast, FID much more closely matches human-

Table 1. Test performance for Data Set 1. *Parameters from a single unidirectional model, thus a secondary model is required for the reverse translation.

Model	Parameters	RGB→NDVI			NDVI→RGB		
		PSNR↑	SSIM↑	FID↓	PSNR↑	SSIM↑	FID↓
PHVAE	38.12M	**30.70**	0.79	171.9	**25.93**	0.73	148.9
ResViT	123.45M^*	30.52	**0.89**	**20.0**	24.15	**0.78**	37.6
pix2pix	54.42M^*	30.54	0.88	22.6	24.68	0.76	41.4
BBDM	237.09M^*	21.99	0.46	65.9	18.46	0.45	86.1
StegoGAN	25.71M	16.38	0.20	49.2	16.13	0.26	**36.7**

Table 2. Test performance for Data Set 2. *Parameters from a single unidirectional model, thus a secondary model is required for the reverse translation.

Model	Parameters	CT→MRI			MRI→CT		
		PSNR↑	SSIM↑	FID↓	PSNR↑	SSIM↑	FID↓
PHVAE	38.11M	**24.86**	**0.85**	173.6	**29.23**	**0.93**	122.8
ResViT	123.45M^*	23.70	0.83	**67.3**	28.57	0.92	81.7
pix2pix	54.41M^*	22.97	0.80	98.2	27.86	0.90	89.7
BBDM	237.09M^*	21.78	0.67	128.9	24.81	0.68	120.7
StegoGAN	25.68M	23.33	0.78	81.7	28.00	0.88	**72.8**

perception for evaluating image quality, however, it does not evaluate the 'correctness'. This is apparent in the StegoGAN outputs for NDVI→RGB, which despite achieving the best FID, exhibit a clear colour-shift (Table 3).

Table 3. PHVAE reconstruction vs translation performances. *VDVAE is constructed with an equivalent shape to the corresponding PHVAE. PHVAE reconstruction differs to VDVAE in that the PHVAE model is trained simultaneously with translation.

Data Set	Task	Modalities		Performance		
		Source	Target	PSNR↑	SSIM↑	FID↓
1	Reconstruction	NDVI	NDVI	40.70	0.96	31.9
	*VDVAE	NDVI	NDVI	37.52	0.93	140.0
	Translation	RGB	NDVI	30.70	0.79	171.9
	Reconstruction	RGB	RGB	39.54	0.96	22.5
	*VDVAE	RGB	RGB	37.97	0.92	108.8
	Translation	NDVI	RGB	25.93	0.73	148.9
2	Reconstruction	MRI	MRI	39.85	0.99	46.4
	*VDVAE	MRI	MRI	35.37	0.95	105.0
	Translation	CT	MRI	24.86	0.85	173.6
	Reconstruction	CT	CT	46.64	0.99	27.9
	*VDVAE	CT	CT	42.95	0.99	174.2
	Translation	MRI	CT	29.23	0.93	122.8

While the predictive performanace for PHVAEs is significantly better than the alternatives, the poorer visual performance for VAEs is clearly still present here. However, exploring the performance in the reconstruction cases for the PHVAE (e.g., CT→CT, MRI→MRI, etc.), the model does manage to achieve significantly better FID scores. While this is to be expected, as reconstruction tasks are significantly less complex, it shows that the model design is capable improving in terms of visual quality.

4.5 Ablations

We perform a number of ablations for evaluating this architecture for image translation tasks. In our results, we strictly ablate on the MRI/CT validation dataset.

The first test compares the hierarchical and non-hierarchical based paired VAEs (Table 4) for various latent-variable scales. High resolution features are important for performance in both types of models, with the models using non-hierarchical VAEs showing a general negative trend as the scales decrease. However, with the hierarchical models, replacing higher resolution latent scales with

Table 4. VAE/HVAE ablation. Comparison of paired, non-hierarchical VAEs, with paired hierarchical VAEs (PHVAE). Each latent variable corresponds to a resolution scale (128, 64, 32, and 16).

Model	Parameters	Latent Depth				CT→MRI			MRI→CT		
		128	64	32	16	PSNR↑	SSIM↑	FID↓	PSNR↑	SSIM↑	FID↓
PVAE	5.7 M	1	–	–	–	21.15	0.68	276.7	26.37	0.80	217.4
	4.0 M	–	1	–	–	21.55	0.68	268.6	26.39	0.80	210.3
	4.3 M	–	–	1	–	20.81	0.65	274.6	25.45	0.76	213.5
	4.6 M	–	–	–	1	20.51	0.61	259.2	24.32	0.71	212.7
PHVAE	6.7 M	8	–	–	–	21.77	0.71	293.1	26.29	0.79	211.8
	5.0 M	4	4	–	–	21.39	0.71	279.5	26.27	0.80	213.9
	5.2 M	3	3	2	–	21.53	0.71	263.3	26.77	0.81	205.7
	5.2 M	2	2	2	2	**22.20**	**0.72**	**258.3**	**27.24**	**0.83**	**197.7**

Table 5. Loss criteria ablation. The 'full' model includes the two-way loss, latent consistency loss, and DMOL reconstruction loss. '1Way' replaces two-way training with one-way, 'NoLCL' removes latent-consistency loss, 'MAE' replaces DMOL reconstruction loss with MAE.

Model	CT→MRI			MRI→CT		
	PSNR↑	SSIM↑	FID↓	PSNR↑	SSIM↑	FID↓
PHVAE-Full	**22.20**	**0.72**	258.3	**27.24**	**0.83**	**197.7**
PHVAE-1Way	21.39	0.66	266.3	25.77	0.78	207.4
PHVAE-NoLCL	21.69	0.69	256.4	26.76	0.81	200.2
PHVAE-MAE	19.74	0.60	**238.5**	25.78	0.76	198.9

lower resolution ones has shown improved performance, thus the best model incorporates all these tested scales.

For the loss function, we evaluate the options discussed previously, such as one/two-way training and paired latent consistency loss. We additionally compare two types of reconstruction losses, Mean-Absolute Error (MAE), and Discrete Mixture of Logistics. As shown in Table 5, the model which uses Two-Way training, includes latent consistency, and DMOL reconstruction loss outperforms on almost all metrics. The largest total effect comes from the use of DMOL.

Finally, as we are modelling this as a stochastic process, we investigate the effect of sampling temperature on performance. Lower temperatures can improve regularity in images, but can come at the cost of diversity. As shown in Fig. 3, the performance gets significantly better for smaller sampling temperatures for both PSNR and SSIM. The FID for CT→MRI could potentially have a local minima around 0.4, however, is insignificant due to the relative scale.

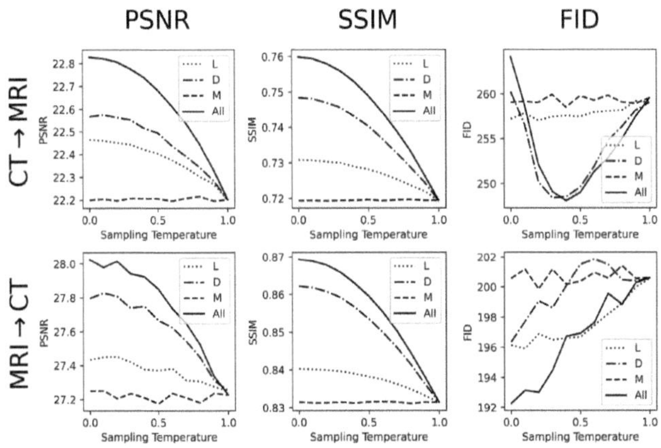

Fig. 3. Sampling temperature ablation. Effect of sampling temperature on translation performance. All VAE models require sampling over the latent variables (labelled L). Models using DMOL additionally require two types of sampling: mixture selection (labelled M), and distribution sampling (D).

5 Conclusion

We examined the use of Paired Hierarchical VAEs in image-to-image translation and cross reconstruction tasks. The novel framework incorporates aspects from previously explored VAE architectures to new areas. As these tasks incorporate both predictive and generative methods, it is difficult to perform quantitative evaluations, as metrics for prediction and visual quality are often at odds.

We explore the performance of these models on two publicly available data sets, and compare against GAN, transformer, and diffusion-based methods. Although these models fall behind the other tested models in terms of visual quality and FID, PHVAEs perform particularly well in predictive metrics such as PSNR and SSIM, while requiring significantly fewer parameters than most alternatives.

Acknowledgments. This research is supported and funded by an Australian Government Research Training Program (RTP) Scholarship.

A Additional Domains

The types of image-to-image translation tasks we have focussed on are those where we suspect preserving semantic information is more important than stylistic information. While our goal is to explore image-to-image translation models in contexts which rely on predictive quality, this only covers a narrow proportion of the potential use-cases.

Because the proposed model relies heavily on collocated features between modalities at high resolutions, we suspect the model will struggle with tasks that do not have such strong structural similarity, such as image-inpainting, style-transfer, or label-to-photo translations.

We have trained the PHVAE model on two additional datasets, to explore broader translation tasks. Cityscapes [3] evaluates label↔photo translations, a difficult task as very little semantic information is preserved in the classification map. We additionally evaluate map↔aerial-photo translations [8], a task similar to style-transfer. The results are provided in Table 6 and Fig. 4.

Table 6. Additional dataset performance. Performance of PHVAE model on two additional datasets.

Dataset	Task	Performance		
		PSNR↑	SSIM↑	FID↓
Cityscapes	Photo→Label	19.90	0.72	131.8
	Label→Photo	16.05	0.32	287.2
Maps	Photo→Map	27.66	0.75	144.0
	Map→Photo	15.40	0.18	259.2

Fig. 4. PHVAE Output samples from additional datasets. Left shows results from Cityscapes dataset, right shows Maps↔Aerial-Photo dataset.

References

1. Aithal, S.K., Maini, P., Lipton, Z.C., Kolter, J.Z.: Understanding hallucinations in diffusion models through mode interpolation. In: Advances in Neural Information Processing Systems, vol. 37, pp. 134614–134644 (2024)
2. Child, R.: Very deep VAEs generalize autoregressive models and can outperform them on images. In: International Conference on Learning Representations (2021)
3. Cordts, M., et al.: The cityscapes dataset for semantic urban scene understanding. In: IEEE Conference on Computer Vision and Pattern Recognition (2016)

4. Dalmaz, O., Yurt, M., Çukur, T.: ResViT: residual vision transformers for multimodal medical image synthesis. IEEE Trans. Med. Imaging **41**(10), 2598–2614 (2022)
5. Harvey, W., Naderiparizi, S., Wood, F.: Conditional image generation by conditioning variational auto-encoders. In: International Conference on Learning Representations (2022)
6. Hazami, L., Mama, R., Thurairatnam, R.: Efficient-VDVAE: less is more (2022)
7. Heusel, M., Ramsauer, H., Unterthiner, T., Nessler, B., Hochreiter, S.: GANs trained by a two time-scale update rule converge to a local nash equilibrium. In: Advances in Neural Information Processing Systems, vol. 30 (2017)
8. Isola, P., Zhu, J.Y., Zhou, T., Efros, A.A.: Image-to-image translation with conditional adversarial networks. In: IEEE Conference on Computer Vision and Pattern Recognition, pp. 5967–5976 (2017)
9. Kingma, D.P., Welling, M.: Auto-encoding variational Bayes. In: International Conference on Learning Representations (2014)
10. Larsen, A.B.L., Sønderby, S.K., Larochelle, H., Winther, O.: Autoencoding beyond pixels using a learned similarity metric. In: International Conference on Machine Learning, vol. 48, pp. 1558–1566 (2016-06-20/2016-06-22)
11. Li, B., Xue, K., Liu, B., Lai, Y.K.: BBDM: image-to-image translation with Brownian bridge diffusion models. In: IEEE Conference on Computer Vision and Pattern Recognition, pp. 1952–1961 (2023)
12. Lu, S., Jayaraman, A.: Pair-variational autoencoders for linking and cross-reconstruction of characterization data from complementary structural characterization techniques. JACS Au **3**(9), 2510–2521 (2023)
13. Nyholm, T., et al.: MR and CT data with multiobserver delineations of organs in the pelvic area-part of the gold atlas project. Med. Phys. **45**(3), 1295–1300 (2018)
14. Parmar, N., et al.: Image transformer. In: International Conference on Machine Learning, vol. 80, pp. 4055–4064 (2018-07-10/2018-07-15)
15. Pawelczyk, M., Broelemann, K., Kasneci, G.: Learning model-agnostic counterfactual explanations for tabular data. In: Proceedings of The Web Conference 2020, pp. 3126–3132 (2020)
16. Ramachandran, P., Zoph, B., Le, Q.V.: Swish: a self-gated activation function (2017)
17. Rezende, D.J., Mohamed, S., Wierstra, D.: Stochastic backpropagation and approximate inference in deep generative models. In: International Conference on Machine Learning, vol. 32, pp. 1278–1286 (2014-06-22/2014-06-24)
18. Rombach, R., Blattmann, A., Lorenz, D., Esser, P., Ommer, B.: High-resolution image synthesis with latent diffusion models. In: IEEE Conference on Computer Vision and Pattern Recognition, pp. 10684–10695 (2022)
19. Saharia, C., et al.: Palette: image-to-image diffusion models. In: ACM SIGGRAPH, pp. 1–10 (2022)
20. Salimans, T., Karpathy, A., Chen, X., Kingma, D.P.: PixelCNN++: improving the PixelCNN with discretized logistic mixture likelihood and other modifications. In: International Conference on Learning Representations (2017)
21. Saxena, D., Cao, J.: Generative adversarial networks (GANs): challenges, solutions, and future directions. ACM Comput. Surv. **54**(3), 1–42 (2022)
22. Sohn, K., Lee, H., Yan, X.: Learning structured output representation using deep conditional generative models. In: Advances in Neural Information Processing Systems, pp. 3483–3491 (2015)

23. Sønderby, C.K., Raiko, T., Maaløe, L., Sønderby, S.K., Winther, O.: Ladder variational autoencoders. In: Advances in Neural Information Processing Systems, vol. 29 (2016)
24. Suekane, K., et al.: Personalized fashion sequential recommendation with visual feature based on conditional hierarchical VAE. In: International Conference on Multimedia Information Processing and Retrieval, pp. 362–365 (2022)
25. Suzuki, M., Nakayama, K., Matsuo, Y.: Joint multimodal learning with deep generative models. In: International Conference on Learning Representations (2017)
26. Vahdat, A., Kautz, J.: NVAE: a deep hierarchical variational autoencoder. In: Advances in Neural Information Processing System, vol. 33, pp. 19667–19679 (2020)
27. Wang, Z., Bovik, A., Sheikh, H., Simoncelli, E.: Image quality assessment: from error visibility to structural similarity. IEEE Trans. Image Process. **13**(4), 600–612 (2004)
28. Wu, S., Others: StegoGAN: leveraging steganography for non-bijective image-to-image translation. In: IEEE Conference on Computer Vision and Pattern Recognition, pp. 7922–7931 (2024)
29. Yaman, M.Y., Kalinin, S.V., Guye, K.N., Ginger, D.S., Ziatdinov, M.: Learning and predicting photonic responses of plasmonic nanoparticle assemblies via dual variational autoencoders. Small **19**(25), 2205893 (2023)
30. Zhang, J., Lang, X., Huang, B., Jiang, X.: VAE-CoGAN: unpaired image-to-image translation for low-level vision. SIViP **17**(4), 1019–1026 (2023)
31. Zhu, J.Y., et al.: Toward multimodal image-to-image translation. In: Advances in Neural Information Processing Systems, pp. 465–476 (2017)

When Language Model Guides Vision: Grounding DINO for Cattle Muzzle Detection

Rabin Dulal[1,2,3](✉), Lihong Zheng[1,2,3], and Muhammad Ashad Kabir[1,2,3]

[1] School of Computing, Mathematics and Engineering, Charles Sturt University, Wagga Wagga, NSW 2678, Australia
{rdulal,lzheng,akabir}@csu.edu.au

[2] Gulbali Institute for Agriculture, Water and Environment, Charles Sturt University, Wagga Wagga, NSW 2678, Australia

[3] Food Agility CRC Ltd, Sydney, NSW 2000, Australia

Abstract. Muzzle patterns are among the most effective biometric traits for cattle identification. Fast and accurate detection of the muzzle region as the region of interest is critical to automatic visual cattle identification. Earlier approaches relied on manual detection, which is labor-intensive and inconsistent. Recently, automated methods using supervised models like YOLO have become popular for muzzle detection. Although effective, these methods require extensive annotated datasets and tend to be trained data-dependent, limiting their performance on new or unseen cattle. To address these limitations, this study proposes a zero-shot muzzle detection framework based on Grounding DINO, a vision-language model capable of detecting muzzles without any task-specific training or annotated data. This approach leverages natural language prompts to guide detection, enabling scalable and flexible muzzle localization across diverse breeds and environments. Our model achieves a mean Average Precision (mAP)@0.5 of 76.8%, demonstrating promising performance without requiring annotated data. To our knowledge, this is the first research to provide a real-world, industry-oriented, and annotation-free solution for cattle muzzle detection. The framework offers a practical alternative to supervised methods, promising improved adaptability and ease of deployment in livestock monitoring applications.

Keywords: Cattle identification · CNN · Transformer · Zero-shot · Grounding DINO · Deep learning

1 Introduction

Accurate cattle identification is essential for biosecurity and livestock management. Traditional methods like RFID tags face issues such as tag loss and tampering [6,10,26]. Biometric approaches using features like the retina, iris, and muzzle offer more reliable alternatives [10]. Muzzle patterns are particularly effective due to their uniqueness and ease of capture.

M. Liu et al. (Eds.): AI 2025, LNAI 16371, pp. 161–173, 2026.
https://doi.org/10.1007/978-981-95-4972-6_13

Deep learning models, especially CNNs and transformers, have improved muzzle-based identification but depend on accurate muzzle detection. Early approaches relied on ink marks and manual cropping, which was labor-intensive and prone to inconsistency. While recent object detectors such as YOLO [23] and its variants have automated this task [7,26], they require annotated datasets and retraining for new cattle [3,21]. Although these models improve accuracy and efficiency, they require annotated datasets with labeled muzzle regions for supervised training [21]. However, collecting such large-scale labeled data is often challenging, expensive, and also requires domain-specific knowledge. To address this, we explore zero-shot muzzle detection as a scalable alternative.

Zero-shot object detection (ZSD) addresses the challenge of detecting objects, such as cattle muzzles, without annotated data [3,21]. It mimics the human ability to generalize from known to unseen categories using shared semantic information. Early ZSD models, such as DeViSE and ALE, used fixed semantic embeddings (e.g., Word2Vec, GloVe) and aligned them with visual features via metric learning. However, their performance was limited in complex visual scenes and lacked contextual understanding.

Recent models like RegionCLIP [34] and Grounding DINO [16] leverage transformer-based architectures and large-scale image-text pairs to learn joint representations. These systems support fine-grained alignment between visual and textual features, improving generalization to unseen classes. In ZSD, models are trained on annotated seen classes with text descriptions and can detect unseen objects at inference by matching visual regions to text embeddings, without retraining [16,21]. State-of-the-art ZSD methods include OWL-ViT [19], ViLD [8], RegionCLIP, OV-DETR [32], DetCLIP [31], OmDet [33], and Grounding DINO.

This research adopts Grounding DINO, a language-guided zero-shot object detection model, for the task of cattle muzzle detection. Inspired by the human ability to generalize from known to unknown concepts by understanding semantics, we explore models that align visual features with semantic descriptions. This approach does not require training the model, as it leverages pre-trained vision-language alignment to perform detection based solely on textual prompts such as "cattle muzzle" or "nose of a cattle". By eliminating the reliance on manually annotated training data and the necessity for model retraining, the proposed method facilitates scalable, adaptable, and efficient muzzle detection in real-world applications. The key contributions of this study are as follows:

- We propose a novel muzzle detection model based on a zero-shot object detection framework, enabling accurate detection without the need for extensive dataset annotation. To the best of our knowledge, this is the first application of a zero-shot approach to the task of cattle muzzle detection.
- We conducted a comprehensive evaluation of seven state-of-the-art zero-shot object detection models that combine language and vision for the task of muzzle detection. This comparison establishes a benchmark to guide and support future research in this domain.

2 Related Work

This section outlines the research efforts focused on the development and evaluation of muzzle detection methods.

Research [11] developed a biometric identification system using Haar cascade classifiers [14] to detect cattle faces, followed by segmentation to isolate the muzzle region. Deep learning further enhanced this process. Research [26] applied YOLOv3 [24] to detect muzzle regions, while research [30] used an improved YOLOv5 [27] model. Similarly, other studies [1,2,12] have effectively utilized YOLO-based models for precise and automated muzzle extraction.

Detecting the muzzle using deep learning models poses challenges due to the need for large amounts of labeled data, which increases the time, cost, and computational resources required for model training and deployment. Unlike existing studies that depend heavily on large, labeled datasets of cattle muzzle images, which can be costly and time-consuming to collect and annotate, this research adopts a zero-shot muzzle detection approach. Zero-shot detection enables the model to identify and localize the muzzle region without requiring any prior labeled examples of muzzle data. This approach reduces the dependency on extensive manual annotation, thereby saving time and resources while maintaining effective detection performance. Most of the prior works employed fine-tuning strategies using pre-trained object detection models such as YOLOv3 [24], YOLOv5 [27], YOLOv7 [29], and YOLOv8 [28], adapting them to specific datasets through additional training. Traditional machine learning methods, such as the Haar cascade classifier, did not involve either fine-tuning or zero-shot learning. In contrast, this study uses a zero-shot detection approach with Grounding DINO. This method does not require retraining on specific tasks and can detect objects without having seen the target classes during training.

3 Methodology

The methodology of this research is illustrated in Fig. 1. It begins with input data, which is a pair of an image and a text prompt. Seven different ZSD models are selected for this research. The outputs are evaluated using appropriate evaluation metrics, such as mean Average Precision (mAP), and the best-performing model is subsequently selected.

3.1 Datasets

This study employs three distinct datasets to ensure diversity in environmental conditions, imaging devices, and cattle breeds. The first dataset, referred to as CSU Data1, was collected at the Global Digital Farm of Charles Sturt University (CSU) in Wagga Wagga, Australia. Image collection was conducted under approved animal ethics protocol A21414, ensuring adherence to animal welfare standards. In addition, two publicly available datasets, the UNE dataset [26] and the NUCES dataset [1], were incorporated to enhance the generalizability

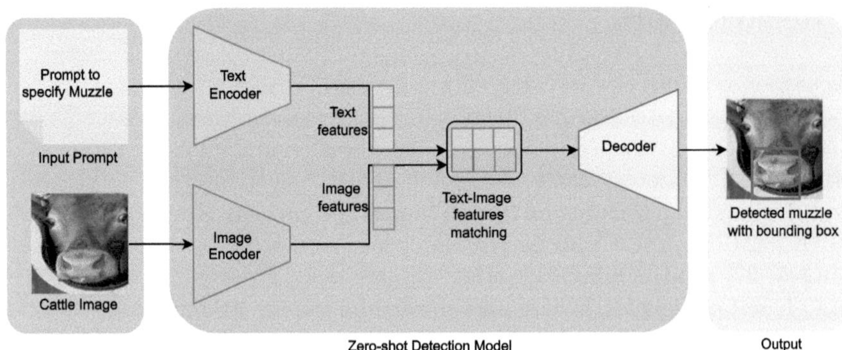

Fig. 1. A general framework for zero-shot muzzle detection using a zero-shot detection model guided by text prompts that describe the muzzle region in cattle images.

and robustness of our approach. Table 1 summarizes the key attributes of each dataset. Notably, the datasets differ in terms of geographical origin, environmental conditions, imaging equipment, and cattle breed diversity, thereby providing a heterogeneous data foundation suitable for evaluating the cross-domain performance of cattle identification models.

Table 1. Summary of datasets used for zero-shot muzzle detection.

Dataset	Image count	Breed	Location	Device
CSU Data1	163	Angus	Wagga Wagga, Australia	Nikon D7200
UNE	2,632	Angus, Simmental, Hereford Charolais	Armidale, Australia	Canon D800
NUCES	2,893	Not mentioned	Lahore, Pakistan	Canon D80, Oppo A76, OnePlus 9 Pro

3.2 Data Preprocessing

To evaluate the performance of ZSD models for muzzle detection, a ground truth dataset was required for objective comparison. Although ZSD models do not require annotated data for the target class, labeled data is essential for model evaluation. In this study, we used Roboflow, a widely adopted annotation and dataset management platform, to draw bounding boxes around the muzzle region of cattle head images. The annotated images were exported in the COCO format, containing image metadata along with the corresponding bounding box coordinates and class labels. These annotations served exclusively for the evaluation, enabling us to assess detection performance.

3.3 Zero-Shot Detection Models

The selected seven ZSD models are OWL-ViT, ViLD, RegionCLIP, OV-DETR, DetCLIP, OmDet, and Grounding DINO. These models were chosen based on two main criteria. First, each model supports text prompts in sentence format, allowing the use of detailed natural language descriptions to guide detection. This is particularly important for detecting specific and subtle features like the cattle muzzle, where a sentence (e.g., "the front part of a cattle's face including the nostrils and mouth") can convey richer semantic information than a single word. Second, all selected models have demonstrated strong performance on standard benchmark datasets such as COCO [15], LVIS [9], and ODinW [13], reflecting their robustness and generalizability across diverse object categories. By leveraging both visual features and semantic text embeddings, these models offer a promising approach to zero-shot muzzle detection tasks.

Table 2 presents a summary of the selected zero-shot detection models, categorizing them based on their base architecture, open-vocabulary strategy, and detection approach. The base architecture refers to the components used for extracting features from both images and texts. It typically includes a visual backbone like ViT [5], Faster R-CNN [25], ConvNeXt [18], DETR [4], and Swin [17] for image feature extraction, and a language encoder like CLIP [22] for processing text descriptions or prompts. The open-vocabulary strategy describes how the model integrates visual features with language supervision to enable the detection of previously unseen categories. Techniques vary from CLIP-guided classification and region-level contrastive learning to query-based grounding and multi-source semantic distillation. The detection approach indicates the model's detection paradigm, whether it requires a region proposal network followed by classification, or end-to-end, where object queries are predicted directly using transformers without separate proposals. Most recent models like OWL-ViT, OV-DETR, DetCLIP, OmDet, and Grounding DINO adopt end-to-end frameworks for better language-vision integration and zero-shot generalization, whereas ViLD and RegionCLIP retain a proposal-based two-stage design for compatibility with traditional detectors.

Based on the methodological framework described above, we conducted a series of experiments to assess the effectiveness of the selected models. The following section reports the results of these evaluations.

4 Experimental Evaluation

All experiments were conducted on an HP Victus laptop equipped with an Intel Core i7 processor and an NVIDIA GeForce RTX 4070 Laptop GPU. The implementation was carried out using Python 3.13 and the PyTorch deep learning framework. All selected models were used with their pre-trained weights for zero-shot detection. We initially conducted a series of experiments using Grounding DINO to identify the most effective text prompt for the muzzle detection task. Multiple prompt variations were tested, and their detection performances were compared. The prompt that yielded the highest accuracy was selected as the

Table 2. Summary of the selected Zero-shot Detection Models.

Model	Base Architecture	Detection Approach	Open-Vocab Strategy
OWL-ViT [19]	ViT+CLIP	End-to-end	CLIP-guided transformer with prompt-based classification
ViLD [8]	Faster R-CNN+CLIP	Proposal-based	Region-level CLIP knowledge distillation
RegionCLIP [34]	Faster R-CNN+CLIP	Proposal-based	Region-text contrastive learning
OV-DETR [32]	DETR+CLIP	End-to-end	CLIP embeddings as query tokens in DETR
DetCLIP [31]	Swin+CLIP	End-to-end	Concept dictionary, multi-source contrastive learning
OmDet [33]	ConvNeXt + CLIP	End-to-end	General open-world detection framework
Grounding DINO [16]	DINO+encoder	End-to-end	Language-guided queries with cross-modal decoder

optimal one. Following this, the selected prompt was uniformly applied across all seven zero-shot learning models to evaluate and compare their detection capabilities under consistent textual guidance. This approach allowed us to isolate model performance from prompt variability and ensured a fair comparison across different architectures. The first two experiments were evaluated by combining the CSU Data1 and UNE datasets. Finally, the performance of the selected ZSD model is compared with other existing muzzle detector models YOLOv3 [26], YOLOv5 [30], YOLOv7 [1], and YOLOv8 [12], with varying training samples of each dataset.

4.1 Selection of Prompts

This section presents an analysis of the different prompts to identify the most effective prompt for muzzle detection. It was observed that customized prompts providing detailed or descriptive information yielded better performance compared to generic prompts commonly used by zero-shot learning models [20]. Based on this observation, we began with a basic, generic prompt and progressively added descriptive phrases or sentences to enhance the specificity of the muzzle. The detection performance of each prompt was evaluated using mAP@0.5, and the results are summarized in Table 3.

In this experiment, we began with a generalized and minimal prompt ("cattle muzzle") and progressively added more descriptive textual elements to enhance the semantic understanding of the target muzzle. Each subsequent prompt built upon the previous one by appending additional context or anatomical details, such as "the nose and mouth of a cattle" and "the area around the nostrils and lips of a cattle". This incremental refinement was aimed at improving the alignment between the visual features and the text input used by the Grounding DINO model. As shown in Table 3, this strategy led to improved detection performance, with the highest mAP@0.5 achieved when using a prompt that balanced specificity and conciseness (Prompt 5). However, adding excessive or overly

Table 3. Detection performance (mAP@0.5) of different prompt variations used with Grounding DINO. The **bold** value represents the best performance.

Prompt No.	Prompt Description	mAP@0.5
1	cattle muzzle	0.242
2	`[Prompt 1]`, the nose and mouth of a cattle	0.386
3	`[Prompt 2]`, the lower front part of a cattle's face	0.536
4	`[Prompt 3]`, the snout of a cattle	0.713
5	`[Prompt 4]`, **the area around the nostrils and lips of a cattle**	**0.768**
6	`[Prompt 5]`, the fleshy soft rounded part of a cattle's face used for eating and smelling	0.755
7	`[Prompt 6]`, cattle's face with visible nasal cavities	0.682

detailed phrases beyond this point (e.g., Prompt 6 and 7) slightly decreased performance, suggesting that overly verbose prompts may dilute the model's focus.

4.2 Performance of ZSD Models

The primary objective of this research is to investigate the feasibility of applying ZSD models for cattle muzzle detection. To facilitate a comparative Performance evaluation, we report the results of selected models using standard metrics. All models were evaluated using their respective base variants to maintain consistency in performance comparison. Specifically, we present the muzzle detection performance in terms of mAP@0.5, mAP@0.75, and mAP@[0.50:0.95]. These results are summarized in Table 4, enabling a comprehensive comparison of each model's accuracy. Among all models, Grounding DINO clearly achieves the best overall performance across all thresholds, with scores of 0.340 ± 0.021 (mAP@0.50:0.95), 0.768 ± 0.025 (mAP@0.5), and 0.180 ± 0.021 (mAP@0.75). This indicates that the model not only detects muzzle regions reliably but also localizes them with high precision. Its superior performance can be attributed to

Table 4. Detection performance (mAP) for ZSD models on cattle muzzle detection. Results show mean ± 95% confidence interval (CI) over five runs. The **bold** value represents the best performance.

Model	mAP		
	0.50:0.95	0.5	0.75
OWL-ViT	0.103 ± 0.019	0.298 ± 0.026	0.032 ± 0.011
ViLD	0.0852 ± 0.014	0.2178 ± 0.018	0.0782 ± 0.016
RegionCLIP	0.160 ± 0.017	0.489 ± 0.028	0.104 ± 0.021
OV-DETR	0.0985 ± 0.012	0.368 ± 0.027	0.0792 ± 0.013
DetCLIP	0.132 ± 0.015	0.386 ± 0.022	0.102 ± 0.020
OmDet	0.227 ± 0.019	0.538 ± 0.030	0.116 ± 0.018
Grounding DINO	**0.340 ± 0.021**	**0.768 ± 0.025**	**0.180 ± 0.021**

the integration of a robust image-text alignment module using CLIP [22] with a transformer-based object detector, enabling fine-grained grounding even without task-specific fine-tuning.

OmDet's performance comes second with scores of 0.227 ± 0.019 (mAP@0.50: 0.95), 0.538 ± 0.030 (mAP@0.5), and 0.116 ± 0.018 (mAP@0.75). While it performs relatively well at lower IoU thresholds, its score at mAP@0.75 indicates a moderate decline in localization accuracy. This gap suggests that OmDet's detection head is less precise in tight object localization compared to Grounding DINO, though its dynamic prompt learning contributes to better generalization.

RegionCLIP and DetCLIP deliver competitive mid-range performance. RegionCLIP achieves 0.160 ± 0.017 (mAP@0.50:0.95), 0.489 ± 0.028 (mAP@0.5), and 0.104 ± 0.021 (mAP@0.75), while DetCLIP scores 0.132 ± 0.015, 0.386 ± 0.022, and 0.102 ± 0.020 respectively. Both models leverage CLIP-based representations enriched with region-level features and contrastive training, resulting in reasonable detection and moderate localization performance. However, their lower mAP@0.75 values reflect a drop in precision for tightly localized predictions.

OV-DETR performs lower with scores of 0.0985 ± 0.012 (mAP@0.50:0.95), 0.368 ± 0.027 (mAP@0.5), and 0.0792 ± 0.013 (mAP@0.75). It outperforms ViLD and OWL-ViT, but its lack of fine-grained localization capabilities, due to the absence of a strong grounding mechanism, limits its usefulness for tasks like muzzle detection.

ViLD and OWL-ViT performed the poorest. ViLD achieves 0.0852 ± 0.014 (mAP@0.50:0.95), 0.2178 ± 0.018 (mAP@0.5), and 0.0782 ± 0.016 (mAP@0.75), while OWL-ViT records 0.103 ± 0.019, 0.298 ± 0.026, and only 0.032 ± 0.011 at mAP@0.75. Their relatively low scores across all thresholds, especially at higher IoU levels, indicate poor localization ability, likely due to their reliance on coarse-level CLIP features and the absence of dedicated detection heads or spatial reasoning modules.

4.3 Performance Comparison with Existing Models

The objective of this experiment is to conduct a comparative study between Grounding DINO and existing muzzle detection models, with a focus on determining the minimum number of labeled muzzle images required for those models to achieve accuracy comparable to that of Grounding DINO. The existing deep learning based muzzle detection models are YOLOv3, YOLOv5, YOLOv7, and YOLOv8. Haar cascade is a classical machine learning model that relies on handcrafted features and does not support automatic learning by extracting features from the input images. Therefore, this study considers existing deep learning muzzle detection models. All three datasets were trained using different numbers of labeled training images: 10, 20, 40, 80, and 160. All experiments were conducted using an image size of 640 × 640 pixels and were trained for 100 epochs with a batch size of 32. All models were evaluated based on the mean Average Precision at IoU threshold 0.5 (mAP@0.5), and the results are presented

Table 5. Comparison of muzzle detection models with varying number of training samples. The **bold** value represents the Grounding DINO's performance. Light green cells highlight cases where other models achieve lower scores than Grounding DINO for the same training sample size.

Model (Approach)	Training Samples	mAP@0.5 on Dataset		
		CSU Data1	UNE	NUCES
YOLOv3 (Fine tuned)	160	0.829	0.985	0.988
	80	0.834	0.887	0.954
	40	0.712	0.697	0.661
	20	0.661	0.635	0.596
	10	0.491	0.365	0.040
YOLOv5 (Fine tuned)	160	0.874	0.975	0.995
	80	0.848	0.975	0.995
	40	0.738	0.655	0.682
	20	0.637	0.629	0.437
	10	0.443	0.464	0.039
YOLOv7 (Fine tuned)	160	0.968	0.995	0.995
	80	0.968	0.995	0.995
	40	0.765	0.729	0.592
	20	0.689	0.581	0.441
	10	0.431	0.419	0.189
YOLOv8 (Fine tuned)	160	0.985	0.995	0.995
	80	0.989	0.995	0.995
	40	0.731	0.728	0.647
	20	0.558	0.493	0.227
	10	0.139	0.256	0.045
Grounding DINO (Zero-shot)	0	**0.753**	**0.789**	**0.758**

in Table 5. Results indicate that Grounding DINO achieves strong zero-shot performance, with mAP@0.5 scores of 0.753 (CSU Data1), 0.789 (UNE), and 0.758 (NUCES), without requiring any fine-tuning or training. In contrast, all YOLO variants require between 40 and 80 labeled training samples to achieve comparable performance levels, highlighting the effectiveness of Grounding DINO in low-data scenarios.

Grounding DINO offers significant practical advantages. Fine-tuning YOLO-based models (YOLOv3, YOLOv5, YOLOv7, YOLOv8) demands a considerable amount of time, computational resources, and manual effort. The process involves dataset preparation, manual annotation, training on GPU-enabled systems, and extensive hyperparameter tuning. Furthermore, the process must be repeated for each target dataset, increasing both the time and cost. In contrast, Grounding DINO provides a cost-effective and scalable alternative, capable of producing competitive results without the need for any supervised training. Its ability to generalize across datasets without re-training makes it particularly suitable for applications with limited data availability or constrained computational budgets.

5 Discussion

As presented in Table 5, the recent studies have utilized supervised deep learning models such as YOLOv3, YOLOv5, YOLOv7, and YOLOv8, achieving high accuracy, up to 99.5% in some cases. However, a major limitation of supervised models is their dependency on the specific dataset they are trained on. These models are typically optimized for particular cattle breeds or controlled environments. If a new dataset is introduced or if the breed changes, the entire dataset must be annotated and retrained from scratch, which is resource-intensive and impractical for real-time or large-scale applications.

In contrast, the zero-shot detection method explored in this study using Grounding DINO does not require task-specific retraining. It leverages natural language prompts (e.g., "cattle muzzle" or "snout of a cattle") to detect the target object, making it more flexible and generalizable across different cattle breeds and unseen environments. Although the accuracy (76.8%) is lower than that of supervised models, this trade-off is offset by the method's adaptability, annotation-free deployment, and potential to scale across breeds and geographies. Therefore, zero-shot detection presents a promising direction for practical muzzle detection applications.

Future work can focus on further improving the accuracy of zero-shot muzzle detection through advanced vision-language models and refined prompt design. Using detailed, domain-specific prompts can lead to more accurate and reliable results. A promising direction is enabling natural language queries, such as "How many cattle have a white muzzle?" or "What is the medication history of cattle with a white muzzle with red marks?", to support intuitive interaction and enhance traceability. This would allow farmers and livestock managers to monitor and search for specific cattle using simple descriptions. Further evaluation across breeds and conditions, along with optimization for edge deployment, will support practical and scalable use in real-world farming environments.

6 Conclusion

In this study, we selected seven ZSD models capable of handling long and descriptive prompts to detect cattle muzzles. We explored various custom natural language prompts to describe muzzle characteristics and identified the most effective ones for detection performance. Through a comprehensive evaluation, we assessed the performance of all seven ZSD models and found that Grounding DINO outperformed the others in terms of accuracy and consistency. Unlike traditional supervised methods, Grounding DINO offers the advantage of detecting muzzles without requiring task-specific training data, making it more adaptable across breeds and environments. These findings demonstrate the potential of prompt-driven ZSD models for practical and scalable livestock monitoring. Nevertheless, certain limitations should be acknowledged. Large language models remain susceptible to hallucination, which may compromise the reliability of generated guidance. Furthermore, the approach is sensitive to prompt design, as variations in phrasing can influence both the consistency and quality of results.

Acknowledgment. This project was supported by funding from Food Agility CRC Ltd, funded under the Commonwealth Government CRC Program. The CRC Program supports industry-led collaborations between industry, researchers, and the community. We thank the travel support from Gulbali Institute for Agriculture, Water and Environment. We also thank Dr Shawn McGrath, Mr. Jonathan Medway, and Prof. Dave Swain for their assistance with the project.

References

1. Ahmed, S.U., Frnda, J., Waqas, M., Khan, M.H.: Dataset of cattle biometrics through muzzle images. Data Brief, 110125 (2024)
2. Anitha, J., Avanthika, R., Kavipriya, B., Vishnupriya, S.: Cattle identification using muzzle images. In: Applied Data Science and Smart Systems, pp. 370–377 (2024)
3. Cao, W., Yao, X., Xu, Z., Liu, Y., Pan, Y., Ming, Z.: A survey of zero-shot object detection. Big Data Min. Analytics **8**(3), 726–750 (2025)
4. Carion, N., Massa, F., Synnaeve, G., Usunier, N., Kirillov, A., Zagoruyko, S.: End-to-end object detection with transformers. In: Proceedings of the European Conference on Computer Vision, pp. 213–229. Springer (2020)
5. Dosovitskiy, A., et al.: An image is worth 16x16 words: transformers for image recognition at scale (2020). arXiv:2010.11929
6. Dulal, R., Zheng, L., Kabir, M.A.: MHAFF: multihead attention feature fusion of CNN and transformer for cattle identification. IEEE Trans. AgriFood Electr., 1–12 (2025)
7. Dulal, R., et al.: Automatic cattle identification using YOLOv5 and mosaic augmentation: a comparative analysis. In: International Conference on Digital Image Computing: Techniques and Applications. pp. 1–8. IEEE (2022)
8. Gu, X., Lin, T.Y., Kuo, W., Cui, Y.: Open-vocabulary object detection via vision and language knowledge distillation (2021). arXiv:2104.13921
9. Gupta, A., Dollar, P., Girshick, R.: LVIS: a dataset for large vocabulary instance segmentation. In: Proceedings of the IEEE Conference on Computer Vision and Pattern Recognition, pp. 5356–5364 (2019)
10. Hossain, M.E., Kabir, A., Zheng, L., Swain, D., McGrath, S., Medway, J.: A systematic review of machine learning techniques for cattle identification: datasets, methods and future directions. Artif. Intell. Agric. (2022)
11. Kusakunniran, W., et al.: Biometric for cattle identification using muzzle patterns. Int. J. Pattern Recognit Artif Intell. **34**(12), 2056007 (2020)
12. Lee, T., Na, Y., Kim, B.G., Lee, S., Choi, Y.: Identification of individual Hanwoo cattle by muzzle pattern images through deep learning. Animals **13**(18) (2023)

13. Li, L.H., et al.: Grounded language-image pre-training. In: Proceedings of the IEEE Conference on Computer Vision and Pattern Recognition, pp. 10965–10975 (2022)
14. Lienhart, R., Maydt, J.: An extended set of Haar-like features for rapid object detection. In: Proceedings. International Conference on Image Processing, vol. 1, pp. I–I. IEEE (2002)
15. Lin, T.Y., et al.: Microsoft COCO: common objects in context. In: Proceedings of the European Conference on Computer Vision, pp. 740–755. Springer (2014)
16. Liu, S., et al.: Grounding DINO: marrying DINO with grounded pre-training for open-set object detection. In: European Conference on Computer Vision, pp. 38–55. Springer (2024)
17. Liu, Z., et al.: Swin transformer: hierarchical vision transformer using shifted windows. In: Proceedings of the IEEE International Conference on Computer Vision, pp. 10012–10022 (2021)
18. Liu, Z., Mao, H., Wu, C.Y., Feichtenhofer, C., Darrell, T., Xie, S.: A convnet for the 2020s. In: Proceedings of the IEEE Computer Vision and Pattern Recognition, pp. 11976–11986 (2022)
19. Minderer, M., et al.: Simple open-vocabulary object detection. In: European Conference on Computer Vision, pp. 728–755. Springer (2022)
20. Nawaz, U., et al.: AgriCLIP: adapting clip for agriculture and livestock via domain-specialized cross-model alignment (2024). arXiv:2410.01407
21. Pourpanah, F., et al.: A review of generalized zero-shot learning methods. IEEE Trans. Pattern Anal. Mach. Intell. **45**(4), 4051–4070 (2022)
22. Radford, A., et al.: Learning transferable visual models from natural language supervision. In: Proceedings of the International Conference on Machine Learning, pp. 8748–8763. PmLR (2021)
23. Redmon, J., Divvala, S., Girshick, R., Farhadi, A.: You only look once: unified, real-time object detection. In: Proceedings of the IEEE/CVF Conference on Computer Vision and Pattern Recognition, pp. 779–788 (2016)
24. Redmon, J., Farhadi, A.: YOLOv3: an incremental improvement (2018). arXiv:1804.02767
25. Ren, S., He, K., Girshick, R., Sun, J.: Faster R-CNN: towards real-time object detection with region proposal networks. In: Advances in Neural Information Processing Systems, vol. 28 (2015)
26. Shojaeipour, A., Falzon, G., Kwan, P., Hadavi, N., Cowley, F.C., Paul, D.: Automated muzzle detection and biometric identification via few-shot deep transfer learning of mixed breed cattle. Agronomy **11**(11), 2365 (2021)
27. Ultralytics: the state-of-the-art yolo model. https://github.com/ultralytics/yolov5. Accessed 15 Jan 2023
28. Ultralytics: the state-of-the-art yolo model. https://ultralytics.com/yolov8. Accessed 15 Jan 2023
29. Wang, C.Y., Bochkovskiy, A., Liao, H.Y.M.: YOLOv7: trainable bag-of-freebies sets new state-of-the-art for real-time object detectors (2022). arXiv:2207.02696
30. Xu, X., et al.: Few-shot cow identification via meta-learning. Inf. Process. Agric. (2024)
31. Yao, L., et al.: DetCLIP: dictionary-enriched visual-concept paralleled pre-training for open-world detection. In: Advances in Neural Information Processing Systems, vol. 35, pp. 9125–9138 (2022)
32. Zang, Y., Li, W., Zhou, K., Huang, C., Loy, C.C.: Open-vocabulary DETR with conditional matching. In: European Conference on Computer Vision. Springer (2022)

33. Zhao, T., Liu, P., Lee, K.: OmDet: large-scale vision-language multi-dataset pre-training with multimodal detection network. IET Comput. Vis. **18**(5) (2024)
34. Zhong, Y., et al.: RegionCLIP: region-based language-image pretraining. In: Proceedings of the IEEE Computer Vision and Pattern Recognition, pp. 16793–16803 (2022)

AFFT: Adapter-Based Few-Shot Fine-Tuning Framework for Remote Sensing Object Detection

Chenguang Zhang[1,2], Yangming Guo[1(✉)], Qianying Tang[3], Pengcheng Zhao[3], and Sicong Zhao[2]

[1] Northwestern Polytechnical University, Xian, China
yangming_g@nwpu.edu.cn
[2] Beijing Aerocim Technology Co., Ltd of CASIC, Beijing, China
[3] Beijing Jiaotong University, Beijing, China

Abstract. Remote Sensing Object Detection (RSOD) presents unique challenges due to arbitrary object orientations, complex backgrounds, and data scarcity. Existing methods often rely on complex network architectures or auxiliary tasks for multi-task optimization, which increase model complexity and require additional data. In this work, we proposed an **A**dapter-based **F**ew-shot **F**ine-**T**uning framework for RSOD, termed **AFFT**. Specifically, to effectively capture global context and extract high-quality features, we adopted a pre-trained hierarchical Swin Transformer based on shifted windows attention mechanism as the backbone. This enables robust feature representation by efficiently modeling both local and global dependencies. Moreover, recognizing the limited availability of annotated remote sensing data, we designed a Group Equivariant Convolutions Multi-cognitive Visual Adapter (GCMONA) for few-shot fine-tuning. This approach significantly enhances model adaptability to new classes with minimal labeled samples, making it particularly well-suited for remote sensing scenarios. It is important that AFFT requires fine-tuning only 5% of the parameters with GCMONA, yet yields competitive or superior performance compared to full fine-tuning methods. Full-parameter training experiments on both DOTA-v1.0 and DOTA-v1.5 demonstrate the superior detection performance of our method in RSOD. Additionally, based on the pre-trained model in DOTA-v1.0, evaluation on our custom few-shot maritime dataset highlights the performance of the proposed few-shot fine-tuning technique when only limited data is available.

Keywords: Remote sensing object detection · Pre-train learning · Fine-tuning technique · Vision transformer

---

Supplementary Information The online version contains supplementary material available at https://doi.org/10.1007/978-981-95-4972-6_14.

M. Liu et al. (Eds.): AI 2025, LNAI 16371, pp. 174–186, 2026.
https://doi.org/10.1007/978-981-95-4972-6_14

1 Introduction

Remote sensing images (RSIs) are characterized by large scale, complex backgrounds, and diverse object orientations [1], while the availability of annotated data is typically limited. These characteristics pose significant challenges for remote sensing object detection (RSOD), severely restricting the practical deployment of intelligent perception systems in critical fields such as urban planning, environmental monitoring and disaster management [2].

Although recent advances in deep learning have significantly improved RSOD performance, most existing methods rely on increasingly complex network architectures [1,3,4] or multi-task optimization strategies [5–7]. These approaches not only require additional annotated data, but also lead to increased model complexity. Such dependence on high-quality, large-scale annotated datasets contradicts real-world applications, where labeled data are often scarce. The scarcity of annotated samples, especially for rare or novel categories, further exacerbates the challenge, making it difficult for traditional supervised and transfer learning approaches to generalize effectively.

Inspired by recent advances in multi-cognitive visual adapters (Mona) [8], which have demonstrated competitive fine-tuning performance and even surpassed full fine-tuning in several challenging vision tasks, we are motivated to explore whether applying the Mona fine-tuning mechanism on pre-trained models to RSOD can overcome the bottlenecks of data scarcity and the complexity associated with full fine-tuning. However, this approach also introduces new challenges: specifically, how to ensure that the Mona fine-tuning method is applicable to RSOD tasks without compromising model performance.

To address these challenges, we proposed an adapter-based few-shot fine-tuning framework for RSOD, termed AFFT. Specifically, AFFT first replaced the original Oriented RCNN [9] backbone with a pre-trained hierarchical Swin Transformer [10] architecture based on shifted windows, endowing the model with powerful multi-scale hierarchical feature extraction capability. This ensures high-quality and less computing cost feature representation even in large-scale RSIs. Furthermore, to substantially reduce the complexity of fine-tuning pre-trained Swin Transformer, we designed a group equivariant convolutions multi-cognitive visual adapter (GCMONA) attached into the backbone, substituting the standard convolution modules in the original MONA with group equivariant convolutions. Group equivariant convolutions effectively enhance parameter efficiency and improve the adapter's capacity to model diverse features, enabling superior fine-tuning performance with a minimal parameter budget. Benefiting from this efficient few-shot fine-tuning approach, AFFT requires fine-tuning only about 5% (2.15M) of the parameters to achieve, or even surpass, the performance obtained by full fine-tuning of 100% (44.75M) parameters.

In summary, our main contributions can be three-fold:

- We propose an adapter-based RSOD few-shot fine-tuning framework, AFFT, which employs the pre-trained Swin Transformer as the feature extraction backbone to efficiently extract high-quality features in large-scale, complex remote sensing scenarios.

– We design a group equivariant convolutions Multi-cognitive Visual Adapter (GCMONA) inserted into the backbone, which replaces the standard convolution modules in the original MONA with group equivariant convolutions, further enhancing the adapter's modeling capability and parameter efficiency.
– We conduct extensive experiments on large-scale RSIs such as DOTA-v1.0 and DOTA-v1.5 [11], demonstrating that AFFT outperforms several state-of-the-art methods. Notably, AFFT achieves comparable or even superior performance to full fine-tuning approaches while requiring fine-tuning of only about 5% of the parameters on a custom few-shot maritime dataset.

2 Related Work

2.1 Remote Sensing Object Detection

Based on oriented rectangular bounding boxes, RRPN [12] and R2CNN [13] generate rotated anchors in advance to provide angular information for horizontal anchors. To tackle the difficulty of feature extraction, many studies have designed sophisticated feature extraction and fusion frameworks. The RoI Transformer [14] learns the transformation from horizontal regions of interest to rotated regions of interest and extracts rotation-invariant features from the rotated regions of interest for subsequent classification and regression subtasks. S2ANet [15] introduces a unique feature alignment module, AlignConv, to optimize anchor distribution and mitigate the misalignment problem. R3Det [16] proposes a Feature Refinement Module (FRM) that utilizes interpolation to precisely obtain the positional information of refined anchors. AO2-DETR [1] further enhances detection performance through three key components based on transformer. Subsequently, based on multi-task learning, SuperYOLO [5] integrates multi-modal data and an auxiliary super-resolution branch to simultaneously optimize both super-resolution and detection tasks. Shi et al. [6] designed a progressive class-aware dual-branch structure to implicitly inject geometric priors into shared features. In recent years, diffusion models have demonstrated powerful capabilities in various vision tasks. DiffusionDet [17] models object detection as a denoising diffusion process and ReDiffDet [18] models it as a rotation-equivariant denoising diffusion process.

However, the aforementioned methods either rely on complex network architectures to obtain high-quality features or introduce additional data and branches for multi-task learning, which significantly increase computational complexity and training costs.

2.2 Fine-Tuning Techniques in Visual Tasks

While full fine-tuning methods can enhance downstream task performance by adjusting all parameters of pre-trained models, their substantial computational costs and training time limit practical applications. To improve fine-tuning efficiency, VPT [19] introduces prompt-based incremental fine-tuning to visual classification tasks. [20] enhances model performance by introducing trainable adapter modules into the backbone network. SSF [21] employs feature scaling and shifting mechanisms for efficient fine-tuning. AdaptFormer [22] constructs a parallel adapter architecture, while KAdaptation [23] optimizes adapter performance through the Kronecker product. LoRand [24] designs a multi-branch low-rank adapter for dense prediction tasks.

Although incremental fine-tuning methods have achieved some success in visual tasks, they often struggle to exceed the performance of full fine-tuning, especially in challenging scenarios. To address this, Mona [8] fine-tuning method introduces multi-cognitive visual filters within the adapter and employs layer normalization to adaptively adjust input feature distributions, which has achieved better performance compared with full fine-tuning on various visual tasks. However, further adapting and optimizing Mona for RSOD to achieve even greater performance gains remains a challenge.

3 Method

3.1 Problem Setup

Assume that given a training set $D_{tr} = \{(x_i, y_i)\}_{i=1}^{n}$, where x_i is an input image and y_i is the ground-truth label with K objects, including the regression labels p_k^* (the Cartesian coordinates of the four vertices of the k-th bounding box) and the classification labels c_k^*. As illustrated in Fig. 1(a), our proposed few-shot fine-tuning method, termed AFFT, consists of a pre-trained Swin Transformer-based backbone inserted with the designed group equivariant convolutions Multi-cognitive Visual Adapter (GCMONA), an oriented RPN and an oriented RCNN head caming from Oriented RCNN. Next, let's introduce more details.

3.2 Overall Framework of AFFT

To take advantage of the powerful and fast feature extraction ability of Swin Transformer, we use the Swin Transformer as the initial backbone. Specifically, for each input image x_i, we first feed it into the Patch Embedding module to obtain the initial feature map P_1. Subsequently, this feature map is processed through multiple SwinBlock Sequences-based backbone to extract multi-scale high-quality features P_2, P_3, P_4, P_5. As illustrated in Fig. 1(a), each sequence primarily consists of two SwinBlocks and a PatchMerge module. More specifically, as shown in Fig. 1(b), each SwinBlock comprises two main input-output streams. For the input feature P_1, the input stream first passes it through a Layer Normalization and a Shifted-Window Multi-head Self-Attention (ShiftWindowMSA) module to reorganize the feature. The output is then element-wise added to the original input through a skip connection. Afterward, GCMONA is applied for few-shot fine-tuning and the specific details will be elaborated subsequently. The fine-tuned feature is then passed through the two linear layers with a GeLU activation to obtain deep semantics, and then combined with itself via another skip

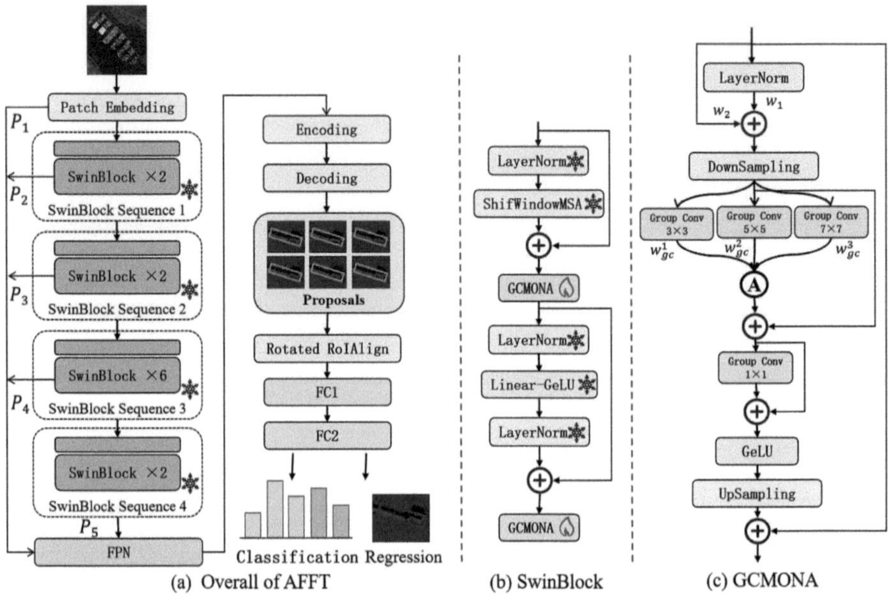

Fig. 1. The architecture of AFFT. (a) is the overall framework of AFFT, (b) is the detailed design of SwinBlock and (c) is the detailed design of GCMONA.

connection. We summarize the feature extraction of the backbone as follows:

$$P_1, P_2, P_3, P_4, P_5 = \begin{cases} P_1 = PE(x_1) \\ P_{j+1} = SBS_j(P_j), \quad j = 1, 2, 3, 4 \end{cases} \tag{1}$$

where $PE(\cdot)$ is the Patch Embedding module and $SBS_j(\cdot)$ is the j-th SwinBlock Sequence in Swin Transformer. Subsequently, the multi-scale features are then fused through a Feature Pyramid Network (FPN) to obtain a fused multi-scale representations $\bar{P_1}, \bar{P_2}, \bar{P_3}, \bar{P_4}, \bar{P_5} = FPN(P_1, P_2, P_3, P_4, P_5)$. Finally, the fused features are fed into the oriented RPN to get a sparse set of oriented proposals for regression and the binary label $I_i^* \in \{0, 1\}$ of each proposal to be a positive sample. We can compute the loss of oriented RPN L_{RPN} as:

$$L_{RPN} = \frac{1}{M}\sum_{i=1}^{M} L_{cls}(q_i, q_i^*) + \frac{1}{M}\sum_{i=1}^{M} I_i^* L_{reg}(\delta_i, t_i^*) \tag{2}$$

where i is the index of the proposal and $M = 256$. q_i^* is the ground-truth label of the i-th proposal, q_i is the predicted probability by the classification branch of the oriented RPN, δ_i is the output offsets of the proposal relative to the anchor by the regression branch of the oriented RPN and t_i^* is the ground-truth offsets. L_{cls}, L_{reg} represent the cross-entropy loss and Smooth L1 loss, respectively.

Finally, we send the fused multi-scale features and the set of oriented proposals into the oriented RCNN head for object classification and regression. For each proposal, we first extract the fixed-size feature from each fused feature. We then feed it into a two fully-connected layers to output the class probability of the proposal belongs to, and the offsets of the proposal. Through post-processing, we can obtain the corresponding predicted detection box p_k and class c_k for each proposal. We can computed the loss of the oriented RCNN head as:

$$L_{HEAD} = \frac{1}{K}\sum_{k=1}^{K} L_{cls}(c_k, c_k^*) + \frac{1}{K}\sum_{k=1}^{K} I_i^* L_{reg}(p_k, p_k^*) \tag{3}$$

Therefore, the total loss of AFFT is computed as follows:

$$L = L_{RPN} + L_{HEAD} \tag{4}$$

3.3 Group Equivariant Convolutions Multi-Cognitive Visual Adapter

Instead of performing full fine-tuning, which updates all parameters in the pre-trained backbone, we propose a group equivariant convolutions Multi-cognitive Visual Adapter (GCMONA) that freezes the pre-trained parameters and updates only the parameters within the adapters. Inspired by MONA, we introduce input optimization and multi-cognitive visual filters, enabling GCMONA to adapt to visual signal processing based on the data distribution of new task. As illustrated in Fig. 1(c), our approach differs from MONA in that we replace the original convolution modules with group equivariant convolution modules.

Specifically, we optimize the input by adding a normalization layer and two learnable hyperparameters w_1, w_2. For each input feature f_0, the optimized feature $\bar{f_0}$ is computed as follows:

$$\bar{f_0} = w_1 \cdot \text{Norm}(f) + w_2 \cdot f_0 \tag{5}$$

where Norm($\cdot$) is the LayerNorm.

To obtain multi-scale features, we design multiple group equivariant convolution-based filters to enhance cognitive dimensions. The motivation for using group equivariant convolutions lies in their ability to capture diverse feature representations, which is particularly beneficial for efficient adaptation in RSOD. Moreover, group equivariant convolutions enhance the robustness of models to geometric transformations by enforcing equivariance under predefined symmetry groups. In practice, the optimized feature is first downsampled as $f_1 = \text{Down}(\bar{f_0})$ and then passed through three group convolution filters with kernel sizes of $3 \times 3, 5 \times 5$ and 7×7 to obtain feature maps $f_1^* = [f_1^1, f_1^2, f_1^3]$ at different scales. These features are weighted by $(w_{gc}^1, w_{gc}^2, w_{gc}^3)$ and added to the original input with a skip connection:

$$\bar{f_1^*} = f_1 + avg(\sum_{i=1}^{3} w_{gc}^i \otimes f_1^i) \tag{6}$$

where $\otimes$ is the element-wise multiplication and $avg(\cdot)$ is the average. Then we use another 1×1 group equivariance convolution with a weight w_{gc}^4 and a skip connection to obtain the aggregated feature:

$$f_2 = \bar{f}_1^* + w_{gc}^4 \otimes \bar{f}_1^* \tag{7}$$

Finally, a GeLU activation is applied for non-linear transformation, and up-projection with a skip connection is used to restore the feature dimensions. The final output is computed as:

$$f_0' = f_0 + \mathrm{Up}(\mathrm{GeLU}(f_2)) \tag{8}$$

With this design, our algorithm achieves efficient few-shot fine-tuning on remote sensing scenarios with limited datasets. By leveraging the rotation-equivariant priors provided by equivariant convolutions, our approach demonstrates superior detection performance compared to conventional methods.

4 Experiment

4.1 Experimental Settings

Dataset and Baseline. The experiments utilize three primary datasets for oriented object detection: DOTA-v1.0, DOTA-v1.5 [11]and a custom maritime few-shot dataset from Maritime [25]. The detailed introduction of the datasets, please refer to Appendix Sect. 1. For DOTA-v1.0, we establish baselines with four representative detectors: Oriented RCNN [9], ROITransformer [14], R3Det [26], and ReDet [27]. The DOTA-v1.5 experiments expand the comparison to include six advanced methods, including EMO2-DETR [28], AO2-DETR [1], Mask R-CNN [29], HTC [30], ReDet [27] and RotatedFCOS [31]. All evaluations follow the DOTA protocol with 1024×1024 image chips generated using an 824-pixel stride. We employ both AP50 and AP75 from COCO metrics to provide a comprehensive performance characterization.

Implementation Details. For fair comparison across all models, we adopt a unified experimental setup implemented within the MMRotate framework. All experiments are conducted on a single NVIDIA RTX 3090 GPU with a batch size of 2. We employ SGD as the optimizer with a momentum of 0.9 and weight decay of 0.0001, training each model for 12 epochs following a step-based learning rate schedule. The learning rate configuration consists of a linear warmup phase for the first 500 iterations with a warmup ratio of 1/3, followed by scheduled decays at the 8th and 11th epochs. For backbone networks, we initialize both ResNet-50 and Swin Transformer with their respective ImageNet pre-trained weights.

4.2 Experimental Results

We conduct extensive quantitative and qualitative experiments to prove the effectiveness of AFFT. For DOTA-v1.0 and DOTA-v1.5, AFFT conducts full

Table 1. Performance Comparison AP50 of Different Methods on DOTA-v1.0 Dataset.

Method	BD	GTF	SV	LV	SH	HC	HA	AP50
Oriented RCNN	0.730	0.697	0.774	0.810	0.873	0.489	0.658	0.727
ROITransformer	0.772	0.673	0.777	0.769	0.871	0.532	0.674	0.731
R3det	0.754	0.683	0.766	0.756	0.850	0.468	0.665	0.715
Redet	0.809	0.685	0.776	0.772	0.874	0.509	0.681	0.735
AFFT	**0.814**	**0.740**	**0.781**	**0.839**	**0.880**	**0.542**	**0.742**	**0.756**

Table 2. Performance Comparison AP75 of Different Methods on DOTA-v1.5 Dataset.

Method	BD	GTF	SV	LV	SH	HC	CC	AP50
EMO2-DETR	0.754	0.582	0.482	0.733	0.803	0.471	0.156	0.617
Mask R-CNN	0.735	0.578	0.513	0.713	0.798	0.578	0.094	0.627
HTC	0.737	0.640	0.515	0.733	0.803	0.559	0.052	0.634
AO2-DETR	0.781	0.612	0.433	0.745	0.796	**0.696**	0.152	0.663
ReDet	**0.828**	0.714	**0.524**	0.757	**0.809**	0.633	0.115	0.669
RotatedFCOS	0.745	0.595	0.505	0.725	0.800	0.565	0.105	0.623
AFFT	0.722	**0.725**	0.520	**0.770**	**0.809**	0.577	**0.239**	**0.676**

fine-tuning to train a powerful pre-trained model. Subsequently, based on this pre-trained model, AFFT freezes the Backbone network and performs GCMONA fine-tuning on the custom maritime few-shot dataset.

Comparison on Dataset DOTA-v1.0. As shown in Table 1, the proposed method achieves superior performance on the DOTA-v1.0 benchmark, with particularly notable improvements in handling oriented objects and complex backgrounds. The integration of Swin Transformer as the backbone network contributes to enhanced feature representation, especially for objects with extreme aspect ratios, like Large Vehicles (LV) and Baseball Diamond(BD), where it achieves 0.839 and 0.814 in AP50, respectively.

Comparison on Dataset DOTA-v1.5. Experimental results on the challenging DOTA-v1.5 benchmark, as shown in Table 2, demonstrate the consistent effectiveness of our approach. The proposed AFFT achieves a competitive 0.676 AP50, outperforming existing methods in several key areas. Notably, our method excels in Container Crane (CC) detection, achieving 0.239 AP50 and significantly surpassing other approaches, highlighting the benefits of our rotation-equivariant design for complex industrial structures. The method also shows robust performance across categories, including 0.520 AP50 for small vehicles and 0.770 AP50 for large vehicles, confirming its ability to handle objects at different scales. Superior results in ground track field detection further validate the architecture's effectiveness for elongated structures. Compared to transformer-based detectors such as EMO2-DETR and AO2-DETR, our method delivers more balanced performance across all categories, with particular advantages in challenging scenarios involving small or densely packed objects.

Fig. 2. Experimental results of our proposed AFFT method on the DOTA-v1.0 benchmark dataset.

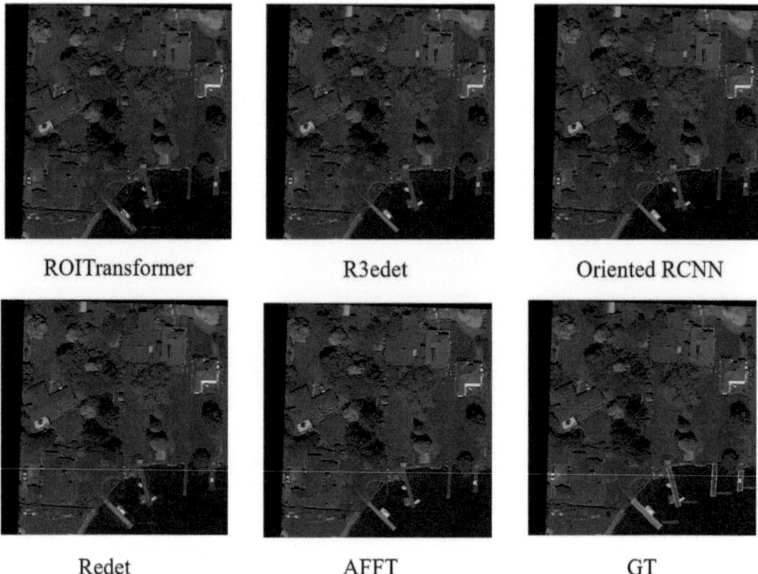

Fig. 3. Visual comparison of detection errors on DOTA-v1.0 dataset. Red boxes indicate missed detections (false negatives), blue boxes show false positives. (Color figure online)

Table 3. Performance Comparison AP50 and AP75 of Different Method Variants on DOTA-v1.0 Dataset.

Method	BD	GTF	SV	LV	SH	HC	HA	AP75
AFFT	**0.814**	**0.740**	0.781	**0.839**	0.880	0.542	**0.742**	**0.463**
w/o GCMONA	0.806	0.727	**0.783**	0.838	**0.881**	**0.565**	0.739	0.462
w/o GCMONA and Swin	0.730	0.697	0.774	0.810	0.873	0.489	0.658	0.429

Table 4. Ablation Study on Few-Shot Maritime Dataset.

Method	AP50	Tuning Params (M)	Total Params (M)
AFFT	**0.394**	2.15	46.90
w/o GC	0.391	**1.41**	46.16
w/o GCMONA	0.390	44.75	**44.75**

Qualitative Analysis. Figure 2 shows visualization results of our method on the DOTA-v1.0 dataset, illustrating stable detection performance across objects of different scales and orientations. Our approach is especially effective at detecting densely packed small objects that are often overlooked by other methods, as well as large, elongated targets that conventional detectors tend to fragment. Moreover, Fig. 3 compares our method with other approaches on the same dataset. Red boxes indicate missed detections, blue boxes denote false positives. Our method achieves higher recall by greatly reducing missed detections, particularly for small and densely arranged objects, while maintaining strong precision with fewer false alarms.

Ablation Study. As shown in Table 3, the experimental results demonstrate progressive improvements through each architectural modification. Replacing the baseline backbone with Swin Transformer yields a notable performance boost, elevating AP75 from 0.429 to 0.462 while achieving significant gains in challenging detection categories. The Baseball-diamond (BD) category achieves an improvement of 7.6% points in AP50, while the Harbor (HA) category shows an increase of 8.1% points in AP50, showcasing Swin Transformer's superior capability in processing complex remote sensing imagery through its hierarchical attention mechanism that effectively captures multi-scale features and spatial relationships. Moreover, introducing the proposed GCMONA module can further improve the performance and balance these trade-offs by introducing geometric awareness, which allows the final architecture to recover initial performance and improve robustness to rotational variations.

Table 4 summarizes the ablation study on the few-shot maritime dataset. AFFT w/o GCMONA uses full-parameter training with the largest number of tunable parameters, yet achieves the lowest detection performance. Replacing GCMONA with the original MONA module (AFFT w/o GC) also leads to a performance drop, indicating that the group equivariant convolutions module benefits RSOD. In contrast, AFFT achieves a 0.4% AP50 improvement with only 100 training samples, demonstrating excellent few-shot adaptation and parameter efficiency. The total parameter increase is just 4.8%, significantly enhancing adaptability in data-scarce scenarios and robustness to orientation variations.

5 Conclusion

In this paper, we presented AFFT, an efficient adapter-based few-shot finetuning framework for remote sensing object detection. AFFT addresses two key challenges: limited annotated data and the difficulty of capturing diverse, complex feature representations. By incorporating the Swin Transformer backbone, the framework effectively models both local and global dependencies, while the proposed Group Equivariant Convolutions Multi-cognitive Visual Adapter (GCMONA) enhances adaptability and parameter efficiency for new tasks with minimal data. Our extensive experiments on DOTA-v1.0, DOTA-v1.5, and a custom few-shot maritime dataset validate the effectiveness of AFFT, demonstrating consistent improvements in detection accuracy and robustness under data-scarce conditions. Moreover, our future work could explore integrating more domain-specific priors, such as handling extreme aspect ratios and complex geometric variations common in remote sensing imagery.

References

1. Dai, L., Liu, H., Tang, H., Wu, Z., Song, P.: AO2-DETR: arbitrary-oriented object detection transformer. IEEE Trans. Circuits Syst. Video Technol. **33**(5), 2342–2356 (2022)
2. Zheng, Z., Zhong, Y., Wang, J., Ma, A., Zhang, L.: Building damage assessment for rapid dis- aster response with a deep object-based semantic change detection framework: from natural disasters to man-made disasters. Remote Sens. Environ. **265**, 112636 (2021)
3. Cai, X., Lai, Q., Wang, Y., Wang, W., Sun, Z., Yao, Y.: Poly kernel inception network for remote sensing detection. In: Proceedings of the IEEE/CVF Conference on Computer Vision and Pattern Recognition, pp. 27706–27716 (2024)
4. Yan, L., He, Z., Zhang, Z., Xie, G.: LS-MambaNet: integrating large strip convolution and mamba network for remote sensing object detection. Remote Sens. **17**(10), 1721 (2025)
5. Zhang, J., Lei, J., Xie, W., Fang, Z., Li, Y., Du, Q.: SuperYOLO: super resolution assisted object detection in multimodal remote sensing imagery. IEEE Trans. Geosci. Remote Sens. **61**, 1–15 (2023)
6. Shi, T., et al.: Progressive class-aware instance enhancement for aircraft detection in remote sensing imagery. Pattern Recogn. **164**, 111503 (2025)
7. Tang, D., et al.: AeroGen: enhancing remote sensing object detection with diffusion-driven data generation. In: Proceedings of the Computer Vision and Pattern Recognition Conference, pp. 3614–3624 (2025)

8. Yin, D., Hu, L., Li, B., Zhang, Y., Yang, X.: 5%¿ 100%: Breaking performance shackles of full fine-tuning on visual recognition tasks. In: Proceedings of the Computer Vision and Pattern Recognition Conference, pp. 20071–20081 (2025)
9. Xie, X., Cheng, G., Wang, J., Yao, X., Han, J.: Oriented R-CNN for object detection. In: Proceedings of the IEEE/CVF International Conference on Computer Vision, pp. 3520–3529 (2021)
10. Liu, Z.: Swin transformer: hierarchical vision transformer using shifted windows. In: Proceedings of the IEEE/CVF International Conference on Computer Vision, pp. 10012–10022 (2021)
11. Xia, G.S., et al.: A large-scale dataset for object detection in aerial images. In: Proceedings of the IEEE Conference on Computer Vision and Pattern Recognition, pp. 3974–3983 (2018)
12. Ma, J., et al.: Arbitrary-oriented scene text detection via rotation proposals. IEEE Trans. Multimedia **20**(11), 3111–3122 (2018)
13. Jiang, Y., et al.: R2CNN: rotational region CNN for orientation robust scene text detection. arXiv preprint: arXiv:1706.09579 (2017)
14. Ding, J., Xue, N., Long, Y., Xia, G.S., Lu, Q.: Learning ROI transformer for oriented object detection in aerial images. In: Proceedings of the IEEE/CVF Conference on Computer Vision and Pattern Recognition, pp. 2849–2858 (2019)
15. Han, J., Ding, J., Li, J., Xia, G.-S.: Align deep features for oriented object detection. IEEE Trans. Geosci. Remote Sens. **60**, 1–11 (2021)
16. Yang, X., Liu, Q., Yan, J., Li, A., Zhang, Z., Yu, G.: R3Det: refined single-stage detector with feature refinement for rotating object. arxiv 2019. arXiv preprint: arXiv:1908.05612 7 (1908)
17. Chen, S., Sun, P., Song, Y., Luo, P.: DiffusionDet: diffusion model for object detection. In: Proceedings of the IEEE/CVF International Conference on Computer Vision, pp. 19830–19843 (2023)
18. Zhao, J., Ding, Z., Zhou, Y., Zhu, H., Du, W.L., Yao, R.: RediffDet: rotation-equivariant diffusion model for oriented object detection. In: Proceedings of the Computer Vision and Pattern Recognition Conference, pp. 24429–24439 (2025)
19. Jia, M., et al.: Visual prompt tuning. In: European Conference on Computer Vision, pp. 709–727. Springer (2022)
20. Chen, Z., et al.: Vision transformer adapter for dense predictions. arXiv preprint: arXiv:2205.08534 (2022)
21. Lian, D., Zhou, D., Feng, J., Wang, X.: Scaling & shifting your features: a new baseline for efficient model tuning. In: Advances in Neural Information Processing Systems, vol. 35, pp. 109–123 (2022)
22. Chen, S., et al.: AdaptFormer: adapting vision transformers for scalable visual recognition. In: Advances in Neural Information Processing Systems, vol. 35, pp. 16664–16678 (2022)
23. He, X., Li, C., Zhang, P., Yang, J., Wang, X.E.: Parameter-efficient model adaptation for vision transformers. In: Proceedings of the AAAI Conference on Artificial Intelligence, vol. 37, pp. 817–825 (2023)
24. Yin, D., Yang, Y., Wang, Z., Yu, H., Wei, K., Sun, X.: 1% vs 100%: parameter-efficient low rank adapter for dense predictions. In: Proceedings of the IEEE/CVF Conference on Computer Vision and Pattern Recognition, pp. 20116–20126 (2023)
25. Sun, X., et al.: FAIR1M: a benchmark dataset for fine-grained object recognition in high-resolution remote sensing imagery. ISPRS J. Photogramm. Remote. Sens. **184**, 116–130 (2022)

26. Yang, X., Yan, J., Feng, Z., He, T.: R3Det: refined single-stage detector with feature refinement for rotating object. In: Proceedings of the AAAI Conference on Artificial Intelligence, vol. 35, pp. 3163–3171 (2021)
27. Han, J., Ding, J., Xue, N., Xia, G.S.: ReDet: a rotation-equivariant detector for aerial object detection. In: Proceedings of the IEEE/CVF Conference on Computer Vision and Pattern Recognition, pp. 2786–2795 (2021)
28. Hu, Z., et al.: EMO2-DETR: efficient-matching oriented object detection with transformers. IEEE Trans. Geosci. Remote Sens. **61**, 1–14 (2023)
29. He, K., Gkioxari, G., Dollár, P., Girshick, R.: Mask R-CNN. In: Proceedings of the IEEE International Conference on Computer Vision, pp. 2961–2969 (2017)
30. Chen, K., et al.: Hybrid task cascade for instance segmentation. In: Proceedings of the IEEE/CVF Conference on Computer Vision and Pattern Recognition, pp. 4974–4983 (2019)
31. Tian, Z., Shen, C., Chen, H., He, T.: FCOS: a simple and strong anchor-free object detector. IEEE Trans. Pattern Anal. Mach. Intell. **44**(4), 1922–1933 (2020)

Depth-Aware Audio Visual Segmentation with Geometry-Heuristic Cross Attention

Hadha Afrisal[1(✉)], Shadi Abpeikar[2], and Francisco Cruz[1,3]

[1] University of New South Wales, Sydney, Australia
hadha.afrisal@unsw.edu.au
[2] University of New South Wales, Canberra, Australia
[3] Universidad Central de Chile, Santiago, Chile

Abstract. The current state-of-the-art in Audio Visual Segmentation (AVS) has demonstrated successful milestones in performing pixel-level sounding object segmentation. However, they faced significant issues of misalignment and performance degradation in complex settings, such as in substandard lighting conditions and cluttered environments. To address those challenges, we introduce Depth-aware Audio Visual Segmentation (Depth AVS) to enhance the capability of current transformer-based AVS. This paper proposes three main contributions: first, a Geometry-Heuristic Cross Attention (GHCA) as a novel method to suppress irrelevant distant features from the main object of interest, thus enhancing the robustness of audio-visual cross-attention fusion in cluttered and inadequate lighting conditions. Second, an Intermediate Fusion module for integrating depth and RGB features to enrich our model's learning with non-redundant visual features. Third, a depth-aware segmentor that outputs not only a binary mask but also a segmented depth mask. We experimented using the S4 AVSBench-Object dataset against the current state-of-the-art in AVS, AVSegFormer. Our experiments demonstrate that our Depth AVS surpasses the performance of the AVS baseline method using only small input sizes. Our Depth AVS also extends the capability of AVS using distance estimation with a small error rate.

Keywords: Audio Visual Segmentation · Multimodal Transformer · Geometry Heuristic Cross Attention · Depth-aware Attention · Sounding Object Distance Estimation

1 Introduction

Although humans and some animals can instinctively perform spatial audio-visual learning, adopting this capability in artificial agents, such as robots, poses theoretical and technical challenges. In the theoretical aspect, there is a limited number of supporting theories on how to optimally and efficiently integrate the audio-visual modalities for robots' spatial learning using a computational approach. The development of audio-visual learning should also be robust and adaptive to noise and non-ideal conditions for deployment in real-world settings.

M. Liu et al. (Eds.): AI 2025, LNAI 16371, pp. 187–199, 2026.
https://doi.org/10.1007/978-981-95-4972-6_15

One of the most vibrant research areas that sheds light on spatial audio-visual learning is Audio Visual Segmentation (AVS). AVS aims to segment fine-grained pixel-based regions of objects that emit sounds. Current AVS state-of-the-art (SOTA) methods successfully demonstrated performance improvements by specifically tackling the arising challenges in AVS, such as suboptimal audio-visual cross-modal fusion [1], overlapping and misalignment of audio-visual matching [11], complex spatio-temporal correspondences [8], and instability in complex audio-visual scenarios [9,10]. The current mainstream AVS methods solely focus on extracting 2D pixel-wise segments of sounding objects, without considering estimating the relative distance of the sounding object.

We identify that the problem of suboptimal cross-modal fusion can be addressed by introducing a geometry prior into the fusion network. This approach is inspired by the way humans focus their attention by combining prior heuristic knowledge about the environment's geometry. To address the issue of AVS's performance degradation under complex conditions, we add extra visual features in the form of depth images into the visual stream of the AVS network, aiming to enrich the model's learning from non-redundant visual features.

Our main contributions can be summarized as follows:

- We implement Geometry-Heuristic Cross Attention (GHCA) to improve the cross-channel fusion in the audio-visual segmentation transformer.
- We add depth features as an extra visual feature that we fuse with RGB features via the Intermediate Fusion module, which enhances the model performance and improves the robustness of transformer-based AVS.
- We extend Audio Visual Segmentation (AVS) to Depth-aware Audio Visual Segmentation (Depth AVS), which segments depth images and estimates the relative distance of sounding objects.

2 Related Works

2.1 Audio Visual Segmentation

Audio-Visual Segmentation (AVS) aims to generate fine-grained segments for objects that emit sounds. This research sub-domain progresses from a branch of Computer Vision that addresses segmentation problems, such as Video Salient Object Detection (VSOD) [18] and Video Object Segmentation (VOS) [4]. While the VSOD focuses on extracting eye-catching visual objects, VOS segments objects of interest. The AVS problem is similar, as it attempts to segment salient objects of interest, which are coincidentally seen and heard. AVS's problem is more complex than sound source localization, as it tries to localize, categorize, and portray the shape of sounding objects in spatiotemporal space.

Zhou et al. [19] presented audio-visual segmentation (AVS) with semantics to generate pixel-wise masks of sounding objects, which consists of a Feature Pyramid Network (FPN) with a Temporal Pixel-level Audio-Visual Interaction (TPAVI) module [20], that incorporates semantic audio features to guide the vision-based segmentation process. The quantitative experiments exhibited a superior performance of mIoU and F1-Score to the preceded SOTA models, such as MSSL [12], SOD-LGVT [18], and VOS-SST [4]. However, the proposed model in [19] struggles to accurately segment multiple overlapping objects that coexist in the same frame. As Gao et al. [5] examined, the model fails to fully optimize the low-level cross-modal correspondences of sound-vision cues.

While many AVS methods focus on improving the performance of the audio branch [8–11] and the visual branch with RGB inputs [4,6], less research focuses on using different visual features for audio-visual segmentation, such as depth images. AVS's methods are uniquely specified to pixel-wise segmentation and hence can provide a sense of the shape of the detected sounding objects. These methods are more intuitive in depicting sound sources embodied in physical objects. However, they only provide spatial information of sounding objects relative to the two-dimensional camera frame coordinates. Unlike the mainstream AVS methods, which produce a binary mask of the audible segment, we aim to extend the capability of AVS to output a depth segment as a mask that contains depth value information for estimating the object location.

2.2 RGBD Semantic Segmentation

Semantic segmentation involves assigning a category label to an image region associated with specific objects. With the rapid progress in deep learning methods, the field has seen considerable improvement. However, interpreting certain complex scenes using only RGB images remains difficult, particularly under poor lighting conditions or when dealing with cluttered environments, as RGB lacks rich textural and a sense of distance. Tackling these challenges, some current research introduced the concept of incorporating depth information to augment RGB-based segmentation. This gives rise to RGB-D semantic segmentation, which has led to numerous studies exploring ways to effectively merge RGB and depth data to enrich scene understanding [17]. Many of the latest approaches for RGB-D semantic segmentation focus on creating interactive modules that integrate RGB and depth features using two separate pretrained neural networks [3]. These strategies significantly enhance the performance of segmentation. However, these milestones still face a major issue: running two full neural networks side by side, one for RGB and one for depth, requires an excessive computing power than using RGB alone. Therefore, in this research, we propose to adopt method of injecting depth and RGB features as geometry prior, inspired by DFormerV2 [17], especially to enhance the cross-attention mixer in our model.

3 Methodology

3.1 Network Design Based on AVSegFormer

Our main research objective is to develop an efficient and light-weight end-to-end approach for audio-visual segmentation with an enhanced capability of sounding object distance estimation. Therefore, for the design of our network, we adopt the basic building block of AVSegFormer [5], which is based on SegFormer [15]

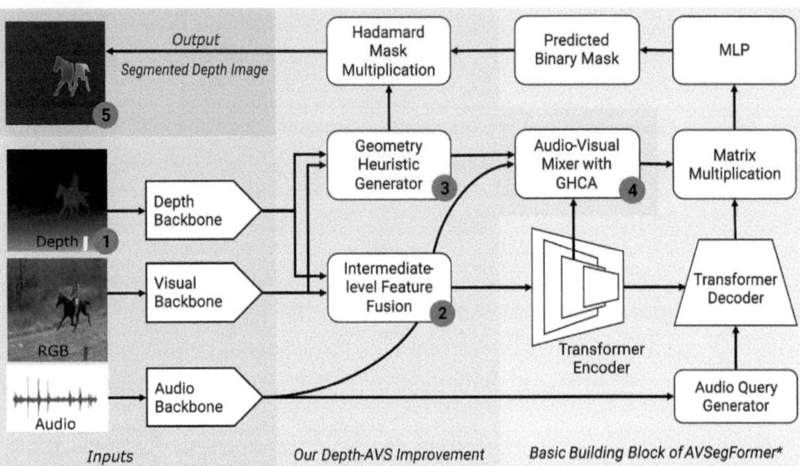

Fig. 1. Our proposed architecture of Depth-aware Audio Visual Segmentation (Depth AVS), inspired by AVSegFormer [5]. Our architecture leverages pre-trained PVT-v2 [14] as the visual and depth backbones, and VGGish [7] as the audio backbone.

and Mask2Former [2]. As visualized in Fig. 1, compared to the original version of AVSegFormer [5] –as the baseline of our performance comparison– the distinctions of our model are as follows: (1) we employ extra depth features to pass as extra visual input for our model together with RGB and audio inputs, (2) we incorporate intermediate fusion via convolution of depth and RGB image features from the backbone to enrich feature learning performance, (3) we design Geometry-Heuristic generator that provide a heuristic value for geometry prior, (4) we apply Geometry-Heuristic Cross Attention in audio-visual channel mixer to enhance attention matrix by suppressing distant features from object of interests, and (5) the output of our system is in the form of depth mask of the audible object, thus the output represent a complete 3D estimated position.

3.2 Geometry-Heuristic Cross Attention

Geometry-Heuristic Cross Attention (GHCA) is comprised of two parts: (1) geometry-heuristic generator and (2) audio-visual mixer with GHCA, as illustrated in Fig. 2 (a) and (b), respectively. We develop the 3D geometry-heuristic prior generator by combining two types of prior represented by Manhattan distance, namely 2D pixel-wise and 1D depth-wise distances. The 2D pixel-wise prior is calculated as the relative distance of every pixel-wise patch position to other patch positions in the image, using Eq. (1)(a), where $P_{xy,x'y'}$ is the representation of pixel-wise Manhattan distance between patches at the position of x, y and x', y'. The 1D depth-wise prior is calculated as depth distances for every patch using Eq. (1)(b), where $D_{xy,x'y'}$ is the representation of depth-wise Manhattan distance between patches at x, y and x', y'.

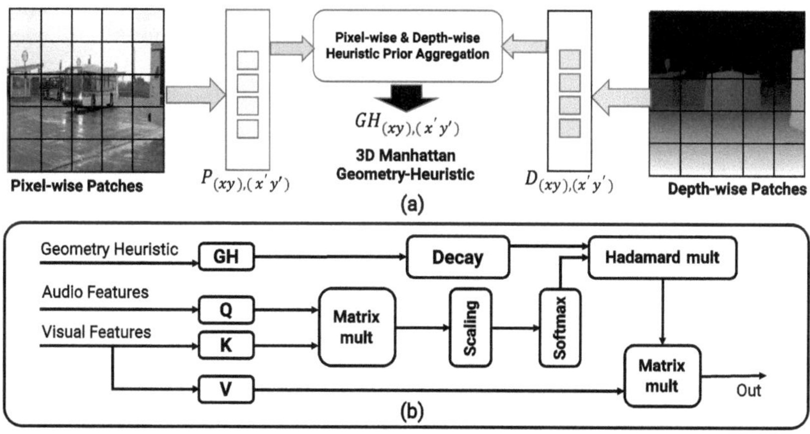

Fig. 2. (a) Geometry-Heuristic generation based on 3D Manhattan distance from pixel-wise and depth-wise patches, as inspired by DFormerv2 [17], and (b) Geometry-Heuristic Cross Attention (GHCA) in audio-visual cross-channel mixer, inspired by Manhattan Self-Attention [13].

$$(a)\ P_{xy,x'y'} = |x - x'| + |y - y'|\ ;\ (b)\ D_{xy,x'y'} = |z_{xy} - z_{x'y'}| \tag{1}$$

Subsequently, given the pixel-wise and depth-wise heuristic distance matrices prior, we calculate the 3D Geometry-Heuristic ($GH_{xy,x'y'}$) by using K_P and K_D, as the balancing learnable weights of pixel-wise and depth-wise heuristic distance matrices, respectively, as presented in Eq. (2).

$$GH_{xy,x'y'} = K_P.P_{xy,x'y'} + K_D.D_{xy,x'y'} \tag{2}$$

Originally, the cross-attention mechanism used in audio-visual mixer of AVSegFormer [5] is as formulated in Eq. (3), where the cross attention mixer is based on query (Q), key (K), and value (V) matrices, embedding dimension (D), and number of heads (n_{head}). In the AVSegFormer, the query matrix is sourced from the audio channel, while the key and value are sourced from the visual channel.

$$\text{CrossAttn}(Q, K, V) = \text{softmax}\left(\frac{QK^T}{\sqrt{D/n_{\text{head}}}}\right)V \tag{3}$$

Differently, we develop the GHCA-based cross-attention audio-visual mixer by injecting geometry-heuristic (GH) using a decay function $e^{\gamma(-GH)}$, where γ is the decay rate, as formulated in Eq. (4). The idea behind this formulation is to attenuate the weight of distant features from the main object of interest using the heuristic approach based on geometrical position composed from GH. In the experiment, we set $\gamma = 1$ as it works best for suppressing distant features.

$$\text{GHCA}(Q, K, V, GH) = \text{softmax}\left(\frac{QK^T}{\sqrt{D/n_{\text{head}}}} \odot e^{\gamma(-GH)}\right)V \tag{4}$$

4 Experiments

4.1 Dataset

We use the Single Sound Source Segmentation (S4) dataset from the AVSBench-Objects dataset [20]. We select this dataset as it has complete annotations for audio-visual pairs, and has been used as a common benchmark in AVS. The dataset comprises short video clips sourced from YouTube, each trimmed to five seconds, thus providing five-second audio segment, with one frame per second images selected for segmentation purposes. The dataset is split into 452 samples for training, 740 samples for validation, and 740 samples for testing, and it consists of 23 categories of sounding objects, such as people, animals, vehicles, and musical instruments, with manually annotated ground truth binary masks that match the activation of sound events in each video. As we aim to extend the objective of AVS for segmenting the depth region and estimating the relative distance of sounding objects, we augmented the dataset with depth videos that we generated by converting RGB videos in the S4 dataset using Depth-Anything-V2-Large [16]. We trimmed the generated depth videos to one frame per second to match the extracted RGB images and audio segments in the dataset.

4.2 Evaluation Metrics

We apply 3 types of evaluation metrics to measure our method's performance:

1. **Intersection over Union (IoU)** $[\frac{TP}{TP+FN+FP}]$. We employ this metric to evaluate audio-visual segmentation performance by quantifying the overlap area of the predicted vs ground truth segments, using $TP = True\,Positive$, $TN = True\,Negative$, $FP = False\,Positive$, and $FN = False\,Negative$.
2. **F-Score** $[\frac{2\times\text{Precision}\times\text{Recall}}{\text{Precision}+\text{Recall}}]$. As audio-visual semantic segmentation can be viewed as pixel-level binary classification, we also use this harmonic mean of $\text{Precision} = \frac{TP}{TP+FP}$ and $\text{Recall} = \frac{TP}{TP+FN}$ to measure the classification of predicted segments against the ground truth segments.
3. **Mean Absolute Error (MAE)** $[\frac{1}{n}\sum_{i=1}^{n}|d_i^G - d_i^P|]$. We use MAE to measure distance estimation errors of the predicted values, with n is the number of objects, d^G is the ground truth distance, and d^P is the predicted distance.

4.3 Experimental Setups

We adopt AVSegFormer [5] as the baseline method for comparison with our developed method—Depth AVS. For a fair performance comparison, we replicate the basic training setup of AVSegFormer, using AdamW optimizer, batch size 2, initial learning rate of 2×10^{-5}, and we train our model using 30 epochs. We follow the setting of the transformer's encoder and decoder of AVSegFormer using 6 layers with embedding dimensions of 256. We also implement AVSegFormer's loss function, $\mathcal{L} = \mathcal{L}_{\text{IoU}} + \lambda\mathcal{L}_{\text{mix}}$, where $\mathcal{L}$ is the total loss, $\mathcal{L}_{\text{IoU}}$ is the IoU loss between the final segmentation mask and the ground truth mask, $\mathcal{L}_{\text{mix}}$ is the AV mixer loss, and λ is the mix loss coefficient that we set to 0.1 as it performed best in AVSegFormer setting. However, during our experimental work, we reduced the input image dimension to 128 × 128 to test the limit of performance robustness under small input sizes. One of our objectives is to develop a lightweight model, therefore, we trained and tested our model using a single NVIDIA GeForce RTX4080 GPU with 16 GB VRAM.

4.4 Performance Comparison and Evaluation

Our Depth AVS aims to combine pixel-level and depth-level segmentation with distance estimation, thus, we evaluate our method's performance as follows:

1. **Audio-visual pixel-level segmentation.** We conduct experiments and an ablation study to quantitatively compare the overall and inter-class performance of our Depth AVS (using variations of applied modules) against the state-of-the-art method of AVSegFormer [5], tested on the same S4 dataset of AVSBench-Objects [20], using mIoU and F-Score evaluation metrics.
2. **Audio-visual depth-level segmentation and distance estimation.** This experiment's objective is to evaluate the performance of our Depth AVS in depth-wise segmentation and distance estimation, via quantitative analysis using MAE and qualitative analysis using feature visualization.

Evaluation of AV Pixel-Level Segmentation: Overall Performance. We perform an ablation study by injecting our GHCA and fusion modules (early and intermediate fusion) on the baseline method of AVSegFormer [5] using a variation of network configurations and inputs, as presented in Table 1. In mode A1, we utilize the original AVSegFormer architecture without any modifications. In mode A2, we replace the RGB images with depth images to be the input for the visual stream, which results in an increase of 0.84% (↑) of mIoU and 0.2% (↑) of F-Score. Introducing depth images to the visual stream has a positive influence on the model's performance. In mode B1, we modify the basic building block of AVSegFormer by inserting the GHCA module into the AV mixer, while

Table 1. Performance comparison using evaluation metric of mIoU and F-Score for the baseline AVSegFormer [5] (mode A), modified AVSegFormer (mode B), and our Depth AVS (mode C) under different module configurations and inputs.

Mode	Methods	Configuration	Inputs [A, V]	mIoU	F-Score
A1	AVSegFormer*	Baseline	Audio, RGB	76.43	86.62
A2	AVSegFormer*	Baseline	Audio, Depth	77.27	86.82
B1	Mdf. AVSegFormer	Baseline+GHCA	Audio, RGB	76.83	86.91
B2	Mdf. AVSegFormer	Baseline+GHCA	Audio, Depth	78.15	86.94
C1	Depth AVS (ours)	Early Fusion (EF)	Audio, RGB+Depth	78.28	86.73
C2	Depth AVS (ours)	Inter. Fusion (IF)	Audio, RGB+Depth	78.87	87.65
C3	Depth AVS (ours)	IF+GHCA	Audio, RGB+Depth	**80.97**	**89.19**

keeping the use of RGB images as the visual input. The GHCA module insertion into the original AVSegFormer improves the model performance by 0.4% (↑) of mIoU and 0.29% (↑) of F-Score. Comparing B2 and A2, using the same depth images as the visual input, the GHCA module insertion consistently increases the performance of the model by 0.88% (↑) of mIoU and 0.12% (↑) of F-Score. This indicates the positive impact of the GHCA module on the model's performance.

In mode C, we use fusion modules combining RGB and depth images as inputs. We observe the impact of injecting this module on the model's performance. We apply the Early Fusion (EF) module without GHCA in mode C1, the Intermediate Fusion (IF) module without GHCA in mode C2, and the Intermediate Fusion (IF) with our GHCA module in mode C3, all using the full-stream visual inputs from RGB and depth images. Compared to the baseline, AVSegFormer, introducing the EF and IF modules increases mIoU by 1.85% (↑) and 2.44% (↑), and F-Score by 0.11% (↑) and 1.03% (↑). This is a positive indicator that our fusion module contributes to our model's performance enhancement. As the IF module offers a better performance improvement than the EF module for both mIoU and F-Score, we employ the IF module to be combined with our GHCA module in our final design of Depth AVS (mode C3). Table 1 shows that our Depth AVS architecture with IF module and full-stream inputs (mode C3) achieves the best overall performance of 80.97% of mIoU and 89.19% of F-Score, enhancing the baseline method by 4.54% (↑) of mIoU and 2.57% (↑) of F-Score. This demonstrates that injecting depth images, the GHCA module, and the Intermediate Fusion module results in a significant improvement for audio-visual segmentation, even in the condition that the input sizes of the visual and audio stream are reduced to only 1/4 of the original ones.

Evaluation of AV Pixel-Level Segmentation: Inter-Class Performance. Figure 3 presents the performance distribution of each class on evaluation metrics of (a) IoU and (b) F-Score, by comparing 3 modes: the baseline method of AVSegFormer (mode A), modified AVSegFormer with GHCA (mode B), and our Depth AVS (mode C). As can be observed in Fig. 3, our Depth AVS (mode C) scores overall better IoU and F-Score in almost all classes against other modes. In addition, we also evaluate our method's robustness under complex settings by selecting specific subsets of classes from the test set containing the major traits of lighting variations and cluttered environments. For variations of lighting conditions, we based our evaluation on three classes: female singing (indoor, low-light), race car, and horse clip-clop (outdoor, overexposure). For cluttered environments, we based our evaluation on three classes: ambulance siren (outdoor, occlusion), helicopter (outdoor, noisy/cluttered), and playing violin (indoor, occlusion). As presented in Fig. 3 (a) and (b), our Depth AVS method demonstrates a higher mean value of IoU and F-Score than the baseline method - AVSegFormer, in the categories of unideal lighting conditions and cluttered scenes, by a significant margin between 2% - 4%. Our method also scores a significantly lower standard deviation for IoU and F-Score, with margin of 1.5% - 4.5%, demonstrating a better consistency in segmentation for various scenes on the subsets, especially for classes with challenging conditions.

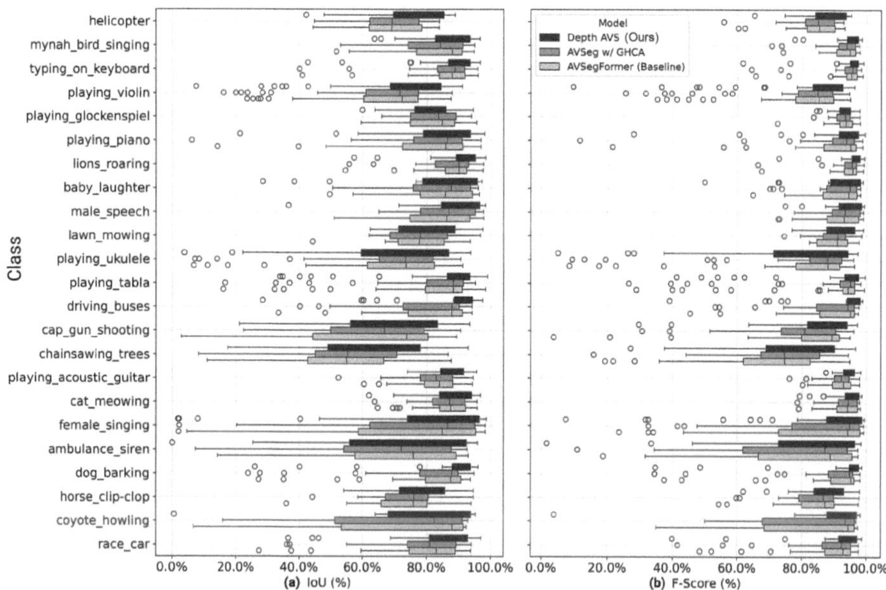

Fig. 3. Comparison of performance distribution of Depth AVS, modified AVSegFormer with GHCA, and AVSegFormer baseline for each class using (a) IoU and (b) F-Score.

AV Depth-Level Segmentation and Distance Estimation. We perform a qualitative comparison on the attention matrix as the output of the audio-visual mixer with and without GHCA. As visualized in Fig. 4, by utilizing GHCA in the audio-visual mixer, we can observe a significant improvement in attention on the object of interest. Our GHCA audio-visual mixer is relatively robust to a variety of scenes and conditions, such as variations in object size, the existence of occlusions, and noisy backgrounds. Our GHCA modules assist the audio-visual mixer in suppressing distant features from the main object of interest. Our Depth AVS extends the capability of the mainstream AVS method by outputting a segmented depth mask of the sounding object. As presented in Fig. 5(a), we qualitatively compare the performance of predicted depth segmentation to our ground truth depth masks. By using the predicted depth segment, we can estimate the relative distance of the sounding-emitting object using a basic averaging technique of the depth value inside the segmented region. We utilize synthetic depth images generated using Depth-Anything-V2 [16] in the form of affine-invariant depth maps for relative pixel-wise distance estimation. Due to the diverse nature of the dataset that contains indoor and outdoor scenes with variations in depth range, we normalize the depth value in the range of 0 to 20 m and represent it

Fig. 4. Qualitative analysis of attention matrix on selected samples of the test set with major lighting variations, such as low-light and overexposure (from left to right: female singing, race car, horse clip-clop) and cluttered scenes, such as existence of occlusions and noises (ambulance siren, helicopter, playing violin), when we apply GHCA and without GHCA compared to the ground truth mask. The GHCA helps our model suppress distant features of the main object of interest, improving performance robustness in complex scenarios.

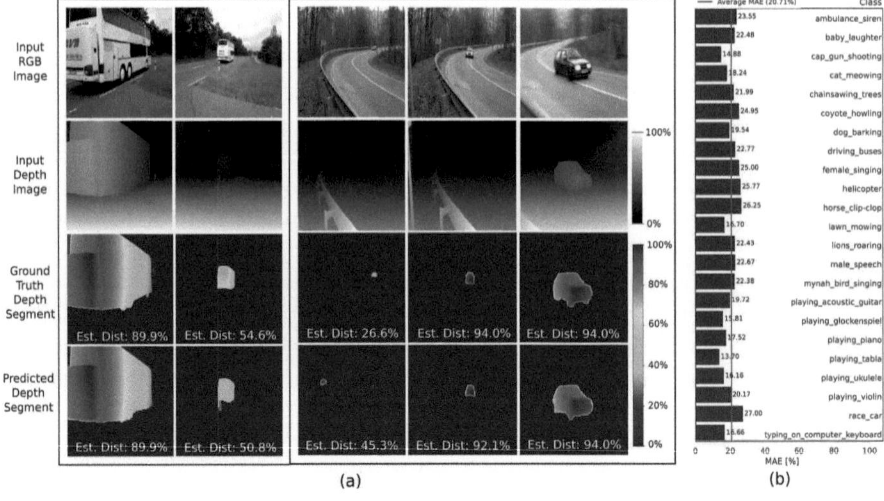

Fig. 5. (a) Qualitative analysis on the result of predicted depth segment from the detected sounding objects compared to the ground truth depth segment, input depth image, and input RGB image, (b) quantitative analysis using Mean Absolute Error (MAE) distribution over class for distance estimation using predicted depth segment compared to the ground truth depth segment.

using a percentage value (100% means close, 0% means far). The Mean Absolute Error (MAE) of inter-class distance estimation is presented in Fig. 5(b). From the experiment, we measure the overall average MAE for our distance estimation, which shows the result of 20.71%.

Computational Cost. The injection of the RGB-depth fusion module in the network and the GHCA module in the audio-visual mixer helps improve the model's performance. However, module insertion results in a small increase of trainable parameters: 0.01 [M] for GHCA and 0.77 [M] for Intermediate Fusion. There is a reduction in inference time of 12–15 ms after adding those modules.

5 Conclusion and Future Work

We extend the mainstream Audio-Visual Segmentation (AVS) to Depth-aware Audio-Visual Segmentation (Depth AVS) that can segment depth images and estimate the relative distance of sounding objects. The key contribution of our proposed method is the novelty of Geometry-Heuristic Cross Attention (GHCA) in the audio-visual mixer. We also process extra visual features of depth images to enrich the visual inputs using the Intermediate Fusion module in our network. We demonstrate that the performance of our Depth AVS is better than the baseline method of AVSegFormer, and more robust to complex scenarios, such as unideal lighting conditions and cluttered environments. However, our Depth AVS method moderately increases computation cost compared to the baseline method. Therefore, a more efficient, lightweight technique will be investigated in the future. In addition, our current method is still limited to predicting relative distance using the training data from synthetic monocular depth estimation with Depth-Anything-V2, by benchmarking using AVSegFormer with the S4 AVSBench-Objects dataset. We plan to employ a depth sensor for absolute distance measurement and perform a more comprehensive evaluation and comparison to diverse state-of-the-art methods on multiple similar datasets.

Acknowledgments. This research is part of Hadha Afrisal's study at UNSW Sydney with a doctoral scholarship from Indonesia Endowment Fund for Education (LPDP).

Disclosure of Interests. The authors have no conflict of interest to declare.

References

1. Chen, Y., et al.: Unraveling instance associations: a closer look for audio-visual segmentation. In: Proceedings of the IEEE/CVF Conference on Computer Vision and Pattern Recognition, pp. 26497–26507 (2024)
2. Cheng, B., Misra, I., Schwing, A.G., Kirillov, A., Girdhar, R.: Masked-attention mask transformer for universal image segmentation. In: Proceedings of the IEEE/CVF Conference on Computer Vision and Pattern Recognition, pp. 1290–1299 (2022)

3. Du, S., Wang, W., Guo, R., Wang, R., Tang, S.: Asymformer: asymmetrical cross-modal representation learning for mobile platform real-time rgb-d semantic segmentation. In: Proceedings of the IEEE/CVF Conference on Computer Vision and Pattern Recognition, pp. 7608–7615 (2024)
4. Duke, B., Ahmed, A., Wolf, C., Aarabi, P., Taylor, G.W.: Sstvos: sparse spatiotemporal transformers for video object segmentation. In: Proceedings of the IEEE/CVF Conference on Computer Vision and Pattern Recognition, pp. 5912–5921 (2021)
5. Gao, S., Chen, Z., Chen, G., Wang, W., Lu, T.: Avsegformer: audio-visual segmentation with transformer. In: Proceedings of the AAAI Conference on Artificial Intelligence, vol. 38, pp. 12155–12163 (2024)
6. Gong, S., et al.: Avs-mamba: exploring temporal and multi-modal mamba for audio-visual segmentation. IEEE Trans. Multimedia (2025). https://doi.org/10.1109/TMM.2025.3542995
7. Hershey, S., et al.: Cnn architectures for large-scale audio classification. In: 2017 IEEE International Conference on Acoustics, Speech and Signal Processing (ICASSP), pp. 131–135 (2017). https://doi.org/10.1109/ICASSP.2017.7952132
8. Li, X., et al.: Qdformer: towards robust audiovisual segmentation in complex environments with quantization-based semantic decomposition. In: Proceedings of the IEEE/CVF Conference on Computer Vision and Pattern Recognition, pp. 3402–3413 (2024)
9. Liu, C., Li, P., Yang, L., Wang, D., Li, L., Yu, X.: Robust audio-visual segmentation via audio-guided visual convergent alignment. In: Proceedings of the Computer Vision and Pattern Recognition Conference, pp. 28922–28931 (2025)
10. Liu, C., et al.: Bavs: bootstrapping audio-visual segmentation by integrating foundation knowledge. IEEE Trans. Multimedia **26**, 10015–10028 (2024)
11. Liu, C., Yang, L., Li, P., Wang, D., Li, L., Yu, X.: Dynamic derivation and elimination: audio visual segmentation with enhanced audio semantics. In: Proceedings of the Computer Vision and Pattern Recognition Conference, pp. 3131–3141 (2025)
12. Qian, R., Hu, D., Dinkel, H., Wu, M., Xu, N., Lin, W.: Multiple sound sources localization from coarse to fine. In: Vedaldi, A., Bischof, H., Brox, T., Frahm, J.-M. (eds.) ECCV 2020. LNCS, vol. 12365, pp. 292–308. Springer, Cham (2020). https://doi.org/10.1007/978-3-030-58565-5_18
13. Wang, H., Xu, L., Yu, Y., Ding, W., Xu, Y.: Manhattan self-attention diffusion residual networks with dynamic bias rectification for bci-based few-shot learning. In: Proceedings of the AAAI Conference on Artificial Intelligence, vol. 39, pp. 14423–14431 (2025). https://doi.org/10.1609/aaai.v39i13.33580
14. Wang, W., et al.: PVT v2: improved baselines with pyramid vision transformer. In: Computational Visual Media, pp. 1–10 (2022). https://doi.org/10.1007/s41095-022-0274-8
15. Xie, E., Wang, W., Yu, Z., Anandkumar, A., Alvarez, J.M., Luo, P.: Segformer: simple and efficient design for semantic segmentation with transformers. Adv. Neural. Inf. Process. Syst. **34**, 12077–12090 (2021)
16. Yang, L., et al.: Depth anything v2. Adv. Neural. Inf. Process. Syst. **37**, 21875–21911 (2024)
17. Yin, B.W., Cao, J.L., Cheng, M.M., Hou, Q.: Dformerv2: geometry self-attention for rgbd semantic segmentation. In: Proceedings of the Computer Vision and Pattern Recognition Conference, pp. 19345–19355 (2025)
18. Zhang, J., Xie, J., Barnes, N., Li, P.: Learning generative vision transformer with energy-based latent space for saliency prediction. Adv. Neural. Inf. Process. Syst. **34**, 15448–15463 (2021)

19. Zhou, J., et al.: Audio-visual segmentation with semantics. Int. J. Comput. Vision 1–21 (2024)
20. Zhou, J., et al.: Audio–visual segmentation. In: European Conference on Computer Vision, pp. 386–403. Springer, Heidelberg (2022). https://doi.org/10.1007/978-3-031-19836-6_22

EyeDentify: A Deep Learning Approach to Non-invasive Biometric Identification from Eye Blink Patterns

Atharva Ratnesh Agate[1(✉)], Shreyash Sharad Wetal[1], Aditya Narvekar[2], and Hazem El-Alfy[1,3]

[1] School of Computer Science, University of Sydney, Sydney, Australia
{aaga0043,swet0507}@uni.sydney.edu.au, hazem.elalfy@sydney.edu.au

[2] Data Science Department, S P Jain School of Global Management, Sydney, Australia
aditya.narvekar@spjain.org

[3] Faculty of Engineering, Alexandria University, Alexandria, Egypt
helalfy@alexu.edu.eg

Abstract. Person identification using behavioural biometrics has been an active area of research. Eye-blink patterns provide a unique set of characteristics that can identify individuals, as they are difficult to imitate. Current blink pattern-based biometric identification approaches use obtrusive head-mounted sensors to accurately capture eye blinks, which limits their daily use. In addition, they rely on one-dimensional temporal signals, such as eye-aspect ratio (EAR), or require fusion with additional biometric data. In contrast, we introduce novel two-dimensional spatio-temporal features, drawing from earlier successful gait recognition templates. In particular, our approach turns eye-blink sequences into discriminative spatiotemporal patterns. We construct and compare two types of "Time Images" to determine the most effective spatio-temporal template: (i) a Colour Outline Time Image (COTI), which uses a red-to-violet hue cycle to convey blink progression; and (ii) a Mean Intensity Time Image (MITI), which is an average intensity image of the periocular region. Our approach uses dynamic features extracted from the EAR and creates superimposed blink frame representations to capture the temporal evolution of eye contours. We used two different Convolutional Neural Network (CNN) architectures to classify these time images. Our method achieves strong results on two publicly available datasets that cover a range of covariates and demonstrates robust identification capability. This research serves as a baseline for future studies and demonstrates the validity of using eye-blink patterns as a standalone biometric feature for person identification.

Keywords: Eye-blink biometrics · Spatio-temporal representation · Eye Aspect Ratio (EAR)

M. Liu et al. (Eds.): AI 2025, LNAI 16371, pp. 200–213, 2026.
https://doi.org/10.1007/978-981-95-4972-6_16

1 Introduction

Biometric identification is widely used for security, yet traditional static traits, such as fingerprints or facial features, are increasingly vulnerable to presentation attacks due to advances in deepfakes [13,22]. In contrast, eye-blink patterns incorporate biological (eye shape) and behavioural (motion) characteristics, making them harder to spoof [18,23]. The neuromuscular processes responsible for eye blinks lead to unique patterns in terms of eye movement, frequency, and duration that vary between individuals [28].

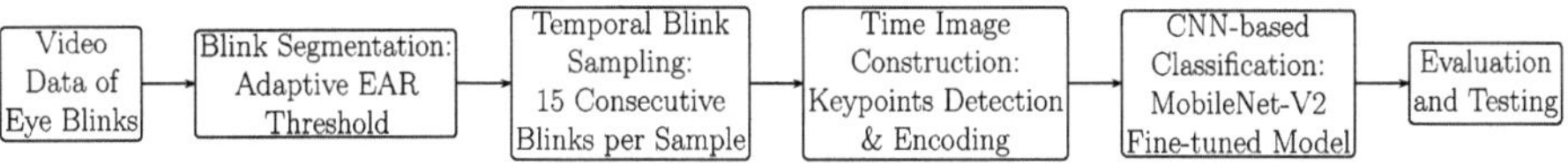

Fig. 1. Pipeline of the Proposed Blink-based Biometric Identification System

Previous eye blink-based identification methods explored simple statistical blink features such as eye opening/closing speed and duration [2,20], more complex EEG-based identification [1,9,29], and eye blink features such as 1D time-series signals [7,21], but often required intrusive sensors or lost spatial detail.

Inspired by the gait recognition methods Chrono-Gait Images (CGI) [25,26] and Gait Energy Images (GEI) [10], we encode similar temporal information of blink patterns in spatial representations; an approach previously unused in periocular biometrics. To the best of our knowledge, our work is the first to innovatively adapt gait-style spatio-temporal templates to blink biometrics.

Subsequently, we establish the first reproducible benchmark for blink-based biometric identification by repurposing the public datasets, NTHU-DDD [27] and UTA-RLDD [8], originally collected for drowsiness detection but well-suited for biometric identification due to their varied lighting, alertness levels, and multi-day recordings. Despite their moderate size, these datasets surpass prior private datasets used in EEG or sensor-based studies [2,17], enabling a fair comparison and future progress.

Our overall system is shown in Fig. 1. It involves breaking the input video of a person's face into sequences of eye blinks, isolating the eyes' region to create time images, which are then used to train a CNN model for multi-class classification. Our proposed time image captures both the spatial features of eye shapes and how they evolve over a blink sequence. The main contributions of this paper are:

1. An adaptive Eye Aspect Ratio threshold calibration technique which improves the robustness of blink detection across subjects by automatically calibrating to the characteristics of the eyes of an individual.
2. A non-invasive biometric identification framework that relies only on eye-blinks from RGB videos with no additional requirements of intrusive sensors, evaluated thoroughly under varying environmental conditions.

3. A novel cross-domain adaptation of gait-style spatio-temporal templates for blink biometrics, introducing the Colour Outline Time Image (COTI) and the Mean Intensity Time Image (MITI) as efficient and principled representations of dynamic blink patterns.

2 Literature Review

Blinking is an involuntary behaviour with biometric potential, as features like blink velocity, duration, and muscle dynamics carry unique information that can distinguish a person [7,12]. Early contributions explored these features using specialised hardware like high-speed cameras to achieve high accuracy (e.g., 99.7% reported in [7]) or asynchronous time-based vision sensors [16]. However, the practical applicability and scientific reproducibility of many of these studies are constrained by several key factors. First, many high-performance methods rely on intrusive hardware, such as electroencephalography (EEG) or electrooculography (EOG) sensors, to capture underlying bio-signals [3,12]. While effective, this limits their use in everyday, non-clinical settings. Second, results are often reported on small and/or private datasets. For example, one study achieved 99.8% accuracy but on a closed dataset of only 10 subjects [3]. Such evaluations on limited data risk learning subject-specific idiosyncrasies rather than robust biometric signatures that generalise to a wider population. The private nature of these datasets also precludes direct, reproducible comparisons, making it difficult to establish a true state-of-the-art benchmark. Our work addresses these gaps by using a non-invasive approach on larger, public datasets.

Some studies have treated characteristics of blinks like time-series signals. For example, gradient intensity around the eye region has been used as a time-series feature with Euclidean distance calculations in [21]. Acceleration, mechanical impulse, and eyelid position have been utilised in multi-subject classification [7]. These approaches discard the spatial information that eye shapes carry during the blinking process, resulting in the loss of valuable discriminative information.

Recent work has also shown that involuntary subtle and micro-expressions like small blinks, mouth movements and eyebrow twitches can improve identification accuracy [15]. They implemented an ensemble-based SlowFast CNN, but the study does not clarify whether the process for isolating the frames containing the micro-expression was automated or manually annotated, introducing subjectivity into the analysis.

2.1 Spatio-Temporal Templates in Behavioural Biometrics

Blinking is essentially a phenomenon that has a temporal aspect, so there is a potential to wrap the time-variant behaviour in a single map. Previous temporal templates from studies of gait recognition serve as a source of inspiration. For example, earlier studies like Gait Energy Image [10] and Chrono-Gait Image [26] proposed to represent the entire sequence of motion in a single image: every silhouette extracted from a walking cycle is colour-coded (e.g., rainbow gradient) based on its time index and then superimposed into a single image. Despite the effectiveness of the chrono-gait image in gait-based identification, early implementations primarily relied on classical techniques such as PCA for dimensionality reduction, followed by classification using LDA and Euclidean distance. While these methods were suitable for their time, contemporary deep learning approaches now provide more robust and scalable solutions for classification tasks.

Although these ideas were initially developed for gait analysis, they can be transferred to blinking, and one can imagine a "Chrono-Blink Image" that superimposes outer shape contours of eyes over time. Applied to pupil tracking [24] or video summarisation [5], such approaches suggest blinking - being temporal - can be similarly encoded.

2.2 Deep Learning for Eye-Based Behavioural Biometrics

The use of deep learning to model blinking behaviours has increased recently, and researchers now commonly employ CNNs, RNNs, or transformers. Although most studies focused on blink detection, rather than identifying people, the underlying architectures remain relevant. CNNs are often used as the foundation to extract spatial features from frames, followed by an LSTM for modelling the blink sequence. Recent studies apply CNN-LSTM hybrids [4] or Vision Transformers [11] for blink detection. While most focus on detection, not identification, they validate the use of deep models for blink behaviour modelling.

In summary, existing blink-based biometric studies often use intrusive sensors, small and private datasets, or reduced 1D features. Spatio-temporal visual templates remain underexplored in non-invasive blink identification. Our work addresses all those gaps.

3 Methodology

This section provides a detailed description of the steps involved in creating time images and using them in a deep learning model for biometric identification.

3.1 Blink Segmentation Using Adaptive EAR

To begin, split the input videos of subjects' faces into subsequences of eye blinks. This is inspired by research in gait analysis, where videos are split into sequences of two steps each [10]. To detect an individual blink instance, we compute Eye Aspect Ratio (EAR) [20] for every frame in an input video stream by using the following formula:

$$\text{EAR} = \frac{\|p_2 - p_6\| + \|p_3 - p_5\|}{2 \cdot \|p_1 - p_4\|} \tag{1}$$

where $p_1, p_2, \ldots, p_6$ are 6 specific facial landmarks around each of the two eyes, identified by MediaPipe's Face Mesh model [14]. Figure 2 illustrates the facial landmark layout from the FaceMesh model.

A sequence of continuous frames is identified as a blink when the EAR drops below a subject-specific threshold, indicating eye closure.

To ensure applicability across subjects with different eye features like shape and size, we implement an adaptive technique that calibrates the aforementioned threshold to each subject's unique eyes. During a 60-s calibration phase (approximately 1800 frames, given that the majority of the videos in the dataset are 30 FPS), we compute a subject-specific adaptive EAR threshold:

$$\mu_{\text{EAR}} = \frac{1}{N}\sum_{i=1}^{N} e_i \quad ; \quad \tau_{\text{adaptive}} = \alpha \cdot \mu_{\text{EAR}}, \quad \text{where } \alpha = 0.7 \tag{2}$$

where e_i is the EAR at the ith frame. The scaling factor $\alpha = 0.7$ was determined empirically through an experimental validation process aimed at balancing the sensitivity and specificity of blink detection across subjects. A lower threshold was found to increase the rate of missed blinks (false negatives), particularly for subjects with naturally narrower eye apertures. Conversely, a higher threshold increased the incidence of false positives, where subtle eye movements or partial eyelid closures were incorrectly classified as full blinks. The 0.7 value thus represents an optimal trade-off, enabling the system to robustly capture genuine blinks across the diverse eye morphologies present in the datasets while minimising detection errors.

3.2 Temporal Blink Sampling

To retain blink dynamics, we sample 15 consecutive blink sequences per time image. Inter-blink frames are excluded. To ensure the temporal continuity while avoiding redundancy, we employ a sliding window with a stride of two blinks between two time images.

For example, the first time image is constructed using blinks 1–15, the second time image is constructed using blinks 3–17, the third time image is constructed using blinks 5–19, and so on. This stride ensures the temporal continuity and increases the diversity of blink mechanics in the dataset.

3.3 Time Image Construction

a. Eye-Opening Keypoints Detection and Normalisation. For each frame in the time image array, 16 keypoints for each eye contour are located using MediaPipe's FaceMesh model [14].

To correct for head pose, landmarks are rotated, scaled, and aligned using two anchor points as shown in Eq. (3).

$$\tilde{\mathbf{x}}_i = \mathcal{S} \cdot \mathcal{R}(\mathbf{x}_i - \mathbf{x}_{\text{anchor}}) + \mathbf{t} \tag{3}$$

where $\mathbf{x}_i$ is the original i^{th} eye keypoint, $\tilde{\mathbf{x}}_i$ is the normalised point after transformation, $\mathbf{x}_{\text{anchor}}$ is the chosen anchor point, $\mathcal{R}$ is a 2×2 rotation matrix that aligns the eye horizontally based on the eye orientation, $\mathcal{S}$ is a scaling factor based on inter-eye distance to normalise scale across subjects, and **t** is a translation vector that centres the aligned keypoints within a standard frame.

Figure 3 demonstrates a face in three distinct head positions. The second row presents the visual depiction of the corresponding eye contours that are normalised to align across all positions, which essentially prevents head position variation from influencing the representation.

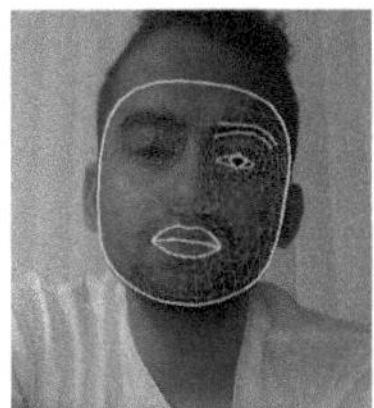

Fig. 2. Keypoints of FaceMesh Model.

b. Colour Outline Time Image (COTI). To represent blink progression over time, we map a colour to each frame in the array of contours using OpenCV's rainbow colourmap:

$$\mathbf{C}_t = \text{colourmap}\left(\frac{t}{T}\right) \tag{4}$$

The final time image is created by superimposing all the coloured contours:

$$\text{COTI}(x, y) = \sum_{t=1}^{T} \delta\left(\mathcal{C}_t(x, y), \mathbf{C}_t\right) \tag{5}$$

Here, t is the frame index, T is the total number of frames in the entire array, $\mathbf{C}_t$ is the colour assigned to the frame, and δ is a function that draws the contour $\mathcal{C}_t$ at pixel location (x, y) using that colour. This produces a colour-coded "Colour Outline Time Image" (COTI) where the hue conveys temporal progression.

c. Mean Intensity Time Image (MITI). MITI is a grayscale image of average pixel intensities across blink frames, as shown in Eq. (6).

$$\text{MITI}(x, y) = \frac{1}{T} \sum_{t=1}^{T} I_t(x, y) \tag{6}$$

where $I_t(x, y)$ is the intensity at pixel (x, y) in frame t. The result is a smooth intensity map which captures the blink motion.

Left and right eye contours are horizontally concatenated in all time images. Figure 4 shows how contours are handled and Fig. 5 depicts visual comparison of COTI and MITI.

3.4 Dataset Preprocessing and Labelling

We evaluate our method on two public datasets, NTHU-DDD and UTA-RLDD. We apply adaptive EAR thresholding to each video to extract blink frames, build time images from 15-blink sequences, and label them per subject. Train/test splits ensure that there is no temporal overlap between the video segments of each person.

3.5 CNN-Based Classification Architecture

Person identification from time-stamped images was tackled using MobileNetV2, a lightweight CNN pre-trained on ImageNet [19]. Its general-purpose feature extraction makes it well-suited for both our datasets. Our architecture uses MobileNetV2 (input: $224 \times 224 \times 3$) followed by a Global Average Pooling layer, a 256-unit ReLU Dense layer, 0.5 Dropout, and a softmax classifier (see Fig. 6). We trained the model in two phases: initially freezing the base layers with a learning rate of 5×10^{-4}, then fine-tuning from layer 100 onward with a smaller learning rate of 1×10^{-5}. Early stopping (patience = 100) and a learning rate scheduler (factor = 0.5, patience = 80) ensured stable convergence. This transfer

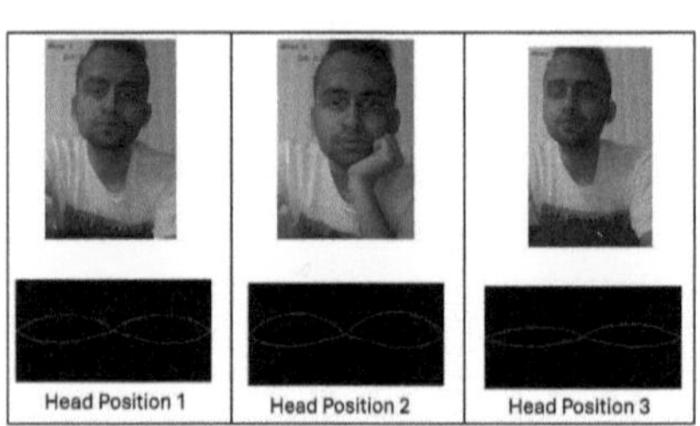

Fig. 3. Demonstration of position invariance in eye contours across different head positions.

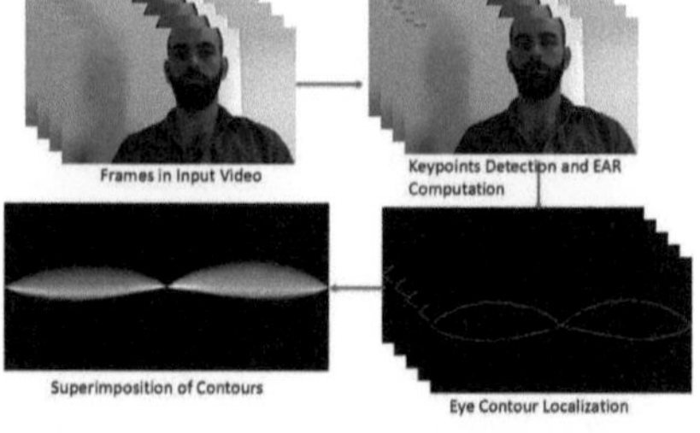

Fig. 4. Methodology for eye-blink-based biometric identification. Eye contours are processed separately for each eye before concatenation.

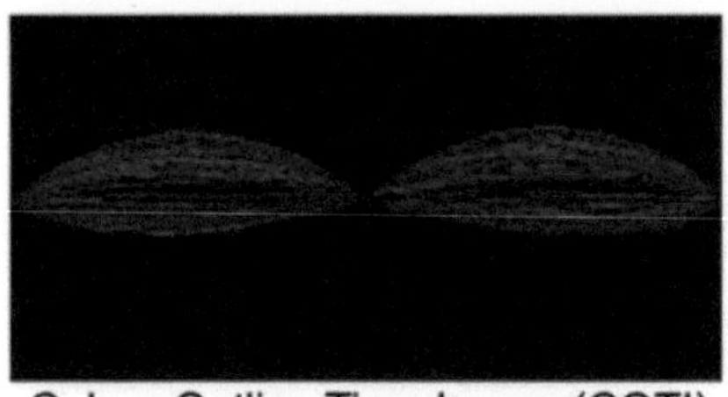

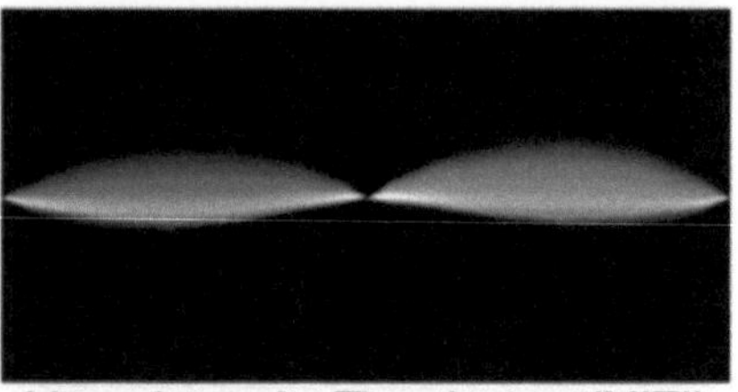

Fig. 5. Visual comparison of the two time image construction methods. (a) Colour Outline Time Image (COTI). (b) Mean Intensity Time Image (MITI)

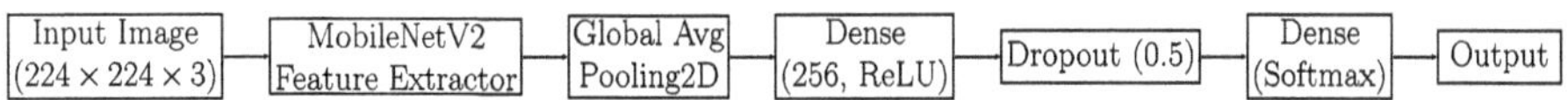

Fig. 6. CNN architecture: MobileNetV2 followed by global average pooling, a 256-unit dense layer, dropout, and softmax output.

learning approach enabled effective classification across both datasets, balancing compact architecture with strong generalisation.

4 Evaluation and Results

4.1 Evaluation

We evaluated our proposed identification framework on two publicly available datasets: NTHU-DDD (approximately 2500 images, 22 persons) [27] and UTA-RLDD (approximately 6000 images, 60 persons) [8], both widely used in computer vision research for analysing eye-related behaviours.

Our evaluation follows the standard closed-set identification protocol, a common practice in biometrics research. In this protocol, all subjects in the testing set (the *probes*) are also present in the training set (the *gallery*) [6]. The system's task is to match a probe sample from a known individual to their corresponding identity in the gallery.

To ensure a rigorous evaluation and prevent data leakage, we employ a strict temporal data split, where the video segment for each subject is partitioned chronologically. The earlier portion of the video (first 70% blinks) is used to generate training samples, and the latter unseen portion (remaining 30% blinks) is used for testing. This methodology is more challenging than a random split and better simulates a real-world scenario. It substantially increases the number of effective samples, enabling a robust evaluation even on datasets with modest subject numbers. Both datasets provide rich eye-blinking sequences critical for assessing the discriminative power of our eye-blink-based identification approach.

Our pipeline comprises four key stages: Keypoints extraction, normalisation, template encoding, and classification. Keypoint extraction uses MediaPipe's FaceMesh to detect landmarks surrounding each eye, computing the EAR as the ratio of vertical to horizontal distances, and capturing blink dynamics. Normalisation involves adaptive and non-adaptive EAR thresholding to segment blinks. Adaptive EAR thresholding calibrates a subject-specific EAR threshold over a 60-s period, accounting for individual variations in eye shape and blink behaviour, while non-adaptive EAR thresholding uses a fixed value, suitable for uniform conditions. Template encoding generates either the Colour Outline Time Image (COTI), an RGB image encoding 15 blinks' contours with a rainbow colourmap, or the Mean Intensity Time Image (MITI), a grayscale image averaging pixel intensities across blinks. Classification employs either MobileNetV2, a deep architecture pre-trained on ImageNet, or a custom 3-layer CNN designed for sparse features, with architectures tailored to the templates' characteristics.

To assess identity recognition, we computed multiple performance metrics. Namely, we used: Accuracy, Precision, Recall, and F1-Score to evaluate the system's ability to correctly identify individuals. Accuracy measures the proportion of correct predictions, precision quantifies true positive identifications relative to all positive predictions, recall assesses the fraction of true positives identified, and F1-Score balances precision and recall for a comprehensive performance indicator. Additionally, we computed Cumulative Match Characteristic (CMC) scores across different ranks to evaluate ranking performance in identification tasks. CMC scores measure the probability that the correct identity appears within the top-k ranked predictions, with Rank-1 indicating the correct identity is the top match. These metrics collectively provide a thorough assessment of our framework's effectiveness in diverse and controlled scenarios.

4.2 Results

This section presents the evaluation results of our eye-blink-based biometric identification system tested on the two datasets, UTA-RLDD and NTHU-DDD. We show our results using accuracy as shown in Fig. 8, precision, recall, and F1-score as shown in Table 1, and losses as shown in Fig. 9. We also outline Cumulative Match Characteristic (CMC) scores in Fig. 7 to evaluate the ranking performance of our system.

Our development process began with MobileNetV2, a deep architecture pretrained on ImageNet, paired with COTI templates, which encode eye contours with a rainbow colourmap to capture temporal blink progression. We hypothesised that contours would effectively represent blink motion. However, the results were suboptimal, particularly on UTA-RLDD, where MobileNetV2 with non-adaptive EAR thresholding and COTI achieved a test accuracy of 0.7867 (training accuracy 0.8105). We attribute this to two factors: (1) COTI's sparse feature set, which lacks the richness needed for robust classification, and (2) the overwriting of earlier contours by later ones due to the high number of contours superimposed in the template. This overwriting, observed during analysis, obscured critical temporal information, limiting discriminative power. On NTHU-DDD, the same configuration performed better, yielding a test accuracy

Table 1. Classification Performance on UTA-RLDD and NTHU-DDD

Dataset	EAR	Template	Model	Test Acc.	F1-Score	Avg. AUC	Rank-1	Rank-5
UTA-RLDD	Adaptive	MITI	MobileNetV2	0.9478	0.9475	0.9939	0.9478	0.9987
UTA-RLDD	Non-Adaptive	COTI	MobileNetV2	0.7867	0.7910	0.9658	0.7867	0.9748
UTA-RLDD	Adaptive	COTI	MobileNetV2	0.8095	0.8145	0.9755	0.8095	0.9763
UTA-RLDD	Adaptive	COTI	3-layer CNN	0.9701	0.9711	0.9745	0.9934	0.9934
NTHU-DDD	Adaptive	MITI	MobileNetV2	0.9904	0.9904	0.9538	0.9904	1.0000
NTHU-DDD	Non-Adaptive	COTI	MobileNetV2	0.9675	0.9701	0.9974	0.9675	1.0000
NTHU-DDD	Adaptive	COTI	MobileNetV2	0.9450	0.9467	0.9936	0.9450	0.9972
NTHU-DDD	Adaptive	COTI	3-layer CNN	0.9850	0.9842	0.9689	0.9850	0.9850

of 0.9675 (training accuracy 0.9644), likely due to the dataset's homogeneous population, which reduced variability in eye shapes and lighting.

Recognising COTI's limitations, we developed the MITI template, a grayscale alternative, preserving richer intensity-based features without overwriting. This shift significantly improved performance. MobileNetV2 with adaptive EAR thresholding and MITI achieved a test accuracy of 0.9478 on UTA-RLDD (training accuracy 0.9036) and an impressive 0.9904 on NTHU-DDD (training accuracy 0.9780). The intensity-based richness of MITI proved particularly effective for UTA-RLDD's diverse conditions, making it suitable for real-world applications where robust feature extraction is critical. The success of MITI led us to hypothesise that COTI's sparse features might be better suited to a shallower architecture. We designed a 3-layer CNN with 32, 64, and 128 filters (3×3 kernels, L2 regularisation), LeakyReLU activation, batch normalisation, 2×2 max-pooling, a 128-unit dense layer with ReLU, 25% dropout, and a softmax output. With adaptive EAR thresholding and COTI, this model achieved a test accuracy of 0.9701 on UTA-RLDD (training accuracy 0.8125) and 0.9850 on NTHU-DDD (training accuracy 0.9865). These results outperformed MobileNetV2 with COTI, confirming that a shallower architecture better matches COTI's sparse features, avoiding overfitting inherent in deeper models. An earlier run on UTA-RLDD with the 3-layer CNN and COTI yielded a lower test accuracy of 0.8095 (training accuracy 0.8125), likely due to suboptimal hyperparameters or data splits, highlighting the importance of careful tuning. To validate our reasoning, we tested the 3-layer CNN with adaptive EAR thresholding and MITI. As expected, performance was poor. Training accuracies were not recorded due to space constraints, but the results align with our hypothesis: MITI's feature-rich nature requires MobileNetV2's depth to extract complex intensity patterns, while the shallow CNN struggles with these demands. This experiment reinforced the importance of matching architecture to template characteristics.

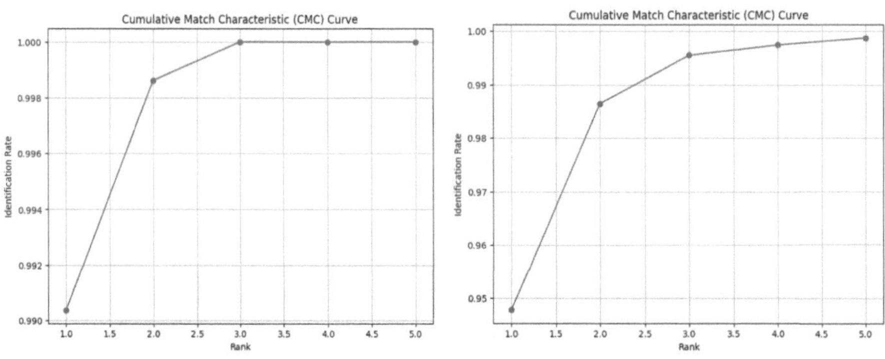

(a) CMC Curve for adaptive EAR thresholding with MITI on NTHU-DDD

(b) CMC Curve for adaptive EAR thresholding with MITI on UTA-DDD

Fig. 7. CMC analysis of adaptive EAR thresholding with MITI.

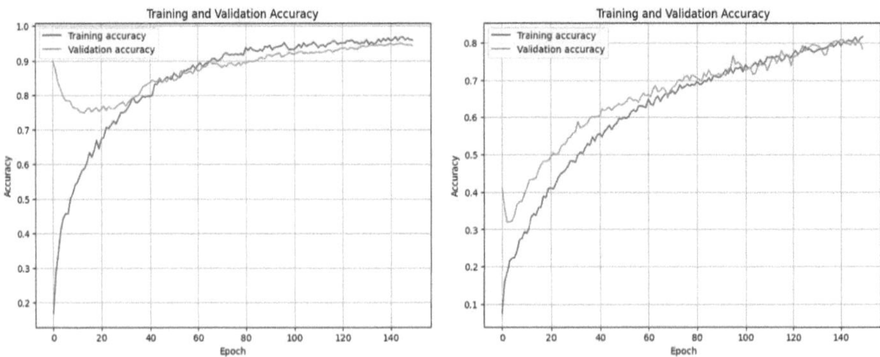

(a) Train/Test Accuracy for Adaptive EAR thresholding with COTI on NTHU-DDD

(b) Train/Test Accuracy for Adaptive EAR thresholding with COTI on UTA-RLDD

Fig. 8. Accuracy of Adaptive EAR thresholding with COTI across datasets.

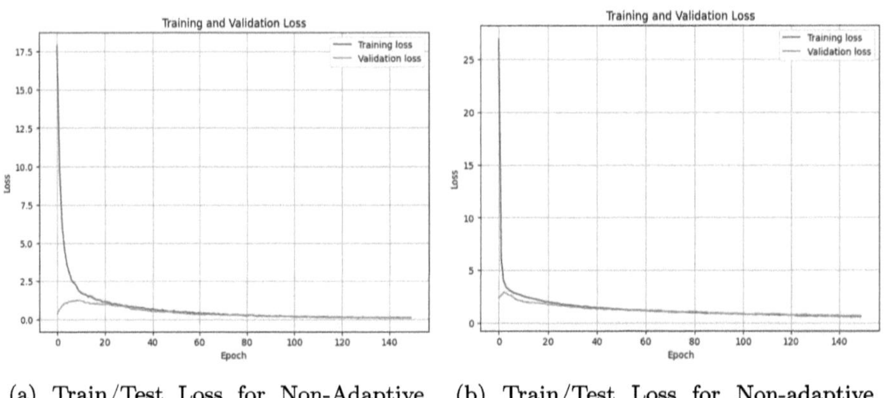

(a) Train/Test Loss for Non-Adaptive EAR thresholding on NTHU-DDD

(b) Train/Test Loss for Non-adaptive EAR thresholding on UTA-RLDD

Fig. 9. Loss analysis for Non-adaptive EAR thresholding with COTI across datasets.

Practically, MITI with MobileNetV2 is ideal for applications requiring high accuracy across varied conditions, such as security systems in uncontrolled environments. The 3-layer CNN with COTI offers a lightweight alternative for embedded devices or settings with consistent lighting and subject demographics, such as workplace access control. Adaptive EAR thresholding enhanced robustness on UTA-RLDD but was less critical for NTHU-DDD's uniformity, suggesting its use depends on deployment context.

5 Conclusion

In this paper, we have presented an innovative framework for biometric identification based on two-dimensional spatio-temporal representation of eye-blink patterns. The novel "time image" representation captures both temporal and spatial features of eye contours, and is later processed through a CNN. This method successfully uses RGB video data recorded by a standard camera, showcasing strong identification capabilities without relying on any extra modalities or specialised hardware. Our results demonstrate the potential of eye-blink patterns for non-obtrusive biometric identification in practical settings.

The proposed method obtained test accuracies of 99.04% and 94.78%, respectively, on the public NTHU-DDD and UTA-RLDD benchmarks. Adaptive EAR thresholding with MITI outperformed COTI, displaying resilience under various conditions and validating the approach for non-obtrusive biometric identification. While the proposed approach demonstrates robust performance using non-invasive dynamic blink patterns, future research could explore combining such dynamic modalities with traditional static biometric cues such as the iris, sclera, and periocular regions. These static features have consistently shown high accuracy, while dynamic traits like blinking offer greater resilience against spoofing. A multimodal fusion of these signals could enhance both accuracy and robustness.

Furthermore, future work could also be done on developing larger datasets that encompass a broader range of subjects. This would assist in evaluating the approach's generalisation across multiple varying factors.

References

1. Abbas, S.N., Abo-Zahhad, M.: Eye blinking EOG signals as biometrics. In: Jiang, R., Al-maadeed, S., Bouridane, A., Crookes, D., Beghdadi, A. (eds.) Biometric Security and Privacy. SPST, pp. 121–140. Springer, Cham (2017). https://doi.org/10.1007/978-3-319-47301-7_5
2. Abo-Zahhad, M., Ahmed, S.M., Abbas, S.N.: A new multi-level approach to eeg based human authentication using eye blinking. Pattern Recogn. Lett. **82**, 216–225 (2016)
3. Abo-Zahhad, M., Ahmed, S.M., Abbas, S.N.: A new eeg acquisition protocol for biometric identification using eye blinking signals. Int. J. Intell. Syst. Appl. **7**(6), 48 (2015)
4. Bennett, R., Joshi, S.H.: A CNN and LSTM network for eye-blink classification from mri scanner monitoring videos. In: 2021 43rd Annual International Conference of the IEEE Engineering in Medicine & Biology Society (EMBC), pp. 3463–3466 (2021)
5. Bilen, H., Fernando, B., Gavves, E., Vedaldi, A.: Action recognition with dynamic image networks (2017)
6. Drozdowski, P., Rathgeb, C., Busch, C.: The watchlist imbalance effect in biometric face identification: comparing theoretical estimates and empiric measurements. In: Proceedings of the IEEE/CVF International Conference on Computer Vision Workshops (ICCVW), pp. 3754–3762. RPRMI: Recognition and Privacy Risks in the Real World, IEEE/CVF (2021). https://doi.org/10.1109/ICCVW54120.2021.00466

7. Espinosa, J., Domenech, B., Vázquez, C., Pérez, J., Mas, D.: Blinking characterization from high speed video records: application to biometric authentication. PLOS ONE **13**(5), 1–19 (2018). https://doi.org/10.1371/journal.pone.0196125
8. Ghoddoosian, R., Galib, M., Athitsos, V.: A realistic dataset and baseline temporal model for early drowsiness detection. In: Proceedings of the IEEE Conference on Computer Vision and Pattern Recognition Workshops (2019)
9. Gupta, E., Agarwal, M., Sivakumar, R.: Blink to get in: biometric authentication for mobile devices using EEG signals. In: ICC 2020-2020 IEEE International Conference on Communications (ICC), pp. 1–6. IEEE (2020)
10. Han, J., Bhanu, B.: Individual recognition using gait energy image. IEEE Trans. Pattern Anal. Mach. Intell. **28**(2), 316–322 (2005)
11. Hong, J., Shin, J., Choi, J., Ko, M.: Robust eye blink detection using dual embedding video vision transformer. In: Proceedings of the IEEE/CVF Winter Conference on Applications of Computer Vision (WACV), pp. 6374–6384 (2024)
12. Iijima, R., Takehisa, T., Ohki, T., Mori, T.: The catcher in the eye: recognizing users by their blinks. In: Proceedings of the 19th ACM Asia Conference on Computer and Communications Security, ASIA CCS '24, pp. 1739–1752. Association for Computing Machinery, New York (2024)
13. Jain, A.K., Ross, A., Prabhakar, S.: An introduction to biometric recognition. IEEE Trans. Circuits Syst. Video Technol. **14**(1), 4–20 (2004)
14. Kartynnik, Y., Ablavatski, A., Grishchenko, I., Grundmann, M.: Real-time facial surface geometry from monocular video on mobile gpus. arXiv preprint arXiv:1907.06724 (2019)
15. Kay, T., Ringel, Y., Cohen, K., Azulay, M.A., Mendlovic, D.: Person recognition using facial micro-expressions with deep learning (2023)
16. Lenz, G., Ieng, S.H., Benosman, R.: Event-based face detection and tracking using the dynamics of eye blinks. Front. Neurosci. **14**, 587 (2020)
17. Mahesh, T.R., Ram, M.S., Ram, N.S.S., Gowtham, A., Swamy, T.V.N.: Real-time eye blinking for password authentication. In: García Márquez, F.P. (ed.) International Conference on Intelligent Emerging Methods of Artificial Intelligence & Cloud Computing, pp. 428–434. Springer, Cham (2022). https://doi.org/10.1007/978-3-030-92905-3_52
18. Pan, G., Sun, L., Wu, Z., Lao, S.: Eyeblink-based anti-spoofing in face recognition from a generic webcamera. In: 2007 IEEE 11th International Conference on Computer Vision, pp. 1–8 (2007)
19. Sandler, M., Howard, A., Zhu, M., Zhmoginov, A., Chen, L.C.: Mobilenetv2: inverted residuals and linear bottlenecks. In: 2018 IEEE/CVF Conference on Computer Vision and Pattern Recognition, pp. 4510–4520 (2018)
20. Soukupova, T., Cech, J.: Eye blink detection using facial landmarks. In: 21st Computer Vision Winter Workshop, Rimske Toplice, Slovenia, vol. 2, p. 4 (2016)
21. Takano, K., Takano, H.: Eye-blink based personal authentication using time-series directional features and waveform features. In: 2019 Asia-Pacific Signal and Information Processing Association Annual Summit and Conference (APSIPA ASC), pp. 231–235 (2019)
22. Tolosana, R., Vera-Rodriguez, R., Fierrez, J., Morales, A., Ortega-Garcia, J.: Deepfakes and beyond: a survey of face manipulation and fake detection (2020)
23. Umashankar, G., et al.: Eye blink based biometric authentication system. In: 2023 International Conference on Bio Signals, Images, and Instrumentation (ICBSII), pp. 1–4 (2023)

24. Villalobos-Castaldi, F.M., Suaste-Gómez, E.: A video-based spatio-temporal biometric template representation of the spontaneous pupillary oscillations: a pilot experiment. In: Carrasco-Ochoa, J.A., Martínez-Trinidad, J.F., Rodríguez, J.S., di Baja, G.S. (eds.) Pattern Recognition, pp. 165–174. Springer, Heidelberg (2013). https://doi.org/10.1007/978-3-642-38989-4_17
25. Wang, C., Zhang, J., Pu, J., Yuan, X., Wang, L.: Chrono-gait image: a novel temporal template for gait recognition. In: Daniilidis, K., Maragos, P., Paragios, N. (eds.) ECCV 2010. LNCS, vol. 6311, pp. 257–270. Springer, Heidelberg (2010). https://doi.org/10.1007/978-3-642-15549-9_19
26. Wang, C., Zhang, J., Wang, L., Pu, J., Yuan, X.: Human identification using temporal information preserving gait template. IEEE Trans. Pattern Anal. Mach. Intell. **34**(11), 2164–2176 (2012)
27. Weng, C.-H., Lai, Y.-H., Lai, S.-H.: Driver drowsiness detection via a hierarchical temporal deep belief network. In: Chen, C.-S., Lu, J., Ma, K.-K. (eds.) ACCV 2016. LNCS, vol. 10118, pp. 117–133. Springer, Cham (2017). https://doi.org/10.1007/978-3-319-54526-4_9
28. Willett, S.M., Maenner, S.K., Mayo, J.P.: The perceptual consequences and neurophysiology of eye blinks. Front. Syst. Neurosci. **17** (2023)
29. Wu, Q., Zeng, Y., Zhang, C., Tong, L., Yan, B.: An EEG-based person authentication system with open-set capability combining eye blinking signals. Sensors **18**(2), 335 (2018)

Augmentation and Transformation for Nighttime Cloud Segmentation in All-Sky Camera Images

Jirui Yang[1(✉)], Charles Gretton[2], and Tony Travouillon[3]

[1] School of Informatics, University of Edinburgh, Edinburgh, UK
s2767383@ed.ac.uk

[2] College of Systems and Society, Australian National University, Canberra, Australia

[3] Research School of Astronomy and Astrophysics, Australian National University, Canberra, Australia

Abstract. For ground-based optical telescopes, automated planning and control systems play a vital role in ensuring efficient acquisition of scientific data. These systems rely on All-Sky Cameras (ASC) to capture real-time sky conditions, including sky obscuration and lunar position. We address the task of assessing sky conditions under ambient nighttime illumination and in the presence of optical artifacts such as lens flares. Seeking to develop a solution based on a neural network model, we are faced with a lack of real nighttime cloud segmentation benchmark datasets. To address this, we propose an **A**ugmentation and **T**ransformation for **N**ighttime **C**loud **S**egmentation in All-Sky Camera Images (AT-NCS) that converts existing daytime and dusk cloud segmentation datasets into a synthetic benchmark simulating realistic nighttime observing conditions. Our approach incorporates key challenges such as night sky glow, and optical artifacts (e.g., lens flare), thereby providing a training and evaluation framework for ASC-based observation systems. This work aims to enhance the reliability and value of autonomous observatory operations under diverse nighttime environments.

1 Introduction

Modern automated observatories require accurate information about the sky conditions for the automated operation of the telescope. All-Sky Camera (ASC) systems are widely employed for sky monitoring and have been deployed across multiple observational facilities [17]. The ASC utilizes a fisheye lens to capture images with a 180-degree field of view. These images enable the extraction of key information, including cloud coverage and lunar position. In recent years, with the development of neural models in computer vision, techniques have been used to extract cloud information from sky images. Multiple cloud detection techniques have been developed for ground-based imagery, including methods proposed by [11,12] and [14].

While these techniques provide valuable insights, automated observatory control systems demand more precise cloud coverage detection. This requires generating a binary mask to classify clouds and clear sky, a task best addressed using semantic

M. Liu et al. (Eds.): AI 2025, LNAI 16371, pp. 214–228, 2026.
https://doi.org/10.1007/978-981-95-4972-6_17

segmentation techniques. Several models have been developed specifically for cloud semantic segmentation, including CloudSegNet by [5], a UNet-based approach by [16], and the Attention and Clustering-based Cloud Segmentation Network (ACLNet) by [13]. These approaches exhibit compelling classification performance in daytime images, where illumination conditions are favourable. Their performance at nighttime remains unknown. Under daylight, clouds are clearly visible and distinct, whereas nighttime observations introduce challenges such as low-light ambiguity. As highlighted in [15], automated cloud detection systems frequently encounter difficulties in differentiating between actual cloud cover and various interference sources, particularly artificial lighting and sky glow. This challenge becomes especially pronounced during nighttime observations. Further, under low-light conditions cloud identification proves significantly more difficult, with distant clouds often failing to register clearly in captured images.

In addition to innate visibility challenges, the most significant obstacle in developing a nighttime cloud classification model is the lack of a benchmark dataset for training and validating performance under real-world nighttime lighting conditions. To address the problem of real-world nighttime cloud segmentation, we sought to leverage existing state-of-the-art daytime cloud segmentation models, adapt them and evaluate the resulting adaptations in nighttime environments. Our main contributions are summarized as follows:

1. **Pipeline Construction and Evaluation.** We developed a novel image augmentation and transformation pipeline to accurately simulate nighttime lighting conditions, ensuring robust model performance across illumination scenarios. The pipeline was evaluated through an ablation study, with model performance assessed across various scenarios.
2. **Pipeline-enhanced Model Validation.** Using cloud masks from Geostationary Orbit (GEO) satellite data as ground truth, we validated our new pipeline-enhanced model with All-Sky Camera data from Mount Stromlo Observatory, Canberra, Australia. Human validation provided direct performance comparison between model outputs and expert assessments.

The remainder of this paper is organized as follows. In Sect. 2, we introduce the research background of cloud segmentation, along with data augmentation and transformation techniques. Section 3 presents the datasets used in our pipeline and details two transformation and two augmentation techniques, along with the specific challenges each technique addresses. Section 4 investigates the effects of our transformation and augmentation pipeline using the state-of-the-art ACLNet model. We evaluate the pipeline through an ablation study and assess the model's performance. Finally, we validate the model's real-world performance using satellite-derived cloud masks and human validation.

2 Related Work

2.1 Deep Learning for Cloud Segmentation

Supervised cloud segmentation relies on labelled datasets. Two key benchmarks are SWIMSEG [4] and SWINSEG [6], covering daytime and dusk conditions (detailed in

Sect. 3.1). CloudSegNet [5] provides a baseline using an encoder-decoder architecture. The UNet-based approach in [16] improves performance by fusing multi-scale features during upsampling. The state-of-the-art ACLNet [13] further improves segmentation accuracy via clustering and attention mechanisms [19].

In astronomy, machine learning has been investigated for cloud detection. The work in [14] achieved 95% accuracy using ResNet and LightGBM, demonstrating the efficiency of feature-based approaches. In [11], the authors applied a multi-color kNN method, achieving 78–95% accuracy, highlighting the role of environmental factors like solar zenith angle. In [1], the authors improved total cloud cover predictions using Multilayer Perceptron (MLP), Random Forest (RF), and gradient boosting machines (GBM) by incorporating additional meteorological data. The authors of [12] leveraged SVR and ANN for continuous cloud cover estimation, outperforming contemporaneous methods. These advances underscore the growing role of machine learning in optimizing astronomical observations through robust cloud segmentation. While existing work has focused on daytime cloud segmentation under ideal lighting conditions, our work considers segmenting nighttime cloud in challenging real-world lighting scenarios by leveraging an existing model enhanced with our proposed transformation and augmentation pipeline.

2.2 Data Transformation and Augmentation for Deep Learning

Collecting labeled datasets for supervised deep learning is often costly, especially for rare or challenging scenarios. To address this, data transformation and augmentation are widely adopted.

Early style transfer methods like Neural Style Transfer [8] applied artistic textures without semantic understanding. Later, pix2pix [9] used conditional Generative Adversarial Networks (cGANs) for structured mappings but required paired data. CycleGAN [21] improved practicality by enabling unpaired translation. Beyond style transfer, fisheye distortion simulation [2] has proven useful for tasks like semantic segmentation on fisheye images. Building on these concepts, our work incorporates style transformation and fisheye distortion to simulate nighttime illumination conditions in the ASC-based observation systems.

While computationally costly methods like CycleGAN and Neural Style Transfer generate realistic data, they demand extensive training. Lightweight parametric augmentations often suffice, as cutout [7] drops random regions to improve occlusion robustness, and motion blur [20] simulates camera movement. Following this approach, our work employs efficient Gaussian blur and lens flare masks to simulate challenging conditions while maintaining low computational cost.

3 Methodology

This section introduces the benchmark dataset used as pipeline input and describes each component of our transformation and augmentation pipeline, along with the specific challenge each component addresses.

3.1 Dataset

The labelled training dataset was compiled from two primary sources: the Singapore Whole Sky IMaging SEGmentation Database (SWIMSEG) [4] and the Singapore Whole Sky Nighttime Image SEGmentation Database (SWINSEG) [6]. These two datasets are as follows:

- **SWIMSEG** comprises $1,013$ daytime sky images acquired under optimal illumination conditions, with each image having a resolution of 600×600 pixels.
- **SWINSEG** includes 115 nighttime images, each at 600×600 pixel resolution. However, the illumination characteristics in these images deviate substantially from authentic nighttime conditions. Rather than true darkness, the scenes exhibit dusk-like qualities characterized by elevated brightness levels and a distinct yellowish hue. This lighting enhances sky visibility and renders cloud structures unusually apparent compared to typical nocturnal observations.

The clouds in these two datasets are under ideal lighting conditions, differing from the real nighttime sky, which has complex illumination and where clouds are usually blurry.

3.2 Data Augmentation and Transformation Pipeline

To enhance the training dataset's realism for ground-based night sky imaging applications, we implemented three key augmentation techniques: *(i)* day-to-night style transfer, *(ii)* night sky glow and lens flare augmentation, *(iii)* and fisheye transformation. The data augmentation and transformation follow the pipeline depicted in Fig. 1. First, a dataset is generated from current existing labeled datasets by using day-to-night transformation. We thereby have a synthetic nightime image with a known ground truth cloudmap. Then, the transformed dataset is augmented with techniques enumerated above as *(ii)* and *(iii)*. The final training dataset combines four distinct components: the original unprocessed dataset, transformed base dataset, separately augmented datasets for each individual technique, and a composite dataset incorporating all augmentation methods simultaneously. Below, we describe each of these techniques in detail, and the full pipeline is visualized in Appendix A.

Day-to-Night Style Transfer. Day-to-night style transfer was performed using the model from [10], based on CycleGAN [21]. Unlike paired translation methods, CycleGAN learns from unpaired images, making it ideal for our task.

Kadam's approach [10] uses urban lighting patterns to handle complex day-to-night transformation, particularly the presence of artificial lights. This method is well-suited for ground-based night sky imagery, where light pollution degrades quality. Built on CycleGAN, the model does not require paired training images. Instead, it learns to convert daytime images into realistic nighttime versions using an unpaired dataset of day and night ground-based sky images from Mount Stromlo Observatory (MSO) as reference. By training on these unpaired MSO images, the CycleGAN effectively transforms daytime sky scenes into their nighttime counterparts.

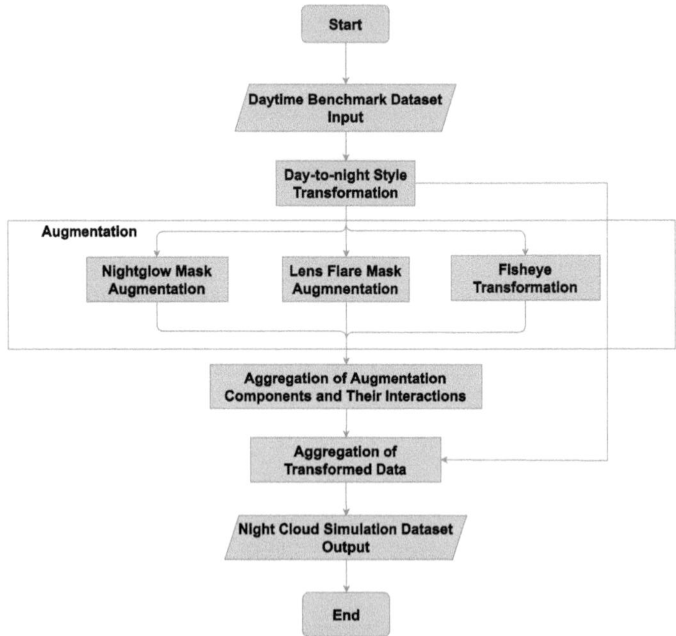

Fig. 1. Flowchart of the data augmentation and transformation pipeline.

All transformed images were sourced from SWIMSEG. While SWINSEG includes nocturnal features such as artificial yellow lighting and white clouds, SWIMSEG offers high-quality cloud samples. To enhance nighttime cloud detection, the two datasets were merged for training. However, this approach still necessitates that the model differentiate between genuine cloud formations and light pollution artifacts.

Night Glow and Lens Flare Augmentation. During manual inspection of MSO site nighttime imagery, we observed two key sources of model misclassification: *(1) nightglow* (artificial urban light reflecting off the atmosphere) and *(2) lens flare* artifacts from lunar illumination. To address these, we implemented two targeted augmentations to both SWINSEG dusk images and SWIMSEG day-to-night transformed images. These augmentations were performed by scripting using GIMP [18]:

- **Night Glow** refers to artificial light pollution reflecting off the atmosphere and clouds. In long-exposure nighttime astronomy images, this effect becomes pronounced, creating irregular brightness patterns, especially near urban areas. The nightglow mask simulates this illumination, helping the model differentiate true clouds from artificial glow artifacts.
- **Lens Flare** refers to localized bright artifacts caused by internal reflections when imaging strong light sources like the moon. While professional astronomy equipment typically minimizes these effects, some flare artifacts remain in the MSO

site's camera images. Our lens flare mask augmentation helps the model distinguish between true cloud formations and flare-induced artifacts.

Fisheye Transformation. The final transformation applied to the labelled dataset is the fisheye transformation. As the ground-based camera is designed to capture a 360-degree view of the sky, the resulting images exhibit fisheye distortion, particularly near the edge which does not exist in the training dataset. According to the findings in [2], which evaluated model performance on fisheye images with fisheye transformation, applying fisheye transformation to augment the data significantly improves the machine learning model's performance in semantic segmentation tasks on fisheye images. Inspired by these results and considering the issue of edge distortion in fisheye images, this transformation was incorporated. The source code and implementation for this work are publicly available on GitHub[1].

4 Experiment Result

In this section, we compare three state-of-the-art models trained on the same simulated night dataset described in Sect. 3.2 using precision and recall metrics. Next, we conduct an ablation study to evaluate the pipeline's effectiveness and assess the pipeline-enhanced models' performance across different challenges. Finally, we validate the pipeline-enhanced model's real-world performance using cloud masks from satellite imagery and human validation.

4.1 Model Comparison

All three models, CloudSegNet, U-Net and ACLNet exhibit strong performance in cloud segmentation under daylight conditions. However, their ability to handle complex lighting conditions will be analyzed. Their performance will be evaluated based on precision and recall performance on the testing dataset in Table 1 to show their relevant performance on handling clouds from other interference.

Table 1. Precision & recall of three models on augmented cloud data (quantiles rounded to 3 decimals).

Model	Metric	Min	Q1	Med.	Q3	Max
CloudSegNet	Precision	0	0.287	0.520	0.754	1.0
	Recall	0.460	0.776	0.951	0.986	1.0
Unet	Precision	0.324	0.713	0.875	0.972	1.0
	Recall	0.678	0.869	0.966	0.996	1.0
ACLNet	Precision	0.658	0.853	0.944	0.983	1.0
	Recall	0.789	0.914	0.976	0.997	1.0

[1] Source Code Available at https://github.com/henry-48/AT-NCS.git.

CloudSegNet shows the highest variability, with precision ranging from 0 to 1 (median 0.5) and a large substantial interquartile range (IQR) of approximately 0.5, indicating inconsistent accuracy. Its median recall matches UNet and ACLNet, but its wider recall IQR implies lower reliability.

UNet has moderately high precision (slightly lower than ACLNet) but a broader IQR and data range, showing greater variability. Its median recall is comparable to ACLNet, yet larger fluctuations reveal less consistency.

ACLNet performs best, with precision and recall medians near 1. Its narrow IQR and limited data range demonstrate exceptional consistency and robust, balanced performance across diverse conditions.

4.2 Pipeline Evaluation

While ACLNet demonstrates superior robustness and cloud classification capabilities compared to previous state-of-the-art models, the absence of a real-world benchmark dataset and the reliance on a synthetic data pipeline (with transformations and augmentations) warrant further investigation. Specifically, it is essential to evaluate:

- **Pipeline Impact**: How each step of the pipeline influences the model's precision and recall.
- **Robustness to Interference**: The model's performance under individual interference conditions (e.g., noise, distortions) and their interactions.

In this section, we first assess the contribution of each pipeline stage to model performance. Subsequently, we evaluate the robustness of the final model, which trained on simulated night-cloud data, against each interference type independently.

Test 1: Augmentation and Transformation Pipeline Ablation Study. This section evaluates each component of the augmentation and transformation pipeline. To assess the pipeline's impact, we train ACLNet, the top performing model identified in Sect. 4.1, on distinct datasets generated at different stages of the pipeline. Specifically, we incrementally train ACLNet using datasets produced by applying each transformation and augmentation step cumulatively.

For the augmentation phase (Fig. 1), the model undergoes sequential training from left to right, beginning with data augmented using only the night glow mask, followed by data augmented with both night glow and lens flare masks independently. This staged approach isolates the individual impact of each augmentation on model performance. A similar incremental evaluation is applied to the fisheye transformation. Finally, all augmentation and transformation techniques are applied simultaneously to create a maximally complex training scenario, challenging the model to adapt to compounded variations.

To assess the knowledge acquired from each augmentation and transformation stage, models trained on different datasets are evaluated on a common testing set. Specifically, the final dataset incorporating all augmentation and transformation techniques (as detailed in Table 2). Additionally, performance is tested on a dataset where all

Table 2. Ablation study: component contributions to performance

Pipeline Component	Precision (%)	Recall (%)
Baseline (Daytime)	69.46	91.87
+ Night Transformation	88.28 (+18.82)	78.02 (-13.85)
+ Sky Glow	92.00 (+3.72)	58.40 (-19.62)
+ Lens Flare	92.80 (+0.80)	90.66 (+32.26)
+ Fisheye	93.61 (+0.81)	62.40 (-28.26)
+ Complex Situation	91.22 (-2.39)	71.58 (+9.18)

augmentations and transformations are applied concurrently (i.e. Complex situation), ensuring a comprehensive evaluation of robustness.

The precision and recall results from models trained on different stages of the Augmentation and Transformation Pipeline show that the day-to-night transformation, night glow mask, and lens flare mask each enhance ACLNet's precision, though with a minor reduction in recall, before the fisheye transformation is applied. Notably, adding the lens flare mask allows the model to maintain recall levels similar to those achieved with the original dataset while increasing precision by 20 %.

The observed recall drop from fisheye transformation is attributed to the mixed test set. Analysis of its performance on fisheye images follows. Finally, the model trained using the complete output dataset from the augmentation and transformation pipeline which are the four dataset mentioned in Table 2 along with the complex situation[2] dataset shows a 9 % improvement in recall at the cost of a slight decrease in precision.

Through this ablation study on the augmentation and transformation pipeline, we observe that integrating day-to-night conversion, night glow masks, and lens flare masks substantially improves model training and boosts overall performance. Conversely, the fisheye transformation adversely affects the recognition of non-fisheye images. However, the impact of fisheye transformation and complex environmental interference on model performance requires further analysis. In the next section, we will compare these effects with the best-performing model (trained with day-to-night transformation, night glow masks, and lens flare masks) as presented in Table 2.

Test 2: Model Performance for Each Interference. This section presents a comparative performance analysis between two models: (1) our pipeline-enhanced model trained on transformed and augmented data, and (2) the current best-performing benchmark model from Table 2. The evaluation focuses on five specific interference conditions: ***(i)*** basic night sky condition, ***(ii)*** night glow effects, ***(iii)*** lens flare artifacts, ***(iv)*** fisheye distortion, and ***(v)*** the complex situation. For testing, we employ both the indi-

[2] The phrase "complex situation" refers to images that have undergone all transformations and augmentations: night background, night glow, lens flare, and fisheye distortion. In the augmentation pipeline, this corresponds to applying Day-to-Night Transformation, Night Glow Mask, Lens Flare Mask, and Fisheye Transformation concurrently to the dataset.

vidual interference-specific datasets documented in Table 2 and an additional composite dataset representing the complex situation scenario.

Table 3. Precision (Prec.) and recall (Rec.) of pipeline-enhanced model and best-performing benchmark model on different interferences

Model	(i) Basic Night Sky		(ii) Night Glow		(iii) Lens Flare	
	Prec.	Rec.	Prec.	Rec.	Prec.	Rec.
Pipeline Model	0.9046	0.9429	0.9231	0.9109	0.8972	0.9351
Benchmark Model	0.9204	0.9605	0.9205	0.9545	0.9184	0.9546

Model	(iv) Fisheye Distortion		(v) Complex Situation	
	Prec.	Rec.	Prec.	Rec.
Pipeline Model	0.8457	0.9266	0.8752	0.8846
Benchmark Model	0.8867	0.8896	0.9405	0.7988

Table 3 compares the performance of the pipeline-enhanced model with the best-performing benchmark model from Table 2 (shown in the fourth row) on the interference dataset. The precision and recall results show that the pipeline-enhanced model performs similarly but slightly worse on the first three interference datasets. However, for fisheye distortion interference, it achieves higher recall while showing lower precision compared to the benchmark model. In more complex scenarios, the pipeline-enhanced model demonstrates a 9 % improvement in recall while maintaining 87 % precision.

These results show our pipeline improves the model's recall, particularly in challenging conditions, enhancing its ability to distinguish clouds from clear sky. Since the model guides an automated telescope system, low recall risks misclassifying clouded areas as clear, which potentially points telescopes toward obscured regions and compromises data integrity. We therefore deploy the pipeline-enhanced model for real-world validation in the next section.

4.3 Model Performance in Real-World

In this section, we evaluate the performance of the pipeline-enhanced ACLNet model on real-world data using two approaches: (1) cloud mask data from the Himawari-AHI satellite [3] and (2) human validation was performed on the predictions for the MSO site; visualization examples are provided in Appendix B. The model's strong performance on real nighttime data confirms its practical applicability, showing that its effectiveness on simulated data generated by our transformation and augmentation pipeline also holds true in real-world scenarios.

Cloud Mask Verification. Our application pertains to the segmentation of nighttime sky images. To the best of our knowledge, no publicly available datasets exist for supervised training of machine learning models specifically tailored to this problem. Therefore, to validate the performance of pipeline-enhanced ACLNet on ground-based night sky images, we have used the cloud mask (CMA) dataset from Himawari-AHI [3].

Validating ACLNet's performance presents a unique challenge since the CMA images lack precise pixel-level masks where each pixel represents a large area. To work around this limitation, we evaluated ACLNet's cloud classification performance by correlating its probability maps (derived from MSO site fisheye images) with ground-truth observations from July–December 2023. Specifically, we compared the model's average cloud probability (within the fisheye image's circular region) to the observed 12×12 pixel averages centered on the MSO site. This approach allows us to assess how well the model's predictions align with real-world conditions despite the coarse resolution of our ground truth data.

The overall correlation coefficient of 0.3657 indicates a moderate positive relationship between ACLNet's predictions and CMA's cloud coverage measurements, confirming the model's ability to identify and segment cloud features, albeit imperfectly. Notably, a comparative analysis reveals a disparity tied to lunar conditions: the correlation weakens significantly during moon-up scenarios (0.3249) compared to moon-down conditions (0.3721). This suggests that lunar illumination adversely impacts ACLNet's segmentation accuracy, likely due to unbalanced brightness and optical artifacts (Table 4).

Table 4. Pearson correlation for overall, moon up, and moon down scenarios

	Overall	Moon Up	Moon Down
Pearson Correlation	0.3657	0.3249	0.3721

Human Validation. In this section, an expert instrument scientist conducted a manual inspection of ACLNet's performance using images from the MSO site. ACLNet's predictions are displayed alongside binary masks, with thresholds applied for clearer visualization. A total of 4,033 images were inspected, offering valuable insights into ACLNet's performance from a human evaluation perspective.

The manual inspection focuses on two key aspects. First, the observer evaluates whether they agree, partially agree, or disagree with ACLNet's segmentation results. Second, the observer assesses the current sky condition by selecting one of four categories: fully cloudy, mostly cloudy, partly cloudy, or clear sky. These classifications allow for a performance analysis of ACLNet under varying weather conditions.

Table 5. Human validation agreement across different weather scenarios

Weather	Agree	Disagree	Partially Agree
Clear Sky	0.898	0.069	0.033
Few Cloud	0.794	0.040	0.166
Most Cloudy	0.542	0.243	0.215
Full Cloudy	0.590	0.240	0.170
Overall	0.753	0.130	0.118

From Table 5, the human validation results demonstrate strong agreement (89.8%) with the pipeline-enhanced model's predictions under clear sky conditions, confirming its effectiveness in ignoring night glow artifacts. Performance remains robust in few cloud scenarios (79.4% agreement), though the higher rate of partial agreement (16.6%) suggests occasional uncertainty, likely due to incomplete cloud segmentation when illumination varies (e.g., moonrise). In mostly cloudy (54.2% agreement) and fully cloudy (59.0% agreement) conditions, performance declines significantly, with disagreement rates nearing 24–24.3%. Overall, the model achieves 75.3% agreement across all weather conditions, but its reliability diminishes with increasing cloud cover.

Manual inspections reveal that while the model captures cloud textures in probability maps, it systematically underestimates darker clouds due to inconsistent sky illumination. This highlights a key limitation: the model's static threshold struggles to adapt to dynamic brightness changes. A dynamic threshold adjustment could address this by scaling detection sensitivity to local lighting, thereby improving robustness.

5 Conclusion

This paper presents a robust solution for nighttime cloud segmentation in automated observatory systems through three key innovations: (1) A novel data augmentation pipeline that transforms daytime imagery into photorealistic nighttime scenes with physically accurate sky glow, lens flare, and fisheye distortion effects; (2) Comprehensive validation showing ACLNet achieves state-of-the-art performance (median 94.4% precision/97.6% recall) when trained with our synthetic data; and (3) Demonstrated operational viability through correlation with satellite cloud masks ($\rho = 0.3657$) and expert validation (75.3% agreement rate).

Our systematic evaluation reveals that the combination of day-to-night transformation with targeted augmentations yields a 9% recall improvement in complex observing conditions compared to the model without fisheye augmentation. That augmentation is particularly valuable for minimizing false negatives in telescope viewing task assignment. The pipeline maintains robust performance (87% precision) across mixed interference scenarios. Human validation confirms exceptional reliability in clear-sky (89.8% agreement) and partially cloudy (79.4%) conditions, while suggesting the need for adaptive illumination handling in overcast scenarios.

Future enhancements will address: (1) Real-time classification threshold adaptation for dynamic lighting, (2) Lunar phase-specific mitigation, (3) State-of-The-Art method on style transfer and (4) Generalization to other nocturnal atmospheric effects. The current implementation can reduce manual intervention requirements by providing a reliable, real-time cloud coverage assessment for autonomous telescope operations.

A Visualization of Augmentation and Transformation Pipeline

(see Figs. 2 and 3).

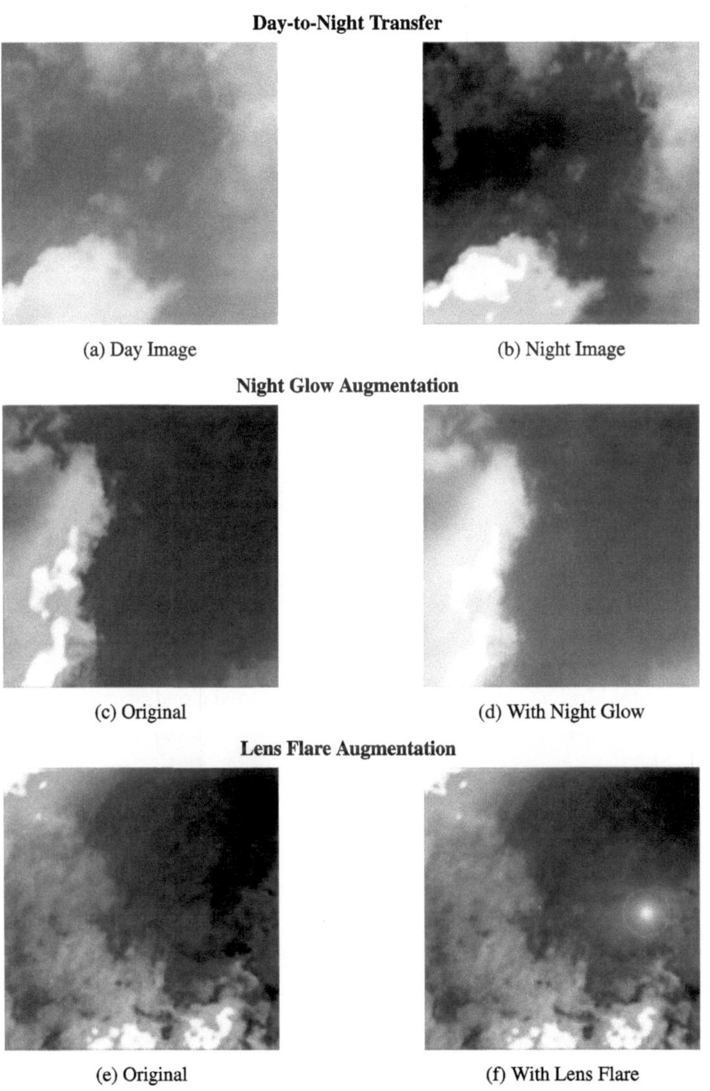

Fig. 2. Visualization of the Augmentation Pipeline.

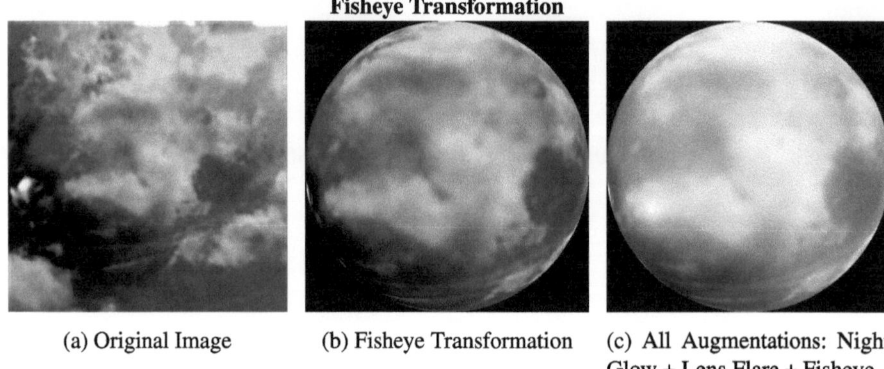

(a) Original Image (b) Fisheye Transformation (c) All Augmentations: Night Glow + Lens Flare + Fisheye

Fig. 3. Visualization of Fisheye Transformation and All Augmentations Applied.

B Visualization of Segmentation Result

(see Fig. 4).

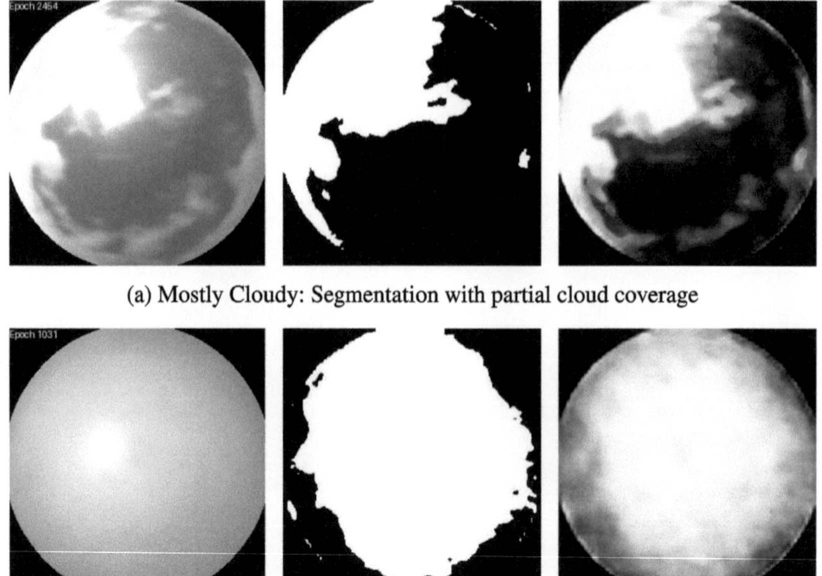

(a) Mostly Cloudy: Segmentation with partial cloud coverage

(b) Fully Cloudy: Segmentation with complete cloud coverage

Fig. 4. Visualization of segmentation performance. Each panel shows the three-component analysis: original input image (left), segmentation result (center), and probability map (right) under different cloud coverage conditions.

References

1. Baran, Á., Lerch, S., Ayari, M.E., Baran, S.: Machine learning for total cloud cover prediction. Neural Comput. Appl. **33**(7), 2605–2620 (2021). https://doi.org/10.1007/S00521-020-05139-4
2. Blott, G., Takami, M., Heipke, C.: Semantic segmentation of fisheye images. In: Leal-Taixé, L., Roth, S. (eds.) ECCV 2018. LNCS, vol. 11129, pp. 181–196. Springer, Cham (2019). https://doi.org/10.1007/978-3-030-11009-3_10
3. Bureau of Meteorology: Himawari 8/9 - cloud mask (CMA). Dataset (2022). https://doi.org/10.25914/az79-5z81. Accessed 13 Aug 2024
4. Dev, S., Lee, Y.H., Winkler, S.: Color-based segmentation of sky/cloud images from ground-based cameras. IEEE J. Sel. Top. Appl. Earth Obs. Remote. Sens. **10**(1), 231–242 (2017). https://doi.org/10.1109/JSTARS.2016.2558474
5. Dev, S., Nautiyal, A., Lee, Y.H., Winkler, S.: Cloudsegnet: a deep network for nychthemeron cloud image segmentation. IEEE Geosci. Remote Sens. Lett. **16**(12), 1814–1818 (2019). https://doi.org/10.1109/LGRS.2019.2912140
6. Dev, S., Savoy, F.M., Lee, Y.H., Winkler, S.: Nighttime sky/cloud image segmentation. In: 2017 IEEE International Conference on Image Processing, ICIP 2017, Beijing, China, 17–20 September 2017, pp. 345–349. IEEE (2017). https://doi.org/10.1109/ICIP.2017.8296300
7. Devries, T., Taylor, G.W.: Improved regularization of convolutional neural networks with cutout. CoRR arXiv:1708.04552 (2017)
8. Gatys, L.A., Ecker, A.S., Bethge, M.: A neural algorithm of artistic style. CoRR abs/1508.06576 (2015). http://arxiv.org/abs/1508.06576
9. Isola, P., Zhu, J., Zhou, T., Efros, A.A.: Image-to-image translation with conditional adversarial networks. In: 2017 IEEE Conference on Computer Vision and Pattern Recognition, CVPR 2017, Honolulu, HI, USA, 21–26 July 2017, pp. 5967–5976. IEEE Computer Society (2017). https://doi.org/10.1109/CVPR.2017.632
10. Kadam, V.: Day-night image translations using cyclegan (2022). https://www.kaggle.com/code/virajkadam/day-night-image-translations-using-cyclegan
11. Kazantzidis, A., Tzoumanikas, P., Bais, A.F., Fotopoulos, S., Economou, G.: Cloud detection and classification with the use of whole-sky ground-based images. Atmos. Res. **113**, 80–88 (2012). https://doi.org/10.1016/j.atmosres.2012.05.005
12. Kim, B.Y., Cha, J.W., Chang, K.H.: Twenty-four-hour cloud cover calculation using a ground-based imager with machine learning. Atmos. Measur. Tech. **14**(10), 6695–6710 (2021). https://doi.org/10.5194/amt-14-6695-2021
13. Makwana, D., et al.: Aclnet: an attention and clustering-based cloud segmentation network. CoRR abs/2207.06277 (2022). https://doi.org/10.48550/ARXIV.2207.06277
14. Mommert, M.: Cloud identification from all-sky camera data with machine learning. Astron. J. **159**(4), 178 (2020). https://doi.org/10.3847/1538-3881/ab744f
15. Price, I., Nielsen, J., Lidman, C., Soon, J., Travouillon, T., Sharp, R.: Converting the ANU 2.3 telescope to fully automated operation. Publications of the Astronomical Society of Australia **41**, e057 (2024). https://doi.org/10.1017/pasa.2024.75, article e057
16. Ronneberger, O., Fischer, P., Brox, T.: U-net: convolutional networks for biomedical image segmentation, vol. abs/1505.04597 (2015). http://arxiv.org/abs/1505.04597
17. Skidmore, W., et al.: Using all sky cameras to determine cloud statistics for the thirty meter telescope candidate sites. **7012**(1), 701224–701229 (2008). https://doi.org/10.1117/12.788141
18. The GIMP Development Team: Gnu image manipulation program (gimp) (2019). https://www.gimp.org

19. Vaswani, A., et al.: Attention is all you need. CoRR abs/1706.03762 (2017). http://arxiv.org/abs/1706.03762
20. Zhang, K., Zuo, W., Zhang, L.: Deep plug-and-play super-resolution for arbitrary blur kernels. arXiv preprint (2019). https://doi.org/10.48550/arXiv.1903.12529
21. Zhu, J.Y., Park, T., Isola, P., Efros, A.A.: Unpaired image-to-image translation using cycle-consistent adversarial networks. In: 2017 IEEE International Conference on Computer Vision (ICCV), pp. 2242–2251. IEEE (2017). https://doi.org/10.1109/ICCV.2017.244

Room Envelopes: A Synthetic Dataset for Indoor Layout Reconstruction from Images

Sam Bahrami(✉) and Dylan Campbell

The Australian National University, Canberra, Australia
{sam.bahrami,dylan.campbell}@anu.edu.au

Abstract. Modern scene reconstruction methods are able to accurately recover 3D surfaces that are visible in one or more images. However, this leads to incomplete reconstructions, missing all occluded surfaces. While much progress has been made on reconstructing entire objects given partial observations using generative models, the structural elements of a scene, like the walls, floors and ceilings, have received less attention. We argue that these scene elements should be relatively easy to predict, since they are typically planar, repetitive and simple, and so less costly approaches may be suitable. In this work, we present a synthetic dataset—Room Envelopes—that facilitates progress on this task by providing a set of RGB images and two associated pointmaps for each image: one capturing the visible surface and one capturing the first surface once fittings and fixtures are removed, that is, the structural layout. As we show, this enables direct supervision for feed-forward monocular geometry estimators that predict both the first visible surface and the first layout surface. This confers an understanding of the scene's extent, as well as the shape and location of its objects.

Keywords: 3D Scene Reconstruction · Indoor Scene Understanding · Layout Estimation · Synthetic Data

1 Introduction

Indoor scene reconstruction remains a challenging problem, particularly when attempting to understand the complete spatial structure of indoor environments from limited visual information. While recent advances in monocular depth estimation and 3D reconstruction have shown promising results, existing approaches often struggle with occluded regions and fail to provide comprehensive geometric understanding of indoor scenes outside of the first visible surface.

Current models used for indoor scene reconstruction typically use depth images [10,25] or layered depth representations [11] for training, but these formats have inherent limitations. Depth maps only capture the first visible surface, losing information about the underlying room structure that might be occluded.

Project website: https://sambahrami.github.io/room_envelopes.

M. Liu et al. (Eds.): AI 2025, LNAI 16371, pp. 229–241, 2026.
https://doi.org/10.1007/978-981-95-4972-6_18

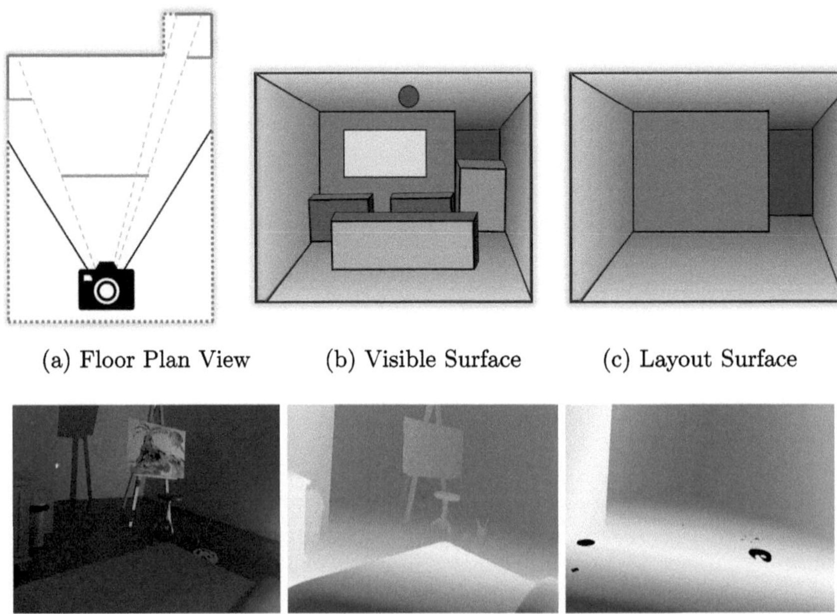

Fig. 1. Room Envelopes dataset overview. Our synthetic dataset provides dual pointmap representations for indoor scene reconstruction. (a) Overhead floor plan showing the view from a camera capturing layout areas in red, the first visible surface in green, and parts shared in both in orange. (b) The visible surface capturing all directly visible geometry including furniture and objects. (c) The layout surface showing structural elements (walls, floors, ceilings, windows, doors) as they would appear without occlusion. This dual representation enables direct supervision for layout reconstruction in occluded regions. (d–f) Example data from our dataset. (d) The original RGB image from Hypersim. (e) The visible surface depth capturing all visible surfaces including furniture and objects (f) The first layout surface depth showing only structural elements (walls, floors, ceiling, windows, doors). (Color figure online)

Layered depth approaches attempt to address occlusion by representing multiple depth layers per pixel, but surfaces that are spatially continuous in 3D space often become fragmented across different depth layers, creating artificial discontinuities and requiring methods that perform layer assignment as well as geometry estimation.

For a given viewpoint, we refer to the union of the first visible surface and the first structural layout surface as the *envelope* of the view, as illustrated in Fig. 1. The structural layout comprises structural elements like walls, floors, ceilings, windows and doors. This is essentially the room with fittings (removable items like furniture, appliances, curtains, carpets) and fixtures (attached items like sinks, baths, lights) removed, which largely coincides with the concept of layout in existing layout estimation works [7,14,15]. Understanding this structural layout is useful for many applications including robotic navigation, augmented reality, and architectural analysis. However, in typical indoor scenes, this layout is often partially occluded by furniture, making it difficult for models to learn robust representations of the underlying room geometry.

Feed-forward 3D reconstruction models face particular challenges at occlusion edges, where they tend to predict averaged depth values rather than making decisive geometric predictions [10,21,25]. This occurs because these models optimise a reconstruction loss by predicting the statistical mean in ambiguous regions, resulting in blurry or noisy outputs at object boundaries. Room layouts however are predominantly continuous and planar, making them well-suited for feed-forward prediction. By focusing on layout estimation, models can leverage the geometric regularity of architectural structures to produce sharp, accurate reconstructions even in partially occluded regions.

To address these limitations, we introduce Room Envelopes, a synthetic dataset that explicitly provides two complementary pointmaps for each RGB image: (1) one representing the first visible surface, capturing all directly visible surfaces including fittings and fixtures, and (2) one representing the first layout surface, capturing the scene's structural elements as they would appear from the same viewpoint if the fittings and fixtures were removed. This dual representation provides several advantages over existing dataset formats. First, the first layout surface provides structural clarity by offering unoccluded views of room boundaries, enabling models to learn robust representations of indoor spatial structure. Second, it facilitates the development of scene reconstructors that have less ambiguity compared to layered depth approaches, eliminating uncertainty about layer assignment. Finally, it allows models to be trained with direct supervision on room layout estimation, rather than relying on indirect multi-view photometric losses.

Our dataset is constructed by combining multiple viewpoints from high-quality synthetic indoor scenes, similar to how real-world datasets are collected through multi-view reconstruction approaches. This multi-view aggregation allows us to build a complete point cloud representation of indoor environments, from which we isolate structural elements and re-render layout surfaces unoccluded. We process these pointclouds into pointmap representations which maintain spatial correspondence with the original RGB images, enable pixel-aligned supervision, and preserve the natural image structure that existing vision neural network architectures are optimised for. This approach produces a dataset that is fundamentally equivalent to what could be achieved with the best possible real-world data collection tools for this task, with the advantage of perfect semantic labelling, perfect camera information, and controlled lighting conditions.

We demonstrate the utility of this dataset by training a model for room layout estimation that can predict the surface envelope of an indoor scene from a single RGB image. The main contributions of this work are

1. Room Envelopes, an indoor synthetic dataset providing an image, a visible surface pointmap, and a layout surface pointmap for each camera pose; and
2. a feed-forward scene reconstruction model demonstrating effective room layout estimation using this dataset.

Table 1. Comparison of dataset features across indoor scene datasets. We show which datasets provide camera poses, point clouds, depth maps (first depth and layout depth), mesh data, semantic labels, and surface normals. Room Envelope is the only dataset that provides both first layer and layout depth representations for comprehensive 3D reconstruction training.

Datasets	#Scenes	Type	Camera Poses	Point Cloud	First Depth	Layout Depth	Mesh	Semantic Labels	Normals
Matterport3D [2]	90	Real	✓	✗	✓	✗	✓	✓	✗
ScanNet++ [29]	1,006	Real	✓	✓	✓	✗	✓	✓	✗
ScanNet [3]	1,513	Real	✓	✗	✓	✗	✓	✓	✗
ARKitScenes [1]	1,661	Real	✓	✓	✓	✗	✓	✗	✗
RealEstate10K [34]	74K	Real	✓	✗	✗	✗	✗	✗	✗
MegaSynth [6]	700K	Synthetic	✓	✗	✓	✗	✗	✗	✗
Structured3D [33]	22K	Synthetic	✓	✗	✓	✗	✗	✓	✓
3D-Front [4]	19K	Synthetic	✗	✗	✗	✗	✓	✗	✗
HyperSim [18]	461	Synthetic	✓	✗	✓	✗	✓	✓	✓
Room Envelopes (Ours)	461	Synthetic	✓	✓	✓	✓	✗	✓	✓

2 Related Work

2.1 Indoor Scene Datasets

Indoor scene understanding relies on both real-world and synthetic datasets, each with distinct advantages and limitations. We provide a detailed comparison of Room Envelopes with existing indoor datasets in Table 1. Real-world datasets [2,3,29,34] face significant collection challenges that limit their scale and quality. Ground truth acquisition is difficult due to depth sensor limitations, including restricted resolution, errors with reflective surfaces, artifacts at object boundaries, and these datasets require extensive manual effort for data capture and post-processing.

Synthetic datasets provide perfect ground truth at scale with reduced manual effort. Recent procedural generation advances enable automatic scene layout generation [17], but often lack natural spatial arrangements that characterise human-designed spaces. Artist-designed synthetic environments offer a compelling middle ground [4,6,18,33], however visual realism remains a key consideration. HyperSim [18] uses V-Ray rendering to achieve photorealistic imagery per view. Room Envelopes builds on this foundation while introducing layout surface maps for each view.

2.2 Scene and Layout Reconstruction

A modern approach to scene reconstruction estimates pixel-aligned pointmaps from images, where each pixel in the output pointmap represents a 3D point in the scene [30]. The pioneering feed-forward pointmap reconstruction method DUSt3R [27] takes pairs of unposed views and estimates pointmaps in the same coordinate system, encoding their geometry and pose. This pairwise representation can be extended to estimate geometry for large scenes with many views, including as part of a SLAM system [16]. The follow-up work MASt3R [10] improved upon this approach by introducing local feature matching and modifying the training methodology to output metric-scale representations. MoGe [25] extended this paradigm to single-view pointmap estimation, with MoGe v2 [26] providing metric-scale output. LaRI [11] adapted the training concept from MoGe to estimate layered pointmap representations, attempting to predict the depth of occluded surfaces with a feed-forward model. VGGT [23] predicts comprehensive 3D scene attributes (such as pointmaps, camera poses, and depth maps) from single or multiple images of a scene without post-processing, achieving state-of-the-art performance in feed-forward 3D reconstruction.

The goal of complete 3D scene reconstruction is to generate plausible representations of entire environments, including both visible and occluded regions, from limited input views. Regression-based methods include 3D Gaussian Splatting [9] feed-forward approaches [20,21,31], as well as encoder–decoder approaches that create intermediate scene representations for novel view synthesis or complete geometric reconstruction [5,8,24]. These methods learn by comparing the rendered predicted representation with other views of the same scene in their ground truth dataset, indirectly supervising the occluded parts of the scene. These methods often produce impressive visual results near the input views, however, they struggle with view extrapolation, leading to poor predictions for areas that were occluded in the source views. In contrast, generative models are able to produce sharper and often more accurate representations of the unseen areas, but compromise speed and require significant training resources [12,19,22,28]. We suspect that generative approaches are less necessary for layout estimation than object reconstruction, since the layout is typically simple, repetitive and often planar. Both approaches, however, require data with layout supervision, which is addressed by the Room Envelopes dataset.

Layout estimation is the task of reconstructing the structure of an indoor scene from an RGB image [14,15]. Traditional approaches often rely on Manhattan world assumptions, where every wall is at right angles to every other wall, modeling scenes as the best fit 3D box defined by simple 3D planes [32]. In contrast, our Room Envelopes approach provides complete geometric details of the first layout surface for each pixel, rather than just 3D scene bounds. This enables a more comprehensive scene reconstruction learning signal that supports scene completion from a single input image, treating layout estimation as a holistic scene completion task rather than only identifying the 3D extents of a room.

3 The Room Envelopes Dataset

In this section, we outline the properties of the Room Envelopes dataset, the construction procedure, and the data format and statistics.

3.1 Hypersim Dataset

The dataset is built upon the Hypersim dataset [18], which provides images of high-quality synthetic indoor environments, containing 77,400 images across 461 indoor scenes with detailed per-pixel annotations. Hypersim includes images generated from fly-through trajectories of each scene, with multiple camera viewpoints per scene providing comprehensive coverage of each environment. Each frame includes precise depth maps, surface normals, ground truth pointmaps providing world coordinates for each pixel location, instance-level semantic segmentations, and detailed lighting information that captures view-dependent effects such as glossy surfaces and specular highlights. The scenes are rendered using V-Ray, a professional physically-based renderer that produces photorealistic imagery. One limitation of this dataset is that the underlying 3D mesh assets are not freely available, and so they cannot be used as part of our pipeline.

3.2 From Multi-view Images to Room Envelopes

Our processing pipeline transforms the multi-view Hypersim data into Room Envelope representations through point cloud aggregation, semantic filtering, and view-specific re-rendering. We extract pointmaps, surface normals, camera poses, and semantic segmentations from each frame in the Hypersim dataset, then aggregate these into point clouds.

Given pointmaps $\mathbf{P}_i \in \mathbb{R}^{H \times W \times 3}$, RGB images $\mathbf{C}_i \in \mathbb{R}^{H \times W \times 3}$, surface normals $\mathbf{N}_i \in \mathbb{R}^{H \times W \times 3}$, and semantic labels $\mathbf{S}_i \in \mathbb{Z}^{H \times W}$ for each frame i, we directly aggregate all world coordinate points with their associated attributes from all frames in the scene

$$\mathcal{P} = \bigcup_{i=1}^{N} \{(\mathbf{p}_{u,v}^{(i)}, \mathbf{c}_{u,v}^{(i)}, \mathbf{n}_{u,v}^{(i)}, s_{u,v}^{(i)}) : \mathbf{p}_{u,v}^{(i)} \in \mathbf{P}_i\}, \tag{1}$$

where $\mathbf{p}_{u,v}^{(i)} \in \mathbb{R}^3$ represents the world coordinate point, $\mathbf{c}_{u,v}^{(i)} \in \mathbb{R}^3$ the RGB colour, $\mathbf{n}_{u,v}^{(i)} \in \mathbb{R}^3$ the surface normal, and $s_{u,v}^{(i)} \in \mathbb{Z}$ the semantic label at pixel (u, v) in frame i, with only valid points included. We then downsample the pointcloud to remove redundant points using a voxel downsampling method with voxel size ρ, while preserving the associated attributes for each point.

Next, we filter the aggregated point cloud $\mathcal{P}$ to retain only layout-relevant points: walls, floors, doors, windows, and ceilings, such that

$$\mathcal{P}_{\text{layout}} = \{\mathbf{p} \in \mathcal{P} : s(\mathbf{p}) \in \mathcal{S}_{\text{layout}}\}, \tag{2}$$

where $s(\mathbf{p})$ denotes the semantic class of point $\mathbf{p}$ and $\mathcal{S}_{\text{layout}}$ is the set of layout-relevant semantic classes. We align the filtered points to each original camera viewpoint by transforming them through a series of coordinate conversions to account for Hypersim's non-standard camera model

$$\tilde{\mathbf{p}}_{\text{cam}} = \mathbf{T}_{\text{rend}} \mathbf{T}_{\text{w2c}} \tilde{\mathbf{p}}_{\text{w}}, \tag{3}$$

where $\tilde{\mathbf{p}}$ represents points in homogeneous coordinates, $\mathbf{T}_{\mathrm{w2c}}$ is the world-to-camera transformation matrix, and $\mathbf{T}_{\mathrm{rend}}$ handles renderer-specific coordinate conversions and camera model corrections. We then rasterize the scene by projecting the 3D points into pixel image space. Since layout elements may project to the same pixel we isolate the closest layout surface using depth thresholding. For each pixel (u, v), we first identify all layout points that project to this pixel,

$$\mathcal{P}_{u,v} = \{\mathbf{p} \in \mathcal{P}_{\text{layout}} : \text{proj}(\mathbf{p}) = (u, v)\}, \tag{4}$$

where $\text{proj}(\mathbf{p})$ is the projection of 3D point $\mathbf{p}$ to image coordinates. We filter these candidate points by depth proximity to select points near the closest surface

$$\mathcal{P}^{\tau}_{u,v} = \{\mathbf{p} \in \mathcal{P}_{u,v} : |z_{\mathbf{p}} - z_{\min}| \leq \tau\}, \tag{5}$$

where $z_{\mathbf{p}}$ is the camera-space depth (z-coordinate) of point $\mathbf{p}$, $z_{\min} = \min_{\mathbf{p} \in \mathcal{P}_{u,v}} z_{\mathbf{p}}$, and τ is the depth threshold. From this candidate set of points we select the point most aligned with the camera ray (the 3D ray from the camera centre through pixel (u,v)), minimising the distance between the point and its projection onto the ray. This process is repeated for each pixel, resulting in a complete layout surface pointmap.

A visualisation of the resulting room envelope is shown in Fig. 1. It inherently contain gaps where layout elements are occluded across all viewpoints, for example, floor partially covered by a rug. A visualisation highlighting these holes is shown in Fig. 2. We observe that missing data of this type would also occur for real-world data collection, for which our processing pipeline could be used with minimal alterations.

3.3 Dataset Format and Statistics

The dataset comprises 77,400 images across 461 indoor scenes. Each data sample consists of an RGB image paired with two pointmaps: a visible surface pointmap capturing all directly observable geometry, and a layout surface pointmap representing structural elements as they would appear without occlusion. Additionally we retain metric scale point clouds of each scene, camera information, and surface normals per pixel. Our approach maintains flexibility for future applications by preserving the completed point clouds, enabling extraction of alternative layers or conversion to other 3D representations as needed.

4 Experiments

4.1 Experimental Setup

We evaluate baseline methods on our Room Envelopes dataset using a comprehensive experimental setup designed to assess layout reconstruction performance. Our dataset is split into 80/10/10 train/validation/test partitions by scene, resulting in 365, 46, and 46 scenes respectively. This scene-level split

(a) RGB Image with occluded layout

(b) Surface layout with missing data

Fig. 2. Missing data and occlusion patterns. (a) Original RGB image showing furniture occluding wall surfaces. (b) Corresponding layout depth image with holes (missing data) where no camera view captured the structural elements.

ensures that test performance reflects the model's ability to generalise to unseen indoor environments rather than memorising specific room configurations.

We fine-tune a pre-trained MoGe v1 model as our base architecture, leveraging its established capabilities in single-view pointmap estimation. Training is conducted on 4 NVIDIA L40S 48GB GPUs with a batch size of 32, fine-tuning for 75,000 training steps over approximately 72 h. We use the AdamW optimiser [13] with a learning rate of 1×10^{-6} for the backbone and 1×10^{-5} for the prediction head, and train with the global pointcloud loss, mask loss, and normal loss from MoGe [25].

4.2 Quantitative Evaluation

Our quantitative evaluation focuses on layout surface reconstruction accuracy across different visibility conditions. For comparison, we evaluate against two other methods. We compare against a pre-trained MoGe v1 model which estimates a pointmap from a single image [25]. Note that MoGe was trained on Hypersim among other datasets. We also compare against a pre-trained LaRI scene estimation model [11], which was trained on layered ray intersection images processed from the 3D-Front dataset [4]. We concatenate the 5 layers of depth predictions from LaRI's output to create a pointcloud, and evaluate the quality of this pointcloud against our Room Envelopes dataset.

For fair comparison, predicted pointclouds are aligned to ground truth pointclouds using scale-shift alignment with least squares optimisation, where a single scale factor is applied to all coordinates and a 3D z translation vector is added. All pointclouds are scaled to metres using the scaling parameters provided in the Hypersim dataset [18]. We evaluate reconstruction quality using Chamfer Distance (CD) and F-Scores (F) similar to previous works [11]. For LaRI evaluation, we take the best one-directional chamfer distance as the evaluation metric, since their method predicts many more points than in the ground truth.

Table 2. Comparison of layout estimation performance on layout pixels seen in the input image, unseen in the input image, and seen or unseen in the input image (overall). CD denotes the chamfer distance and F denotes the F-score.

Method	Seen			Unseen			Overall		
	CD ↓	F@0.1↑	F@0.05↑	CD↓	F@0.1↑	F@0.05↑	CD↓	F@0.1↑	F@0.05↑
MoGe [25]	**0.229**	**0.657**	0.507	1.037	0.114	0.059	**0.330**	0.565	0.431
LaRI [11]	0.948	0.412	0.286	1.522	0.400	0.283	0.925	0.410	0.285
Ours	0.362	0.647	**0.512**	**0.753**	**0.512**	**0.390**	0.401	**0.633**	**0.499**

Table 3. Surface Normal Likelihood Analysis. We construct a kernel density estimator using von Mises–Fisher kernels from normals in seen layout regions and evaluate the likelihood of normals in unseen regions under this distribution. We compare our method against uniformly distributed random normals (Baseline-uniform) and normals fixed to the most frequent direction in seen regions (Baseline-dominant). Higher average likelihood values indicate more geometrically consistent normal predictions in unseen areas.

Method	Average Likelihood ↑
Baseline-uniform	0.0795
Baseline-dominant	0.8788
Ours	0.4093

Our evaluation distinguishes between "seen", "unseen", and "overall" layout regions. Seen regions correspond to layout elements that are fully visible in the source image, while unseen regions represent areas occluded by furniture and other objects but reconstructed using information from other viewpoints. Overall regions include all layout elements, regardless of visibility. In our test set, unseen pixels comprise 23.57% of all layout pixels, representing a substantial reconstruction challenge. The quantitative results are summarised in Table 2, demonstrating our method's effectiveness particularly in reconstructing occluded layout geometry.

We evaluate the geometric consistency of our predicted layout surfaces by analysing how well normals in occluded regions conform to the distribution of normals from visible regions. Normal vectors are computed from the estimated point cloud using local surface fitting. We randomly sample 5,000 normal vectors from the seen layout regions to construct a kernel density estimator using von Mises–Fisher kernels, where each sampled normal serves as a kernel centre with equal weight, and concentration parameter $\kappa = 15$. We then evaluate the likelihood of 5,000 randomly sampled normals in unseen layout regions under this kernel density estimator. This process is repeated for each test image and the resulting likelihood values are averaged across the entire test split. We compare against two baselines: uniformly distributed random normals on the unit sphere, and normals fixed to the most frequent direction in the seen regions of each test

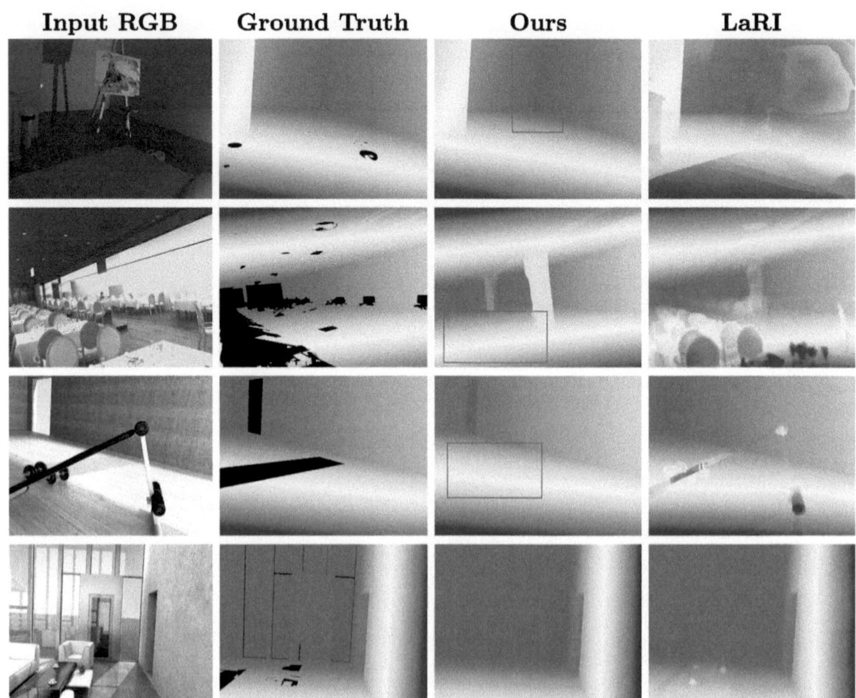

Fig. 3. Qualitative comparison of 3D first surface layout results as depth images. From left to right: input RGB image, ground truth layout geometry, our fine-tuned model trained on Room Envelopes, LaRI [11]. Our method shows superior reconstruction of occluded layout elements. Red boxes highlight regions where our method successfully reconstructs layout geometry that is completely occluded in the input image.

sample. Table 3 shows that our estimated normals in occluded regions are substantially more likely under the seen region distribution than uniformly random normals, though not as consistent as simply using the dominant normal direction from visible regions. This indicates that our method produces both geometrically plausible point distributions in occluded areas and maintains geometric variation rather than defaulting to a single dominant orientation.

4.3 Qualitative Results

Figure 3 shows qualitative comparisons of our layout reconstruction against LaRI [11] on several test set scenes. Our Room Envelopes approach demonstrates superior reconstruction of layout elements, particularly in occluded regions where furniture blocks the direct view of walls and structural elements. For LaRI, we visualise the best possible permutation of their 5 predicted depth images that best aligns to the ground truth. Additional in-the-wild qualitative results on real indoor images are provided in Appendix A.

5 Conclusion

We have introduced Room Envelopes, a synthetic dataset that addresses fundamental limitations in indoor scene reconstruction by providing dual pointmap representations: visible surfaces and layout surfaces. This dual representation enables direct supervision for layout reconstruction tasks, eliminating the ambiguity inherent in layered depth approaches and providing clear geometric targets for feed-forward models. This work opens several directions for future research, including extension to complete scene geometry reconstruction within camera frustums, integration with real-world data collection pipelines, and application to downstream tasks such as robotic navigation and augmented reality.

A In-the-Wild Qualitative Results

We present additional qualitative results on real-world indoor images captured from indoor environments in Fig. 4. These examples show how our layout estimation model performs on real indoor environments with different lighting conditions and realistic furniture arrangements. We convert the estimated pointmap into a depth image, and estimate surface normals through local surface fitting on the pointmap. The results illustrate the model's ability to reconstruct plausible

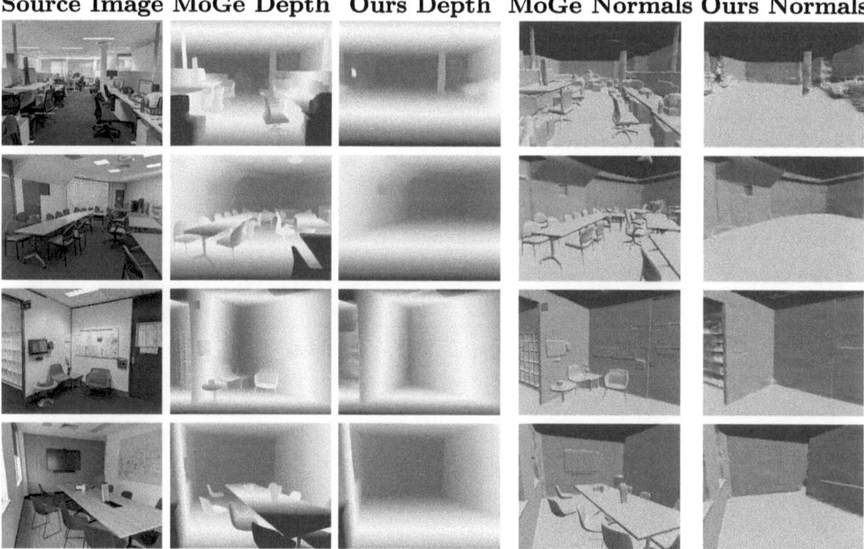

Fig. 4. Qualitative comparison on in-the-wild images captured by a phone camera comparing MoGe and our layout trained model on real indoor images. Normals are estimated using local surface fitting and colourised by mapping the x, y, z components to red, green, blue channels respectively. Please zoom in to see finer details. (Color figure online)

layout geometry even when applied to real images that differ from the synthetic training data, highlighting the value of our layout supervision approach for practical applications.

References

1. Baruch, G., et al.: ARKitScenes: a diverse real-world dataset for 3D indoor scene understanding using mobile RGB-D data. arXiv preprint arXiv:2111.08897 (2021)
2. Chang, A., et al.: Matterport3D: learning from RGB-D data in indoor environments. arXiv preprint arXiv:1709.06158 (2017)
3. Dai, A., Chang, A.X., Savva, M., Halber, M., Funkhouser, T., Nießner, M.: ScanNet: Richly-annotated 3D reconstructions of indoor scenes. In: CVPR (2017)
4. Fu, H., et al.: 3D-front: 3D furnished rooms with layouts and semantics. In: CVPR, pp. 10933–10942 (2021)
5. Jiang, H., et al.: RayZer: a self-supervised large view synthesis model. arXiv preprint arXiv:2505.00702 (2025)
6. Jiang, H., et al.: MegaSynth: scaling up 3D scene reconstruction with synthesized data. In: CVPR, pp. 16441–16452 (2025)
7. Jiang, Z., Liu, B., Schulter, S., Wang, Z., Chandraker, M.: Peek-a-boo: occlusion reasoning in indoor scenes with plane representations. In: CVPR, pp. 113–121 (2020)
8. Jin, H., et al.: LVSM: a large view synthesis model with minimal 3D inductive bias. In: ICLR (2025). https://openreview.net/forum?id=QQBPWtvtcn
9. Kerbl, B., Kopanas, G., Leimkühler, T., Drettakis, G.: 3D Gaussian splatting for real-time radiance field rendering. ACM TOG **42**(4), 139–1 (2023)
10. Leroy, V., Cabon, Y., Revaud, J.: Grounding image matching in 3D with MASt3R. In: ECCV (2024)
11. Li, R., Zhang, B., Li, Z., Tombari, F., Wonka, P.: LaRI: layered ray intersections for single-view 3D geometric reasoning. arXiv preprint arXiv:2504.18424 (2025)
12. Liu, R., Wu, R., Van Hoorick, B., Tokmakov, P., Zakharov, S., Vondrick, C.: Zero-1-to-3: zero-shot one image to 3D object. In: ICCV, pp. 9298–9309 (2023)
13. Loshchilov, I., Hutter, F.: Decoupled weight decay regularization. In: ICLR (2019)
14. Mathew, B.: Review on room layout estimation from a single image. Int. J. Eng. Res. Technol. **V9** (2020). https://doi.org/10.17577/IJERTV9IS060820
15. Meng, M., Zhu, Y., Zhao, Y., Li, Z., Zhu, Z.: 3D indoor scene geometry estimation from a single omnidirectional image: a comprehensive survey. Comput. Vis. Media **11**(3), 431–464 (2025). https://doi.org/10.26599/CVM.2025.9450438
16. Murai, R., Dexheimer, E., Davison, A.J.: MASt3R-SLAM: real-time dense SLAM with 3D reconstruction priors. In: CVPR, pp. 16695–16705 (2025)
17. Raistrick, A., et al.: Infinigen indoors: photorealistic indoor scenes using procedural generation. In: CVPR, pp. 21783–21794 (2024)
18. Roberts, M., et al.: Hypersim: a photorealistic synthetic dataset for holistic indoor scene understanding. In: ICCV, pp. 10912–10922 (2021)
19. Shi, Y., Wang, P., Ye, J., Long, M., Li, K., Yang, X.: MVDream: multi-view diffusion for 3D generation. arXiv preprint arXiv:2308.16512 (2023)
20. Smart, B., Zheng, C., Laina, I., Prisacariu, V.A.: Splatt3R: zero-shot Gaussian splatting from uncalibrated image pairs (2024). https://arxiv.org/abs/2408.13912
21. Szymanowicz, S., et al.: Flash3D: feed-forward generalisable 3D scene reconstruction from a single image. In: 3DV (2025)

22. Szymanowicz, S., et al.: Bolt3D: generating 3D scenes in seconds. arXiv:2503.14445 (2025)
23. Wang, J., Chen, M., Karaev, N., Vedaldi, A., Rupprecht, C., Novotny, D.: VGGT: visual geometry grounded transformer. In: CVPR (2025)
24. Wang, Q., Zhang, Y., Holynski, A., Efros, A.A., Kanazawa, A.: Continuous 3D perception model with persistent state. In: CVPR (2025)
25. Wang, R., et al.: MoGe: unlocking accurate monocular geometry estimation for open-domain images with optimal training supervision. In: CVPR (2025)
26. Wang, R., et al.: MoGe-2: accurate monocular geometry with metric scale and sharp details (2025). https://arxiv.org/abs/2507.02546
27. Wang, S., Leroy, V., Cabon, Y., Chidlovskii, B., Revaud, J.: DUSt3R: geometric 3D vision made easy. In: CVPR (2024)
28. Xiang, J., et al.: Structured 3D latents for scalable and versatile 3D generation. In: CVPR (2025)
29. Yeshwanth, C., Liu, Y.C., Nießner, M., Dai, A.: ScanNet++: a high-fidelity dataset of 3D indoor scenes. In: ICCV, pp. 12–22 (2023)
30. Zhang, J., et al.: Advances in feed-forward 3D reconstruction and view synthesis: a survey (2025). https://arxiv.org/abs/2507.14501
31. Zhang, K., et al.: GS-LRM: large reconstruction model for 3D Gaussian splatting. In: ECCV, pp. 1–19. Springer, Cham (2024)
32. Zhang, W., Qiao, Y., Liu, Y., Song, R., Zhang, W.: Fast 3D room layout estimation based on compact high-level representation. IEEE TIP (2025)
33. Zheng, J., Zhang, J., Li, J., Tang, R., Gao, S., Zhou, Z.: Structured3D: a large photo-realistic dataset for structured 3D modeling. In: ECCV (2020)
34. Zhou, T., Tucker, R., Flynn, J., Fyffe, G., Snavely, N.: Stereo magnification. ACM TOG **37**(4), 1–12 (2018)

PLNet-12: A Vision-Language Benchmark for Zero-Shot Physical Literacy Analysis Across 12 Fundamental Movements

Tianchen Guo[1,2], Peter Anthony Logan[2], Thomas Wackwitz[1,2], and David Martin[1(✉)]

[1] The University of Queensland, Brisbane, Australia
{tianchen.guo,t.wackwitz,d.martin1}@uq.edu.au
[2] FOLLOW ME AI, Sydney, Australia
peterlogan@fome.ai

Abstract. Physical literacy, encompassing movement skills, coordination, and confidence, is a critical yet underexplored domain in both sports and computer vision. In this paper, we introduce **PLNet-12**, a novel video dataset consisting of 12 fundamental movement skills designed to support the evaluation of physical literacy through multi-task vision-language models (VLMs). Unlike existing datasets, PLNet-12 includes fine-grained annotations for temporal action detection, repetition counting, displacement and distance measurement, duration estimation, and physical literacy scoring. We focus on zero-shot evaluation using large-scale VLMs without task-specific fine-tuning to investigate their generalization ability in real-world settings. To ensure annotation quality, we employ a semi-automated, rule-based pipeline verified by motion capture data and domain experts. We benchmark several state-of-the-art VLMs under different temporal and spatial settings and reveal challenges including fine-grained temporal understanding, action confusion, and temporal hallucination. Our findings highlight critical limitations of current VLMs in structured physical tasks and suggest directions for future multi-modal reasoning research. PLNet-12 offers a lightweight yet challenging platform for advancing temporal video understanding in human movement analysis.

Keywords: Physical Literacy Analysis · Vision-Language Models · Zero-Shot Temporal Action Understanding

1 Introduction

Physical literacy (PL) [5,7] encompasses the ability, confidence, and motivation to engage in physical activity throughout life. It has become a cornerstone in physical education [8], sports science [1], and public health [24], reflecting a holistic integration of physical, cognitive, psychological, and social competencies. Despite increasing global recognition, definitions of PL remain inconsistent.

The original version of the chapter has been revised. The Chapter 19 Figure 2 has been corrected. A correction to this chapter can be found at
https://doi.org/10.1007/978-981-95-4972-6_40

M. Liu et al. (Eds.): AI 2025, LNAI 16371, pp. 242–254, 2026.
https://doi.org/10.1007/978-981-95-4972-6_19

However, a shared emphasis across frameworks is the development of fundamental movement skills (FMS) [2,3]–such as running, jumping, throwing, and balancing–which serve as the foundation for more complex physical activities [1].

Analyzing FMS through observational assessments, motion capture, or performance metrics provides valuable insights into an individual's physical development, helping identify strengths, limitations, and areas for intervention. A fundamental prerequisite in this analysis is the ability to accurately localize and classify distinct action segments. More broadly, human analysis has been studied across diverse domains beyond physical literacy. For example, affective behavior analysis [21,22,25,26] aims to automatically interpret emotions, expressions, and social signals from visual and audio cues. In parallel, research on human authenticity has grown rapidly, with deepfake voice detection [9] emerging as a key area for safeguarding trust in speech-based interaction. Other domains such as human activity recognition [4] and gait analysis [13] similarly highlight the wide spectrum of challenges in understanding human behavior. Positioning physical literacy within this broader landscape emphasizes both the uniqueness of our contribution and its connection to ongoing efforts in holistic human-centered analysis.

Zero-shot temporal action detection (ZS-TAD), also known as zero-shot temporal action localization (ZS-TAL), aims to localize and classify action segments in untrimmed videos without access to labeled examples of the target classes during training [19] [16]. Current ZS-TAD research faces two key challenges: limited datasets and the need for model adaptation. Many existing benchmarks, such as THUMOS14, lack clear taxonomies, fine-grained temporal boundaries, and semantic detail–e.g., broad classes like "billiards" span multiple sub-actions without precise definitions, making it difficult for vision-language models to align textual prompts with visual segments. While some newer datasets offer large-scale, hierarchically annotated videos, they are often restricted to narrow domains like elite sports [17] or egocentric cooking [6], making them unsuitable for studying general physical literacy in everyday settings. Moreover, even state-of-the-art methods often rely on complex pipelines and extensive tuning but still achieve low accuracy, limiting fair evaluation of new approaches. Methodologically, most VLMs are trained on static images or short clips and require domain adaptation to handle spatiotemporal content. While fine-tuning improves performance, it demands detailed annotations that are costly and impractical in real-world applications like physical literacy assessment.

To address these gaps, we introduce **PLNet-12**, a novel dataset comprising 12 fundamental movement skills specifically designed to evaluate physical literacy through video analysis. Unlike prior domain-specific datasets, PLNet-12 captures everyday actions performed by ordinary individuals, with each movement clearly defined using simple, interpretable instructions to minimize ambiguity and improve annotation quality. The dataset provides diverse annotation types to support multiple computer vision tasks, with a particular focus on zero-shot temporal action detection (ZS-TAD) using large-scale vision-language models (VLMs). By intentionally avoiding task-specific fine-tuning and complex optimization strategies, PLNet-12 enables rigorous evaluation of VLM generalization in realistic, real-world settings, and offers a strong foundation for advancing research in physical literacy assessment and general-purpose video understanding.

Our contributions are summarized as follows:

- We propose **PLNet-12**, the first dataset dedicated to assessing fundamental movement skills within the context of physical literacy.
- We provide high-quality, multi-task annotations using a semi-automated pipeline, validated through expert review and consistency checks.
- We benchmark multiple state-of-the-art vision-language models on the ZS-TAD task, offering a detailed analysis of their strengths, limitations, and generalization performance.

We hope PLNet-12 will serve as a valuable resource for advancing research in physical literacy analysis, zero-shot action detection, and robust video-language understanding. Due to the page limitation, the related work section can be found in Appendix A.

2 PLNet-12 Dataset

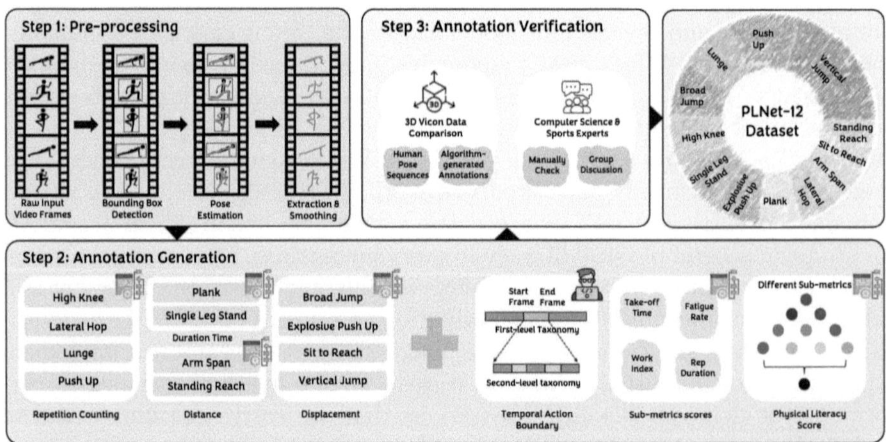

Fig. 1. Overview of the dataset construction pipeline. Pose sequences are first extracted from raw videos and processed using a white-box algorithm to generate initial annotations. Temporal action boundaries are then manually labeled. Finally, rigorous verification is conducted through comparison with 3D Vicon motion capture data and expert review by a panel of sports professionals, resulting in a high-quality, fully validated dataset.

Our objective is to construct a dataset designed for fundamental movement skills measurement with fine-grained action classes and dense annotations to support a range of computer vision understanding tasks Fig. 1. Compared to large-scale, complex datasets, our dataset enables more efficient and time-saving exploration of advanced techniques, simplifies failure case analysis, and allows for fairer comparisons of model performance and generalization. This avoids the problem of uninformative or misleading results due to extremely low scores on overly complex benchmarks. Although this paper primarily focuses on temporal action detection, our dataset is not limited to that task. It also supports a variety of hybrid computer vision tasks, including repetition counting and distance measurement. These features aim to foster a more comprehensive exploration of model capabilities and contribute to the development of human physical literacy research. In this section, we describe the dataset preparation process, including the action taxonomy, video collection, annotation methodology, dataset statistics, and key dataset properties.

2.1 Task Definitions

Our dataset supports a range of tasks, including temporal action detection (TAD), repetition counting, distance measurement, displacement estimation, time duration calculation, and human physical literacy scoring. In this paper, we primarily focus on benchmarking the first five tasks using several state-of-the-art vision-language models (VLMs).

Temporal Action Detection Temporal action detection can be formulated as follows: given an untrimmed video $\mathbf{X} \in \mathbb{R}^{3\times H\times W\times T}$, where H and W are the height and width of each frame, and T is the frame number. The temporal action boundary annotations can be denoted as $\Psi_g = \{\varphi_i = (t_s, t_e, c)\}_{i=1}^{N}$, where t_s, t_e, c are the start, end time and category of action instance φ_i, and N is the number of ground truth actions. ZS-TAD aims to predict a candidate proposal set to cover the annotations under for unseen data $\Psi_p = \{\hat{\varphi}_i = (\hat{t}_s, \hat{t}_e, \hat{c})\}_{i=1}^{M}$ to cover Ψ_g.

2.2 Dataset Annotation

Our dataset includes multiple layers of annotation to support a range of computer vision tasks, including temporal action detection (TAD), repetition counting (for exercises involving repeated movements), and a novel task we introduce: physical literacy scoring. The physical literacy score is a biomechanical age index, a quantitative metric that reflects an individual's functional physical capacity. It is derived from movement quality, biomechanical dynamics, and force generation, offering insight into biological age relative to chronological age. This metric is particularly valuable for detecting age-related changes in physical literacy and identifying early signs of physical decline.

Annotation for Temporal Action Detection. To annotate temporal boundaries, we first provide annotators with detailed definitions for each action. Annotators are trained with these guidelines and tasked with marking action start and end points. Spot checks and consistency reviews are conducted throughout to ensure annotation quality.

Table 1. Complete organizational taxonomy of our dataset, comprising 12 first-level categories and multiple second-level categories. #Videos indicates the number of video samples per category. #Second-level refers to the number of second-level categories. #Sub-metrics denotes the number of sub-metrics used to compute the human physical literacy score. Additional annotations include all labels beyond the temporal action boundaries.

First-level	#Videos	#Second-level	#Sub-metrics	Additional Annotations
Arm Span	41	3	-	Arm Span Distance
Broad Jump	58	3	-	Jump Displacement
Exp Push Up	52	2	6	Push Up Displacement
High Knee	56	3	9	Repetition Count
Lateral Hop	44	1	-	Repetition Count
Lunge	60	3	-	Repetition Count
Plank	49	1	3	Time Duration
Push Up	82	3	6	Repetition Count
Single Leg Stand	56	1	1	Time Duration
Sit to Reach	30	1	-	Reach Distance
Standing Reach	28	3	-	Reach Distance
Vertical Jump	145	4	5	Jump Displacement
Total	701	28	30	-

First-level annotations (i.e., full action spans) are manually labeled following this protocol. For second-level annotations (i.e., fine-grained action phases), we consult domain experts, including Ph.D. researchers in sports science and kinesiology, to define each phase with clear operational criteria. These phase-level annotations are generated using a rule-based algorithm informed by domain knowledge, allowing both consistency and scalability. The complete two-level taxonomy is presented in Table 1.

Table 2. Comparison with related datasets. Source indicates whether videos were collected online or recorded using mobile devices from real-life subjects. Hand specifies if the videos were captured using handheld cameras. Age denotes whether the dataset includes participants of diverse age groups (youth, adults, and elderly). For Participant, A represents athletes and N represents non-athletes.

Dataset	#Class	#Video	Source	Hand	Age	Participant		Environment	
						A	N	Indoor	Outdoor
THUSMOS14	20	413	online	✗	✗	✓	✗	✓	✗
MPII Cooking	65	45	real	✗	✗	✗	✓	✓	✗
FineGym V1.0	530	303	online	✗	✗	✓	✗	✓	✗
Ours	12	701	real	✓	✓	✓	✓	✓	✓

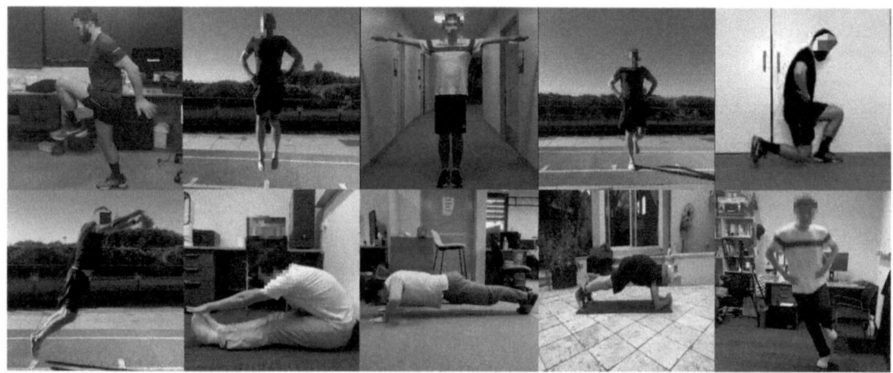

Fig. 2. Visualization examples from our dataset, showcasing diverse environments including both indoor and outdoor settings, as well as participants from different age groups (young, adults, and elderly). Some images also depict subjects wearing the Vicon system for motion capture.

2.3 Dataset Properties

Dataset Statistics. To provide a comprehensive overview (see Fig. 2) of our proposed dataset, we summarize key statistics in Table 1, Table 2 and Table 5. The dataset comprises 12 first-level action categories and 30 s-level action phases. Most videos include a single first-level action, accompanied by 0 to 9 s-level phases, depending on the complexity and structure of the movement.

Video durations vary considerably across different action types. For instance, actions such as *plank* and *high knee* tend to have longer durations due to their sustained or repetitive nature, whereas *arm span* and *standing reach* are among the shortest due to their simple and brief execution. We also report the average duration for each action class, highlighting the temporal variability across exercises.

Overall, the dataset includes six distinct annotation types, as illustrated in Fig. 1. These rich and diverse annotations enhance the dataset's versatility, enabling a wide range of downstream tasks such as zero-shot temporal action detection, repetition counting, displacement measurement, and physical literacy scoring.

3 Experiments

In this section, we systematically evaluate both the quality of our proposed dataset and the performance of several state-of-the-art vision-language models (VLMs). Our objectives are threefold: (1) to demonstrate the high quality of our annotations using multiple verification approaches; (2) to highlight the unique challenges inherent to physical literacy analysis and the current limitations of VLMs in this domain; and (3) to showcase the lightweight nature of our dataset for zero-shot temporal action detection (ZS-TAD), making it a suitable testbed for exploring a variety of strategies. All experiments are conducted using two NVIDIA RTX 4090 GPUs.

3.1 Benchmark Results

Table 3. Benchmark results of the Qwen series models, including Qwen2.5-VL-7B-Instruct and Qwen2.5-VL-32B-Instruct-AWQ. For the TAD task, mAP refers to mean Average Precision, MAE denotes the Mean Absolute Error between predictions and ground-truth annotations, and UR represents the Uniqueness Rate, calculated as the number of unique predicted actions divided by the total number of predictions.

VLM	Action Type	mAP(%)↑					MAE↑	UR↑
		0.3	0.5	0.7	0.9	Avg.		
Qwen2.5-VL-7B-Instruct	arm span	87.18	56.41	33.33	10.26	55.90	13.25	0.13
	broad jump	7.14	0.00	0.00	0.00	3.57	68.29	0.09
	exp push up	13.95	11.63	0.00	0.00	7.91	13.14	0.05
	high knee	96.23	94.34	88.68	66.04	88.68	71.19	0.13
	lateral hop	0.00	0.00	0.00	0.00	0.00	12.25	0.14
	lunge	76.67	73.33	71.67	31.67	67.67	9.08	0.13
	plank	95.12	82.93	80.49	56.10	82.93	51.72	0.27
	push up	53.25	44.16	38.96	29.87	44.16	9.74	0.09
	single leg stand	40.38	34.62	34.62	26.92	36.54	16.65	0.17
	sit to reach	90.00	63.33	26.67	0.00	56.00	14.10	0.03
	standing reach	78.57	28.57	10.71	3.57	41.43	28.34	0.04
	vertical jump	14.71	7.35	4.41	1.47	9.41	25.48	0.01
	average	**54.43**	**41.39**	**32.46**	**18.82**	**36.78**	-	**0.11**
Qwen2.5-VL-32B-Instruct-AWQ	arm span	89.74	66.67	28.21	2.56	56.41	7.89	0.08
	broad jump	46.43	10.71	0.00	0.00	25.71	70.34	0.14
	exp push up	58.14	41.86	18.60	6.98	40.00	9.72	0.07
	high knee	94.34	94.34	88.68	73.58	89.06	55.51	0.36
	lateral hop	4.55	4.55	4.55	0.00	3.64	16.52	0.11
	lunge	90.00	90.00	90.00	71.67	86.33	6.9	0.13
	plank	80.49	80.49	80.49	63.41	77.07	33.67	0.46
	push up	77.92	75.32	74.03	53.25	71.69	7.6	0.14
	single leg stand	44.23	44.23	40.38	38.46	42.31	10.88	0.44
	sit to reach	70.00	46.67	16.67	3.33	43.33	15.02	0.13
	standing reach	50.00	39.29	17.86	10.71	35.71	42.26	0.11
	vertical jump	30.15	13.97	5.15	0.74	19.41	13.66	0.07
	average	**61.33**	**50.67**	**38.72**	**27.06**	**44.45**	-	**0.19**

We present our benchmark results in Table 3, focusing on two state-of-the-art vision-language models from the Qwen series. Due to hardware memory constraints, we select the 7B and 32B variants for fair comparison. All experiments are conducted at a temporal sampling rate of 2 FPS, which follows the default configuration recommended in the original Qwen technical report and is commonly adopted in other vision-language benchmarks.

As expected, the 32B model generally outperforms the 7B model across most tasks, consistent with findings reported in the official Qwen documentation. However, despite the relatively straightforward nature of the physical literacy tasks (e.g., basic daily exercises), several notable failure cases emerge.

First, VLMs struggle to distinguish between visually similar actions, such as lateral hop and high knee. A particularly high error rate is observed for lateral hop, which is frequently misclassified either as high knee or as unrelated actions. Second, for tasks requiring fine-grained temporal understanding, such as identifying the moment of take-off and landing in broad jump, VLMs perform poorly. This indicates a clear limitation in their ability to integrate spatial and temporal cues jointly.

Third, fast-motion activities further challenge the models' perception, leading to degraded performance. When evaluating auxiliary tasks, such as displacement/distance measurement, repetition counting, and duration estimation, we observe noticeably poor results. One major advantage of our dataset is the ability to directly analyze such errors in a transparent manner, without relying on multiple-choice question-answering schemes. Unlike conventional VQA setups [11] [15] where models may guess the correct answer from a limited set, our regression-based formulation requires true reasoning and temporal understanding.

Interestingly, we observe that the models often fail to perceive or interpret these tasks altogether. In many cases, the model outputs nearly identical predictions across different input videos, indicating an over-reliance on prior knowledge rather than visual grounding. For example, in the repetition counting task, the model tends to prefer rounded numbers divisible by 5 or 10, regardless of the actual content of the video. We interpret this phenomenon as a form of **temporal hallucination**, wherein the model fabricates plausible yet inaccurate outputs in the absence of genuine understanding. We encourage the community to explore this challenge further and develop solutions that improve temporal reasoning in VLMs.

3.2 Ablation Study

To gain deeper insights into the factors that influence model performance, we conduct a series of rigorous ablation studies on our dataset. Specifically, we examine the effects of prompt design, video frame rate (FPS), and spatial input resolution on several downstream tasks (Table 4).

Prompt Design. We begin by evaluating the impact of prompt informativeness. In the baseline setting, we provide the model with detailed textual descriptions of each action. We then compare this with a simplified version where only a list of action labels is provided, omitting detailed descriptions or context. The results show a dramatic performance drop, primarily due to increased wrong

Table 4. Comparison of different frame rates (FPS), resolutions, and prompt types on the selected evaluation set. This analysis evaluates how variations in prompt design, video frame rate, and resolution impact the final performance results.

Prompt	FPS	Min pixels	Max pixels	mAP(%)↑					UR↑
				0.3	0.5	0.7	0.9	Avg.	
✗	2	64*28*28	128*28*28	36.95	23.73	16.56	9.33	21.64	0.08
✓	2	64*28*28	128*28*28	45.40	30.58	20.16	8.21	26.09	0.08
✓	2	128*28*28	256*28*28	45.39	28.10	21.69	13.25	27.11	0.08
✓	4	64*28*28	128*28*28	46.00	30.42	20.02	10.19	26.66	0.09
✓	8	64*28*28	128*28*28	53.02	36.48	26.05	13.23	**32.20**	0.09

label errors. This observation underscores the importance of well-crafted, informative prompts–especially in zero-shot settings, where the model lacks task-specific fine-tuning. This insight may also extend to traditional TAD datasets like ActivityNet and THUMOS14. However, challenges remain: ActivityNet contains a large number of action classes, making it difficult to include comprehensive descriptions for all in a single prompt; THUMOS14 has fewer classes, but the actions are often ambiguous and boundaries are poorly defined, even for human annotators. These findings highlight the need for future research into prompt engineering strategies that effectively communicate action semantics and boundaries, especially in multi-label or overlapping action scenarios that require fine-grained spatial and temporal understanding.

Temporal and Spatial Resolution. We further study the impact of FPS and spatial resolution. Both factors positively influence performance, but increasing FPS yields a substantially larger performance gain compared to increasing spatial resolution. This suggests that access to more temporal frames allows the model to capture subtle motion patterns, which are critical for distinguishing fine-grained temporal changes. In zero-shot TAD tasks, a practical trade-off, such as using higher FPS with lower resolution, can be beneficial under hardware memory constraints. Additionally, these findings open opportunities to explore more sophisticated frame sampling strategies. Current approaches apply uniform sampling (i.e., increasing FPS), but given the limited frame capacity of most VLMs, intelligent sampling strategies that select temporally salient frames could further improve performance, especially for fast and fine-grained actions.

Model Biases and Temporal Hallucination. Despite these adjustments, we observe persistent model biases, particularly in tasks that require unique and grounded outputs. Even across different configurations, the models tend to produce repetitive or identical predictions, leading to extremely low Unique Response (UR) (see Table 1 and Table 5) scores compared to the diversity observed in human annotations. A deeper analysis reveals several interesting patterns:

Table 5. Dataset statistics, including the average video length for each action type and the uniqueness rate of the annotations. A uniqueness rate of 1.0 indicates that all annotations are unique, while a value of 0.5 suggests that approximately two videos share the same annotation. For example, performing the same number of repetitions.

Action Type	Average Length(s)	Annotation UR
Arm Span	7.89	1.00
Broad Jump	7.42	1.00
Exp Push Up	15.72	1.00
High Knee	92.64	0.81
Lateral Hop	22.79	0.57
Lunge	56.47	0.48
Plank	127.15	0.97
Push Up	53.23	0.46
Single Leg Stand	51.64	0.76
Sit to Reach	8.27	0.86
Standing Reach	6.50	1.00
Vertical Jump	6.97	0.95

- **Displacement Measurement.** Models often fail to reason about before-and-after body positions, even under 8 FPS settings. They are unable to infer relative displacements, e.g., jump distance, by comparing temporal frames in relation to human height or background reference. This task demands an integrated understanding of spatial reasoning, temporal memory, and metric estimation, which current VLMs lack.
- **Repetition Counting.** Fast, cyclic motions, such as high knees, pose significant challenges. Even with increased FPS, models struggle to count accurately due to the absence of clear temporal gaps between repetitions. Moreover, VLMs often cannot distinguish between left and right leg movements, leading to unordered and inconsistent counting.

These findings point to a broader issue of *temporal hallucination*, where models generate plausible but unfounded outputs without genuine temporal understanding. This underscores the importance of developing new benchmarks and model architectures that better integrate spatial-temporal reasoning and memory in vision-language models.

4 Conclusion

We present PLNet-12, a benchmark for evaluating physical literacy through video-based, multi-task analysis, focusing on zero-shot capabilities of vision-language models (VLMs). The dataset covers 12 fundamental movements with expert-verified annotations for tasks such as temporal action detection, repetition counting, and biomechanical scoring. Experiments on two state-of-the-art VLMs reveal strengths and limitations in modeling fine-grained, temporally grounded actions. Ablation studies highlight the impact of prompt design and resolution, while exposing persistent temporal hallucination in counting and displacement tasks. PLNet-12 provides a compact yet challenging platform for advancing temporal reasoning, physical behavior analysis, and robust video-language modeling.

Acknowledgements. This research is supported by Advanced Queensland Industry Research Project. We thank all anonymous reviewers and ACs for their constructive suggestions.

A Related Work

A.1 Zero-Shot Temporal Action Detection

Zero-shot temporal action detection (ZS-TAD), also known as zero-shot temporal action localization (ZS-TAL), aims to jointly localize action segments and recognize action classes that are absent during training. This task is inherently challenging due to the presence of multiple unseen categories and the need to generalize beyond labeled data. The core idea is to transfer shared knowledge from seen classes to unseen ones using learned visual-semantic representations. EffPrompt [12] uses a pre-trained detector to generate the action proposals, followed by classification using CLIP [23]. However, this structure introduces error propagation between stages. To address this limitation, STALE [19] proposes a one-stage architecture consisting of two concurrent streams for classification and localization, leveraging large-scale pretrained vision-language models (VLMs).

Despite these developments, current methods are heavily relies on elaborately hyperparameter settings for different dataset and CLIP is relatively old with a low performance compared to several latest large-scale vision language models. To bride this gap, in our work we aim to expore the performance of these latest large-scale vision language models on our realistic PLNet-12 dataset to evualte these model's ability on real-world hand-collected videos.

A.2 Vision-Language Models

Recent years have witnessed significant advancements at the intersection of computer vision and natural language processing, particularly in tasks such as video captioning [27], visual question answering [10], and video grounding [18]. A central focus in this domain is the development of pre-trained vision-language models (VLMs) [14], which aim to improve the generalization ability of visual models–especially when dealing with unseen object classes or action categories.

T3AL [16], for instance, employs test-time adaptation to update the model using an unlabeled stream of videos, offering a practical solution for handling out-of-distribution (OOD) data. ZEETAD [20] proposes a dual-localization mechanism and incorporates adapter modules into an end-to-end VLM-based architecture, achieving strong performance without extensive training data.

References

1. Bailey, R., Glibo, I., Koenen, K., Samsudin, N.: What is physical literacy? An international review and analysis of definitions. Kinesiol. Rev. **12**(3), 247–260 (2023)
2. Barnett, L.M., et al.: Fundamental movement skills: an important focus. J. Teach. Phys. Educ. **35**(3), 219–225 (2016)
3. Basman, A.J.: Assessment criteria of fundamental movement skills for various age groups: a systematic review. J. Phys. Educ. Sport **19**(1), 722–732 (2019)
4. Bukht, T.F.N., Rahman, H., Shaheen, M., Algarni, A., Almujally, N.A., Jalal, A.: A review of video-based human activity recognition: theory, methods and applications. Multimed. Tools Appl. **84**(17), 18499–18545 (2025)
5. Carl, J., Barratt, J., Wanner, P., Töpfer, C., Cairney, J., Pfeifer, K.: The effectiveness of physical literacy interventions: a systematic review with meta-analysis. Sports Med. **52**(12), 2965–2999 (2022)
6. Damen, D., et al.: The epic-kitchens dataset: collection, challenges and baselines. IEEE Trans. Pattern Anal. Mach. Intell. **43**(11), 4125–4141 (2020)
7. Jean de Dieu, H., Zhou, K.: Physical literacy assessment tools: a systematic literature review for why, what, who, and how. Int. J. Environ. Res. Public Health **18**(15), 7954 (2021)
8. Essiet, I.A., et al.: Exploring Australian teachers' perceptions of physical literacy: a mixed-methods study. Phys. Educ. Sport Pedagog. **29**(1), 18–37 (2024)
9. Guo, T., Du, H., Huo, H., Liu, B., Yu, X.: Who is being impersonated? Deepfake audio detection and impersonated identification via extraction of id-specific features. In: International Conference on Algorithms and Architectures for Parallel Processing, pp. 301–320. Springer, Cham (2024)
10. Hartsock, I., Rasool, G.: Vision-language models for medical report generation and visual question answering: a review. Front. Artif. Intell. **7**, 1430984 (2024)
11. Hong, W., et al.: Motionbench: benchmarking and improving fine-grained video motion understanding for vision language models. In: Proceedings of the Computer Vision and Pattern Recognition Conference, pp. 8450–8460 (2025)
12. Ju, C., Han, T., Zheng, K., Zhang, Y., Xie, W.: Prompting visual-language models for efficient video understanding. In: European Conference on Computer Vision, pp. 105–124. Springer, Cham (2022)
13. Khan, A., Galarraga, O., Garcia-Salicetti, S., Vigneron, V.: Deep learning for quantified gait analysis: a systematic literature review. IEEE Access (2024)
14. Laurençon, H., Tronchon, L., Cord, M., Sanh, V.: What matters when building vision-language models? Adv. Neural. Inf. Process. Syst. **37**, 87874–87907 (2024)
15. Li, K., et al.: Mvbench: a comprehensive multi-modal video understanding benchmark. In: Proceedings of the IEEE/CVF Conference on Computer Vision and Pattern Recognition, pp. 22195–22206 (2024)
16. Liberatori, B., Conti, A., Rota, P., Wang, Y., Ricci, E.: Test-time zero-shot temporal action localization. In: Proceedings of the IEEE/CVF Conference on Computer Vision and Pattern Recognition, pp. 18720–18729 (2024)
17. Liu, Y., Wang, L., Wang, Y., Ma, X., Qiao, Y.: Fineaction: a fine-grained video dataset for temporal action localization. IEEE Trans. Image Process. **31**, 6937–6950 (2022)
18. Mu, F., Mo, S., Li, Y.: Snag: scalable and accurate video grounding. In: Proceedings of the IEEE/CVF Conference on Computer Vision and Pattern Recognition, pp. 18930–18940 (2024)

19. Nag, S., Zhu, X., Song, Y.Z., Xiang, T.: Zero-shot temporal action detection via vision-language prompting. In: European Conference on Computer Vision, pp. 681–697. Springer, Cham (2022)
20. Phan, T., Vo, K., Le, D., Doretto, G., Adjeroh, D., Le, N.: Zeetad: adapting pre-trained vision-language model for zero-shot end-to-end temporal action detection. In: Proceedings of the IEEE/CVF Winter Conference on Applications of Computer Vision, pp. 7046–7055 (2024)
21. Qiu, F., et al.: Learning transferable compound expressions from masked autoencoder pretraining. In: Proceedings of the IEEE/CVF Conference on Computer Vision and Pattern Recognition, pp. 4733–4741 (2024)
22. Qiu, F., et al.: Language-guided multi-modal emotional mimicry intensity estimation. In: Proceedings of the IEEE/CVF Conference on Computer Vision and Pattern Recognition, pp. 4742–4751 (2024)
23. Radford, A., et al.: Learning transferable visual models from natural language supervision. In: International Conference on Machine Learning, pp. 8748–8763. PMLR (2021)
24. Young, L., O'Connor, J., Alfrey, L., Penney, D.: Assessing physical literacy in health and physical education. Curriculum Stud. Health Phys. Educ. **12**(2), 156–179 (2021)
25. Zhang, W., et al.: An effective ensemble learning framework for affective behaviour analysis. In: Proceedings of the IEEE/CVF Conference on Computer Vision and Pattern Recognition, pp. 4761–4772 (2024)
26. Zhang, W., et al.: Affective behaviour analysis via integrating multi-modal knowledge. arXiv preprint arXiv:2403.10825 (2024)
27. Zhou, X., et al.: Streaming dense video captioning. In: Proceedings of the IEEE/CVF Conference on Computer Vision and Pattern Recognition, pp. 18243–18252 (2024)

TrashTracer: Enabling Efficient Real-Time Detection of Underwater Marine Debris

Yifan Yin[1], Xiufeng Liu[2], Xu Cheng[2], Hua Zuo[1], Ling Chen[1], Tianqing Zhu[3], and Huan Huo[1](✉)

[1] School of Computer Science, University of Technology Sydney, Sydney, Australia
Yifan.Yin-2@student.uts.edu.au, {Hua.Zuo,Ling.Chen,Huan.Huo}@uts.edu.au
[2] Department of Technology, Management and Economics Energy Markets and Finance, Technical University of Denmark, Kongens Lyngby, Denmark
{xiuli,xuche}@dtu.dk
[3] Faculty of Data Science, Macau University of Science and Technology, Macau, China
tqzhu@cityu.edu.mo

Abstract. Human-induced marine pollution has led to increasing underwater debris, threatening ecosystems. Although vision-based robots like AUVs and ROVs assist in seafloor monitoring, effective detection remains difficult due to poor visibility and degraded image quality in underwater environments. In this paper, we propose **TrashTracer**, a novel and lightweight framework for real-time underwater debris detection. To address visibility degradation, we introduce CG-Enhance, an online image enhancement method combining CLAHE and Gamma correction, which improves image quality without adding inference overhead. We further design an Efficient Reparameterized Backbone (ERB) utilizing the Diverse Branch Block (DBB) to enhance feature learning during training and simplify computation during inference. Additionally, we propose the Reparameterized Adaptive Spatial Fusion Head (RASF-Head) to adaptively integrate multi-scale features for robust detection. Extensive experiments on the TrashCan dataset demonstrate that our model achieves superior detection accuracy while maintaining low computational complexity and real-time performance, highlighting its strong potential for practical deployment in underwater debris monitoring applications.

Keywords: Underwater debris detection · Real-time object detection · Image enhancement · Reparameterized neural networks

1 Introduction

Marine pollution has become a pressing global issue, with plastic debris, discarded fishing gear, and industrial waste accumulating on the seafloor. This threatens ecosystems, navigation, and the marine economy [1]. Recent advances in vision-based robots allow Autonomous Underwater Vehicles (AUVs) and Remotely Operated Vehicles (ROVs) to detect seafloor debris using deep learning, enabling recognition from underwater imagery.

M. Liu et al. (Eds.): AI 2025, LNAI 16371, pp. 255–266, 2026.
https://doi.org/10.1007/978-981-95-4972-6_20

Deep learning-based object detection methods, two-stage detectors (e.g., Faster R-CNN [2]) and one-stage detectors (e.g., YOLO [3]), achieve strong results on datasets like COCO [4]. However, underwater detection remains challenging due to low visibility, light scattering, and poor image quality [5], which limit feature extraction and reduce detection accuracy. To address this, researchers have explored data augmentation [6] and integrated image enhancement with detection [5], as well as improved feature extractors [7]. Despite progress, existing methods often rely on complex image enhancement as a preprocessing step or employ heavy end-to-end models that are impractical for AUVs with limited computational resources.

To overcome these issues, we propose **TrashTracer**, a high-performance underwater debris detection framework that balances accuracy, efficiency, and real-time capability. Our contributions are:

- We design **TrashTracer**, a lightweight yet accurate marine debris detection network.
- We introduce **CG-Enhance**, a real-time underwater image enhancement method based on CLAHE and Gamma correction, which improves visibility without adding parameters or inference time.
- We optimize the backbone with the **Efficient Reparameterized Backbone (ERB)** and propose the **Reparameterized Adaptive Spatial Fusion Head (RASF-Head)**, which enhance feature representation while simplifying inference.
- We evaluate TrashTracer on the TrashCan dataset [8], demonstrating competitive accuracy, lightweight design, and real-time performance compared with state-of-the-art detectors.

2 Related Work

2.1 Underwater Image Enhancement (UIE)

In this study, we focus on non-physics-based UIE methods, which enhance underwater images by adjusting pixel values for better contrast and color. Common techniques for enhancing contrast and brightness include Gamma Correction, Histogram Equalization (HE) [9], and Contrast Limited Adaptive Histogram Equalization (CLAHE) [10]. For color correction, typical methods include White Balance (WB), the Gray-World Assumption (GWA), and the Gray-Edge Assumption (GEA). However, due to the lack of adequate lighting in underwater environments, relying on a single method may amplify internal image noise. Therefore, fusion has proven to be an effective strategy for UIE—combining multiple processing results with weighted mappings to produce images with improved contrast and finer details [11].

2.2 Real-Time Detection Methods

One-stage detectors (e.g., YOLO) provide faster inference than two-stage approaches, making them more suitable for engineering tasks. Speed-up strategies include pruning [12], quantization [13], and compact convolution designs such as depthwise separable or group convolutions [14]. Although effective, these can lead to fragmented architectures that slow real-time inference. Reparameterization techniques [15] address this by using multi-branch structures during training and collapsing them into single convolutions at inference, improving both speed and accuracy.

2.3 Feature Fusion for Multi-Scale Detection

Detecting objects of varying scales is a core challenge. SSD [16] first introduced multi-scale features, while FPN [17] and PANet [18] improved fusion through top-down and bottom-up paths. Advanced approaches like NAS-FPN [19] and ASFF [20] further enhance multi-scale representation. However, many fusion methods increase latency. In this work, we integrate reparameterization into ASFF and design the **Reparameterized Adaptive Spatial Fusion Head (RASF-Head)** to achieve accurate feature fusion without sacrificing inference efficiency.

3 Methods

Our goal is to build a detection framework optimized for low-visibility seafloor environments, enabling timely marine debris removal and extending the operational capabilities of AUVs. We present TrashTracer, a real-time detector with four key components (Fig. 1): (1) Online Data Preprocessing Module, responsible for generating enhanced underwater images; (2) Efficient Reparameterized Backbone (ERB), which enriches feature extraction using a multi-branched block and applies the reparameterization trick during inference without incurring additional time costs; (3) Neck, which fuses multi-scale features, enabling the network to focus on targets of varying sizes simultaneously; and (4) Reparameterized Adaptive Spatial Fusion Head (RASF-Head), which adaptively fuses features from different Neck layers by assigning weights; additionally, by replacing the original convolution with a multi-branched block for channel transformation and size alignment, the method optimizes inference time and enhances detection accuracy.

3.1 Online Underwater Image Enhancement

Deep learning-based enhancement is often too slow for real-time detection and limited by the lack of paired training samples. To maintain speed and accuracy, we integrate two traditional methods—**Contrast Limited Adaptive Histogram Equalization and Gamma Correction (CG-Enhance)**—directly into the detector, avoiding extra parameters or computation.

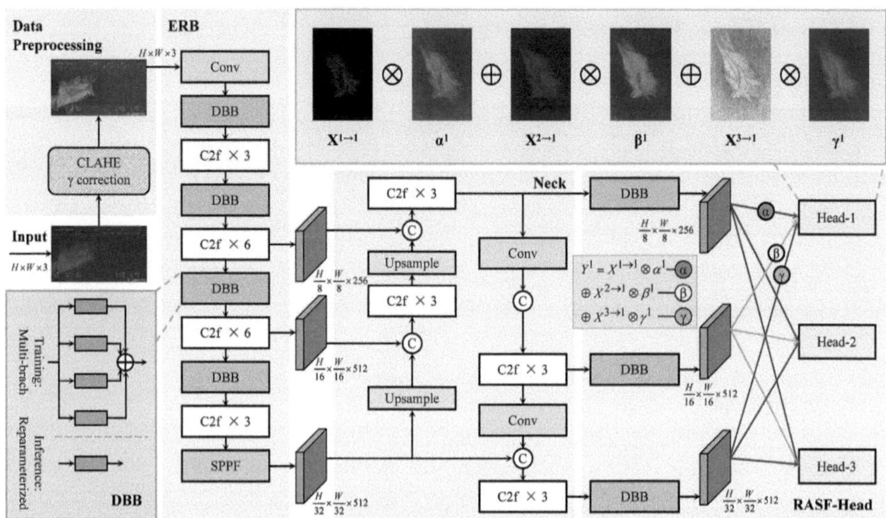

Fig. 1. Overview of the proposed framework with key innovations: (1) Online Preprocessing enhances underwater images; (2) ERB uses multi-branched blocks and reparameterization for efficient, rich feature extraction; (3) Neck fuses multi-scale features; (4) RASF-Head adaptively fuses features across scales; The reparameterized multi-branch design boosts both inference speed and detection accuracy.

Contrast Limited Adaptive Histogram Equalization (CLAHE): This study employs the CLAHE [10] method to enhance image brightness by improving local contrast while suppressing noise. CLAHE divides the image into small regions and applies histogram equalization within each block. A clipping threshold is used to limit peak values in the histogram, with excess counts redistributed to prevent over-enhancement. Finally, bilinear interpolation is applied between adjacent blocks to ensure smooth transitions in the output image.

Gamma Correction: In image processing, gamma correction is a nonlinear transformation that adjusts pixel values to produce a more natural-looking image. The transformation is defined as:

$$I_{out}(x,y) = 255 \cdot (\frac{I_{in}(x,y)}{255})^r \tag{1}$$

Here, I_{in} denotes the brightness value of the input image at pixel position (x, y), while I_{out} represents the output after transformation. The parameter γ controls the degree of nonlinearity in the transformation. When $\gamma > 1$, the image brightness is suppressed, resulting in a darker appearance; conversely, when $\gamma < 1$, the brightness is enhanced, making the image appear lighter.

Combined Strategy: Combining CLAHE and gamma correction creates an effective enhancement approach for underwater imagery. CLAHE improves local contrast and detail, while gamma correction adjusts overall brightness for a natural look. Together, they enhance contrast and maintain balanced illumination, making the images clearer for detection in challenging underwater conditions.

3.2 Optimizing Backbone Efficiency with Diverse Branch Block (DBB)

Diverse Branch Block (DBB). The DBB [15] is a multi-branch convolutional module that increases feature representation during training and collapses into a single convolution at inference, yielding both accuracy and speed. DBB adopts a four-branch topology during training, consisting of the following: a 1×1 convolutional branch, a sequential 1×1 - $K \times K$ convolutional branch, an average pooling branch, and a $K \times K$ convolutional branch. Their outputs are summed, then mathematically fused into one $K \times K$ convolution for deployment.

Fundamental Form of Convolution: Let $\mathbf{I} \in \mathbb{R}^{C\times H\times W}$ represent the input, $\mathbf{O} \in \mathbb{R}^{D\times H'\times W'}$ the output, $\mathbf{F} \in \mathbb{R}^{D\times C\times K\times K}$ the kernel, and $\mathbf{b} \in \mathbb{R}^{D}$ the bias term. The convolution operation is then defined as:

$$\mathbf{O} = \mathbf{I} \circledast \mathbf{F} + \mathbf{REP}(\mathbf{b}) \tag{2}$$

Here, $\circledast$ denotes the convolution operation, and $\mathbf{REF}(\mathbf{b}) \in \mathbb{R}^{D\times H'\times W'}$ refers to a replicated version of the bias term for the subsequent addition. Additionally, the linear properties of the convolution operation, including homogeneity and additivity, can be expressed as follows:

$$\mathbf{I} \circledast (p\mathbf{F}) = p(\mathbf{I} \circledast \mathbf{F}), \forall p \in \mathbb{R} \tag{3}$$

$$\mathbf{I} \circledast \mathbf{F}^{(1)} + \mathbf{I} \circledast \mathbf{F}^{(2)} = \mathbf{I} \circledast (\mathbf{F}^{(1)} + \mathbf{F}^{(2)}) \tag{4}$$

Transform I: Fusing BN into Conv. A convolution followed by BatchNorm (BN) can be merged into a single convolution. For the j-th channel:

$$\mathbf{O}_{\mathbf{j},:,:} = ((\mathbf{I} \circledast \mathbf{F})_{j,:,:} - \mu_j)\frac{\gamma_j}{\sigma_j} + \beta_j \tag{5}$$

Here, μ_j and σ_j are the mean and standard deviation from batch normalization (BN), while γ_j and β_j are the learnable scale and bias. Leveraging the linearity of convolution, the BN and convolution layers can be merged into a single equivalent convolution with transformed kernel $\mathbf{F}'$ and bias $\mathbf{b}'$ as shown in Eq. (6), enabling efficient inference.

$$\mathbf{F}'_{j,:,:,:} \leftarrow \frac{\gamma_j}{\sigma_j}\mathbf{F}_{j,:,:,:}, \quad \mathbf{b}'_j \leftarrow -\frac{\mu_j\gamma_j}{\sigma_j} + \beta_j \tag{6}$$

Transform II: Fusing Branches into Conv. By convolution additivity, branches with the same setup merge into one:

$$\mathbf{F}' \leftarrow \mathbf{F}^{(1)} + \mathbf{F}^{(2)}, \quad \mathbf{b}' \leftarrow \mathbf{b}^{(1)} + \mathbf{b}^{(2)} \tag{7}$$

Transform III: Fusing Sequential Convolutions into Conv. A sequence of 1×1 convolution - BN - $K \times K$ convolution - BN can be merged into a single $K \times K$ convolution. Let the kernel of the 1×1 convolution has a shape of $D \times C \times 1 \times 1$, and the kernel of the $K \times K$ convolution has a shape of $E \times D \times K \times K$. According to **Transform I**, the parameters of the BN layers following both the 1×1 and $K \times K$ convolutions can be fused into the respective convolutions, resulting in the updated parameters $\mathbf{F}^{(1)} \in \mathbb{R}^{D \times C \times 1 \times 1}$, $\mathbf{b}^{(1)} \in \mathbb{R}^{D}$ and $\mathbf{F}^{(2)} \in \mathbb{R}^{E \times D \times K \times K}$, $\mathbf{b}^{(2)} \in \mathbb{R}^{E}$. The output is:

$$\begin{aligned} \mathbf{O}' &= (\mathbf{I} \circledast \mathbf{F}^{(1)} + \mathbf{REP}(\mathbf{b}^{(1)})) \circledast \mathbf{F}^{(2)} + \mathbf{REP}(\mathbf{b}^{(2)}) \\ &= \mathbf{I} \circledast \mathbf{F}^{(1)} \circledast \mathbf{F}^{(2)} + \mathbf{REP}(\mathbf{b}^{(1)}) \circledast \mathbf{F}^{(2)} + \mathbf{REP}(\mathbf{b}^{(2)}) \end{aligned} \tag{8}$$

Our goal is to derive a single convolution with $\mathbf{F}'$ and bias $\mathbf{b}'$ that satisfy the following equation:

$$\mathbf{O}' = \mathbf{I} \circledast \mathbf{F}' + \mathbf{REP}(\mathbf{b}') \tag{9}$$

For the input $\mathbf{I}$, the 1×1 convolution performs channel-wise linear combinations without spatial aggregation, making its output effectively a new weight matrix for the $K \times K$ convolution. This can be expressed as:

$$\mathbf{F}' \leftarrow \mathbf{F}^{(2)} \circledast \mathbf{TRANS}(\mathbf{F}^{(1)}) \tag{10}$$

Here, $\mathbf{TRANS}(\mathbf{F}^{(1)}) \in \mathbb{R}^{C \times D \times 1 \times 1}$ denotes the transposition of the 1×1 weights to align with the $K \times K$ convolution.

Since the bias of a 1×1 convolution, when passed through a $K \times K$ convolution, results in a constant matrix, its value can be determined using the following equation:

$$\hat{\mathbf{b}}_{\mathbf{j}} \leftarrow \sum_{d=1}^{D} \sum_{u=1}^{K} \sum_{v=1}^{K} \mathbf{b}_d^{(1)} \mathbf{F}_{j,d,u,v}^{(2)}, \quad 1 \leq j \leq E \tag{11}$$

It is then combined with the bias of the $K \times K$ convolution as follows:

$$\mathbf{b}' \leftarrow \hat{\mathbf{b}} + \mathbf{b}^{(2)} \tag{12}$$

Transform IV: Fusing Depth Concatenation into Conv. Branches can also be merged by concatenating their kernels along the channel dimension. Given $\mathbf{F}^{(1)} \in \mathbb{R}^{D_1 \times C \times K \times K}$, $\mathbf{b}^{(1)} \in \mathbb{R}^{D_1}$ and $\mathbf{F}^{(2)} \in \mathbb{R}^{D_2 \times C \times K \times K}$, $\mathbf{b}^{(2)} \in \mathbb{R}^{D_2}$, they can be equivalently combined into $\mathbf{F}' \in \mathbb{R}^{(D_1+D_2) \times C \times K \times K}$, $\mathbf{b}' \in \mathbb{R}^{D_1+D_2}$ through channel-wise concatenation:

$$CONCAT(\mathbf{I} \circledast \mathbf{F}^{(1)} + \mathbf{REP}(\mathbf{b}^{(1)}), \mathbf{I} \circledast \mathbf{F}^{(2)} + \mathbf{REP}(\mathbf{b}^{(2)})) = \mathbf{I} \circledast \mathbf{F}' + \mathbf{REP}(\mathbf{b}') \tag{13}$$

Transform V: Fusing Average Pooling into Conv. An average pooling operation with a kernel size of K, a stride of s, and C input channels can be equivalently represented by a convolution $\mathbf{F}' \in \mathbb{R}^{C \times C \times K \times K}$ with the same kernel size and stride:

$$\mathbf{F}'_{d,c,:,:} = \begin{cases} \frac{1}{K^2}, & if \quad d = c, \\ 0, & elsewise. \end{cases} \tag{14}$$

Transform VI: Fusing multi-scale convolutions into Conv. In fact, a $k_h \times k_w$ ($k_h \le K, k_w \le K$) kernel is equivalent to a $K \times K$ kernel with zero padding, allowing the $k_h \times k_w$ kernel to be transformed into a $K \times K$ kernel by adding zero terms.

DBB-Enhanced Backbone. As shown in Fig. 1, we design a new backbone called the **Efficient Reparameterized Backbone (ERB)**, built upon the YOLOv8 backbone. Specifically, we replace all standard convolutional layers (Conv) in the original backbone with DBB modules, except for the first convolutional layer. During training, the DBB boosts the network's feature extraction capability through its multipath structure. In the inference phase, these branches are efficiently reparameterized into a single-layer $K \times K$ convolution, enhancing target detection performance while maintaining high computational efficiency. Experimental results indicate that ERB enhances performance without impacting inference speed.

3.3 Improved Detection Head: RASF-Head

Underwater targets vary greatly in scale, making feature fusion challenging. Deep layers in CNNs provide global semantics, while shallow layers preserve spatial detail for small targets. Directly merging them can create conflicts. To address this, we design the **Reparameterized Adaptive Spatial Fusion Head (RASF-Head)**, which resolves cross-scale inconsistencies and uses reparameterization to cut inference cost. RASF-Head has two parts: Feature Resizing and Adaptive Fusion.

Feature Resizing: Effective fusion requires aligned resolutions and channel numbers. YOLOv8 feature maps differ across layers, so we use DBB (Sect. 3.2) for upsampling, downsampling, and channel compression. A 1×1 DBB with interpolation handles upsampling, while a 3×3 DBB (stride 2) or a max-pooling + DBB handles downsampling. During inference, DBB branches collapse into one convolution, reducing cost.

Adaptive Fusion: In RASF-Head, each layer adaptively participates in fusion through dynamic weighting. At each location (i, j), the feature vector $\mathbf{x}_{ij}^{n\to l}$ resized from level n to level l is assigned adaptively learned weights α_{ij}^l, β_{ij}^l, and γ_{ij}^l, which capture the spatial importance of features at different levels. These weighted features are then combined to produce the output feature vector:

$$\mathbf{y}_{ij}^l = \alpha_{ij}^l \cdot \mathbf{x}_{ij}^{1\to l} + \beta_{ij}^l \cdot \mathbf{x}_{ij}^{2\to l} + \gamma_{ij}^l \cdot \mathbf{x}_{ij}^{3\to l} \tag{15}$$

Weights satisfy $\alpha_{ij}^l + \beta_{ij}^l + \gamma_{ij}^l = 1$ and are normalized by softmax:

$$\alpha_{ij}^l = \frac{e^{\lambda_{\alpha_{ij}}^l}}{e^{\lambda_{\alpha_{ij}}^l} + e^{\lambda_{\beta_{ij}}^l} + e^{\lambda_{\gamma_{ij}}^l}}. \tag{16}$$

1×1 convolutions generate the scalar maps $\lambda_{\alpha_{ij}}^l, \lambda_{\beta_{ij}}^l, \lambda_{\gamma_{ij}}^l$, which are learned via backpropagation. This produces output feature maps $\{y_1, y_2, y_3\}$ for detection.

4 Experiments

4.1 Experimental Setup

Implementation Details. We adopt YOLOv8s as the baseline for its balance of accuracy and efficiency, suitable for AUV deployment. Input images are resized to 480 × 480. Training uses a batch size of 4, SGD optimizer (momentum = 0.937, weight decay = 5e-4), and an initial learning rate of 0.01 for 300 epochs with early stopping to prevent overfitting. Other hyperparameters follow YOLOv8 defaults. All experiments run on NVIDIA Quadro RTX 6000 GPUs using PyTorch 1.9.0 and Python 3.8.0. For baseline and comparison models, we retain original training settings without hyperparameter tuning.

Dataset. We use TrashCan [8], a large-scale underwater debris dataset containing 7,212 labeled RGB images sourced mainly from JAMSTEC's ROV footage. It provides two annotation schemes: TrashCan-Material, which categorizes debris into 16 material-based classes with 5,008 training, 1,000 validation, and 1,204 testing images; and TrashCan-Instance, which classifies debris into 22 object-based classes with 5,065 training, 1,000 validation, and 1,147 testing images.

4.2 State-of-the-Art Comparison

Comparison with Generic Object Detectors. We first benchmark TrashTracer against representative generic detectors: YOLO series [21–23], DETR [24], and Deformable DETR [25]. YOLO models offer different speed-accuracy trade-offs, while DETR and Deformable DETR bring transformer-based designs but with higher computational cost.

Table 1. Comparison of the detection performance of generic object detectors.

Method	Precision		mAP_{50}		Params	GFLOPs	Time (ms)
	Inst	Mat	Inst	Mat			
YOLOv5m	74.55	**74.29**	64.41	59.51	20.93M	48.1	6.8
YOLOv5l	74.32	73.44	66.07	59.90	46.21M	108.0	**6.7**
YOLOv8m	71.41	69.43	65.76	59.97	25.85M	78.8	6.9
YOLOv8l	73.47	69.65	66.22	60.16	43.63M	165.2	8.8
YOLOv10m	73.94	71.16	63.63	58.15	**16.48M**	63.5	8.7
YOLOv10l	72.44	71.67	62.73	58.13	25.75M	126.5	9.3
DETR	-	-	60.29	55.77	41.28M	99.82	219.1
Deformable DETR	-	-	56.23	54.48	39.83M	209.18	397.6
TrashTracer	**74.79**	73.47	**66.27**	**60.17**	16.57M	**36.9**	7.5

Table 1 shows TrashTracer achieves the highest detection accuracy (mAP_{50}: 66.27% Instance / 60.17% Material) and the best instance-level precision (74.79%). Despite having only 16.57M parameters and 36.9 GFLOPs, TrashTracer surpasses larger YOLO variants (e.g., YOLOv8l, YOLOv10l) and runs at 7.5 ms per image—far faster than DETR models (200 ms+).

Comparison with Underwater Debris Detectors. We further compare TrashTracer with recent underwater debris detection methods, including GCC-Net [5], Mask R-CNN-based work by Deng et al. [26], ERL-Net [27], YOLO-TrashCan [7], Aqua-DETR [28], and a RetinaNet-based approach by Ali et al. [29].

Table 2. Comparison of the detection performance of underwater debris detectors.

Method	mAP_{50}		Params	GFLOPs	Size(MB)
	Instance	Material			
GCC-Net [5]	61.10	57.87	38.34M	70.97	427M
Literature [26]	65.00	-	-	-	-
ERL-Net [27]	-	58.90	-	-	-
YOLOTrashCan [7]	65.01	58.66	-	-	214M
Aqua-DETR [28]	63.00	-	50.36M	78.33	-
Literature [29]	63.09	56.70	-	-	-
TrashTracer	**66.27**	**60.17**	**16.57M**	**36.9**	**56M**

Table 2 shows TrashTracer achieves the highest mAP_{50} on both datasets (66.27% Instance / 60.17% Material) with only 16.57M parameters and 36.9 GFLOPs, far smaller and more efficient than GCC-Net (38.34M/70.97 GFLOPs) and Aqua-DETR (50.36M/78.33 GFLOPs). The model's footprint is just 56 MB, compared to 427 MB for GCC-Net and 214 MB for YOLOTrashCan, making TrashTracer highly suitable for edge deployment.

4.3 Ablation Studies

Table 3. The check table for module ablations on the TrashCan-Instance and TrashCan-Material datasets in terms of mAP_{50}.

CG-Enhance	ERB	RASF-Head	mAP_{50}	
			Instance	Material
			63.79	58.19
✓			64.30	58.44
✓	✓		65.36	58.99
✓	✓	✓	**66.27**	**60.17**

Table 3 presents the variation in mAP_{50} on the TrashCan-Instance and TrashCan-Material datasets as different modules are incrementally integrated

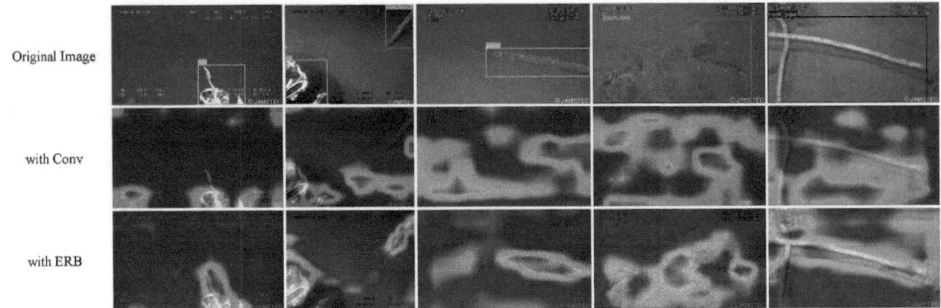

Fig. 2. Visual comparison of heatmaps for the backbone equipped with Conv and ERB.

into the network. Compared to the baseline model without any modules, the introduction of CG-Enhance yields improvements of 0.51 and 0.25 in mAP_{50} on the two datasets, respectively. Further gains are observed with the successive addition of ERB and RASF-Head, resulting in significant performance boosts of 2.48 and 1.98 over the baseline.

From the ablation experiments, it is evident that CG-Enhance contributes to improved image quality by enhancing contrast and balancing luminance distribution. The ERB module, through its multi-path structure, introduces diverse receptive fields and increased complexity, thereby enriching the feature representation. Importantly, the reparameterization technique used during inference ensures that ERB retains the computational efficiency of a single-branch structure, avoiding any additional overhead. To further analyze the effectiveness of ERB, we visualize the activation maps using Grad-CAM [30], as shown in Fig. 2. The model with standard convolution tends to produce widespread and less focused activations, often highlighting irrelevant background areas. In contrast, the ERB-equipped backbone yields more concentrated and accurate responses around key objects, indicating that ERB enhances the model's ability to focus on salient features and suppress background noise.

Finally, RASF-Head enhances detection accuracy by adaptively learning spatial fusion weights across different feature scales, effectively filtering out conflicting information.

5 Conclusion

In this study, we present TrashTracer, an efficient and lightweight framework for underwater debris detection, designed to tackle challenges such as poor visibility and varying object scales. The framework incorporates three key components: CG-Enhance for real-time image enhancement using CLAHE and Gamma Correction; the Efficient Reparameterized Backbone (ERB) for enriched feature representation without added inference cost; and the Reparameterized Adaptive Spatial Fusion Head (RASF-Head) for adaptive multi-scale feature fusion. Extensive evaluations of the TrashCan dataset show that our method outperforms generic, lightweight, and specialized detectors in accuracy, model size, and real-time performance. With its strong balance of precision and efficiency, TrashTracer is well-suited for deployment in resource-limited underwater scenarios, such as AUV-based debris removal.

References

1. Politikos, D.V., Fakiris, E., Davvetas, A., Klampanos, A., Papatheodorou, G.: Automatic detection of seafloor marine litter using towed camera images and deep learning. Marine Pollution Bulletin, 164:111974, 2021
2. Ren, S., He, K., Girshick, R., Sun, J.: Faster R-CNN: towards real-time object detection with region proposal networks. IEEE Trans. Pattern Anal. Mach. Intell. **39**(6), 1137–1149 (2016)
3. Redmon, J.: Yolov3: an incremental improvement. arXiv preprinthttp://arxiv.org/abs/1804.02767arXiv:1804.02767 (2018)
4. Lin, T.-Y., et al.: Microsoft COCO: common objects in context. In: Fleet, D., Pajdla, T., Schiele, B., Tuytelaars, T. (eds.) ECCV 2014. LNCS, vol. 8693, pp. 740–755. Springer, Cham (2014). https://doi.org/10.1007/978-3-319-10602-1_48
5. Dai, L., Liu, H., Song, P., Liu, M.: A gated cross-domain collaborative network for underwater object detection. Pattern Recogn. **149**, 110222 (2024)
6. Huang, H., Hao Zhou, X., Zhang, Y.L., Qi, L., Zang, A.-Y.: Faster r-CNN for marine organisms detection and recognition using data augmentation. Neurocomputing **337**, 372–384 (2019)
7. Zhou, W., Zheng, F., Yin, G., Pang, Y., Yi, J.: Yolotrashcan: a deep learning marine debris detection network. IEEE Trans. Instrum. Meas. **72**, 1–12 (2022)
8. Hong, J., Fulton, J., Sattar, J.: Trashcan: a semantically-segmented dataset towards visual detection of marine debris. arXiv preprinthttp://arxiv.org/abs/2007.08097arXiv:2007.08097 (2020)
9. Hummel, R.: Image enhancement by histogram transformation. Unknown (1975)
10. Zuiderveld.: Contrast limited adaptive histogram equalization. In: Graphics gems IV, pp. 474–485 (1994)
11. Ancuti, C., Ancuti, C. O., Haber, T., Bekaert, P.: Enhancing underwater images and videos by fusion. In: 2012 IEEE conference on computer vision and pattern recognition, pp. 81–88. IEEE (2012)
12. Han, S., Mao, H., Dally, W. J.: Deep compression: compressing deep neural networks with pruning, trained quantization and huffman coding. arXiv preprinthttp://arxiv.org/abs/1510.00149arXiv:1510.00149 (2015)
13. Rastegari, M., Ordonez, V., Redmon, J., Farhadi, A. :Xnor-net: imagenet classification using binary convolutional neural networks. In: European conference on computer vision, pp. 525–542. Springer (2016)
14. Howard, A.G.: Mobilenets: efficient convolutional neural networks for mobile vision applications. arXiv preprinthttp://arxiv.org/abs/1704.04861arXiv:1704.04861 (2017)
15. Ding, X., Zhang, X., Han, J., Ding, G.: Diverse branch block: building a convolution as an inception-like unit. In: Proceedings of the IEEE/CVF conference on computer vision and pattern recognition, pp. 10886–10895 (2021)
16. Liu, W., et al.: SSD: single Shot MultiBox Detector. In: Leibe, B., Matas, J., Sebe, N., Welling, M. (eds.) ECCV 2016. LNCS, vol. 9905, pp. 21–37. Springer, Cham (2016). https://doi.org/10.1007/978-3-319-46448-0_2

17. Lin, T., Dollár, T., Girshick, R., He, K., Hariharan, B., Belongie, S.: Feature pyramid networks for object detection. In: Proceedings of the IEEE conference on computer vision and pattern recognition, pp. 2117–2125 (2017)
18. Liu, S., Qi, L., Qin, H., Shi, J., Jia, J.: Path aggregation network for instance segmentation. In: Proceedings of the IEEE conference on computer vision and pattern recognition, pp. 8759–8768 (2018)
19. Ghiasi, G., Lin, T., Le, Q.V.: NAS-FPN: learning scalable feature pyramid architecture for object detection. In: Proceedings of the IEEE/CVF conference on computer vision and pattern recognition, pp. 7036–7045 (2019)
20. Liu, S., Huang, D., Wang, Y.: Learning spatial fusion for single-shot object detection. arXiv preprinthttp://arxiv.org/abs/1911.09516arXiv:1911.09516 (2019)
21. Ultralytics. yolov5 (2020). https://github.com/ultralytics/yolov5. GitHub repository
22. Ultralytics. Yolov8. https://github.com/ultralytics/ultralytics, 2023. GitHub repository
23. Wang, A., et al.: Yolov10: real-time end-to-end object detection. Adv. Neural. Inf. Process. Syst. **37**, 107984–108011 (2024)
24. Carion, N., Massa, F., Synnaeve, G., Usunier, N., Kirillov, A., Zagoruyko, S.: End-to-end object detection with transformers. In: European conference on computer vision, pp. 213–229. Springer (2020)
25. Zhu, X., Su, W., Lu, L., Li, B., Wang, X., Dai, J.: Deformable detr: deformable transformers for end-to-end object detection. arXiv preprinthttp://arxiv.org/abs/2010.04159arXiv:2010.04159 (2020)
26. Deng, H., Ergu, D., Liu, F., Ma, B., Cai, Y.: An embeddable algorithm for automatic garbage detection based on complex marine environment. Sensors **21**(19), 6391 (2021)
27. Dai, L., Liu, H., Song, P., Tang, H., Ding, R., Li, S.: Edge-guided representation learning for underwater object detection. CAAI Trans. Intell. Technol. **9**(5), 1078–1091 (2024)
28. Xiao, Z., Li, Z., Li, H., Li, M., Liu, X., Kong, Y.: Multi-scale feature fusion enhancement for underwater object detection. Sensors **24**(22), 7201 (2024)
29. Ali, M., Khan, S.: Underwater object detection enhancement via channel stabilization. In: 2022 International Conference on Digital Image Computing: Techniques and Applications (DICTA), pp. 1–8. IEEE (2022)
30. Selvaraju, R. R., Cogswell, M., Das, A., Vedantam, R., Parikh, D., Batra, D.: Grad-cam: visual explanations from deep networks via gradient-based localization. In: Proceedings of the IEEE international conference on computer vision, pp. 618–626 (2017)

AI for Healthcare

A-PriDiff: Anatomical Prior-Guided Conditional Diffusion for Ultrasound Spine Image Synthesis

Chen Zhang[1], Jian Li[1], Wei Guo[2], Weidong Cai[2], Yong-Ping Zheng[3], and Sai Ho Ling[1(✉)]

[1] School of Electrical and Data Engineering, University of Technology Sydney, NSW, Australia

[2] School of Computer Science, University of Sydney, Sydney, NSW, Australia

Steve.Ling@uts.edu.au

[3] Department of Biomedical Engineering, Hong Kong Polytechnic University, Hong Kong, China

Abstract. The scarcity of medical images poses major challenges for downstream tasks and AI deployment. While generative models like VAEs, GANs, and diffusion models offer potential solutions, they often fail to capture the fine-grained anatomical details essential in medical imaging, particularly for scoliosis. This highlights the need for a specialized model that preserves structural and lesion-specific features.

We propose A-PriDiff, a diffusion-based generative model tailored for ultrasound scoliosis imaging. It adopts a two-phase framework: a pre-trained segmentation U-Net extracts anatomical features, which are refined by an Anatomical Prior Encoding Module (APEM) to guide a diffusion module in synthesizing high-fidelity images.

Tested on a dataset of 1,170 raw and 109 segmented ultrasound images, A-PriDiff achieves superior performance (FID: 9.69, KID: 0.013), outperforming DDPM and GAN baselines. It also yields the lowest mean curve angle error ($1.80 \pm 0.90°$). Ablation results confirm APEM's critical role in enhancing anatomical consistency.

Keywords: Ultrasound-Based Scoliosis · Diffusion-based network · Image Generation · Medical Image Synthesis

1 Introduction

Recent advances in generative AI have expanded its applications to healthcare, where it addresses challenges such as data scarcity, privacy concerns, and ethical limitations [1,11,22,41]. In medical imaging, generative models enable synthetic data creation for tasks like segmentation, classification, and disease prognosis [6,25,38,39]. Traditional models like GANs and VAEs, while widely used,

C. Zhang, J. Li, and W. Guo—Equal contribution.

M. Liu et al. (Eds.): AI 2025, LNAI 16371, pp. 269–280, 2026.
https://doi.org/10.1007/978-981-95-4972-6_21

suffer from training instability, mode collapse, or blurry outputs [5,18]. Diffusion models have emerged as a superior alternative, offering stable training and high-quality, diverse outputs by gradually denoising noise [19,27]. They have demonstrated strong performance in generating realistic MRI/CT images and allow controllable synthesis through segmentation maps or text prompts [10,15], making them especially valuable for medical image augmentation and privacy-preserving clinical applications.

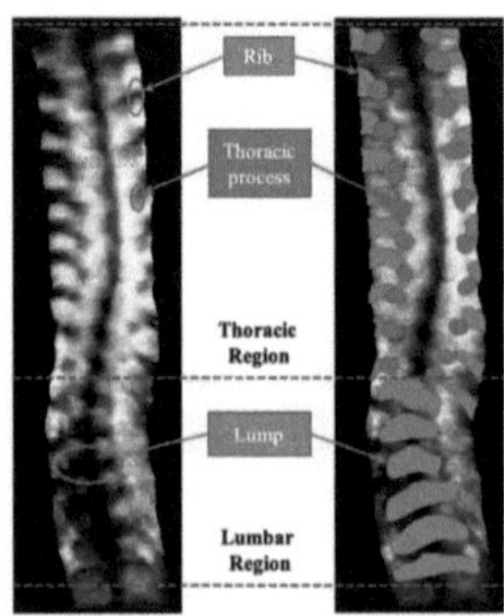

Fig. 1. A pair of data, the image on the left side is the original ultrasound image, while on the right side is the ground truth label overlapping the raw data.

Ultrasound-based scoliosis assessment poses unique challenges for image synthesis due to the modalityâĂŹs inherent limitations—ambiguous anatomical boundaries, low contrast, and the irregular shapes of thoracic and lumbar bony features (TBF/LBF) [20,26,34,42], as shown in Fig. 1. These issues complicate the generation of anatomically accurate spinal curves, which are critical for diagnostic tasks. While diffusion models have shown promise in medical imaging, standard Denoising Diffusion Probabilistic Models (DDPMs) often fail to preserve fine-grained bony structures and subtle curvature features [32].

To address this, we propose A-PriDiff, a diffusion-based generative model tailored for scoliosis ultrasound imaging. It consists of two phases: (1) a pre-trained Segmentation UNet extracts anatomical structures, and (2) an Anatomical Prior Encoding Module (APEM) generates smoothed class-specific heatmaps as conditional priors. These priors are fused with Gaussian noise and fed into a UNet-based diffusion model (Denoising UNet) to synthesize anatomically consistent ultrasound images.

Evaluated on a dataset of 1,170 raw images and 109 paired segmentation maps, A-PriDiff outperforms both DDPM and GAN-based baselines (e.g., MEDGAN, WGAN), achieving the lowest FrÃĺchet Inception Distance (FID: 9.69) and Kernel Inception Distance (KID: 0.013). It also delivers the most accurate clinical estimation, with a mean Ultrasound Curve Angle (UCA) error of $1.80° \pm 0.90°$. Ablation studies confirm the effectiveness of APEM, with its removal degrading both FID (from 9.69 to 12.4) and Structural Similarity Index

Measure (SSIM) (from 0.85 to 0.79). FID of 9.69 and KID of 0.013, indicating superior image quality. Moreover, clinical feature assessment revealed that A-PriDiff attained the smallest UCA error of 1.80° ± 0.90°, reflecting its ability to capture critical anatomical details. An ablation study further validated the importance of our Anatomical Prior Encoding Module, as its removal resulted in degraded performance (FID increasing to 12.4 and SSIM dropping from 0.85 to 0.79). Our main contributions are as follow:

- A flexible generative architecture supporting both conditional and unconditional modes, enabling diverse outputs while leveraging anatomical priors.
- Integration of a pre-trained Segmentation UNet to extract structural information, guiding diffusion for anatomically consistent synthesis.
- Proposal of APEM to refine segmentation maps into informative priors, improving structural fidelity, UCA estimation, and overall image quality.

2 Related Works

2.1 Significant in Scoliosis Assessment

Scoliosis, characterized by lateral spinal curvature, is traditionally diagnosed using X-rays [8,35], but radiation risks limit frequent use [23]. Ultrasound imaging has emerged as a radiation-free, cost-effective alternative for scoliosis detection [37], with high-quality synthetic ultrasound images potentially improving AI diagnostic accuracy and efficiency [4]. However, ultrasound has limitations, including low spatial resolution, limited field of view, poor tissue penetration for bone imaging, and operator dependency that reduces accuracy, especially in obese patients [24,28,42].

Generating synthetic ultrasound images addresses these challenges by expanding and balancing training datasets, particularly for rare cases [16,33]. This enhances model recognition of various scoliosis degrees, reduces overfitting, and improves classification accuracy through controlled parameters like scoliosis angle and noise level [40]. Synthetic images also enable real-time processing for dynamic assessments and serve as valuable training tools for clinicians [14].

2.2 Medical Image Generation

Generative models have recently transformed medical image synthesis, addressing data scarcity and privacy concerns. Applications include BrainGAN for brain MRI generation [2], conditional GANs for pseudo liver lesion images in CT [17], and diffusion priors for liver anomaly inpainting [30]. Diffusion models outperform GANs in 3D MRI and CT generation, offering greater diversity, accuracy, and segmentation improvements [19]. Advanced designs include Med-DDPM with semantic conditioning [9], DiffuMask for text-guided localization [36], segmentation-guided anatomical conditioning [21], and MCDN for morphological consistency [43].

Nevertheless, generative models encounter significant constraints. Unbalanced generatorâĂŞdiscriminator updates cause training instability, which is made worse by small datasets [7,31]. Overfitting and repeated images result from mode collapse, which limits output variety [29]. Above all, anatomical inconsistency continues to be a significant obstacle: medical synthesis necessitates precise structural and pathological fidelity across modalities, which calls for specialized models that retain anatomy while capturing lesion-specific characteristics, in contrast to natural images.

3 Method

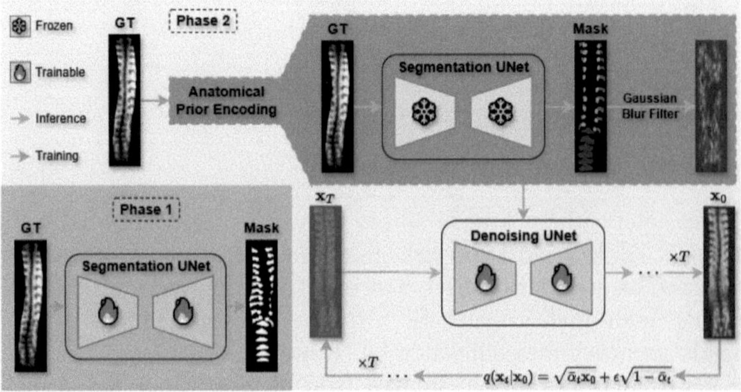

Fig. 2. Overview of our two-phase framework for anatomical condition diffusion. Phase 1: pretrain a Segmentation UNet to generate an anatomical mask. Phase 2 uses the frozen, pretrained Segmentation UNet to encode an anatomical prior (optionally blurred), resulting in a heatmap that guides the Denoising UNet during the diffusion process, ensuring anatomical accuracy while preserving essential structures.

In this study, we employ a two-phase approach to enhance image generation. In Phase 1, we begin with a pre-trained Segmentation UNet model, which is active and trainable during this phase. This allows the model to effectively learn segmentation features. The Segmentation UNet is then frozen, and its learned features are utilized to generate an anatomical heatmap. In Phase 2, this heatmap serves as a conditional input for a Denoising UNet, which is active and trainable. The Denoising UNet learns to denoise images conditioned on the anatomical information provided by the heatmap, facilitating improved image generation. Both the Segmentation UNet and Denoising UNet use a basic UNet as the backbone, with added residual connections and a self-attention mechanism to enhance their capability. This structured approach leverages the strengths of both models, enabling effective segmentation and denoising through sequential training phases. The network structure is shown in Fig. 2.

3.1 Diffusion Process

In diffusion models, the noise-adding process is implemented as a T-step Markov chain, where Gaussian noise is gradually introduced to the data in a controlled manner. At each step, a carefully scaled amount of noise is added to the original data $\mathbf{x}_0$. This process can be expressed as (Eq. (1)):

$$q(x_t \mid x_0) = \sqrt{\bar{\alpha}_t}\, x_0 + \sqrt{1 - \bar{\alpha}_t}\, \epsilon \tag{1}$$

where $\sqrt{\bar{\alpha}_t}$ and $\sqrt{1 - \bar{\alpha}_t}$ control the contributions of the original data $\mathbf{x}_0$ and the noise term ϵ, respectively.

Additionally, to guide the image generation process, we incorporate a heatmap condition h, which is integrated into the diffusion process as shown in the following equation (Eq. (2)):

$$q(x_t \mid x_0, h) = \sqrt{\bar{\alpha}_t}\, \mathbf{x}_0 + \sqrt{1 - \bar{\alpha}_t}\, (\epsilon_t + h) \tag{2}$$

Here, ϵ_t is the noise term at time step t, which is modified to include temporal information, as described by (Eq. (3)):

$$\epsilon_t = \epsilon + \mathrm{SinusoidalEmbed}(t) \tag{3}$$

The sinusoidal embedding function SinusoidalEmbed(t) injects temporal information into the model by encoding the time step t into a high-dimensional embedding vector. This process is defined by the following equations:

$$\mathrm{SinusoidalEmbed}(t) = (f_i(t))_{i=1}^{d} \tag{4}$$

$$f_i(t) = \begin{cases} \sin(\omega_i t), & \text{if } 1 \leq i \leq d/2, \\ \cos(\omega_{i-d/2} t), & \text{if } d/2 + 1 \leq i \leq d. \end{cases} \tag{5}$$

In these equations (Eqs. (4) and (5)), ω_i represents the frequency associated with the i-th component of the sinusoidal embedding, which is spaced logarithmically. For the first half of the embedding vector (i.e., for i from 1 to $d/2$), the sine function $\sin(\omega_i t)$ is used, while for the second half (i.e., for i from $d/2 + 1$ to d), the cosine function $\cos(\omega_{i-d/2} t)$ is applied. The dimensionality d defines the size of the embedding vector, and the frequencies ω_i are chosen to be logarithmically spaced, which helps capture both fine-grained and coarse temporal patterns.

3.2 Anatomical Prior Encoding

To ensure data diversity, we introduce anatomical features into the model by randomly integrating them at multiple scales (as Fig. 3 shows). This enriches the modelâĂŹs capacity to capture diverse anatomical patterns while reducing overfitting to any specific scale. Given an input image $x \in \mathbb{R}^{B \times 1 \times H \times W}$, where B is the batch size, H is the height, and W is the width, anatomical conditions

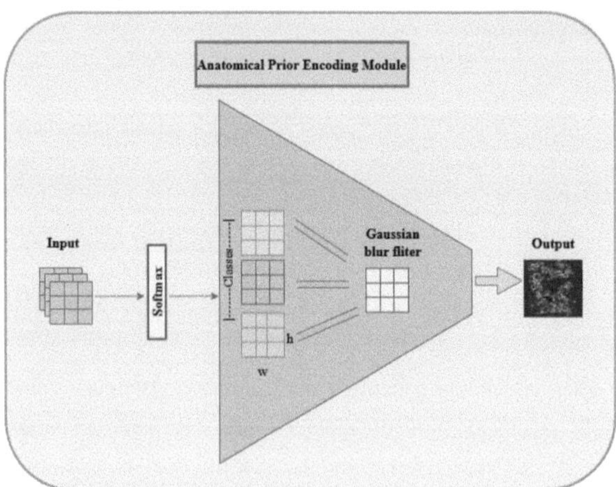

Fig. 3. The Anatomical Prior Encoding Module processes input data through a Softmax layer, followed by multi-channel grid representation and Gaussian blur filtering to produce anatomically informed feature maps.

are extracted using a pre-trained segmentation model S. This model outputs pixel-wise class probabilities, which are transformed into smoothed multi-channel heatmaps representing anatomical structures.

The heatmap for each class c is computed as:

$$\text{Heatmap}_c = \frac{1}{2\pi\sigma^2} \sum_{i',j'} \exp\left(- \frac{(i-i')^2 + (j-j')^2}{2\sigma^2} \right) \cdot \text{Probability Map}_{c,i',j'}. \tag{6}$$

This equation performs a Gaussian smoothing operation over the probability map for each anatomical class c. Each pixel in the heatmap is computed by applying a Gaussian kernel centered at (i, j), weighted by the corresponding probability values at neighboring pixels (i', j'). The result is a soft, spatially coherent representation of anatomical features that preserves local structure and facilitates stable conditioning in the diffusion process.

To ensure data diversity, we introduce anatomical features into the model by randomly integrating them at multiple scales. This approach enriches the model's ability to capture various anatomical details while avoiding overfitting to any specific scale. As illustrated in Fig. 3, the APEM generates multi-scale, anatomically-informed feature maps that are integrated into the model to enhance its ability to capture and preserve anatomical structures. Given an input image $x \in \mathbb{R}^{B\times 1\times H\times W}$, where B is the batch size, H is the height, and W is the width, the anatomical condition is derived using a pre-trained segmentation model S. The segmentation model outputs logits for each pixel, which are processed to generate a multi-channel heatmap.

3.3 Dataset and Experimental Details

This research utilizes input images acquired from the Scolioscan system [42], which employs 3D ultrasound imaging techniques to generate 3D Volume Projection Images (VPI) from data collected from 109 patients. After slicing each 3D sample, there are 1170 2D images.

Ultrasound images were preprocessed by converting to tensors, normalizing to the range $[-1, 1]$ (mean = std = 0.5), and resizing to 448×448. For Phase 1, segmentation, a 4-class classification was used: 0 (Background), 1 (Thoracic), 2 (Rib), and 3 (Lumbar). For Phase 2, the model was trained for 2000 epochs and 1000 timesteps with an initial learning rate of 2×10^{-4}, dynamically adjusted by a scheduler for stable convergence.

To align with real ultrasound resolution, generated images were upsampled to 2580×640 using bicubic interpolation before being saved for further analysis.

3.4 Result

Synthetic Similarity. The table presents a comparison of four generative models based on their FID (FrÃľchet Inception Distance) and KID (Kernel Inception Distance) scores. These models include two diffusion-based models (A-PriDiff and DDPM [13]) and two GAN-based models (MEDGAN [12] and WGAN [3]). The FID and KID scores are used to evaluate the quality of the generated images with the original images, with lower values indicating better performance, as shown in Table 1.

Table 1. Summary of different models with their respective FID and KID values.

Model Type	Model Name	FID	KID
A-PriDiff	Diffusion-based	**9.69**	**0.013**
DDPM	Diffusion-based	10.6	0.028
MEDGAN	GAN	27.38	0.043
WGAN	GAN	25.88	0.048

A-PriDiff performs the best with the lowest FID (9.69) and KID (0.013), indicating high-quality image generation. DDPM has slightly higher scores (FID 10.6, KID 0.028), suggesting slightly lower image quality. MEDGAN and WGAN have higher FID and KID scores, indicating lower quality images, with WGAN having the highest KID (0.048).

Clinical Feature Assessment. For each model, we computed the Ultrasound Curve Angle (UCA) for each output [4]. The mean error reflects the accuracy of the segmentation-based UCA estimation. Its magnitude is largely determined by the precision of landmark identification—if key bony structures are unclear

Table 2. Comparison of Range Values and Mean Error between each model for Thoracic and Lumbar Regions(°)

	Thoracic Range	Lumbar Range	Mean error (Std)
A-PriDiff	2.10 – 15.50	4.00 – 20.30	**1.80 ± 0.90**
DDPM	1.80 – 18.60	2.50 – 22.40	3.20 ± 2.10
MEDGAN	1.90 – 19.20	3.00 – 23.50	3.50 ± 2.30
WGAN	2.00 – 17.80	3.50 – 21.80	2.90 ± 2.80

or misaligned due to poor image quality, accurate landmark detection becomes challenging, leading to a higher mean error.

The results, presented in Table 2, highlight the performance differences among the models. A-PriDiff achieves the lowest mean error (1.80 ± 0.90) and produces the most stable images. In contrast, DDPM, WGAN, and MEDGAN exhibit higher mean errors, indicating lower reliability.

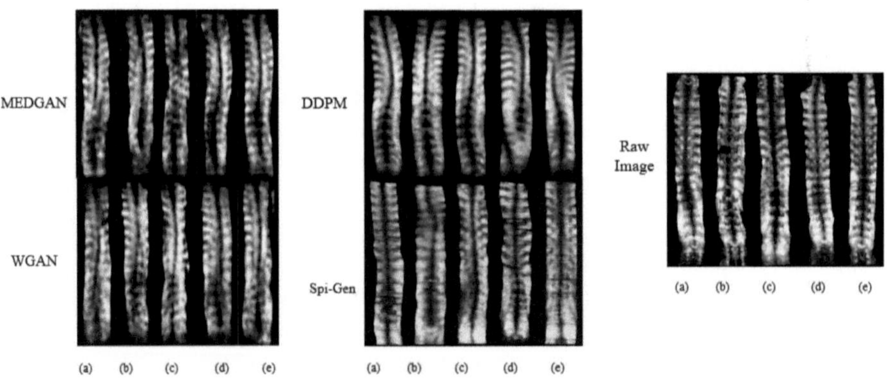

Fig. 4. A pair of generated data. a,b,c,d,e are five images generated by each model.

Qualitative Review. We conducted a clinical evaluation of all generative models in collaboration with medical experts. As shown in Fig. 4, the generated results are compared with raw training images.

Both models show strong clinical utility, accurately capturing key anatomical structures. Thoracic and lumbar bones are clearly defined with preserved contrast, aiding visual interpretation. Most outputs also feature a continuous and distinct spinal centerline, supporting reliable diagnostic use.

However, based on clinical feedback, certain generated images appear to be theoretically unreliable. As shown in Fig. 5, we observed a few instances of failed generation for MedGAN, WGAN, and DDPM, where the spinal centerline is visibly discontinuous. This outcome contradicts established anatomical principles:

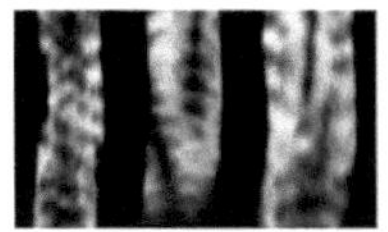

Fig. 5. Three clinical failure results, with uncontinuous spinal structure.

although severe scoliosis can result in a markedly curved spine, the vertebral column itself remains continuous. Such unrealistic discontinuities raise concerns regarding the clinical applicability of the generated images. If left unaddressed, these artifacts could mislead clinical assessments and compromise diagnostic accuracy. Nevertheless, the A-PriDiff model tends to produce images with clear organization, distinct structural contrasts, and properly displayed key anatomical landmarks, suggesting it may be more reliable for clinical applications.

Ablation Study. To evaluate the Anatomical Prior Encoding Module (APEM), we compare the full A-PriDiff model with two variants: one without APEM and another using binary masks instead of refined priors. As shown in Table 3, removing APEM degrades FID and KID, highlighting its role in preserving anatomical consistency. The binary mask variant performs better than the no-prior version but remains inferior to the full model.

Table 3. Ablation study results for the Anatomical Prior Encoding Module. The first one is Full A-PriDiff (with Anatomical Prior Encoding Module), the second one is Without Anatomical Prior Encoding Module, and the third one is with binary masks.

Model Variant	FID	KID
Setting 1	**9.69**	**0.013**
Setting 2	12.4	0.024
Setting 3	11.2	0.019

4 Discussion and Conclusion

We propose A-PriDiff, a diffusion-based generative model for ultrasound scoliosis imaging that addresses data scarcity and poor image quality. By integrating anatomical priors via an Anatomical Prior Encoding Module and a two-phase training strategy, A-PriDiff enhances anatomical consistency and supports medical image augmentation.

Experiments show that A-PriDiff generates high-fidelity, clinically plausible images, outperforming DDPM and GAN-based models, reducing UCA errors (1.80 vs. 3.20) and achieving superior image quality (FID: 9.69, KID: 0.013).

This work demonstrates the potential of diffusion models in ultrasound diagnostics. Limitations include a small dataset (109 patients); future work will expand to multi-center datasets, diverse devices, and semi-supervised preprocessing with image outpainting to improve segmentation and diagnosis.

References

1. Achiam, J., et al.: Gpt-4 technical report. arXiv preprint arXiv:2303.08774 (2023)
2. Alrashedy, H.H.N., Almansour, A.F., Ibrahim, D.M., Hammoudeh, M.A.A.: Braingan: brain mri image generation and classification framework using gan architectures and cnn models. Sensors **22**(11), 4297 (2022)
3. Arjovsky, M., Chintala, S., Bottou, L.: Wasserstein generative adversarial networks. In: International Conference on Machine Learning, pp. 214–223. PMLR (2017)
4. Banerjee, S., et al.: Automatic assessment of ultrasound curvature angle for scoliosis detection using 3-d ultrasound volume projection imaging. Ultrasound Med. Biol. (2024)
5. Chatterjee, S., Maity, S., Bhattacharjee, M., Banerjee, S., Das, A.K., Ding, W.: Variational autoencoder based imbalanced covid-19 detection using chest x-ray images. N. Gener. Comput. **41**(1), 25–60 (2023)
6. Chen, Y., Esmaeilzadeh, P.: Generative ai in medical practice: in-depth exploration of privacy and security challenges. J. Med. Internet Res. **26**, e53008 (2024)
7. Chen, Y., et al.: Generative adversarial networks in medical image augmentation: a review. Comput. Biol. Med. **144**, 105382 (2022)
8. Cheng, J.C., et al.: Adolescent idiopathic scoliosis. Nat. Rev. Dis. Primers. **1**(1), 1–21 (2015)
9. Dorjsembe, Z., Pao, H.K., Odonchimed, S., Xiao, F.: Conditional diffusion models for semantic 3d brain MRI synthesis. IEEE J. Biomed. Heal. Inf. (2024)
10. Graf, R., et al.: Denoising diffusion-based MRI to CT image translation enables automated spinal segmentation. European Radiol. Experiment. **7**(1), 70 (2023)
11. Guo, D., et al.: Deepseek-coder: when the large language model meets programming–the rise of code intelligence. arXiv preprint arXiv:2401.14196 (2024)
12. Guo, K., et al.: Medgan: an adaptive GAN approach for medical image generation. Comput. Biol. Med. **163**, 107119 (2023)
13. Ho, J., Jain, A., Abbeel, P.: Denoising diffusion probabilistic models. Adv. Neural. Inf. Process. Syst. **33**, 6840–6851 (2020)
14. Hu, X., et al.: Standalone ultrasound-based highly visualized volumetric spine imaging for surgical navigation. Sci. Rep. **15**(1), 4922 (2025)
15. Hung, A.L.Y., et al.: Med-cdiff: conditional medical image generation with diffusion models. Bioengineering **10**(11), 1258 (2023)
16. Hwang, B.Y., et al.: Ultrasound in traumatic spinal cord injury: a wide-open field. Neurosurgery **89**(3), 372–382 (2021)
17. Ikeda, Y., Doman, K., Mekada, Y., Nawano, S.: Lesion image generation using conditional GAN for metastatic liver cancer detection. J. Image Graph. **9**(1), 27–30 (2021)
18. Jiang, L., Zhang, M., Wei, R., Liu, B., Bai, X., Zhou, F.: Reconstruction of 3d CT from a single x-ray projection view using cvae-gan. In: 2021 IEEE International Conference on Medical Imaging Physics and Engineering (ICMIPE), pp. 1–6. IEEE (2021)

19. Khader, F., et al.: Denoising diffusion probabilistic models for 3d medical image generation. Sci. Rep. **13**(1), 7303 (2023)
20. Kim, H., et al.: Scoliosis imaging: what radiologists should know. Radiographics **30**(7), 1823–1842 (2010)
21. Konz, N., Chen, Y., Dong, H., Mazurowski, M.A.: Anatomically-controllable medical image generation with segmentation-guided diffusion models. In: International Conference on Medical Image Computing and Computer-Assisted Intervention, pp. 88–98. Springer (2024)
22. Lambert, B., Forbes, F., Doyle, S., Dehaene, H., Dojat, M.: Trustworthy clinical AI solutions: a unified review of uncertainty quantification in deep learning models for medical image analysis. Artifi. Intell. Med. 102830 (2024)
23. Levy, A.R., Goldberg, M.S., Mayo, N.E., Hanley, J.A., Poitras, B.: Reducing the lifetime risk of cancer from spinal radiographs among people with adolescent idiopathic scoliosis. Spine **21**(13), 1540–1547 (1996)
24. Luo, Y., et al.: Investigating a novel non-surgical management for adolescent and adult scoliosis using digitalized 3d correction and ultrasound assessment: a preliminary feasibility study. In: International Research Society of Spinal Deformities (IRSSD) Scientific Meeting 2024 (2024)
25. Ning, Y., et al.: Generative artificial intelligence and ethical considerations in health care: a scoping review and ethics checklist. Lancet Digit. Heal. **6**(11), e848–e856 (2024)
26. Noble, J.A., Boukerroui, D.: Ultrasound image segmentation: a survey. IEEE Trans. Med. Imaging **25**(8), 987–1010 (2006)
27. Rombach, R., Blattmann, A., Lorenz, D., Esser, P., Ommer, B.: High-resolution image synthesis with latent diffusion models. In: Proceedings of the IEEE/CVF Conference on Computer Vision and Pattern Recognition, pp. 10684–10695 (2022)
28. Rykkje, A., Carlsen, J.F., Nielsen, M.B.: Hand-held ultrasound devices compared with high-end ultrasound systems: a systematic review. Diagnostics **9**(2), 61 (2019)
29. Shahbazi, M., Danelljan, M., Paudel, D.P., Van Gool, L.: Collapse by conditioning: training class-conditional GANs with limited data. arXiv preprint arXiv:2201.06578 (2022)
30. Shi, Y., Niu, C., Simpson, A.L., De Man, B., Do, R., Wang, G.: Ct-based anomaly detection of liver tumors using generative diffusion prior. arXiv preprint arXiv:2408.00092 (2024)
31. Singh, N.K., Raza, K.: Medical image generation using generative adversarial networks: a review. Health Informatics: a Computational Perspective in Healthcare, pp. 77–96 (2021)
32. Song, J., Meng, C., Ermon, S.: Denoising diffusion implicit models. arXiv preprint arXiv:2010.02502 (2020)
33. Tawfik, N.A., Ahmed, A.T., El-Shafei, T.E., Habba, M.R.: Diagnostic value of spinal ultrasound compared to MRI for diagnosis of spinal anomalies in pediatrics. Egyptian J. Radiol. Nuclear Med. **51**, 1–11 (2020)
34. Trac, S., Zheng, R., Hill, D.L., Lou, E.: Intra-and interrater reliability of cobb angle measurements on the plane of maximum curvature using ultrasound imaging method. Spine Deformity **7**, 18–26 (2019)
35. Weinstein, S.L., Dolan, L.A., Cheng, J.C., Danielsson, A., Morcuende, J.A.: Adolescent idiopathic scoliosis. Lancet **371**(9623), 1527–1537 (2008)
36. Wu, W., Zhao, Y., Shou, M.Z., Zhou, H., Shen, C.: Diffumask: synthesizing images with pixel-level annotations for semantic segmentation using diffusion models. In: Proceedings of the IEEE/CVF International Conference on Computer Vision, pp. 1206–1217 (2023)

37. Yang, D., et al.: Semi-automatic ultrasound curve angle measurement for adolescent idiopathic scoliosis. Spine Deformity, 1–9 (2022)
38. Zhang, C., Deng, X., Ling, S.H.: Next-gen medical imaging: U-net evolution and the rise of transformers. Sensors **24**(14), 4668 (2024)
39. Zhang, C., Zheng, Y., McAviney, J., Ling, S.H.: Ssat-swin: Deep learning-based spinal ultrasound feature segmentation for scoliosis using self-supervised swin transformer. Ultrasound Med. Biol. **51**(6), 999–1007 (2025)
40. Zhang, J., Cui, X., Chen, S., Dai, Y., Huang, Y., Zhang, S.: Ultrasound-guided nusinersen administration for spinal muscular atrophy patients with severe scoliosis: an observational study. Orphanet J. Rare Dis. **16**, 1–8 (2021)
41. Zhang, L., Rao, A., Agrawala, M.: Adding conditional control to text-to-image diffusion models. In: Proceedings of the IEEE/CVF International Conference On Computer Vision, pp. 3836–3847 (2023)
42. Zheng, Y.P., et al.: A reliability and validity study for scolioscan: a radiation-free scoliosis assessment system using 3d ultrasound imaging. Scoliosis Spinal Disorders **11**, 1–15 (2016)
43. Zhou, Y., et al.: Morphological-consistent diffusion network for ultrasound coronal image enhancement. arXiv preprint arXiv:2409.16661 (2024)

MammoMix: Leveraging Mixture of Experts for Robust Mammogram Breast Detection

Dinh Tan Nguyen[1], Hoang Quan Dang[3], Chen Zhang[2], and Sai Ho Ling[3](✉)

[1] School of Biomedical Engineering, University of Technology Sydney, Sydney, Australia
dinhtan.nguyen@uts.edu.au

[2] School of Computer Science, University of Technology Sydney, Sydney, Australia
chen.zhang@uts.edu.au

[3] School of Electrical and Data Engineering, University of Technology Sydney, Sydney, Australia
hoangquan.dang@student.uts.edu.au, steve.ling@uts.edu.au

Abstract. Breast lesion detection in mammography remains a challenging task due to variations in image quality, lesion appearance, and population demographics across datasets. While current object detectors such as YOLO and DETR achieve strong results on individual datasets, their performance often degrades when trained on or applied across heterogeneous sources. To address this, we propose MammoMix, a novel framework based on Mixture-of-Experts (MoE) paradigm for robust and generalizable lesion detection. In MammoMix, each expert model is trained on a specific domain, allowing it to specialize in distinct characteristics of its source data. A gating mechanism adaptively weighs contributions from each expert based on input image, combining their outputs to enable domain-adaptive inference. To improve reliability, we further incorporate a calibration module, MoCAE, which adjusts confidence scores to reflect true predictive uncertainty. We evaluate MammoMix on 3 public mammography datasets: CSAW, DDSM, and DMID, covering diverse clinical settings. Results show that MammoMix outperforms baseline detectors in both average precision and reliability, particularly on datasets with greater variability. Our findings demonstrate that expert specialization and calibrated ensemble fusion significantly enhance model generalization and robustness. MammoMix offers a promising step toward dependable AI-assisted breast cancer screening across real-world clinical domains. Code is available at: https://github.com/tommyngx/MammoMix

Keywords: Mixture-of-Experts · Object Detection · Breast Cancer

1 Introduction

Breast cancer screening via mammography is a critical application of computer vision, where accurate lesion detection significantly influences early diagnosis

M. Liu et al. (Eds.): AI 2025, LNAI 16371, pp. 281–294, 2026.
https://doi.org/10.1007/978-981-95-4972-6_22

and outcomes [7]. Deep learning-based object detectors such as Faster R-CNN, YOLO, and DETR have shown strong performance on automated mammogram analysis, particularly when trained and tested on a single dataset [2,6]. In such controlled settings, these models often match or even exceed expert radiologist performance, highlighting their potential for computer-aided diagnosis. However, challenges emerge when models are applied across diverse clinical environments. Real-world mammography data vary significantly across populations, imaging devices, and lesion types. When multiple datasets are combined for training, performance often degrades, even if models perform well on each dataset individually. This drop in generalization is well-documented: detectors that perform strongly on their development set frequently suffer substantial losses in AUC and sensitivity on unseen domains. A unified model struggles to reconcile conflicting data distributions, resulting in unstable training and reduced effectiveness.

To overcome this, we propose MammoMix, a breast lesion detection framework based on a Mixture-of-Experts (MoE) architecture [12]. Instead of a single detector, MammoMix trains multiple expert models, each specialized to a dataset or domain. This divide-and-conquer approach allows each expert to learn domain-relevant features without interference from other distributions. A gating mechanism learns to weigh and combine expert predictions on a per-image basis, allowing the system to dynamically adapt to varying input domains. This specialization improves robustness and generalization across heterogeneous datasets.

Additionally, MammoMix incorporates a confidence calibration module, MoCAE (Mixture of Calibrated Experts), to ensure that expert outputs reflect their actual reliability. Without calibration, overconfident experts may dominate the ensemble, degrading performance. MoCAE calibrates each expert's confidence score to align predictions with actual precision, enabling fairer fusion. This ensures that the gating mechanism reflects true expertise rather than raw confidence. Together, MoE and MoCAE allow MammoMix to deliver reliable, generalizable lesion detection across diverse mammography data.

2 Related Work

Lesion detection in mammograms has attracted significant research interest, with deep learning models increasingly used to support radiologists. Two-stage detectors like Faster R-CNN have shown strong results in identifying masses and calcifications on datasets such as DDSM [9]. One-stage YOLO models, valued for their real-time speed, have also demonstrated competitive performance. More recently, Transformer-based detectors like DETR [1] have been introduced, though their application to mammography remains limited [2]. DETR introduces self-attention mechanisms to better capture global dependencies, which are essential for detecting subtle lesions in complex mammogram backgrounds. It treats object detection as a set prediction problem, eliminating the need for anchor boxes or NMS by using bipartite matching to align predictions with ground truth [1]. On the COCO dataset, DETR-ResNet50 achieves 42.0% mAP, but its slow convergence and computational demands have limited its adoption in real-time medical applications, where faster inference is crucial for clinical workflows. Variants like RT-DETRv2 [11] address these issues by optimizing for real-time performance, achieving 53.1% mAP on COCO at 108 FPS on a T4 GPU, outperforming YOLO models or even DETR in both speed and accuracy, and showing promise in medical imaging for tasks like lesion detection in mammograms due to its hybrid CNN-transformer design. However, while these models often achieve competitive accuracy when trained and tested on a single dataset, their performance tends to be domain-specific. Most studies report results confined to the dataset used for training, with accuracy dropping considerably when applied to different sources without adaptation.

Mammography datasets differ by population, imaging equipment, and annotation standards, creating conflicting data distributions that challenge generalization. Multi-center studies [5] have shown models trained on one dataset perform poorly on others, even with similar imaging protocols. Combining datasets can even reduce sensitivity, as noted by [5]. In response, domain adaptation methods such as transfer learning and adversarial alignment have been proposed, though they often struggle with overfitting or feature collapse. More recent domain generalization approaches aim to learn transferable features without relying on target data. However, most still rely on a single model stretched across varied data, limiting robustness and motivating a multi-expert strategy.

Mixture-of-Experts (MoE) models [12] assign specialized networks to different sub-tasks or domains, with a gating mechanism selecting the most relevant expert per input. This architecture has re-emerged in modern deep learning as an effective solution for handling diverse data. In medical imaging, MoEs remain underexplored, though ensemble classifiers are sometimes used in breast cancer detection. However, standard ensembles lack learnable gating and treat all models equally, limiting adaptability across domains. MammoMix is among the first to apply MoE specifically for mammographic lesion detection. Recent work on the Mixture of Calibrated Experts (MoCaE) demonstrates that calibration is crucial for effective ensembling [13]. MoCaE improves detector fusion by aligning each model's confidence with its actual performance, avoiding dominance by overconfident predictors. MammoMix builds on this idea by integrating calibration directly into MoE pipeline. Unlike MoCaE's post-hoc fusion, our experts are trained for domain specialization, optimized jointly with gating and calibration. This enables MammoMix to better accommodate distributional shifts, offering improved accuracy and reliability for multi-dataset breast lesion detection.

3 Dataset Formulation

To evaluate the effectiveness of our MoE approach, we employed 3 publicly available mammography datasets: CSAW, DDSM, and DMID. Each dataset brings unique imaging protocols, patient demographics, and lesion types, reflecting the heterogeneity found in real-world screening programs. Training expert models on these distinct subsets allowed us to explore how domain-specific specialization can enhance overall ensemble performance, reduce dataset-specific biases, and improve generalizability across clinical settings.

3.1 Dataset Description

The **Cohort of Screen-Age Women - Case Control (CSAW-CC)** [2] contains high-resolution mammograms with detailed pixel-level annotations. Designed to support AI-based breast cancer screening, it includes both cancer and control cases. For this study, we used 472 image-mask pairs, with original image resolutions of 3328×4096 pixels cropped to 1493×3023 to focus on breast regions and improve segmentation efficiency.

The **CBIS-DDSM (Curated Breast Imaging Subset of DDSM)** [10] offers 2,620 mammograms from 1,566 patients, paired with pathology reports. We selected 1,244 patients (2,142 images) and focused on 1,333 image-mask pairs involving mass lesions. Images (originally 2986×5491) were cropped to 2134×4627 to standardize input size and enhance training consistency.

The **DMID (Digital Mammography Dataset)** [14] comprises 510 images curated for diagnostic research, with 269 including segmentation masks. Original images (4751×6000) were cropped to 2602×5010 pixels. The dataset provides high-quality DICOM/TIFF images, reports, and pixel-level annotations, supporting breast mass segmentation and model evaluation.

Our expected outcome is enhanced detection accuracy, particularly for underrepresented lesion types, as models learn to adapt to inter-dataset variations without assuming homogeneity. The goal of using these datasets instead of a single merged one is to allow our MoE variants to learn specialized representations, simulating real-world clinical scenarios where X-ray images come from varied sources. This mirrors "federated learning" paradigms in healthcare, where data silos are common due to privacy or institutional constraints, aiming to achieve models that are not only accurate but also adaptable to new clinical environments. Quantitatively, the combined dataset size (2,006 images) enables robust statistical evaluation; however, the imbalance (DDSM dominating, fewer positives in DMID) highlights the need for MoE to prevent overfitting to the largest sub-dataset, motivating our designs to pool knowledge across subsets.

3.2 Data Preprocessing and Augmentation

The preprocessing began with image resizing and standardization. All images were resized to a dimension of 640 × 640 pixels while preserving aspect ratios through padding, ensuring compatibility with Transformer-based architectures like YOLOS, which benefit from square inputs for efficient attention mechanisms. Following resizing, pixel values were normalized to the range ([0, 1]). Subsequently, we utilized a composition of both geometric and intensity-based augmentations using the albumentations[1] library to simulate real-world variability in mammograms, such as differences in patient positioning, imaging conditions, and tissue density. The goal is to improve model generalization and reduce overfitting by exposing the model to diverse image variations, addressing the imbalance across sub-datasets (e.g., DDSM's dominance) and the scarcity of positive cancer samples. Our geometric augmentations included:

[1] https://albumentations.ai.

- *Elastic Transformations*: Simulate breast tissue deformation (alpha = 50, sigma = 5), helping the model learn deformation-invariant features, especially beneficial for dense breasts.
- *Perspective Transform*: Adds mild distortions (scale 0.05âĂŞ0.1) to mimic imaging angle variations and dataset differences (e.g., DDSM vs. DMID), improving viewpoint robustness.
- *Horizontal Flip*: Mirrors images to enforce left-right symmetry learning, enhancing orientation invariance across breast views.
- *Rotate*: Small random rotations ($\pm 10°$) simulate patient misalignment while preserving lesion shape, aiding generalization.
- *Random Scale*: Varies image size by ± 0.2 to mimic distance and lesion size variability, supporting scale-invariant detection.
- *Affine Transformations*: Apply combined scaling, translation (± 0.1), rotation ($\pm 10°$), and shear ($\pm 6°$) to simulate geometric distortions.
- *Random Brightness and Contrast*: Adjusted within ± 0.2 to simulate X-ray exposure variation and enhance robustness to contrast inconsistencies.
- *Gaussian Noise*: Added with standard deviation 0.05 to mimic scanner noise and low-quality scans, helping models generalize under real-world artifacts.
- *Gaussian Blur*: Slight blur applied to replicate mild defocus, encouraging the model to rely on robust edge features rather than sharp contours alone.

All augmentations were implemented with bounding box awareness, ensuring labels were adjusted accordingly using VOC format ($[x_{min}, y_{min}, x_{max}, y_{max}]$) and clipped to image boundaries. Boxes with area less than 25 pixels or visibility below 10% post-augmentation were discarded to maintain annotation quality. If augmentations resulted in no valid boxes, we re-sampled the image to ensure at least one valid detection target, addressing the risk of losing positive samples in small datasets like DMID. Augmentations were applied only during training without altering validation or test data to reflect real-world inference.

4 Mixture of Approaches: MoMo, MoE, and MoCAE

We designed and evaluated 3 approaches using YOLOS as the backbone: MoMo, Simple MoE, and MoCaE. Each variant explores a distinct strategy for leveraging the sub-datasets' heterogeneity, aiming to enhance detection accuracy and robustness by combining the strengths of multiple models or training paradigms. The rationale behind these variants is to exploit dataset-specific patterns while mitigating issues like domain shifts and sub-dataset imbalance. Their designs were motivated by the hypothesis that leveraging dataset diversity through specialized models could outperform a unified model. Our goal is to achieve a detection system that maximizes mAPs across varied clinical scenarios, particularly for small and medium-sized lesions, which are critical for early diagnosis.

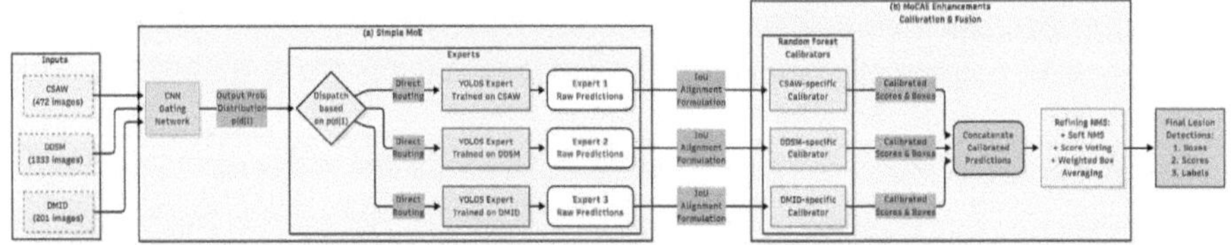

Fig. 1. Overview of the MammoMix framework: (a) Expert models specialized on individual datasets got selected by a CNN gate (b) Calibration module aligns image embeddings and confidence scores with IoU by employing the predictive reliability from Random Forest; then fusion via Soft NMS and Score Voting.

4.1 Expert Selection

We focused on Transformer-based architectures and the core of our detection framework is the YOLOS model [4], a Vision Transformer (ViT) [3] object detector that serves as the backbone for all approaches evaluated in this study: MoMo, Simple MoE, and MoCaE. We selected YOLOS over DETR [1], despite both being Transformer-based, for several reasons informed by our experiments and literature. First, YOLOS's purely ViT-based backbone eliminates the need for a convolutional backbone (e.g., ResNet in DETR), reducing architectural complexity of Transformer, which is advantageous for training on moderate-sized datasets like ours (2006 total images). This choice was evidenced by our initial experiments, showing YOLOS outperforming DETR in terms of convergence speed and accuracy on our datasets. YOLOS's novelty in medical imaging, underexplored compared to CNNs or DETR, further motivated its selection, aiming to contribute a new benchmark for Transformer-based detection in mammography.

Unlike traditional CNN-based detectors like YOLO variants that rely on local receptive fields, YOLOS leverages self-attention mechanisms to model long-range dependencies without the inductive biases of convolutions, which is particularly advantageous for medical imaging where contextual relationships, such as the surrounding tissue density influencing lesion detectability, are crucial. The YOLOS architecture applied a simple ViT directly to image patches and treated them as a sequence of tokens, augmented with learnable detection tokens and processed through Transformer layers to predict bounding boxes and class probabilities. Specifically, an input image $I \in R^{H \times W \times C}$ is divided into non-overlapping patches of size $P \times P$, flattened into a sequence of N tokens, where $N = (H \cdot W)/P^2$. In other words, all images in our datasets were reshaped to $640 \times 640 \times 3$, resulting in $N = \lfloor \frac{640}{16} \rfloor^2 = 1600$ patches (patch size 16×16), each embedded into a $d = 768$-dimensional vector. These patch embeddings were augmented with positional encodings and concatenated with a set of learnable [DET] tokens (fixed at 100), equivalent to "object queries" in DETR, forming the input sequence to the Transformer encoder. The encoder processes this sequence through 12 Transformer layers of multi-head self-attention and feed-forward networks, producing a sequence of contextualized features. The detection heads then predict up to 100 boxes and their class probabilities (in our case, for "cancer" class, plus a "no-object" class) from detection tokens' outputs.

The output of YOLOS is a set of predictions $\{(b_i, s_i, l_i)\}_{i=1}^{100}$, where $b_i \in [0,1]^4$ represents the normalized bounding box coordinates $[x_{\text{center}}, y_{\text{center}}, w, h]$, $s_i \in [0,1]$ is the confidence score, and $l_i \in \{0,1\}$ is the class label (0 for "cancer", 1 for "no-object"). These predictions are matched to ground truth boxes using bipartite matching, optimized via a combination of classification loss (cross-entropy) and box regression loss (L1 and generalized IoU loss). This architecture's simplicity with no anchor boxes or Non-maximum Suppression (NMS) required during single-model inference made it an ideal candidate for adaptation into our MoE frameworks where multiple model outputs must be combined.

4.2 MoMo: Combined Dataset Training

The MoMo (Monolithic Model) approach serves as a baseline to assess the efficacy of combining all sub-datasets into a unified source of mammography images without distinguishing their origins. In this method, a single YOLOS model is trained on the aggregated dataset comprising CSAW, DDSM, and DMID. The rationale for MoMo was to test whether a single, general-purpose exposed to a larger, more diverse dataset could generalize across the varied imaging conditions without requiring explicit specialization. By pooling the data, we aimed to maximize training sample size, potentially mitigating the data scarcity issue in smaller datasets like DMID while allowing the model to learn shared features or a robust representation of cancerous lesions across diverse imaging modalities (e.g., masses, calcifications) without the need for specialized training. The goal was to achieve robust performance across all test sets, particularly for DDSM, which dominates the combined dataset. However, we anticipated potential drawbacks: the model might dilute sub-dataset-specific patterns and overfit to DDSM's characteristics, such as its lower-resolution digitized film images, potentially reducing sensitivity to the high-contrast lesions in CSAW or subtle calcifications in DMID. The MoMo approach thus provides a point of comparison to evaluate whether specialized models offer significant advantages over naive data aggregation.

Predictions in MoMo are generated directly by the single YOLOS model, with no additional combination or calibration steps. The output bounding boxes and confidence scores for the "cancer" class are post-processed using standard non-maximum suppression (NMS) to eliminate redundant detections, with an IoU threshold of 0.5 to balance precision and recall. The anticipated outcome is a model that performs adequately across all sub-datasets but may underperform on tasks requiring fine-grained adaptation to dataset-specific nuances, particularly for the smaller DMID dataset, where specialized knowledge is advantageous.

4.3 Simple MoE: Simple Ensemble of Experts

The Simple MoE approach trains 3 separate YOLOS models, each as an "expert" on one of the sub-datasets, and combines their predictions using a learned gating mechanism. Unlike traditional MoE implementations that use a learned gating function trained jointly with experts, we trained a separate CNN on the combined dataset to classify an image's dataset origin or to select the most relevant expert for each image. This gating network, denoted as $G : \mathcal{X} \rightarrow \{1, 2, 3\}$, where $\mathcal{X}$ is the image space and the output indices correspond to CSAW, DDSM, or DMID experts, assigns each test image to the expert most likely to perform well based on its dataset origin. Formally, for an input image x, the gating network outputs a probability distribution over the 3 datasets, $p(d|I) \in [0, 1]^3$, where d is the dataset and I is the input image.

During inference, we select the expert corresponding to the dataset with the highest probability, $\arg\max_d p(d|I)$, and use its predictions as the final output $\hat{y}_d = f_d(x)$, where f_d is the YOLOS model trained on the d-th sub-dataset. This approach was chosen to simplify the combination process and exploit the specialization of each expert while introducing a dynamic selection mechanism to route images to the most appropriate model, expecting improved performance over MoMo by prioritizing dataset-relevant predictions. However, it risks suboptimal performance if the gating network misclassifies the dataset or if multiple experts produce overlapping or complementary detections, motivating the more sophisticated MoCaE approach.

4.4 MoCaE: Mixture of Calibrated Experts

The MoCAE approach, inspired by [13], extends the Simple MoE by incorporating calibration of expert predictions and a sophisticated prediction fusion strategy known as Refining NMS to address the limitations of naive ensemble methods. This approach addressed the critical issue of miscalibration in ensemble methods, where raw confidence scores from different experts may be misaligned due to variations in training data, leading to biased combinations where a confident but inaccurate expert dominates. Calibration aligns confidence scores with the true likelihood of correct detection, measured as the maximum IoU with ground truth boxes. Formally, for a prediction with confidence score s, the calibrator $C : s \rightarrow \hat{s}$ should map s to an expected IoU value, such that $\hat{s} \approx E[\mathrm{IoU}(b, g)|s]$, where b is the predicted box and g is the ground truth. We aim to maximize the complementary strengths and ensure fair contribution from each expert, expecting improved overall mAP, particularly for challenging cases like small and medium-sized lesions where dataset-specific expertise is critical.

For each YOLOS expert, we constructed a calibration dataset using validation images from all 3 sub-datasets. This was motivated by the need to capture both in-domain and out-of-domain performance, as experts may encounter images from other datasets during testing, reflecting real-world scenarios where image sources are mixed. For an expert e trained on dataset D_e (e.g., CSAW), we processed validation images from all datasets to collect:

1. Image embeddings extracted using a pre-trained ResNet-18 backbone (without its classification head), producing a 512-dimensional feature vector per image, capturing high-level visual characteristics like texture and contrast.
2. Confidence scores from the expert's predictions for the "cancer" class.

3. Intersection over Union (IoU) values: For in-domain images from D_e, we computed predictions $\{(b_i^e, s_i^e, l_i^e)\}$ and their maximum IoUs with ground truth boxes, $\text{IoU}_{\max} = \max_k \text{IoU}(b_i^e, g_k)$, where g_k is a ground truth box. For out-of-domain images from other datasets $(D_k, k \neq e)$, we assign an IoU of 0, as these images lack ground truth annotations relevant to the expert's training domain, reflecting the expert's likely poor performance due to domain mismatch. The confidence score $s_i^e \in [0, 1]$ and image embeddings $f \in \mathbb{R}^{512}$ form the input, resulting in a 513-dimensional vector and paired with target $\text{IoU}_{\max} \in [0, 1]$ for in-domain or 0 for out-of-domain samples.

3 separate calibrators were trained, one per expert, to account for dataset-specific biases and ensure each expert's predictions are adjusted independently. Here, we employed 3 Random Forest regressors as 3 calibrators to minimize the Mean Squared Error between predicted and actual IoU values: $\text{MSE} = \frac{1}{N}\sum_{k=1}^{N}\left(C([f_k, s_k]) - \text{IoU}_{\max,k}\right)^2$, where N is the number of validation samples, and C is the calibrator. The Random Forest was chosen over isotonic regression (used in the original MoCaE [13]) due to its robustness to high-dimensional inputs and ability to model non-linear relationships between embeddings, scores, and IoU, outperforming simpler methods when incorporating image features. This calibration strategy ensures the calibrator learns to assign low confidence to out-of-domain predictions, preventing overconfident errors. We aimed to mimic clinical settings where image sources are not pre-identified, enhancing the generalizability and reliability of confidence scores in mixed-domain testing scenarios.

During inference, all 3 experts process each test image, producing predictions $\{(b_i^e, s_i^e, l_i^e)\}_{i=1}^{100}$. Confidence scores are calibrated using the respective Random Forests, yielding $\tilde{s}_i^e \approx \text{IoU}(b_i^e, g_k)$. These predictions are concatenated into a single set, filtered for the "cancer" class ($l_i^e = 0$), and processed through Refining NMS, which acted as a routing mechanism to handle overlapping predictions from multiple experts with 2 stages:

- Soft NMS: Unlike traditional NMS, which discards overlapping boxes based on a fixed IoU threshold, Soft NMS reduces the scores of overlapping boxes using a Gaussian decay. For a box b_i with calibrated score $\tilde{s}_i$, and another box b_j with $\text{IoU}(b_i, b_j) > \theta$, the score is updated as $\tilde{s}_i \leftarrow \tilde{s}_i \cdot \exp\left(-\frac{\text{IoU}(b_i,b_j)^2}{\sigma_{\text{nms}}}\right)$, where σ_{nms} controls the influence of nearby boxes. This soft suppression, inspired by the MoCAE framework in [13], was chosen to reduce false negatives and avoid hard suppression of valid detections, particularly for dense or clustered lesions common in mammography.
- Score Voting: After Soft NMS, surviving boxes are refined by weighted averaging based on their calibrated scores and pairwise IoUs, where the weight for box b_i is computed as $w_{ij} = \tilde{s}_i \cdot \exp\left(-\frac{(1-\text{IoU}(b_i,b_j))^2}{\sigma_{\text{nms}}}\right)$, excluding self-influence $(i \neq j)$. The refined box is $b_i' = \sum_{j \neq i} w_{ij} \cdot b_j / \sum_{j \neq i} w_{ij}$. This fusion improved box localization by aggregating information from all experts to refine their coordinates, leveraging their complementary detections strengths

5 Experimental Setup

YOLOS with base version serves as the unified backbone across all MoE approaches, ensuring consistency in feature extraction while allowing each variant to explore different training and combination strategies. Optimization used AdamW with a learning rate of 5×10^{-5}, weight decay of 10^{-4}, and a cosine scheduler with warmup ratio 0.05 and one restart cycle. The batch size was set to 8, with gradient accumulation over 2 steps to simulate larger effective batches on a 40GB memory NVIDIA A100 GPU. We trained for 200 epochs, monitoring validation mAP to enable early stopping if no improvement occurred for 20 epochs, preventing overfitting on small datasets like DMID. This extended epoch count was selected to allow sufficient adaptation of the transformer layers, which often require longer training than CNNs due to their data-hungry nature. For the MoCAE variant, post-training calibration was integrated using a Random Forest regressor with 300 estimators to predict maximum IoU values, producing calibrated scores that better reflect true detection quality to enable more effective combination with a $\sigma_{\text{nms}} = 0.08$ in the ensemble phase.

The splitting strategy was chosen to allocate approximately 64% of data for training, 16% for validation (to tune hyperparameters and calibrate models), and 20% for testing (to evaluate generalization), as shown in Table 1. The validation set is particularly important for MoCAE, where it is used to fit 3 calibrators associated with 3 datasets used, ensuring that confidence scores are aligned with actual detection quality.

Table 1. Data distribution across training, validation, and test sets.

Subset	Train	Validation	Test	Total
CSAW	301	76	95	472
DDSM	852	214	267	1333
DMID	128	32	41	201

In evaluating the performance of our methods, several key metrics are used to measure accuracy and overlap between predicted and true regions. Among these metrics, the Mean Average Precision (mAP) [8] is a commonly used measure, especially in object detection and segmentation tasks. The mAP values are provided for different IoU thresholds to capture the performance across various levels of overlap criteria. The IoU=0.50 (average) measures precision when the IoU between predicted and ground truth regions is at least 0.50. The IoU=0.50:0.95 (average) calculates average precision across multiple IoU thresholds, offering a comprehensive performance overview. The IoU=0.50:0.95 (large) focuses on larger objects, while IoU=0.50:0.95 (medium) targets medium-sized objects, and IoU=0.50:0.95 (small) evaluates performance for smaller objects, providing insights into detection and segmentation accuracy for different sizes.

6 Results and Discussion

Our preliminary experiments confirmed that YOLOS-base outperformed DETR on mAP@50 across datasets, likely due to its simpler design and fewer modifications needed for fine-tuning on Table 2. On the CSAW dataset, individually trained YOLOS experts achieved the highest mAP@50-95 (0.3292), outperforming DETR (0.2763), RT-DETRv2 (0.2260), and even the ensemble-based approaches. While Simple MoE reached comparable performance (0.3287), the calibration-enhanced MoCAE showed slightly lower overall mAP (0.3044), yet it achieved the best performance for small lesions (0.3119). This suggests that calibration improves model reliability in detecting subtle and low-contrast abnormalities, which are typically more challenging in clinical settings.

Table 2. Comparison of mAPs for different models across 3 datasets

Subset	mAP	DETR baseline		YOLOS			
		DETR	RT-DETRv2	Individually	All (MoMo)	Simple MOE	MoCAE
CSAW	50-95	0.2763	0.2260	**0.3292**	0.2458	0.3287	0.3044
	50	0.6491	0.4687	**0.7185**	0.6321	**0.7185**	0.6854
	75	**0.1616**	0.1250	0.1524	0.1149	0.1481	0.1161
	small	0.2673	0.1181	0.2106	0.1641	0.2106	**0.3119**
	medium	0.2683	0.2000	**0.3210**	0.2568	**0.3210**	0.2978
	large	0.3942	0.4325	**0.4932**	0.2966	0.4881	0.4447
DDSM	50-95	**0.1995**	0.0720	0.1973	0.1710	0.1973	0.1357
	50	0.5027	0.1813	**0.5103**	0.4699	0.5047	0.3508
	75	**0.1042**	0.0410	0.0989	0.0810	0.0972	0.0562
	small	**0.2901**	0.1765	0.2066	0.2024	0.2316	0.1975
	medium	**0.2036**	0.0673	0.1986	0.1658	0.1986	0.1318
	large	0.1892	0.0965	**0.1998**	0.1975	0.1927	0.1576
DMID	50-95	0.1965	0.2030	**0.3364**	0.2660	0.3348	0.3204
	50	0.4180	0.3456	**0.6249**	0.5630	**0.6249**	0.6016
	75	0.1360	0.2210	0.2409	0.2299	0.2409	**0.2464**
	small	0.1010	0.3535	0.0000	**0.4040**	0.0000	0.0505
	medium	0.1519	0.1225	0.2437	0.1488	0.2437	**0.2504**
	large	0.3297	0.3086	**0.4805**	0.4169	0.4763	**0.4910**

For the DDSM dataset, the DETR baseline outperformed all other approaches in terms of mAP@50-95 (0.1995), with YOLOS individually (0.1973) and Simple MoE (0.1973) closely following. This indicates that for more homogeneous datasets like DDSM, a strong unified model such as DETR can still generalize effectively without requiring domain-specific specialization. Nevertheless, MoE variants remained competitive across lesion sizes, particularly for large lesions, maintaining stable performance without significant degradation.

In contrast, on the DMID dataset, which is smaller and more diverse, the MoE-based models demonstrated clear advantages. The YOLOS individually trained experts reached the highest mAP@50-95 (0.3364), while both Simple MoE (0.3348) and MoCAE (0.3204) followed closely. Notably, MoCAE achieved the highest mAP for large lesions (0.4910) and partially recovered detection capability for small lesions (0.0505), where YOLOS individually failed (0.0000). This highlights MoCAE's ability to provide more balanced and reliable predictions across lesion scales.

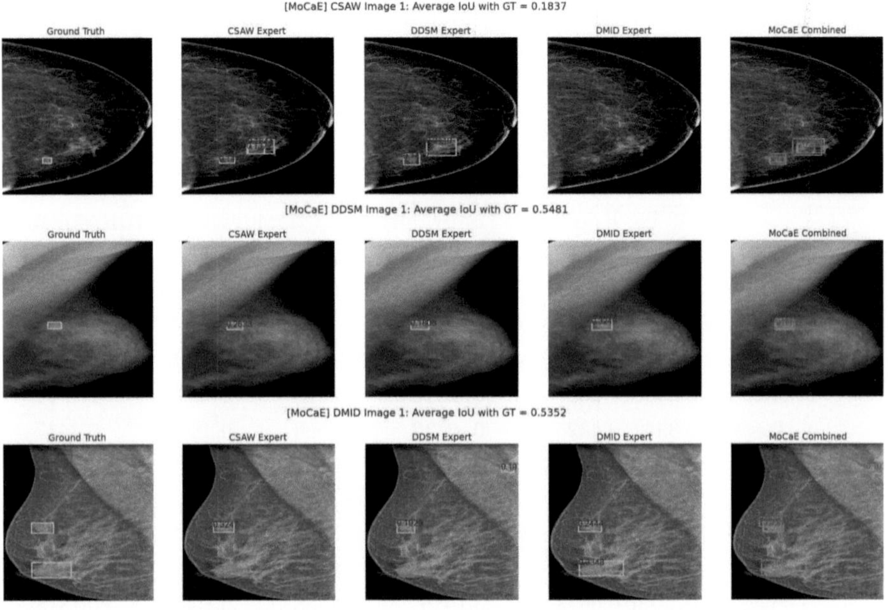

Fig. 2. Inference Breakdown Across Multiple Expert Models accross dataset.

Figure 2 illustrates how the MoE-based combination (MoCaE Combined) consistently outperforms individual experts in terms of localization accuracy across diverse datasets. In each row, we observe that while individual experts (trained on CSAW, DDSM, or DMID) produce fragmented or misaligned detections—often with low IoU scores—the combined MoCaE output yields more complete and precise bounding boxes that better match the ground truth. For instance, in the CSAW image (top row), none of the individual experts align well with the lesion, whereas the MoCaE fusion successfully captures the region with higher spatial coverage. Similarly, in the DDSM and DMID examples (second and third rows), the MoCaE output consolidates partial detections from different experts into a unified and more accurate prediction. These qualitative results support the advantage of expert specialization and calibrated fusion in handling domain-specific variance and improving overall lesion detection.

While our framework improved robustness in breast lesion detection across diverse mammography datasets, several limitations warrant consideration. First, computational costs remain a concern. Training multiple experts (3 YOLOS-base models, each with 127.73M parameters, over 200 epochs) obviously requires substantial resources comparing to a single one, with total training time exceeding 4 GPU-hours on a single NVIDIA A100, which will be potentially longer on larger dataset. Inference runtime, while moderately efficient at around 8 s per image for the ensemble, is 2âĂŞ3 times slower than a single YOLOS model due to the calibration and fusion steps (e.g., Soft NMS and Score Voting), potentially hindering deployment in resource-limited clinical settings where real-time processing is essential. Moreover, clinical interpretability poses another challenge. Although MoCAE enhances confidence calibration, the "black-box" nature of YOLOS transformers limits radiologists' ability to understand decision-making processes, such as how attention mechanisms prioritize subtle lesions. Without integrated explainable AI (XAI) tools like Grad-CAM, the system may face resistance in clinical adoption, where transparency is critical for trust and error tracing. Dataset biases also further complicate generalization. Our subsets (CSAW, DDSM, DMID) exhibit imbalances (e.g., DDSM's 1,333 samples vs. DMID's 201), potentially amplifying demographic or imaging protocol biases common in mammography data, such as underrepresentation of dense breasts or diverse ethnic groups. This could lead to reduced performance on underrepresented cases, as evidenced by MoCaE's lower mAP on DDSM large objects, reflecting domain shifts not fully addressed by our calibration.

Overall, these results support the effectiveness of the Mixture-of-Experts approach in addressing domain heterogeneity, particularly in datasets with diverse characteristics like CSAW and DMID. While unified models may suffice for cleaner, single-domain datasets, MoE models—especially when enhanced with confidence calibration via MoCAE—offer improved robustness, generalization, and reliability across varied clinical scenarios. This makes them promising candidates for real-world deployment in breast cancer screening workflows.

7 Conclusion

In this work, we introduced MammoMix, a Mixture-of-Experts framework designed to improve breast lesion detection across heterogeneous mammography datasets. By training specialized experts on individual data domains and integrating their outputs through a learned gating mechanism, MammoMix effectively addresses domain-specific variability. We further incorporated a calibration module, MoCAE, to ensure reliable and well-calibrated predictions. Experimental results across three diverse datasets, CSAW, DDSM, and DMID, demonstrated that MammoMix consistently improves detection accuracy and robustness, particularly in challenging or mixed-domain settings. These findings highlight the potential of MoE-based architectures in advancing generalizable, trustworthy AI tools for clinical breast cancer screening. Our future iterations on this matter could incorporate lightweight architectures or pruning to reduce costs, while integrating XAI for better interpretability.

Acknowledgments. We acknowledge the support from the Research team of the University of Anonymous for providing storage and facilities to conduct this study.

Disclosure of Interests. (None. Submission to AJCAI2025 with original research.)

References

1. Carion, N., Massa, F., Synnaeve, G., Usunier, N., Kirillov, A., Zagoruyko, S.: End-to-end object detection with transformers. In: European Conference on Computer Vision, pp. 213–229. Springer (2020). https://doi.org/10.1007/978-3-030-58452-8_13
2. Dembrower, K., Liu, Y., Azizpour, H., Eklund, M., Smith, K., Lindholm, P., Strand, F.: Comparison of a deep learning risk score and standard mammographic density score for breast cancer risk prediction. Radiology **294**(2), 265–272 (2020)
3. Dosovitskiy, A., et al.: An image is worth 16x16 words: transformers for image recognition at scale. arXiv preprint arXiv:2010.11929 (2020). https://doi.org/10.48550/arxiv.2010.11929
4. Fang, Y., et al.: You only look at one sequence: rethinking transformer in vision through object detection. Adv. Neural Inf. Process. Syst. **34**, 26183–26197 (2021). https://doi.org/10.48550/arxiv.2106.00666
5. Garrucho, L., Kushibar, K., Jouide, S., Diaz, O., Igual, L., Lekadir, K.: Domain generalization in deep learning based mass detection in mammography: a large-scale multi-center study. Artif. Intell. Med. **132**, 102386 (2022)
6. Giaquinto, A.N., et al.: Breast cancer statistics, 2022. CA Cancer J. Clinicians **72**(6), 524–541 (2022)
7. Gøtzsche, P.C., Jørgensen, K.J.: Screening for breast cancer with mammography. Cochrane Database Syst. Rev. **1**(6) (2013)
8. Hamed, G., Marey, M.A.E.R., Amin, S.E.S., Tolba, M.F.: Deep learning in breast cancer detection and classification. In: Proceedings of the International Conference on Artificial Intelligence and Computer Vision (AICV2020), pp. 322–333. Springer (2020)
9. Kim, H.E., et al.: Changes in cancer detection and false-positive recall in mammography using artificial intelligence: a retrospective, multireader study. Lancet Digit. Heal. **2**(3), e138–e148 (2020)
10. Lee, R.S., Gimenez, F., Hoogi, A., Miyake, K.K., Gorovoy, M., Rubin, D.L.: A curated mammography data set for use in computer-aided detection and diagnosis research. Sci. Data **4**(1), 1–9 (2017)
11. Lv, W., Zhao, Y., Chang, Q., Huang, K., Wang, G., Liu, Y.: Rt-detrv2: improved baseline with bag-of-freebies for real-time detection transformer. arXiv preprint arXiv:2407.17140 (2024), https://doi.org/10.48550/arxiv.2407.17140
12. Masoudnia, S., Ebrahimpour, R.: Mixture of experts: a literature survey. Artif. Intell. Rev. **42**(2), 275–293 (2014)
13. Oksuz, K., Kuzucu, S., Joy, T., Dokania, P.K.: Mocae: mixture of calibrated experts significantly improves object detection. arXiv preprint arXiv:2309.14976 (2023). https://doi.org/10.48550/arxiv.2309.14976
14. Oza, P., et al.: Digital mammography dataset for breast cancer diagnosis research (dmid) with breast mass segmentation analysis. Biomed. Eng. Lett. **14**(2), 317–330 (2024)

TripletResNet: A Deep Metric Learning Approach for mTBI Diagnosis from 3D CT

Hanem Ellethy[1(✉)], Shekhar S. Chandra[1], and Viktor Vegh[2,3]

[1] School of Electrical Engineering and Computer Science, University of Queensland, Brisbane, QLD, Australia
h.elwaseif@uq.edu.au

[2] The Centre for Advanced Imaging, Australian Institute for Bioengineering and Nanotechnology, University of Queensland, Brisbane, QLD, Australia

[3] ARC Training Centre for Innovation in Biomedical Imaging Technology, Brisbane, QLD, Australia

Abstract. Mild Traumatic Brain Injury (mTBI) is a prevalent neurological condition with significant diagnostic challenges. Accurate and timely diagnosis is critical to guiding effective treatment strategies and improving patient outcomes. Traditional diagnostic techniques often suffer from limited accuracy and sensitivity, leading to potential misdiagnoses. To overcome these limitations, we propose **TripletResNet**, a novel residual neural network trained using metric learning with triplet loss. Our method effectively distinguishes mTBI cases from healthy individuals by embedding 3D Computed Tomography (CT) scans into a discriminative feature space. Optimizes representations by maximizing distances between dissimilar cases and minimizing them between similar ones, thereby highlighting diagnostic features. Our TripletResNet achieved superior diagnostic performance, demonstrating an average accuracy of 94.3%, sensitivity of 94.1%, and specificity of 95.2% through five-fold cross-validation. Notably, compared to our previously introduced Residual Convolutional Neural Network (RCNN), TripletResNet exhibits significant performance improvements, a 22.5% increase in specificity, 16.2% higher accuracy, and 11.3% enhanced sensitivity. Additionally, TripletResNet requires fewer memory resources, offering a practical and efficient solution. The integration of occlusion sensitivity maps further enhances the interpretability of our model, providing visual explanations that build clinician trust and facilitate transparent informed decision-making.

Keywords: Mild Traumatic Brain Injury (mTBI) · 3D Computed Tomography (CT) · Deep Learning · Metric Learning · Triplet Loss

1 Introduction

Mild Traumatic Brain Injury (mTBI) represents a pervasive and complex public health concern [1]. Given the potential for both short-term and persistent cognitive impairments, accurate and timely diagnosis of mTBI is crucial for optimal patient outcomes. However, traditional diagnostic methods, including clinical assessments and neuroimaging,

M. Liu et al. (Eds.): AI 2025, LNAI 16371, pp. 295–308, 2026.
https://doi.org/10.1007/978-981-95-4972-6_23

often lack the sensitivity needed to reliably detect subtle yet clinically important abnormalities [2, 3]. Misdiagnoses or delayed diagnoses can compromise patient recovery, underscoring the urgent need for more precise diagnostic technologies.

Computed Tomography (CT) remains the primary neuroimaging modality for evaluating traumatic brain injury (TBI) [4] due to its widespread availability, cost-effectiveness, and rapid acquisition [5]. Despite these advantages, standard CT scans may fail to detect subtle structural changes typical of mTBI, limiting diagnostic accuracy [6].Three-dimensional (3D) CT imaging provides comprehensive volumetric data, significantly enhancing the visualization of subtle brain abnormalities often missed in traditional 2D slices, and thus holds considerable promise for improving mTBI diagnostic precision [7].

Deep learning (DL), particularly Convolutional Neural Networks (CNNs), has revolutionized medical imaging analysis by automating the extraction of discriminative features from complex imaging data [8]. Although DL has shown substantial promise in various medical imaging tasks, including lesion detection and classification, its specific application to mTBI diagnosis using 3D CT imaging remains relatively unexplored [9]. Notably, Residual CNN architectures (RCNN) have exhibited excellent feature extraction capabilities in medical imaging, making them particularly suitable for identifying subtle abnormalities associated with mTBI [10].

Complementing CNN architectures, metric learning techniques aim to optimize feature representations by explicitly enhancing the separability of data points in a learned feature space [11, 12]. This approach can significantly improve diagnostic accuracy by ensuring extracted features capture clinically relevant distinctions [13–16]. Metric learning presents a new approach for improving mTBI diagnosis, particularly when coupled with RCNN architectures, by offering extraction of discriminative and clinically relevant features from medical 3D scans [17]. Within metric learning, Triplet Loss (TL) has shown exceptional effectiveness, guiding models to learn representations that minimize intra-class variation while maximizing inter-class differences, thus boosting discriminative power [13].

Building upon these insights, we propose **TripletResNet**, a novel ResNet architecture that integrates metric learning via triplet loss to enhance the extraction of diagnostic features from 3D CT scans. This approach improves the sensitivity and specificity of mTBI diagnosis by embedding CT data into a discriminative and clinically meaningful feature space.

Specifically, our study makes the following key contributions:

- **Novel DL model:** We introduce TripletResNet, metric-learning-enhanced residual network designed to improve mTBI diagnosis using 3D CT imaging. By optimizing feature representations, the model supports better case differentiation and aids in informed clinical decision-making.
- **Superior diagnostic performance:** We demonstrate that integrating triplet loss into the residual architecture significantly outperforms a conventional RCNN model, achieving a 16.2% increase in accuracy, an 11.3% improvement in sensitivity, and a 22.5% gain in specificity, all while using memory resources more efficiently.

- **Improved interpretability:** We apply Occlusion Sensitivity Maps (OSM) to visualize the model's decision-making process. These visual explanations enhance transparency, support clinical validation, and help build trust in the model's outputs.

Overall, TripletResNet represents a meaningful advancement in mTBI diagnostics, offering both technical innovation and clinical relevance through faster, more accurate, and interpretable diagnostic support.

2 Methods

2.1 Study Population

This study utilized data from the TRACK-TBI Pilot Study, a prospective, multi-centre cohort initiative designed to investigate various aspects of traumatic brain injury (TBI). The pilot study enrolled 600 patients aged over seven years, all of whom sustained a TBI and underwent head CT scans in the emergency department within 24 h of injury. Data collection was conducted across four U.S. sites: University of California San Francisco, University Medical Centre Brackenridge, University of Pittsburgh Medical Centre, and Mount Sinai Rehabilitation Centre, between April 2010 and June 2012 [18].

The TRACK-TBI dataset includes demographic and clinical information, head CT scans obtained within 24 h of injury, MRI scans within 14 days, and functional outcome assessments at three- and six-months post-injury. A key objective of the TRACK-TBI Pilot Study was to evaluate the feasibility of implementing TBI Common Data Elements (TBI-CDEs). At the time of our access, the database contained 589 records (out of the original 600 enrolled participants). Of these, 323 patients exhibited mTBI-related abnormalities on CT.

We obtained access to the dataset through the Federal Interagency Traumatic Brain Injury Research (FITBIR) repository, with approval granted by the Data Access and Quality (DAQ) Committee. Our study protocol was approved by the University of Queensland Human Research Ethics Committee (Approval No. 2020002583).

For this study, we focused on patients who underwent non-contrast head CT scans within 48 h of sustaining closed head injuries and had a Glasgow Coma Scale (GCS) score ≥ 13, indicative of mild TBI. After applying these criteria, we analysed a total of 296 de-identified cases from the publicly available TRACK-TBI dataset.

2.2 Data Pre-processing

To ensure high-quality and consistent input for deep learning (DL) models, we implemented a structured pre-processing pipeline for all CT scans, as illustrated in Fig. 1.

We first converted the original DICOM files to the Neuroimaging Informatics Technology Initiative (NIfTI) format using the dcm2nii tool [10]. This step included a broad field-of-view alignment to standardize spatial orientation across all scans (Fig. 1a).

Next, we applied standardized brain window settings (window width = 80 Hounsfield Units (HU), window level = 40 HU) [18] to enhance soft tissue contrast and improve the visibility of potential mTBI-related abnormalities (Fig. 1b).

To normalize spatial resolution and reduce computational load, we resampled all images using the Spline Interpolated Zoom (SIZ) algorithm [19], producing isotropic volumes with dimensions 128 × 128 × 64 voxels at 1 mm3 resolution. Z-axis interpolation was applied to ensure uniform slice spacing [20] (Fig. 1c).

Finally, non-brain regions and high-intensity artifacts (e.g., scanner bed) were removed to focus the model exclusively on relevant brain structures (Fig. 1d). A subset of the pre-processed volumes was visually inspected using MRIcron (https://www.nitrc.org/projects/mricron), a neuroimaging viewer bundled with dcm2nii, to verify anatomical consistency, appropriate cropping, and absence of processing artifacts.

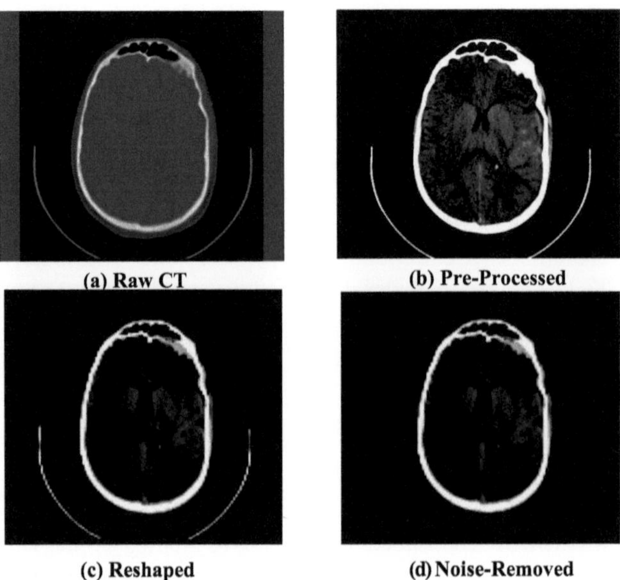

Fig. 1. The **CT slice pre-processing steps.** (a) The original, unprocessed CT slice before any pre-processing steps were applied. (b) The same CT slice after undergoing initial pre-processing, which includes standardizing the dynamic range to enhance the visibility of brain anatomical structures. (c) The CT slice after being reshaped, which is a crucial part of the pre-processing steps. (d) The CT slice has been further processed to remove high-intensity noise artifacts such as the scanner bed.

2.3 Study Design

This study investigates the impact of incorporating metric learning into a modified residual RCNN to improve diagnostic accuracy for mTBI from 3D CT images. First, we implement a baseline RCNN model. Then, we integrate the architecture with triplet loss to create **TripletResNet**, our proposed metric-learning-based model. The performance of both models is compared quantitatively and qualitatively, with visual interpretability provided via Occlusion Sensitivity Maps (OSM) [21].

TripletResNet. TripletResNet extends our previously published RCNN model [10] by integrating metric learning through triplet loss. RCNN architecture, which serves as the baseline in this study, comprises an encoder with six residual blocks and a multi-layer perceptron classifier. TripletResNet builds upon this foundation by mapping 3D CT scans into a discriminative embedding space, where samples from the same class are positioned closer together and those from different classes are pushed farther apart. As illustrated in Fig. 2, the TripletResNet architecture consists of four main modules: a **Sampler** that ensures balanced batch construction, an **Embedder** that generates low-dimensional representations, a **Miner** that identifies informative training pairs, and a **Classifier** that performs the final binary prediction. The full training workflow is outlined in Algorithm 1.

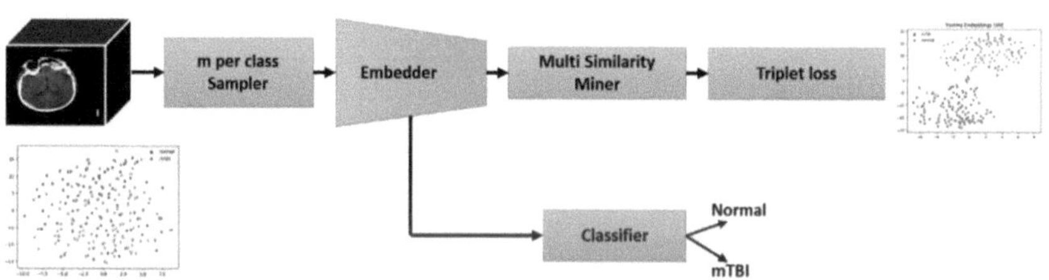

Fig. 2. Network architecture of TripletResNet.

Embedder. The embedder maps each 3D CT volume to a 512-dimensional vector in the embedding space. Initially, we adopted the RCNN encoder architecture as the embedder. However, through systematic experimentation and hyperparameter tuning, we developed a simplified version that outperformed the original in both efficiency and classification accuracy.

As illustrated in Fig. 3, the embedder consists of four residual stages with channel depths of 32, 64, 128, and 256, respectively. Each stage contains three 3D convolutional residual units, followed by a 1×1 pointwise convolution for dimensionality adjustment. Instance normalization and parametric ReLU (PReLU) activations are applied within each convolutional layer. The encoder output is flattened and passed through three fully connected layers with dimensions $5000 \rightarrow 3000 \rightarrow 1024 \rightarrow 512$, producing a compact 512-dimensional embedding vector. Compared to the baseline RCNN encoder with six residual blocks, this streamlined four-stage embedder reduces intermediate feature map size and depth, lowering memory demand during training. This design mitigates overfitting on a dataset of limited size, stabilizes training convergence, and still extracts highly discriminative features essential for effective metric learning.

Miner. The Multi-Similarity Miner (MSM) plays a critical role in selecting informative training pairs by identifying hard positive and hard negative examples within each batch [22]. In our implementation, w a dynamic MSM was used to select training pairs based on their similarity relative to the hardest opposing pair in the batch [23]. Specifically, a positive pair (samples from the same class) is selected if its similarity is less than the similarity of the hardest negative pair plus a margin ($\varepsilon = 0.1$). Conversely, a negative pair (samples from different classes) is selected if its similarity is greater than the similarity of the hardest positive pair minus the margin. This margin-based selection strategy enables the model to focus on the most challenging and informative pairs, those that are difficult but not outliers, thereby promoting the learning of a more discriminative and robust embedding space.

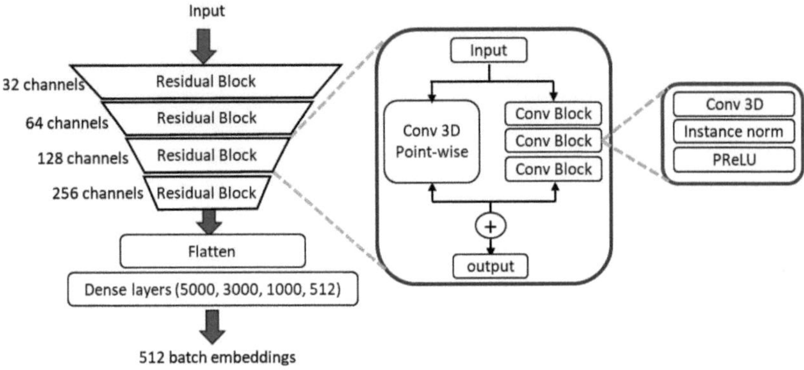

Fig. 3. Embedder architecture of TripletResNet.

Triplet Loss. Triplet loss is a core component of our model, designed to promote the learning of discriminative feature representations in the embedding space. In the context of metric learning, the loss function ensures that samples from the same class are mapped closer together, while those from different classes are separated by a specified margin. The standard triplet loss formulation is given by:

$$L_{\text{triplet}} = \max\{0,\ d(\text{a, p}) - \text{d(a, n)} + \text{margin}\} \quad (1)$$

where:

- d (a, p) denotes the distance between the anchor and a positive sample (same class),
- d (a, n) denotes the distance between the anchor and a negative sample (different class),
- margin is a hyperparameter that enforces a minimum separation between positive and negative pairs. In our experiments, the margin was empirically set to 0.4 based on validation performance.

Algorithm 1: Learning Discriminative Embeddings with TripletResNet

Input: A dataset of labelled 3D CT images.

Output: Trained Embedder, Trained TripletResNet model, Visualized Embeddings, OSM.

Steps:

1. Initialize the TripletResNet model.
2. For each training epoch:
   - 2.1. Use the m-per-class sampler to form balanced batches.
     - 2.1.2. Randomly sample "m" images from each class
     - 2.1.3. Repeat until the total batch size is reached.
   - 2.2. For each batch:
     - 2.2.2. Compute embeddings using the Embedder.
     - 2.2.3. Calculate pairwise L2 distances between embeddings.
     - 2.2.4. Apply MSM to mine for pairs where:
       - $d(a\text{-}n)_{selective} > d(a\text{-}p)_{hardest} - \varepsilon$
       - $d(a\text{-}p)_{selective} < d(a\text{-}n)_{hardest} + \varepsilon$
   - 2.3. Compute triplet loss:

     $$L_{triplet} = \max\ \{0, d\ (a, p) - d\ (a, n) + margin\}$$
   - 2.4. Backpropagate the loss and update the Embedder weights.
3. Visualize the learned embeddings using t-Distributed Stochastic Neighbour Embedding (t-SNE).
4. Freeze the trained Embedder.
5. Attach the MLP Classifier to the frozen embedder.
6. Train the classifier on the fixed embeddings.
7. Evaluate the complete TripletResNet model on the test set.
8. Generate the OSM for interpretability.

In practice, loss becomes zero when the anchor-positive pair is sufficiently close compared to the anchor-negative pair—specifically, when d (a, p) + margin $\leq d$ (a, n). This condition indicates that the model has learned an adequate separation between similar and dissimilar samples. If the condition is not met—meaning the negative is too close to the anchor—the loss becomes positive. The model is then penalized and encouraged to update the embeddings by pulling the anchor closer to the positive and pushing it away from the negative, thus improving the discriminative power of the learned space.

In our implementation, distances are computed using the Euclidean (L2) distance between L2-normalized embeddings, which enhances training stability and performance. Additionally, to improve robustness, we allow for adaptive switching between the anchor and positive samples when the margin violation is more significant in the positive-negative relationship. This strategy helps the model focus better on difficult relationships during training.

Classifier. The classifier is a fully connected neural network composed of three hidden layers with 256, 64, and 32 units, respectively, followed by a sigmoid output layer for binary classification. This architecture enables the model to learn hierarchical representations from the 512-dimensional embedding vectors produced by the trained embedder. The classifier is responsible for distinguishing between mTBI and normal CT scans based on the learned embeddings, completing the end-to-end diagnostic pipeline.

3 Results and Discussion

The proposed model consists of two main branches: the triplet loss branch and the classification branch. In this section, we present and discuss the results obtained from each branch, highlighting their respective contributions to the overall model performance.

3.1 Triplet Loss Branch and Embeddings

The first branch leverages the triplet loss function to train the model for producing highly discriminative embeddings. These embeddings capture rich information about the underlying structure and semantic relationships within the data. A key outcome of this branch is the improved quality of the learned embedding space, which plays a critical role in boosting classification performance. Through rigorous training, the model successfully learned to map samples from the same class closer together while distancing those from different classes. This resulted in a compact and well-separated embedding space that facilitates more effective classification.

To evaluate the learned embeddings, we applied t-distributed Stochastic Neighbor Embedding (t-SNE) [12] for visualization. As shown in Fig. 4, the resulting 2D projections of the embeddings exhibit clear and well-defined clusters for each class, with minimal overlap between mTBI and normal cases. This visual clustering demonstrates that the model effectively minimized intra-class distances and maximized inter-class separation, a result attributed to the combined effect of the triplet loss and the hard mining strategy.

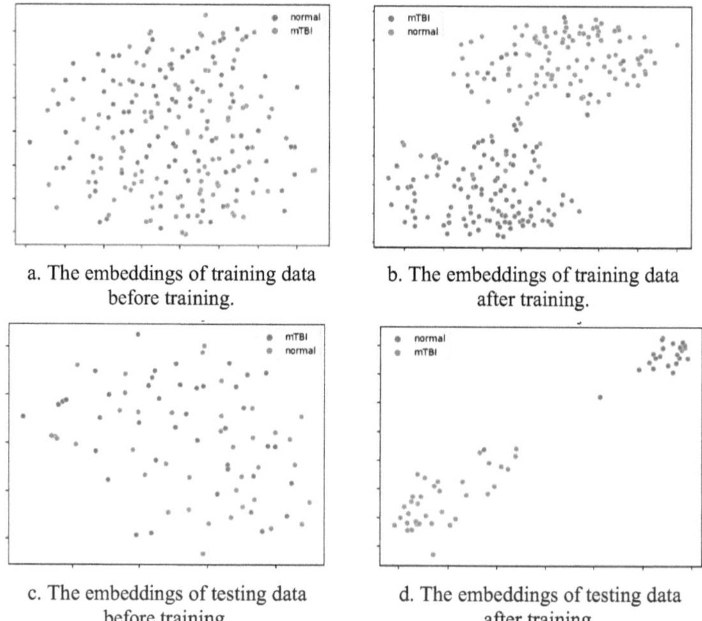

a. The embeddings of training data before training.

b. The embeddings of training data after training.

c. The embeddings of testing data before training.

d. The embeddings of testing data after training.

Fig. 4. Visualization of the embedding distributions for training and testing datasets before and after training using t-SNE. The top row shows the untrained embeddings, where samples from mTBI and normal cases are not well separated. The bottom row displays the trained embeddings generated by TripletResNet, highlighting distinct and well-separated clusters for each class, demonstrating effective metric learning and improved class separability.

3.2 Classification Branch

The second branch of our model utilizes the learned embeddings for binary classification, yielding notable improvements in accuracy, sensitivity, and specificity. These embeddings, produced by TripletResNet, effectively capture subtle and clinically relevant patterns within 3D CT scans, enabling more accurate differentiation between mTBI and normal cases. As shown in Table 1, the proposed model achieved a classification accuracy of 94.3 $\pm$ 4.0%, with a sensitivity of 94.1 $\pm$ 6.7% and a specificity of 95.2 $\pm$ 3.4%, outperforming both the baseline RCNN and the commonly used DenseNet121 [24] architecture. These results highlight the significant impact of integrating metric learning and optimized embeddings in enhancing diagnostic precision.

The significant gain in specificity—the ability to correctly identify non-mTBI cases (true negatives)—indicates a substantial reduction in false positives. At the same time, the increased sensitivity reflects the model's improved ability to correctly detect mTBI cases (true positives), minimizing the risk of missed diagnoses. This well-balanced performance profile underscores the potential of TripletResNet as a reliable and clinically valuable tool for supporting accurate mTBI diagnosis from 3D CT imaging.

Table 1. Performance comparison of DenseNet121, RCNN, and TripletResNet on mTBI classification. p-values (one-sided t-test) indicate the statistical significance of performance improvements between RCNN and TripletResNet.

Metric	Accuracy ($\mu \pm \alpha$ in %)	Sensitivity ($\mu \pm \alpha$ in %)	Specificity ($\mu \pm \alpha$ in %)
Densenet121 [24]	76.5 ± 2.6	77.5 ± 5.9	75.3 ± 5.3
RCNN [10]	78.1 ± 6.6	82.8 ± 5.1	72.7 ± 8.7
TripletResNet	**94.3 ± 4.0** ($p = 0.001$)	**94.1 ± 6.7** ($p = 0.020$)	**95.2 ± 3.4** ($p < 0.001$)

In addition to its diagnostic performance, TripletResNet demonstrates superior computational efficiency. Despite having more parameters than the baseline RCNN, it achieves substantially lower memory consumption during forward and backward passes (Table 2). This apparent paradox is explained by the distribution of parameters versus activations. In CNNs, memory usage during training is dominated not by stored weights but by activations (feature maps) that must be retained for backpropagation. RCNN employs six residual stages with wider feature maps, which significantly increases activation storage requirements. By contrast, TripletResNet streamlines the encoder to only four residual stages with narrower channels (32–256), substantially reducing intermediate feature map size.

While TripletResNet introduces additional parameters through its fully connected embedding head ($5000 \rightarrow 3000 \rightarrow 1024 \rightarrow 512$), these dense weights add little to memory load compared to 3D convolutional activations. Furthermore, optimizing embeddings via triplet loss emphasizes geometric relations in a compact 512-dimensional space, reducing redundancy in training signals. As a result, TripletResNet combines strong predictive performance with a resource-efficient profile, a property especially advantageous for real-world clinical deployment where GPU memory and processing budgets are often constrained.

Table 2. Memory utilization of DenseNet121, RCNN, and TripletResNet models. MB = Megabytes.

	Number of Parameters	Forward/backward pass size (MB)	Parameters size (MB)	Total Memory Consumption (MB)
Densenet121 [24]	**11, 243, 649**	1342.45	**42.89**	1389.34
RCNN [10]	35, 335, 490	1875.94	134.79	2014.73
TripletResNet	66, 741, 964	**448.58**	254.60	**707.18**

3.3 Visual Assessment

To further interpret model behavior, we applied Occlusion Sensitivity Maps (OSM) to visualize the regions of an image that most influence a model's predictions. OSM systematically occludes parts of the input and monitors prediction changes, revealing the model's attention and supporting validation and trust [21]. We selected OSM over alternative techniques, such as Grad-CAM, based on our previous comparative study [10], which demonstrated that OSM provides more reliable and informative explanations for 3D CT data in the context of mTBI diagnosis.

Figure 5 illustrates OSM examples for a correctly classified mTBI case using both the RCNN and TripletResNet models. The color gradient distinguishes between highly influential (blue) and less important (red) regions in the image. Notably, TripletResNet's OSM highlights a broader distribution of relevant areas, indicating a more comprehensive spatial awareness and greater contextual understanding. In contrast, RCNN's attention appears more narrowly focused, potentially overlooking peripheral but informative regions.

This difference underscores TripletResNet enhanced capacity to integrate spatial information across the 3D CT scan, contributing to its improved diagnostic performance. Such behavior may contribute to its superior performance and enhanced generalizability, further reinforcing its suitability for clinical decision support in analyzing complex 3D CT data.

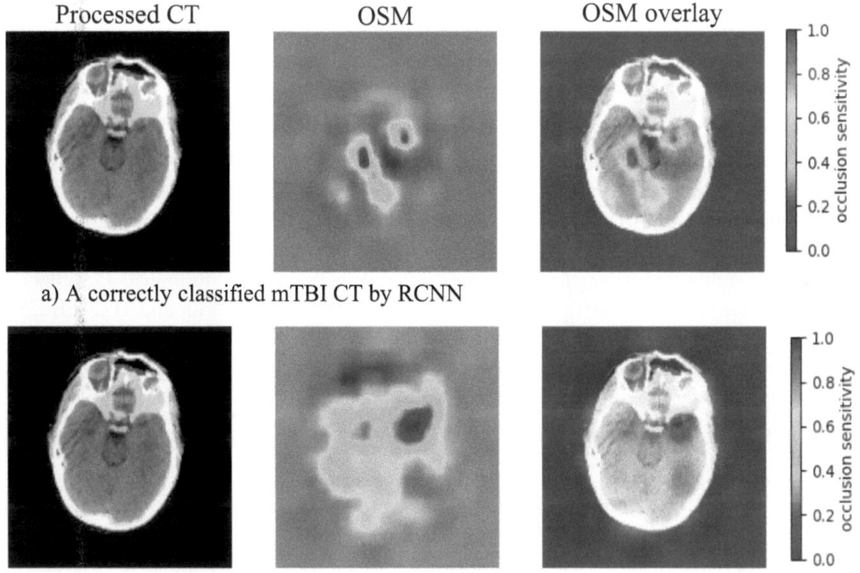

Fig. 5. Occlusion sensitivity map (OSM) for a correctly classified mTBI CT scan by RCNN and TripletResNet. Each row shows the CT, OSM, and overlay. Blue regions indicate critical areas influencing predictions. TripletResNet shows a broader focus, suggesting greater contextual awareness.

4 Conclusion

This study presents a significant advancement in the field of mild traumatic brain injury (mTBI) diagnosis by introducing TripletResNet—a novel deep learning framework that integrates 3D CT imaging with triplet loss-based metric learning. Compared to existing methods, TripletResNet achieved substantial performance gains, including a 16.2% increase in accuracy, a 22.5% improvement in specificity, and an 11.3% boost in sensitivity. In addition to its diagnostic accuracy, the model demonstrates lower memory consumption, making it a resource-efficient solution suitable for real-world clinical applications. Furthermore, our use of occlusion sensitivity maps (OSM) enhances the model's interpretability by revealing critical image regions influencing predictions. This contributes to greater transparency and trust in AI-assisted decision-making.

Despite these promising outcomes, several limitations warrant attention. The model's performance depends on access to high-quality, accurately labeled 3D CT datasets—resources that can be scarce and costly to curate. Variability in data quality and limited diversity may impact generalizability. To address these challenges, future work should consider integrating complementary clinical data (e.g., patient history, symptoms) to build more holistic diagnostic models. Additionally, evaluating TripletResNet on alternative imaging modalities or datasets beyond CT can help assess its generalizability and adaptability across broader diagnostic contexts. Future work could also include human-in-the-loop studies or usability testing with clinicians to ensure practical relevance and adoption. Exploring alternative neural architectures or self-supervised learning strategies may further enhance robustness and scalability. Continued efforts in these directions will be vital to advancing automated mTBI diagnosis and improving patient care.

Acknowledgement. NITRC, NITRC-IR, and NITRC-CE have been funded in whole or in part with Federal funds from the Department of Health and Human Services, National Institute of Biomedical Imaging and Bioengineering, the National Institute of Neurological Disorders and Stroke, under the following NIH grants: 1R43NS074540, 2R44NS074540, and 1U24EB023398 and previously GSA Contract No. GS-00F-0034P, Order Number HHSN268200100090U. Moreover, we would like to acknowledge the principal investigators of the TRACK TBI Pilot research program, the sub-investigators and research teams that contributed to TRACK TBI Pilot, and the patients who participated.

Disclosure of Interests. The authors have no competing interests.

References

1. Gerberding JL, B.S.: Report to congress on mild traumatic brain injury in the United States: steps to prevent a serious public health problem. Centers Dis. Control Prev. **45** (2003)
2. Korley, F.K., Kelen, G.D., Jones, C.M., Diaz-Arrastia, R.: Emergency department evaluation of traumatic brain injury in the United States, 2009–2010. J. Head Trauma Rehabil. (2016). https://doi.org/10.1097/HTR.0000000000000187
3. Ellethy, H., Chandra, S.S., Nasrallah, F.A.: The detection of mild traumatic brain injury in paediatrics using artificial neural networks. Comput. Biol. Med. **135**, 104614 (2021). https://doi.org/10.1016/J.COMPBIOMED.2021.104614

4. Lee, B., Newberg, A.: Neuroimaging in traumatic brain imaging. NeuroRx **2**(2), 372–383 (2005)
5. Schweitzer, A.D., Niogi, S.N., Whitlow, C.T., Tsiouris, A.J.: Traumatic brain injury: imaging patterns and complications. Radiographics **39**(6), 1571–1595 (2019). https://doi.org/10.1148/rg.2019190076
6. Easter, J.S., Haukoos, J.S., Meehan, W.P., Novack, V., Edlow, J.A.: Will neuroimaging reveal a severe intracranial injury in this adult with minor head trauma?: The rational clinical examination systematic review. JAMA **314**(24), 2672–2681 (2015). https://doi.org/10.1001/jama.2015.16316
7. Srikrishna, M., et al.: Comparison of two-dimensional-and three-dimensional-based U-Net architectures for brain tissue classification in one-dimensional brain CT. Front. Comput. Neurosci. **15**, 785244 (2022)
8. Litjens, G., et al.: A survey on deep learning in medical image analysis. Med. Image Anal. (2017). https://doi.org/10.1016/j.media.2017.07.005
9. Ellethy, H., Chandra, S.S., Nasrallah, F.A.: Deep neural networks predict the need for CT in Pediatric mild traumatic brain injury: a corroboration of the PECARN rule. J. Am. Coll. Radiol. **19**(6), 769–778 (2022). https://doi.org/10.1016/j.jacr.2022.02.024
10. Ellethy, H., Vegh, V., Chandra, S.S.: Interpretable 3d multi-modal residual convolutional neural network for mild traumatic brain injury diagnosis. In: Liu, T., Webb, G., Yue, L., Wang, D. (eds.) BT - AI 2023: Advances in Artificial Intelligence. Springer Nature Singapore, Singapore, pp. 483–494 (2024)
11. Deepak, S., Ameer, P.M.: Brain tumour classification using siamese neural network and neighbourhood analysis in embedded feature space. Int. J. Imaging Syst. Technol. **31**(3), 1655–1669 (2021). https://doi.org/10.1002/ima.22543
12. Yan, Y., Hao, H., Xu, B., Zhao, J., Shen, F.: Image clustering via deep embedded dimensionality reduction and probability-based triplet loss; image clustering via deep embedded dimensionality reduction and probability-based triplet loss. IEEE Trans. Image Process. **29**, 2020 (2020). https://doi.org/10.1109/TIP.2020.2984360
13. Huang, Z., Zhou, Q., Zhu, X., Zhang, X.: Batch similarity based triplet loss assembled into light-weighted convolutional neural networks for medical image classification. Sensors **21**(3), 764 (2021). https://doi.org/10.3390/S21030764
14. Osama, S., Zafar, K., Sadiq, M.U.: Predicting clinical outcome in acute ischemic stroke using parallel multi-parametric feature embedded Siamese network. Diagnostics **10**(11) (2020). https://doi.org/10.3390/DIAGNOSTICS10110858
15. Alaverdyan, Z., Jung, J., Bouet, R., Lartizien, C.: Regularized Siamese neural network for unsupervised outlier detection on brain multiparametric magnetic resonance imaging: application to epilepsy lesion screening. Med. Image Anal. **60**, 101618 (2020). https://doi.org/10.1016/J.MEDIA.2019.101618
16. Alaverdyan, Z., Chai, J., Lartizien, C.: Unsupervised feature learning for outlier detection with stacked convolutional autoencoders, Siamese networks and wasserstein autoencoders: application to epilepsy detection. In: Lect. Notes Comput. Sci. (including Subser. Lect. Notes Artif. Intell. Lect. Notes Bioinformatics), LNCS, vol. 11045, pp. 210–217 (2018). https://doi.org/10.1007/978-3-030-00889-5_24/FIGURES/3
17. Onga, Y., Fujiyama, S., Arai, H., Chayama, Y., Iyatomi, H., Oishi, K.: Efficient feature embedding of 3D brain MRI images for content-based image retrieval with deep metric learning. In: 2019 IEEE International Conference on Big Data (Big Data), pp. 3764–3769 (2019). https://doi.org/10.1109/BigData47090.2019.9006364
18. Lolli, V., Pezzullo, M., Delpierre, I., Sadeghi, N.: MDCT imaging of traumatic brain injury. Br. J. Radiol. **89**(1061), 20150849 (2016)

19. Zunair, H., Rahman, A., Mohammed, N., Cohen, J.P.: Uniformizing techniques to process CT scans with 3D CNNs for tuberculosis prediction. In: International Workshop on PRedictive Intelligence in Medicine, pp. 156–168. Springer (2020)
20. Goenka, N., Tiwari, S.: AlzVNet: a volumetric convolutional neural network for multiclass classification of Alzheimer's disease through multiple neuroimaging computational approaches. Biomed. Signal Process. Control **74**, 103500 (2022). https://doi.org/10.1016/j.bspc.2022.103500
21. Rieke, J., Eitel, F., Weygandt, M., Haynes, J.-D., Ritter, K.: Visualizing convolutional networks for MRI-based diagnosis of Alzheimer's disease. In: Understanding and Interpreting Machine Learning in Medical Image Computing Applications: First International Workshops, MLCN 2018, DLF 2018, and iMIMIC 2018, Held in Conjunction with MICCAI 2018, Granada, Spain, 16–20 September 2018, Proceedings 1, pp. 24–31. Springer (2018)
22. Wang, X., Han, X., Huang, W., Dong, D., Scott, M.R.: Multi-similarity loss with general pair weighting for deep metric learning. In: Proceedings of the IEEE/CVF Conference on Computer Vision and Pattern Recognition, pp. 5022–5030 (2019)
23. Musgrave, K., Belongie, S.J., Lim, S.-N.: PyTorch metric learning. ArXiv, vol. abs/2008.0 (2020)
24. Huang, G., Liu, Z., Van Der Maaten, L., Weinberger, K.Q.: Densely connected convolutional networks. In: Proceedings of the IEEE Conference on Computer Vision and Pattern Recognition, pp. 4700–4708 (2017)

Efficient Craniofacial Microsomia Detection via Edge-Focused 3D Point Cloud Network

Qichen Zhao, Wei Emma Zhang(✉), Sarbin Ranjitkar, Robin Viltoriano, Zonghan Xie, and Peter J. Anderson

The University of Adelaide, Adelaide, Australia
qichen.zhao@student.adelaide.edu.au,
{wei.e.zhang,sarbin.ranjitkar,zonghan.xie,peter.anderson,
robin.viltoriano}@adelaide.edu.au

Abstract. Automated classification of the severity of hemifacial microsomia (HFM) from craniofacial scans remains challenging due to limited medical imaging data, challenging anatomical features, and computational constraints. We propose an efficient and effective classification framework that leverages lightweight STL meshes and a streamlined processing pipeline, including mesh cleaning, edge-focused point-cloud processing strategy and intra-class MixUp augmentation. Specifically, the proposed edge-focused strategy is tailored explicitly to select points near anatomical boundaries critical for HFM diagnosis, preserving essential diagnostic features while dramatically reducing data complexity. To overcome data scarcity, we implement intra-class MixUp, significantly improving model robustness by synthesizing anatomically realistic additional samples. Evaluations on a clinically labeled HFM dataset confirm that our optimized model, built upon the Pretrained PointNet backbone, achieves superior accuracy and Macro-F_1 compared to state-of-the-art alternatives. Our results demonstrate that this edge-focused, lightweight, augmentation-enhanced approach effectively balances diagnostic accuracy with computational efficiency, advancing automated craniofacial severity classification.

Keywords: Hemifacial Microsomia · 3D Point Cloud · Data Augmentation · Lightweight 3D Analysis

1 Introduction

Hemifacial Microsomia (HFM) is a common craniofacial anomaly characterized by asymmetrical mandibular and facial development, necessitating accurate and objective severity grading for clinical management [17]. Traditional assessment methods often depend heavily on subjective evaluation of radiographs or resource-intensive CT imaging with manual segmentation [2], limiting their clinical scalability. While recent deep learning-based methods address aspects of automated HFM detection, they typically either perform binary classification or landmark detection rather than comprehensive severity grading [24,35]. And

M. Liu et al. (Eds.): AI 2025, LNAI 16371, pp. 309–321, 2026.
https://doi.org/10.1007/978-981-95-4972-6_24

current clinical workflows rely on high-resolution DICOM volumes, which are computationally demanding [31,37].

In the intersection of medical imaging and computer vision, point cloud sampling strategies have begun to incorporate boundary awareness to preserve critical anatomical structures [16,36]. However, this has not yet been applied to HFM detection, where conventional sampling remains common [35]. Introducing boundary-focused sampling into HFM analysis is therefore both novel and necessary to capture clinically relevant morphological features. Moreover, most studies depend on high-resolution DICOM data, with limited work exploring lightweight STL-based point clouds. To our knowledge, no prior work has applied edge-aware sampling on STL meshes for HFM severity classification, making our approach a novel contribution in this area.

To bridge these gaps, we present an efficient deep learning framework explicitly tailored for automated severity classification of HFM. First, we employ a largest-component filtering step to remove scanning artifacts and surgical debris from the original STL meshes, preserving a clean, anatomically consistent surface model. Subsequently, our edge-focused sampling strategy selects points concentrated along clinically critical anatomical boundaries, effectively capturing the high-curvature mandibular features essential for accurate severity discrimination. To further address data scarcity, we integrate intra-class MixUp augmentation, synthesizing additional anatomically realistic training samples. This targeted approach significantly expands the dataset size, enhances intra-class variability, and promotes smoother decision boundaries in the classification model.

Our main contributions include:

- Developing an anatomically-driven edge-focused processing strategy specifically optimized for HFM mandibles, efficiently capturing diagnostically critical morphological features.
- Implementing intra-class MixUp augmentation to effectively mitigate the constraints of limited medical data availability.
- Employing a lightweight STL-based data processing pipeline involving mesh cleaning and targeted point-cloud generation, significantly reducing computational overhead. To our knowledge, this is among the first attempts to perform HFM classification solely based on STL data.

2 Related Work

2.1 Approaches in 3D Dental Analysis

Early severity schemas such as Pruzansky-Kaban [9], OMENS [32], and SAT [4] relied heavily on 2D radiographs or basic 3D CT reconstructions, leading to subjective interpretation and variability among examiners [2]. Although Qiu et al. [24] quantified mandibular asymmetry using bone-volume ratios from 3D CT scans, their approach required extensive manual segmentation, limiting clinical applicability and scalability.

Recent advances in deep-learning-based methods demonstrated some progress in detecting HFM. Nazira et al. [20] employed convolutional neural networks for binary classification to distinguish HFM patients from controls, whereas Xu et al. [35] focused specifically on automated craniofacial landmark detection. However, these approaches had not addressed automated severity classification, essential for detailed clinical assessment. For medical applications involving limited and imbalanced data, pre-trained neural networks remained widely adopted to mitigate data scarcity and facilitate robust feature extraction [8].

Sampling strategies also play a significant role in craniofacial point cloud analysis. Recent dental research identified tooth crown size and mandibular border morphology as critical features for assessing HFM severity [25]. Nevertheless, conventional uniform sampling methods often lost crucial diagnostic details at highly curved anatomical edges, potentially compromising the precision of severity classification. Specifically in the field of HFM, existing point-cloud sampling strategies typically did not emphasize anatomical boundaries [35], resulting in potential loss of valuable morphological information.

Additionally, most existing AI-based 3D medical analyses relied on resource-intensive CT DICOM data [12,18,31,37]. Despite their effectiveness, these approaches typically required specialized imaging infrastructure, significant computational resources, and might not have been widely accessible in clinical environments.

Unlike existing methods, our approach efficiently utilized lightweight STL meshes with the edge-focused processing strategy, preserving critical diagnostic features while significantly reducing computational demands and data complexity.

2.2 CV Techniques for 3D Point Cloud Analysis

Conventional 3D analysis often relied on surrogate representations. Multiview methods rendered multiple 2D projections and applied convolutional neural networks (CNNs) [22,28], while voxel- or octree-based approaches discretized space into grids for volumetric convolutions [7,27,33]. Despite their effectiveness, these methods typically required significant computational resources and might have sacrificed fine-grained surface details due to occlusions or limited resolution.

In point-cloud learning, the raw XYZ coordinates were processed directly. Uniform samplers, such as Farthest Point Sampling (FPS) [19], often discarded high-curvature points that were most informative for shape interpretation [5]. Boundary-aware studies tackled this problem: Hu et al. [11] introduced a dual branch with edge loss to sharpen borders; Tang et al. [30] made down-sampling contrastive by pulling boundary features of different classes apart; Gong et al. [6] aggregated neighborhood cues to predict boundaries while limiting cross-class leakage; and attention-weighted sampling further increased the priority of curved regions [34]. Guided by these ideas, we designed a sampler customized for the mandibular rims and dental arches, which matched the clinical needs in Hemifacial Microsomia diagnosis.

3 Methodology

Our methodology involves cleaning raw anatomical meshes, selecting edge-focused points to preserve critical morphological information, and applying intra-class MixUp augmentation tailored for our HFM dataset. We adapt and optimize the Pretrained PointNet released by Griffiths [8], through an optimized model-selection strategy based explicitly on validation set accuracy.

3.1 Mesh Cleaning

STL files are typically much smaller than the original DICOM or CBCT datasets because DICOM stores full volumetric grayscale information, whereas STL only represents surface geometry [13]. As a result, after segmenting specific anatomical regions and applying mesh simplification, the size of the STL file is often substantially reduced compared to the source imaging data. The raw skull and mandible STL files often include small fragments of scanning artifacts or surgical hardware. Each mesh is therefore decomposed into connected components, after which only the surface with the largest area is retained and all other pieces are discarded. This single step removes virtually all non-anatomical debris while preserving an intact anatomical model for subsequent analysis.

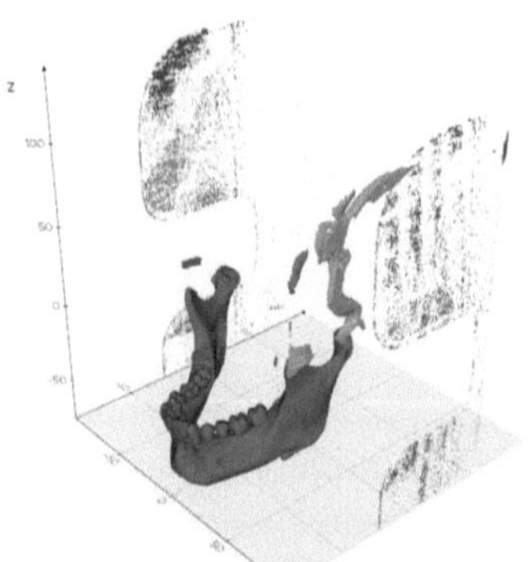

(a) Raw STL scan before cleaning.

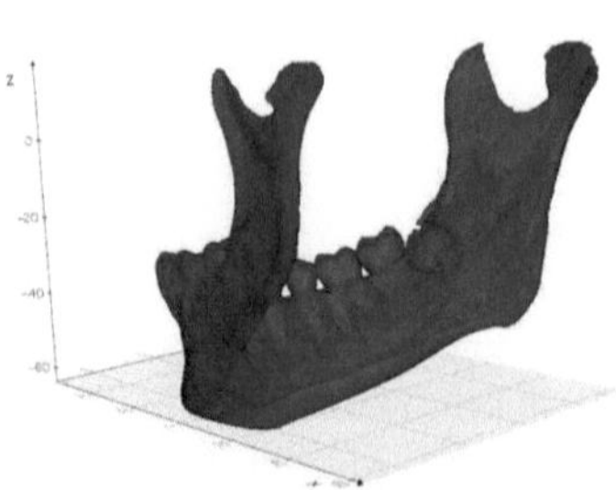

(b) After largest-component filtering.

Fig. 1. Mandible surface before and after cleaning.

3.2 Edge-Focused Point Sampling Strategy

We propose an edge-focused processing strategy that selectively prioritizes points near anatomical boundaries with significant curvature, such as the mandibular rim and dental arches. This approach effectively preserves essential diagnostic features of HFM while substantially reducing the number of points, thereby enabling more efficient processing without compromising clinically relevant morphological information.

Dense Point Cloud Generation. The cleaned STL mesh of each subject is first converted to a dense point set $\mathbf{P} = \{\mathbf{v}_i \in \mathbb{R}^3\}_{i=1}^N$ by retaining all surface vertices; if $N > 20{,}000$ we draw a uniform random subset of size 20,000. This cloud provides sufficient spatial coverage for robust local geometry estimation.

Neighbourhood and Normal Estimation. To accurately identify anatomical boundaries, this step estimates the local surface geometry by computing normals for each point on its spatial neighborhood. These normals serve as the basis for detecting regions of high curvature, essential for subsequent edge scoring. For every point $\mathbf{v}_i$ we query its k nearest neighbours, via a KD–tree search [1]. The neighbourhood covariance matrix is

$$\mathbf{C}_i = \frac{1}{k} \sum_{\mathbf{v}_j \in \mathcal{N}_i} (\mathbf{v}_j - \bar{\mathbf{v}}_i)(\mathbf{v}_j - \bar{\mathbf{v}}_i)^\top, \tag{1}$$

where $\bar{\mathbf{v}}_i$ is the local centroid. Let $\lambda_1 \le \lambda_2 \le \lambda_3$ be its eigenvalues and $\mathbf{v}_1$ the eigenvector of λ_1; $\mathbf{n}_i := \mathbf{v}_1 / \|\mathbf{v}_1\|$ is taken as the unit normal at $\mathbf{v}_i$ following the PCA normal-estimation scheme of Hoppe *et al.* [10].

Normal–Variation Edge Score. To quantify surface irregularity, we compute a curvature-based score that reflects the variation in local normals. This edge score highlights regions of rapid normal change, which typically correspond to anatomical boundaries. Normal diversity in $\mathcal{N}_i$ is quantified as

$$s_i = 1 - \frac{1}{k} \sum_{\mathbf{v}_j \in \mathcal{N}_i} \left| \mathbf{n}_i^\top \mathbf{n}_j \right|, \quad 0 \le s_i \le 1, \tag{2}$$

which is analogous to the normal-variation criterion used in point-cloud edge detection [21]. A higher s_i indicates stronger orientation change and thus a greater likelihood of lying on a ridge or border.

Vertex Ranking and Selection. This step filters the point cloud to retain only those points with the highest curvature variation, focusing computational resources on regions most relevant for HFM severity discrimination. All scores are z-normalised and sorted in descending order. The top $M{=}2{,}048$ indices $\mathcal{S} = \arg\operatorname{top}_M\{s_i\}$ form the edge cloud $\mathbf{P}_{\text{edge}} = \{\mathbf{v}_i\}_{i \in \mathcal{S}}$. This removes $>90\%$ of vertices yet preserves the mandibular rim and dental-arch details necessary for severity classification.

As shown in Fig. 2, compared to uniform sampling via Farthest Point Sampling (FPS) [19] (gray), our proposed edge-focused sampling (red) concentrates more points near structural boundaries (top zoom-in), while avoiding redundancy in flat, less informative regions (bottom zoom-in). This leads to a more efficient and informative point representation.

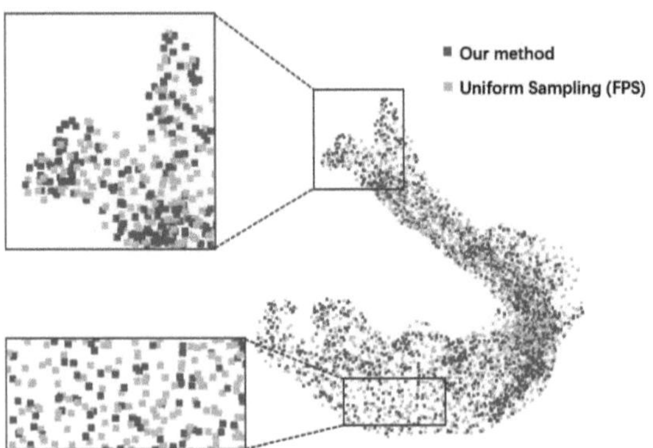

Fig. 2. Edge-focused vs. Uniform Sampling (2,048 pts).

3.3 Data Augmentation

In medical research, limited data availability remains a significant challenge. After evaluating several 3D augmentation strategies (Sect. 4.2), MixUp demonstrates the greatest improvement on our HFM dataset. This superior performance attributed to: (i) the inherent smooth and continuous morphological variation within each severity category of HFM, making linear interpolation anatomically realistic; and (ii) MixUp's capability to effectively enhance intra-class variability without introducing additional label noise.

Intra-class MixUp on HFM Point Clouds. To augment our training set and encourage smoother decision boundaries, we applied an intra-class MixUp strategy specifically adapted for our HFM point cloud data. Consider two point clouds, $\mathbf{P}_i, \mathbf{P}_j \in \mathbb{R}^{N\times3}$, representing mandibles from patients belonging to the *same* severity category (control / mild / moderate–severe). We sampled a mixing coefficient $\mu \sim \mathrm{Beta}(\alpha, \alpha)$ to generate a new synthetic sample through linear interpolation:

$$\mathbf{P}_{\mathrm{mix}} = \mu\mathbf{P}_i + (1-\mu)\mathbf{P}_j. \tag{3}$$

As both source point clouds originate from the same severity category, the augmented point cloud $\mathbf{P}_{\mathrm{mix}}$ retains the same category label. In our implementation, we first group all training mandible samples according to their severity category. Within each severity category, we retain all original samples and exhaustively generate additional augmented samples by applying MixUp to all unordered pairs and form the training set.

We set $\alpha = 0.2$ to ensure that most augmented samples remain close to their original counterparts, achieving effective regularization without compromising the integrity of class-specific morphological characteristics.

4 Experiment

4.1 Setting

Dataset. Our anonymised dataset consisted of mandibular STL scans from patients diagnosed with Hemifacial Microsomia (HFM). The dataset was obtained from an Australian public hospital. Each scan was clinically labeled by dentists according to disease severity: 20 samples were labeled as S1, representing mild cases, and 10 samples as S2, representing moderate to severe cases. Additionally, we incorporated 38 STL scans of mandibles from healthy individuals as a control group (i.e., healthy cases), labeled as S0. We partitioned the 68 samples into training, validation, and test subsets in a 60/20/20 ratio, yielding 40, 14, and 14 cases, respectively. Given the scarcity of publicly available medical imaging data for HFM and the practical difficulties of acquiring additional cases, we applied data augmentation techniques in subsequent experiments to enhance the performance (Sect. 3.3).

Evaluation Metrics. Model performance was assessed with two complementary scores: overall Accuracy and class-balanced Macro-F_1.

Accuracy. Let N_{corr} be the number of test scans whose predicted severity matches the ground truth and N_{tot} the total number of scans. The accuracy is simply

$$\text{Accuracy} = \frac{N_{\text{corr}}}{N_{\text{tot}}}. \tag{4}$$

Macro-F_1. For each severity category $t \in \mathcal{T}$ we computed precision P_t, recall R_t, and the class-wise F_1 score:

$$P_t = \frac{\text{TP}_t}{\text{TP}_t + \text{FP}_t}, \qquad R_t = \frac{\text{TP}_t}{\text{TP}_t + \text{FN}_t}, \qquad F1_t = \frac{2\,P_t R_t}{P_t + R_t}, \tag{5}$$

where TP_t, FP_t, and FN_t denoted true positives, false positives, and false negatives for class t, respectively. Averaging the class-wise scores gave the macro value:

$$\text{Macro-}F1 = \frac{1}{|\mathcal{T}|} \sum_{t \in \mathcal{T}} F1_t. \tag{6}$$

Accuracy reflected overall correctness, whereas Macro-F_1 treated each severity category equally, making it was especially informative under class imbalance.

4.2 Results

Comparison with Baseline Models. To determine the most suitable network architecture for our HFM craniofacial classification task, we compared three prominent point cloud deep learning methods: PointNet [23], PointCNN [15], and the Pretrained PointNet [8]. Considering the limited size of our dataset, the pretrained variant demonstrated clear advantages over the other two models by leveraging previously learned visual feature representations, thereby mitigating the risks associated with training deep networks from scratch on smaller datasets.

To further optimize the pretrained PointNet specifically for our HFM data characteristics, we introduced several key enhancements. First, we refined the feature transformation network within PointNet to robustly handle variations in orientation and scale present in craniofacial point clouds. Additionally, we employed an improved model-selection strategy based explicitly on validation set accuracy, facilitating better generalization to unseen data.

Moreover, by combining these modifications with our proposed edge-focused point selection strategy and data augmentation, we significantly boosted the network's sensitivity to subtle yet critical geometric features that distinguish HFM severity levels. This integrated approach led to substantial improvements, achieving the highest overall accuracy and Macro-F_1 score among all the evaluated models as shown in Table 1.

Table 1. Performance comparison of different models (with 2,048 point cloud points input).

Model	Accuracy	Macro-F_1
PointNet	0.36	0.30
PointCNN	0.50	0.46
Pretrained PointNet	0.57	0.42
Our Model	**0.79**	**0.76**

Ablation Study. We performed an ablation study to evaluate the contributions of our proposed edge-focused sampling and data augmentation methods individually and jointly. Results were summarized in Table 2. Data augmentation increased the size of our training set substantially, from 40 to 352 samples, significantly enhancing intra-class variability and model generalization. Our edge-focused sampling was specifically compared against the widely-used Farthest Point Sampling (FPS) method [5], with both methods used point clouds of 2,048 points. As illustrated in Fig. 2, our sampling strategy better emphasizes critical anatomical boundary features of HFM mandibles compared to FPS, thus providing more informative inputs for severity classification. The integration of edge-focused sampling and intra-class MixUp augmentation achieved the best performance, underscoring their complementary roles in effectively addressing data scarcity while preserving essential morphological details.

Choice of k. Neighbourhood size k directly influenced the estimation of local normals and the accuracy of edge scoring, significantly affecting downstream classification performance. A smaller k was sensitive to fine local structures but susceptible to noise, while a larger k reduced noise but might have oversmoothed crucial anatomical edges.

Table 2. Ablation Study

Model Configuration	Accuracy	Macro-F_1
Optimised Pretrained PointNet (baseline)	0.57	0.54
Optimised Pretrained PointNet + Data Augmentation	0.64	0.59
Optimised Pretrained PointNet + Edge-Focused Strategy	0.71	0.60
Optimised Pretrained PointNet + Edge-Focused + Augmentation	**0.79**	**0.76**

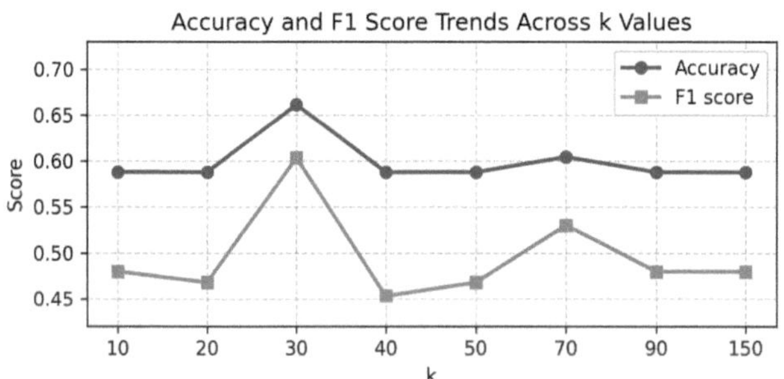

Fig. 3. Results of 5-fold cross-validation: selection of the optimal k based on Accuracy and F1 Score.

To select an optimal k, we performed a range of candidate values ($k \in \{10, 20, 30, 40, 50, 70, 90, 150\}$). As shown in Fig. 3, the model achieved peak performance at $k = 30$, balancing sensitivity to critical mandibular features with robustness against noise. Thus, we adopted $k = 30$ in all subsequent experiments.

Table 3. Dense versus edge-focused point clouds (w/o augmentation). The proposed edge-focused method outperforms the dense cloud given significantly less runtime and the size of file to process.

Representation	Size (per mandible)	Runtime	Accuracy	Macro-F_1
Dense (20 000 pts)	234.4 kB	1 h 51 m 33 s	0.56	0.49
Edge-focused (2 048 pts)	24.0 kB	1 m 53 s	0.71	0.60

Edge-Focused Vs. Dense Point Clouds. To assess the efficiency and diagnostic value of our edge-focused processing strategy, we compared two point-cloud representations extracted from the same cleaned STL mesh: a Dense cloud of 20,000 uniformly sampled points and an edge-focused cloud of 2,048 points selected by our curvature-driven method. To isolate the effect of the point-selection strategy itself, no data augmentation was applied here; both models were trained on the same 40 original mandibles under identical setting.

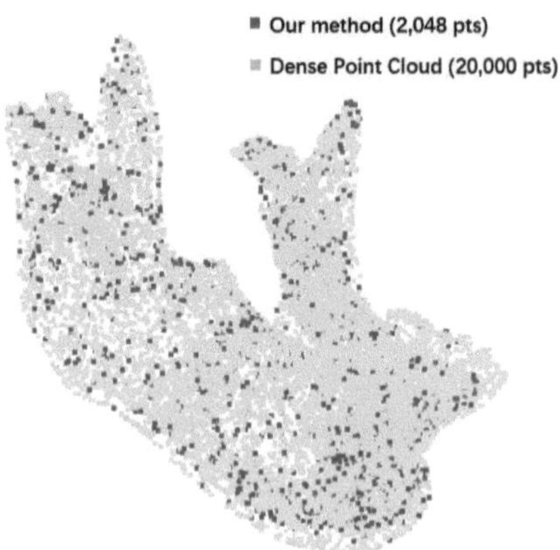

Fig. 4. Comparison of edge-focused (2,048 pts) and dense point clouds (20,000 pts) showing efficient preservation of anatomical features.

As shown in Table 3, the dense representation did not perform well despite its higher point count - probably because a large proportion of low-curvature, low-information points introduced noise and diluted salient boundary cues. In contrast, our edge-focused representation concentrated learning on anatomically informative regions, producing superior results with a 90% reduction in storage and a 60× speed-up in training time. This shows that emphasizing the mandibular boundaries is both efficient and effective for the classification of HFM severity. Figure 4 further depicts that the edge-focused cloud preserves key anatomical structure (e.g., gonial angle, condylar head) while reducing the point count by an order of magnitude.

Table 4. Effectiveness of Data Augmentation Strategies.

Augmentation	Accuracy	Macro-F_1
None (baseline)	0.71	0.60
MixUp	**0.79**	**0.76**
Random scaling	0.71	0.66
Gaussian jitter	0.57	0.52
Random rotation	0.57	0.42

Data Augmentation. To improve generalization and reduce overfitting, we employed four point-cloud augmentation strategies during training: random scaling [14], Gaussian jitter [26], in-plane rotation [29], and MixUp [3]. In our experiments, evaluated their individual impact on classification performance and aimed to retain only the most effective strategy for our model.

- Random scaling: uniformly scale each point cloud by a factor drawn from $\mathcal{U}(0.8, 1.2)$.
- Gaussian jitter: add zero-mean Gaussian noise $\mathcal{N}(0, 0.01^2)$ to each coordinate (clipped to ± 0.06).
- Random rotation: rotate each cloud by 30° around the x-axis and 45° around the y-axis.
- MixUp: combine two clouds of the same class via convex mixing $\tilde{P} = \mu P_i + (1 - \mu)P_j$, with $\mu \sim \text{Beta}(0.2, 0.2)$.

Among all tested augmentation strategies, MixUp yielded the largest performance gain as shown in Table 4. This is because HFM mandibles within a given severity grade vary smoothly. Convex interpolation therefore produced anatomically plausible samples that enlarge the training manifold without altering class semantics. In contrast, simple geometric or noise-based transforms such as scaling, jittering, and fixed-axis rotation either introduce distortions that blur key boundary cues or add variation that is not clinically informative, offering only marginal or even negative benefit for this small and highly structured dataset.

5 Conclusion

In this study, we proposed an efficient deep-learning framework for automated severity classification of HFM. By leveraging lightweight STL meshes and edge-focused sampling, we addressed key challenges including limited data, computational burden, and anatomically complex craniofacial structures. Intra-class MixUp augmentation further enhanced data diversity and model robustness. Our optimized pretrained PointNet-based model achieved higher accuracy and Macro-F_1 scores than the baseline methods, and greatly reduces storage and runtime costs. These results support the clinical potential of our lightweight, anatomically informed approach in resource-constrained settings.

References

1. Bentley, J.L.: Multidimensional binary search trees used for associative searching. Commun. ACM **18**(9), 509–517 (1975)
2. Birgfeld, C.B., et al.: A phenotypic assessment tool for craniofacial microsomia. Plast. Reconstr. Surg. **127**(1), 313–320 (2011)
3. Chen, Y., et al.: PointMixup: augmentation for point clouds. In: Vedaldi, A., Bischof, H., Brox, T., Frahm, J.-M. (eds.) ECCV 2020. LNCS, vol. 12348, pp. 330–345. Springer, Cham (2020). https://doi.org/10.1007/978-3-030-58580-8_20
4. David, D.J., et al.: Hemifacial microsomia: a multisystem classification. Plast. Reconstr. Surg. **80**(4), 525–533 (1987)
5. Eldar, Y., et al.: The farthest point strategy for progressive image sampling. IEEE Trans. Image Process. **6**(9), 1305–1315 (1997)
6. Gong, J., et al.: Boundary-aware geometric encoding for semantic segmentation of point clouds. In: AAAI 2021

7. Graham, B., et al.: 3D Semantic segmentation with submanifold sparse convolutional networks. In: CVPR (2018)
8. Griffiths, D.: keras-io/pointnet (2022). https://huggingface.co/keras-io/PointNet, Accessed 2024
9. Hopkins, E., et al.: Fundamentals of hemifacial microsomia, pp. 317–322 (2024)
10. Hoppe, H., et al.: Surface reconstruction from unorganized points. SIGGRAPH Comput. Graph. **26**(2), 71–78 (1992)
11. Hu, Z., et al.: JSENet: joint semantic segmentation and edge detection network for 3d point clouds. In: ECCV (2020)
12. Järnstedt, J., et al.: Reproducibility analysis of automated deep learning based localisation of mandibular canals on a temporal CBCT dataset. Sci. Rep. **13**, 14159 (2023)
13. Kamio, T., et al.: DICOM segmentation and STL creation for 3D printing: a process and software package comparison for osseous anatomy. 3D Print. Med. **6**, 17 (2020)
14. Li, R., et al.: PointAugment: an auto-augmentation framework for point cloud classification. In: CVPR 2020
15. Li, Y., et al.: PointCNN: convolution on x-transformed points. In: NeurIPS 2018
16. Liu, Y., et al.: Edge-oriented point-cloud transformer for 3d intracranial aneurysm segmentation. In: MICCAI (2022)
17. Makinde, P.M., et al.: Hemifacial microsomia: a scoping review on progressive facial asymmetry due to mandibular deformity. Oral Maxillofac. Surg. **28**, 1441–1455 (2024)
18. Miki, Y., et al.: Classification of teeth in cone-beam CT using deep convolutional neural network. Comput. Biol. Med. **80**, 24–29 (2017)
19. Moenning, C., et al.: Fast marching farthest point sampling for point clouds and implicit surfaces. Citeseer (2003)
20. Nazira, F.A., et al.: Hemifacial microsomia detection using convolutional neural networks. In: ICCA (2022)
21. Pauly, M., et al.: Multi-scale feature extraction on point-sampled surfaces. Comput. Graph. Forum **22**(3), 281–290 (2003)
22. Pontes, J.K., et al.: Image2Mesh: a learning framework for single image 3d reconstruction. In: ACCV (2018)
23. Qi, C.R., et al.: PointNet: deep learning on point sets for 3d classification and segmentation. In: CVPR (2017)
24. Qiu, X., et al.: Classification of hemifacial microsomia based on bone volume ratio of the healthy and affected sides of the mandible: a retrospective self-control experiment. Chin. J. Plast. Reconstr. Surg. **2**(1), 35–50 (2020)
25. Ranjitkar, S., et al.: Computed tomography assessment of hypodontia and crown size in hemifacial microsomia. Arch. Oral Biol. **147**, 105633 (2023)
26. Ren, N., et al.: PointNet-based jitter decomposition on point cloud of jitter histogram. In: SCAS (2021)
27. Riegler, G., et al.: OctNet: learning deep 3d representations at high resolutions. In: CVPR (2017)
28. Su, H., et al.: Multi-view convolutional neural networks for 3d shape recognition. In: ICCV (2015)
29. Suzuki, S., et al.: Scene generalized multi-view pedestrian detection with rotation-based augmentation and regularization. In: ICIP (2024)
30. Tang, L., et al.: Contrastive boundary learning for point cloud segmentation. In: CVPR (2022)
31. Tian, Y., et al.: Automatic jawbone structure segmentation on dental CBCT images via deep learning. Clin. Oral Investig. **28**(12), 663 (2024)

32. Vento, A.R., et al.: The O.M.E.N.S. Classification of Hemifacial Microsomia. Cleft Palate Craniofac. J. **28**(1), 68–76 (1991)
33. Wang, P.S., et al.: O-CNN: octree-based convolutional neural networks for 3d shape analysis. ACM Trans. Graph. **36**(4), 72:1–72:11 (2017)
34. Wu, C., et al.: Attention-based point cloud edge sampling. In: CVPR (2023)
35. Xu, M., et al.: Using a new deep learning method for 3d cephalometry in patients with hemifacial microsomia. Ann. Plast. Surg. **91**, 381–384 (2023)
36. Yang, X., et al.: IntrA: 3D intracranial aneurysm dataset for deep learning. In: CVPR (2020)
37. Zhang, Z., et al.: Quantitative structural analysis of hemifacial microsomia mandibles in different age groups. Front. Pediatr. **11**, 1157607 (2023)

Privacy-Centric Seizure Detection Using Surface Normals, Pose and Segmentation Masks

Kishore Gandhi Ponnambalam[1], Talha Ilyas[2,4], Shobi Sivathamboo[3], Zongyuan Ge[4], Patrick Kwan[3], Levin Kuhlmann[1], and Deval Mehta[1(✉)]

[1] Department of Data Science and AI, Faculty of IT, Monash University, Melbourne, Australia
deval.mehta@monash.edu

[2] Faculty of Engineering, Monash University, Melbourne, Australia

[3] Department of Neuroscience,Faculty of Medicine, Nursing and Health Sciences,, Monash University, Melbourne, Australia

[4] AIM for Health Lab, Faculty of IT, Monash University, Melbourne, Australia

Abstract. Continuous video-based seizure detection remains clinically challenging owing to occlusions, environmental variations, and subtle seizure manifestations. We introduce a privacy-centric, non-invasive video-based seizure detection system that leverages dense surface normals to encode geometric features. This approach achieves superior generalization as the features remain invariant to patient appearance. This work explores the first application of surface normal analysis to seizure detection, demonstrating that geometry-based features not only preserve privacy but also outperform traditional pose-based methods. Through a rigorous evaluation on 821 clinical video clips from 7 patients, we systematically compare surface normals against pose estimation, semantic segmentation, and multi-modal fusion approaches under both patient-dependent (5-fold CV) and patient-independent (LOPO CV) validation protocols. Raw surface normals achieve 89.4% accuracy, significantly outperforming pose estimation (82.8%) with a 26.8% relative improvement in F1-score. Critically, it maintains exceptional robustness in scenarios where semantic methods catastrophically fail, including multi-person interactions and severe occlusion, where its F1-score is robustly maintained at 0.833 while the pose-based F1-score drops to 0.0. Our proposed geometry-based, privacy-centric approach enables continuous monitoring in both clinical and home settings without compromising patient privacy.

Keywords: Privacy-Preservation · Patient Monitoring · Seizure Detection · Surface Normals

1 Introduction

Continuous seizure detection constitutes a critical challenge in epilepsy treatment, as detection delay is directly linked to patient morbidity and in severe

L. Kuhlmann, D. Mehta—Equal senior authorship

M. Liu et al. (Eds.): AI 2025, LNAI 16371, pp. 322–334, 2026.
https://doi.org/10.1007/978-981-95-4972-6_25

cases, mortality [14,18]. While EEG is clinically the gold standard, invasiveness, cost, and interpretability limit its practicality for continuous monitoring [15]. Video-based methodologies present a promising non-contact, scalable option, yet they struggle to bridge the semantic gap, the inherent difficulty in translating raw, complex visual data into clinically meaningful interpretations of pathological motion, particularly under real-world conditions [1]. Furthermore, a formidable barrier to the clinical translation of these systems is the inherent compromise of patient privacy [20]. Methods that operate on raw video raise significant ethical and data security concerns, severely limiting their practical adoption.

Current deep learning methods for video-based seizure detection have predominantly focused on end-to-end learning utilising RGB frames or optical flow [1,7]. These methods, while achieving high accuracy in controlled settings, often fail when confronted with clinical distribution shifts, such as variations in patient presentation, environmental noise like caregiver interactions, blankets, and voluntary movements.

Semantic features such as **2D Pose Estimation** and **Semantic Segmentation** have been widely adopted in broader human action recognition, due to their intuitive mapping of human body parts and actions, and successes in areas such as sports analytics and surveillance [13]. Their application to seizure detection aims to capitalize on this direct human-centric understanding. However, the nuanced, subtle, and highly inconsistent motion characteristics of seizures, compounded by frequent occlusions (e.g., by blankets) and peculiar viewpoints in clinical settings, expose the limitations of relying solely on these high-level representations [2,13]. Challenging this approach is significant because the intrinsic complexities of seizure events may mandate features that are more robust to appearance changes and less dependent on consistent, clear visibility of the entire body.

We introduce a new paradigm for video-based seizure detection, shifting the focus from appearance to privacy-centric geometry. For the first time, we employ dense **surface normals**, demonstrating that this geometric approach decisively outperforms established semantic features like pose-estimation and segmentation under real-world clinical conditions. We posit that geometry, not appearance, provides a more robust and generalizable signal for this task. Surface normals, which encode the 3D orientation of a surface independent of texture or illumination, can capture subtle deformations indicative of underlying pathological movement even through occluding objects like a blanket. This geometric approach inherently preserves patient privacy by abstracting away identifiable features. Despite surface normals being fundamental to 3D reconstruction in CGI and robotics, their utilization in clinical motion analysis, specifically for distinguishing seizure tremors from other movements, is largely untapped [5].

Our methodology directly addresses the deficiencies of the aforementioned methods by (1) leveraging dense surface information capable of capturing subtle deformations (e.g., clothing or blanket movement) indicative of seizures even when the patient is partially occluded, and (2) conducting a thorough patient-independent evaluation (LOPO) to gauge real-world generalizability. Our main contributions are: (i) A systematic framework for privacy-centric epilepsy seizure detection using dense surface normals, segmentation masks, and human pose information from RGB video. (ii)Rigorous qualitative and quantitative generalization analysis of dense surface normals, pose, segmentation and masked normals for unseen patient conditions. (iii) Comprehensive empirical analysis revealing modality-specific failure modes, multi-modality fusion, and robustness to different clinical context settings. This work establishes geometric deep learning as a powerful paradigm for privacy-centric clinical monitoring.

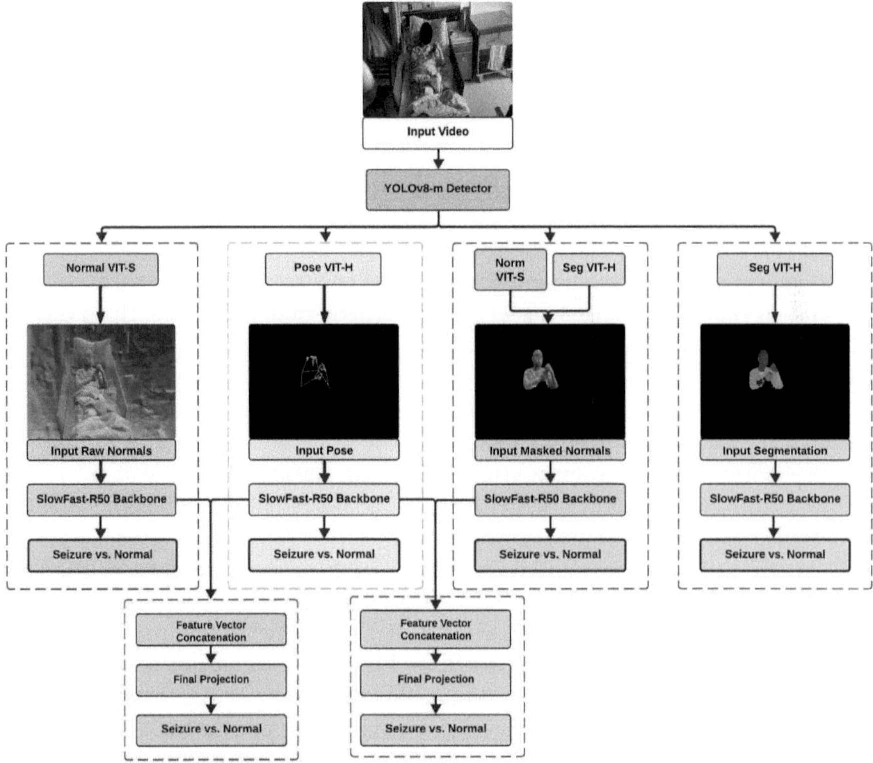

Fig. 1. Pipeline for privacy-centric seizure detection leveraging Raw Surface Normals (geometry-invariant), 2D Pose (17 keypoints), Semantic Segmentation, and Masked Normals. Features are independently processed by SlowFast-R50, with late fusion combining normals & pose and masked normals & pose for robust geometric abstraction.

2 Related Work

Video-based seizure detection has evolved from handcrafted motion descriptors and optical flow to deep learning approaches using raw RGB video and semantic representations. Early methods relied heavily on optical flow and spatio-temporal descriptors to identify rhythmic seizure movements, reporting sensitivities up to 97% in controlled scenarios [7]. However, these techniques were constrained by manual threshold tuning and failed to generalize across varied environmental conditions. The advent of CNNs enabled end-to-end learning of seizure patterns from raw video, achieving inflated performance metrics, as they may learn subject-specific features that don't apply to new individuals. Patient-independent evaluation reveals a significant accuracy drop, exposing the brittleness of appearance-based models under real-world distribution shifts [1].

Human pose estimation abstracts video into skeletal joint trajectories, offering interpretable, compact features aligned with clinical semiology. Early 3D video-EEG systems tracked limb kinematics to quantify convulsive patterns [4]. Modern graph convolutional networks on skeleton sequences demonstrate significant improvements, VSViG achieved 94.1% accuracy (5.9% error) after fine-tuning on 580 annotated seizure frames [21]. Pose reduces background noise, but fails on subtle tremor or occlusion, as off-the-shelf estimators produce missing or erroneous keypoints [1].

Despite clinical interest, optical flowâĂŞbased seizure detection remains inherently sensitive to any pixel-level movement, generating high false-positive rates from non-seizure movements (e.g., caregiver interventions). Even a recent state-of-the-art privacy-centric framework, achieving a promising accuracy of 83.9% through progressive knowledge distillation, inherits optical flow's susceptibility to environmental disturbances [11].

The brittleness of skeleton-based models stems from their two-stage pipeline, which introduces two primary failure points. First, noisy joint estimates, whose accuracy degrades sharply under occlusion, propagate errors directly into the recognition model. Second, the sparse keypoint representation omits the critical shape and appearance cues necessary to decode subtle seizure semiology.

3 Dataset

We utilized the publicly available tonic-clonic seizure dataset from 7 adult EMU patients introduced in VSViG [21], comprising 13 video recordings at 1920×1080 resolution, 30 FPS. See [21] for collection details. We uniformly down sampled all the videos to 5 frames per second (FPS), resulting in a total of $13,426$ frames across all sessions. We spatially resized the frames to 1024×768 pixels to conform to the input specifications of the Sapiens architecture. The frame rate was carefully selected to balance the need for adequate temporal resolution to capture seizure dynamics against the limitations imposed by computational resources. We then employed a sliding-window extraction method with a window length of 32 frames and a stride of 16 frames, generating 821 temporally overlapping clips of ($6.4s$ each), yielding $2685.2s$ of annotated data.

3.1 Class Distribution and Heterogeneity

Our dataset exhibits the characteristic imbalance of clinical seizure monitoring: while aggregate statistics show 38.0% Normal vs. 62.0% Seizure activity across 2685.2s of annotated video, this masks extreme inter-patient variability. Session-level seizure prevalence ranges from 15.17% to 83.8%, a 5.5-fold difference reflecting true epilepsy phenotype heterogeneity (Fig. 2 (b)). This natural distribution, combined with session durations varying from 58.2s to 475.6s across 7 patients (Fig. 2 (a)), creates a challenging testbed where traditional sparse representations (pose, segmentation) fail catastrophically under distribution shift, while our dense geometric features maintain robustness.

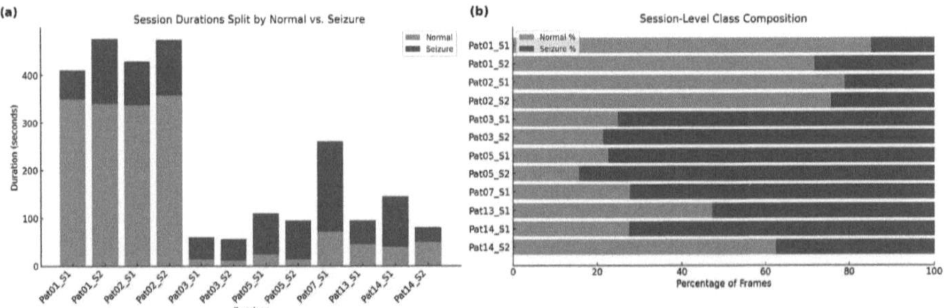

Fig. 2. Dataset heterogeneity among 7 EMU patients: (a) Diverse session durations (58.2 to 475.6s); (b) Seizure prevalence variability (15.17 to 83.8%) reflecting real-world clinical imbalance and challenges in generalization.

3.2 Annotation and Labelling

The dataset provided clinical seizure onset timestamps without explicit termination markers, necessitating a weakly supervised labeling pipeline [10,17].

First, we performed **frame-level temporal alignment** at 5 fps, converting each onset timestamp into a frame index and labeling all subsequent frames as Seizure (1) and preceding frames as Normal (0). Second, we applied clip-level aggregation over 32-frame (6.4s) windows, assigning a seizure label whenever $\geq$ 50% of frames were Seizure, a threshold chosen to maximize early detection sensitivity while containing false positives. Although this majority-vote scheme can occasionally extend the Seizure label into the post-seizure period, it aligns with clinical priorities for prompt warning over exact offset delineation [19].

The resulting labels reveal extreme inter-patient variability, seizure clip prevalence spans 15.17% to 83.8% which, while challenging for model training, provides a rigorous testbed for patient-independent (LOPO) evaluation of true generalization (Fig. 2 (b)).

4 Methodology

In this section, we present our unified framework for video-based seizure detection, detailing the feature-extraction pipeline built on Meta's Sapiens models, the spatiotemporal representation learning architecture, and the validation protocols employed to ensure robust evaluation. A high-level overview of the entire pipeline is illustrated in Fig. 1.

4.1 Feature Extraction Pipeline

Our central hypothesis posits that dense, geometry-informed features outperform sparse semantic cues for seizure detection, particularly under clinical distribution shifts. To test this, we first detected human instances using YOLOv8-m($\tau = 0.25$). We discarded bounding boxes covering $< 50\%$ of frame height and applied a 50-pixel padding to preserve motion context, producing patient-centric ROIs. Then, we leveraged the Sapiens architecture [9], a state-of-the-art human-centric vision model pretrained on over 300 million images as our feature extraction backbone. Sapiens achieves field-leading performance with 61.1 AP for whole-body pose estimation and 11.84 mean angular error for surface normal prediction, while crucially extracting geometric primitives rather than appearance features, thereby inherently preserving patient privacy. We extracted **four** complementary visual modalities from each video frame using the Sapiens family of Vision Transformer (ViT) models:

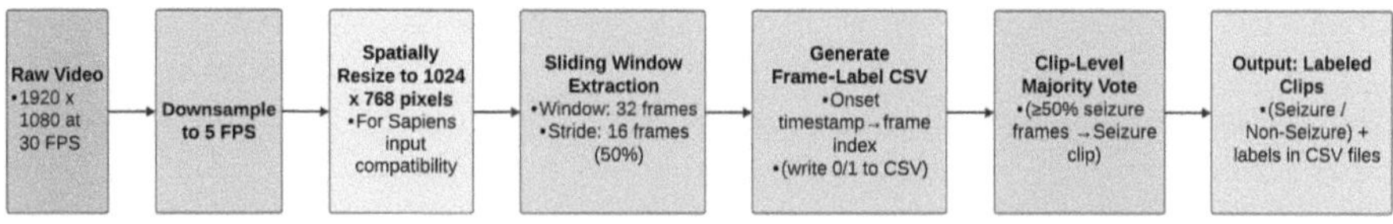

Fig. 3. Temporal pre-processing pipeline: EMU videos are downsampled (5 FPS), resized for Sapiens compatibility, and segmented into 821 overlapping clips. Clinical timestamps are frame-aligned and aggregated into binary clip labels (majority voting), balancing early seizure detection with false-positive reduction.

1. **Raw Surface Normals:** Sapiens-2B (ViT-Huge) computes dense, per-pixel normals, capturing subtle 3D deformations invariant to texture and lighting, crucial for detecting seizure tremors under variable clinical conditions. Surface normals offer geometry-aware features invariant to illumination and texture, which are critical for clinical environments with volatile lighting [5].
2. **2D Keypoint Pose :** Sapiens-1B predicts 17 COCO-format keypoints. Hierarchical spatial pooling encodes both local joint detail and global posture, while learned anatomical priors enhance resilience to occlusion.
3. **Semantic Segmentation :** Sapiens-1B also generates a 28-class human segmentation mask, a pixel-level map that isolates the patient's silhouette.

4. **Masked Normals:** We used the Segmentation Masks to isolate patient-specific geometry by filtering the dense surface normal maps. However, this approach is vulnerable to cascading errors, where segmentation failures under occlusion can catastrophically attenuate the underlying geometric signal.

4.2 Spatiotemporal Representation Learning

We leveraged SlowFast R50's dual-pathway architecture [6] to capture multiscale temporal dynamics inherent in seizure semiology that captures the multiscale temporal dynamics inherent in seizure manifestations, from rapid myoclonic jerks to slower tonic-clonic progressions. The Slow pathway ($\tau = 4$ temporal downsampling) extracts spatial-semantic features through 3D ResNet-50 blocks, for capturing sustained seizure patterns, while the Fast pathway maintains full temporal resolution ($5fps$) with lightweight convolutions ($\beta = 1/8$ channel capacity) to detect fine-grained tremors and sudden onset dynamics. This asymmetric fusion preserves pathway-specific representations while facilitating selective incorporation of high-frequency motion signals, making it particularly suited for detecting both subtle pre-seizure tremors and pronounced convulsive movements characteristic of epileptic seizures.

4.3 Transfer Learning and Domain Adaptation

Given our limited clinical dataset (821 clips from 7 patients), we initialized networks with Kinetics-400 pre-trained weights [8] to leverage robust motion priors. Despite the domain gap between voluntary actions and seizure movements, low-level motion features, particularly geometric deformations transfer effectively to seizure detection. We fine-tuned all network parameters (rather than freezing layers) to enable comprehensive adaptation to seizure-specific spatiotemporal patterns, replacing only the final 400-class head with a binary classifier for seizure/non-seizure discrimination.

4.4 Multi-Modal Fusion Architecture

To exploit complementary information across visual modalities, we implemented a late fusion architecture combining features from independent SlowFast R50 streams. Each modality-specific network processes inputs in parallel, generating specialized 2048-dimensional representations that are concatenated (yielding 4096-dimensional features) and passed through a learned fusion classifier for final seizure/non-seizure discrimination. This strategy, grounded in established multimodal learning principles [3], preserves modality-specific feature development. This allows the network to learn high-level cross-modal combinations, providing crucial robustness when individual streams fail, such as segmentation degradation under patient occlusion.

4.5 Implementation Details

We trained models using RMSprop optimization with OneCycleLR scheduling and addressed class imbalance through focal loss $\gamma = 1.385$ [16] combined with weighted random sampling. We optimized hyperparameters via Bayesian search (Optuna , 50 trials) over learning rate (1×10^{-6} to 1×10^{-3}), weight decay (1×10^{-6} to 1×10^{-2}.), momentum (0.7 to 0.99), and focal loss gamma (1.0 âĂŞ3.0), maximizing validation accuracy on a 20% held-out split. We employed mixed-precision (FP16/FP32) with batch sizes of 4 (single-stream) and 2 (fusion models). Due to computational constraints, we conducted hyperparameter search on the masked normals configuration and adapted empirically for other modalities.

Feature extraction required 17.6 hours (Table 1), and aggregate training at 20 epochs consumed 1,556 hours across 5-fold and LOPO protocols (Table 2), for a total of 1,573.6 hours ($\approx$ 65.6 GPU-days) on a single NVIDIA A40 GPU.

Table 1. Feature Extraction Cost and Throughput

Modality	Time (h)	Rate
Raw Surface Normals	5.6	1.50 ± 0.09 s/frame
Pose Estimation	0.04	9.9 ± 0.1 ms/frame
Semantic Segmentation	3.2	0.85 ± 0.11 s/frame
Masked Normals	8.8	2.35 ± 0.2 s/frame
Total	17.6	-

Table 2. Model Training Cost by Protocol and Model Type

Patient-Mixed Validation (5-Fold CV)				
Model Type	Count	Time/model(h)	Fold	Subtotal (h)
Single Modality	6	20	-	120
Fusion	2	32	-	64
Patient-Independent Validation (LOPO CV)				
Model Type	Count	Time/model (h)	Fold	Subtotal (h)
Single Modality	6	22	7	924
Fusion	2	32	7	448
Total				1556

4.6 Validation Protocols

We evaluated our approach using two complementary protocols: (i) **Stratified 5-fold cross-validation** on 821 clips to establish performance upper bounds and enable architecture comparison, and (ii) **Leave-One-Patient-Out (LOPO) cross-validation** across 7 patients as our primary metric, ensuring zero patient overlap between training and test sets to assess true generalization capability [12]. The stark performance gap between these protocols (Table 3 vs. Table 4) quantifies the challenge of inter-patient heterogeneity in clinical deployment.

5 Results

5.1 Patient-Mixed Validation (5-Fold CV)

Under 5-fold cross-validation (Table 3), all methods achieve strong performance with Fusion:Raw Normals & Pose reaching near-ceiling sensitivity (0.974±0.020) and specificity (0.993 ± 0.004). The performance hierarchy:Raw Normals ($F1 = 0.959$) > Pose ($F1 = 0.915$) > Masked Normals ($F1 = 0.847$) aligns with representational richness, while variance patterns reveal modality-specific stability. The fusion approaches maintain tight bounds ($\sigma \leq 0.010$) whereas segmentation-based methods exhibit 4 × higher variance ($\sigma \leq 0.047$), foreshadowing their brittleness under distribution shift.

Table 3. 5-Fold CV Performance: Fusion Raw Normals & Pose achieves the highest accuracy (98.6%) and sensitivity (97.4%). Individual modalities show descending performance: Raw Normals > Pose > Masked Normals. Fusion approaches demonstrate lowest variance, highlighting consistent robustness.

Model	Accuracy (±SD)	Sensitivity (±SD)	Specificity (±SD)	Precision (±SD)	F1-Score (±SD)
F: Raw Normals & Pose	0.986 ± 0.006	0.974 ± 0.020	0.993 ± 0.004	0.988 ± 0.006	0.981 ± 0.010
Raw Normals	0.968 ± 0.014	0.982 ± 0.017	0.959 ± 0.023	0.938 ± 0.029	0.959 ± 0.014
F: Masked Normals & Pose	0.953 ± 0.019	0.943 ± 0.036	0.960 ± 0.011	0.934 ± 0.019	0.938 ± 0.020
Pose	0.934 ± 0.032	0.920 ± 0.032	0.943 ± 0.034	0.911 ± 0.047	0.915 ± 0.032
Segmentation	0.885 ± 0.019	0.845 ± 0.042	0.909 ± 0.040	0.856 ± 0.040	0.849 ± 0.017
Masked Normals	0.878 ± 0.024	0.888 ± 0.033	0.872 ± 0.025	0.811 ± 0.026	0.847 ± 0.026

5.2 Patient-Independent Generalization Assessment (LOPO CV)

LOPO evaluation unmasks the fundamental limitations of semantic features in real-world clinical deployment (Table 4). While Fusion:Raw Normals & Pose maintains sensitivity (0.937 ± 0.094), specificity collapses catastrophically ($99.3\% \rightarrow 72.2\%, \Delta = 27.1pp$), exposing how pose estimation's vulnerability propagates through fusion. Pose-only exhibits extreme variance (sensitivity=0.628 ± 0.408) as keypoint detection fails under blanket occlusion while generating false positives from caregiver movements. Segmentation-based methods show similar brittleness: Masked Normals' specificity (0.748 ± 0.217) reflects segmentation's binary failure modes, either missing occluded patients entirely or misclassifying caregiver movements as seizures.

In stark contrast, Raw Normals maintains balanced performance (Sensitivity = 0.894 ± 0.090, Specificity = 0.810 ± 0.247) by capturing motion through subtle geometric deformations of occluding surfaces, a signal entirely missed by keypoint-based methods. This robustness extends to precision, while Fusion raw normals & pose drops 16.4%($0.988 \rightarrow 0.827 \pm 0.080$) due to misinterpreted caregiver interactions, Raw Normals declines only 2.9%($0.938 \rightarrow 0.909 \pm 0.062$), confirming that dense geometric features distinguish patient from non-patient motion even through blankets.

The specificity degradation hierarchy crystallizes modality-specific failure mechanisms:Fusion Raw normals & Pose exhibits catastrophic collapse ($-27.1pp$), driven primarily by pose estimation's vulnerability to occlusion and multi-person scenarios. While pose estimation shows severe degradation ($-15.3pp$) with only partial recovery through visible extremities (Fig4(c)), Raw Normals demonstrate remarkable robustness ($-5.0pp$). This differential performance confirms that continuous surface deformations, rather than discrete anatomical landmarks provide the invariant representation necessary for real-world clinical monitoring. Segmentation similarly fails ($-18.8pp$), establishing that dense geometric features fundamentally outperform both sparse keypoints and binary masks in deployment scenarios.

Table 4. LOPO Performance: Raw Surface Normals maintain superior robustness (Accuracy: 89.4%, F1: 0.899), significantly outperforming pose and segmentation modalities, especially under severe occlusion and multi-person interference scenarios. Semantic methods suffer specificity collapse, indicating vulnerability to environment challenges.

Model	Accuracy(±SD)	Sensitivity(±SD)	Specificity(±SD)	Precision(±SD)	F1-Score(±SD)
Raw Normals	0.894 ±0.095	0.894 ±0.0899	0.810±0.247	0.909±0.062	0.899±0.063
F:Raw Normals and Pose	0.883 ±0.034	0.937±0.0943	0.722±0.188	0.826±0.079	0.877±0.080
Masked Normals	0.843 ±0.077	0.737±0.339	0.748±0.217	0.788±0.265	0.736±0.289
Segmentation	0.828 ±0.043	0.821±0.094	0.721±0.024	0.797±0.151	0.804±0.113
Pose	0.828 ±0.043	0.628±0.408	0.790±0.345	0.637±0.4133	0.631±0.407
F:Masked Normals and Pose	0.8022 ±0.143	0.806±0.140	0.822±0.168	0.813±0.176	0.798±0.141

5.3 Qualitative Analysis of Challenging Clinical Scenarios

To elucidate the underlying causes of performance degradation observed in LOPO evaluation, we conducted frame-level analysis on exemplar patients representing distinct clinical challenges (Fig 4).

Pat02 (Severe Occlusion): Under heavy blanket occlusion, segmentation produced complete null outputs, which propagated through Masked Normals as

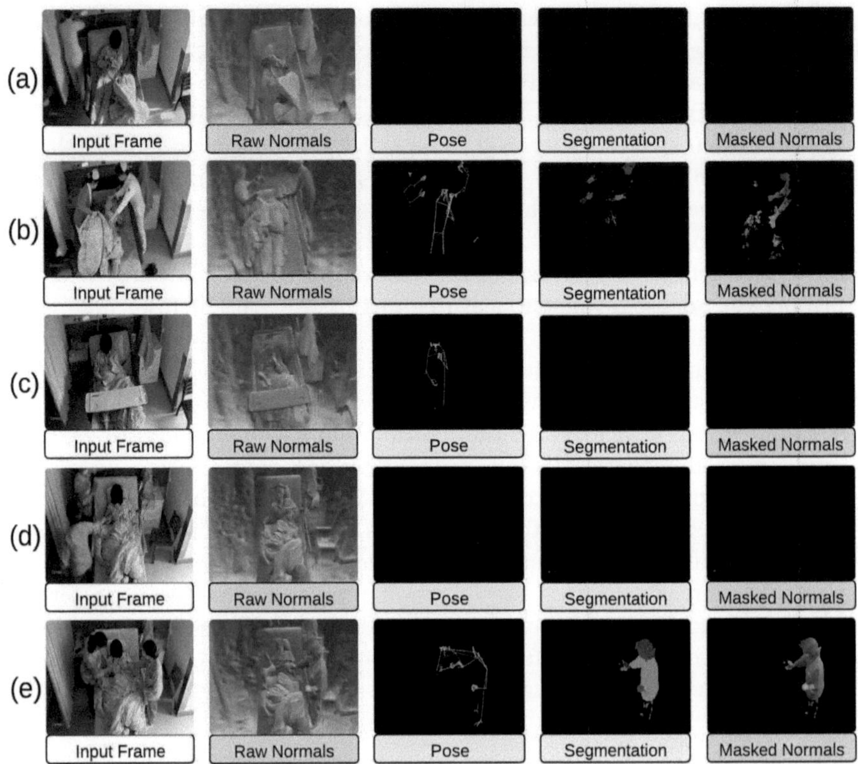

Fig. 4. Robustness analysis under clinical failure scenarios: (aâĂŞb) Severe blanket occlusion and caregiver interventions causing semantic failures: pose estimation misses keypoints or misattributes caregiver movements, and segmentation yields empty masks. (c) Pose estimation partially recovers through visible extremities (ankles, hands). (dâĂŞe) Multi-person caregiver interventions further degrade semantic accuracy (pose specificity=0.0; segmentation specificity=0.25). In contrast, surface normals consistently distinguish patient-generated movements through fine-grained geometric differences, maintaining perfect specificity.

zero-valued feature maps despite Raw Normals successfully detecting motion, explaining the 25% performance gap between these geometrically-related modalities (Fig. 4 (a, d)). Pose estimation partially recovered through visible ankles and hands (Fig. 4(c)), achieving $F1 = 0.58$ versus Raw Normals' $F1 = 0.83$, which captured seizure dynamics through subtle blanket surface deformations invisible to keypoint-based methods.

Pat05 (Multi-Person Interference): During caregiver restraint, as soon as the patient experienced a seizure(Fig. 4(d)), both pose estimation and segmentation catastrophically failed (specificity: 0.0 and 0.25 respectively) due to heavy blanket occlusion (Fig. 4(c))combined with the multi-person scenario. Subsequent nurse arrival (Fig. 4(e)) compounded these failures, with both modalities detecting predominantly caregiver/nurse motion rather than patient seizures(4(b)).Âă Raw Normals' perfect specificity (1.000) demonstrates its unique ability to detect patient motion through blanket deformations even in complex multi-person environments, leveraging dense geometric features that remain informative when semantic understanding fails due to occlusion and scene complexity.

These failure modes reveal a fundamental principle: robustness scales with representational density. Segmentation's binary masks and pose's sparse keypoints lack the information capacity to disambiguate overlapping motion sources or infer through occlusion. Raw Normals' dense, continuous surface representation maintains discriminative power across diverse clinical challenges. This mechanistic understanding explains the high variance observed in LOPO metrics (e.g., Pose: specificity $\sigma = 0.345$, F1 $\sigma = 0.407$, sensitivity $\sigma = 0.408$) and validates surface normals as the optimal representation for real-world deployment where such conditions are routine rather than exceptional.

6 Conclusion

This work establishes dense surface normals as a superior, privacy-centric feature primitive for automated seizure detection. Our geometry-based approach achieves a patient-independent accuracy of 89.4% and, critically, maintains high robustness ($F1 = 0.833$) in clinical scenarios such as severe occlusion and multi-person interaction, where traditional pose-based methods fail entirely ($F1 = 0.0$). By shifting from appearance to geometry, we overcome fundamental limitations of semantic features in clinical environments, our method's 26.8% relative improvement over state-of-the-art demonstrates that geometric invariance, not anatomical tracking, provides the robust signal necessary for real-world deployment. The clinical impact extends beyond improved accuracy; surface normal analysis enables 24/7 monitoring across clinical and home settings without EEG constraints, while its sensitivity to subtle deformations opens possibilities for detecting non-convulsive seizures. Future work should pursue real-time optimization ($< 100ms$ latency), extension to broader neurological movement disorders and external validation on different cohorts. This study provides a blueprint for developing clinical vision systems that preserve privacy while handling the inherent complexity of patient care environments.

References

1. Ahmedt-Aristizabal, D., et al.: Deep learning approaches for seizure video analysis: a review. Epilepsy Behav. **154**, 109735 (2024)
2. Arabboev, M., Begmatov, S., Rikhsivoev, M., Saydiakbarov, S., Nosirov, K.: Advances in pose estimation: techniques, applications, and challenges across domains
3. Baltrušaitis, T., Ahuja, C., Morency, L.P.: Multimodal machine learning: a survey and taxonomy. IEEE Trans. Pattern Anal. Mach. Intell. **41**(2), 423–443 (2018)

4. Cunha, J.P.S., et al.: Neurokinect: a novel low-cost 3dvideo-EEG system for epileptic seizure motion quantification. PLoS ONE **11**(1), e0145669 (2016)
5. Eigen, D., Fergus, R.: Predicting depth, surface normals and semantic labels with a common multi-scale convolutional architecture. In: Proceedings of the IEEE International Conference on Computer Vision (ICCV) (2015)
6. Feichtenhofer, C., Fan, H., Malik, J., He, K.: Slowfast networks for video recognition. In: Proceedings of the IEEE/CVF international conference on computer vision, pp. 6202–6211 (2019)
7. Geertsema, E.E., et al.: Automated video-based detection of nocturnal convulsive seizures in a residential care setting. Epilepsia **59**, 53–60 (2018)
8. Kay, W., et al.: The kinetics human action video dataset. arXiv preprint arXiv:1705.06950 (2017)
9. Khirodkar, R., et al.: Sapiens: foundation for human vision models. In: European Conference on Computer Vision, pp. 206–228. Springer (2024)
10. Kuehne, H., Arslan, A., Serre, T.: The language of actions: recovering the syntax and semantics of goal-directed human activities. In: Proceedings of the IEEE conference on computer vision and pattern recognition, pp. 780–787 (2014)
11. Mehta, D., Sivathamboo, S., Simpson, H., Kwan, P., O'Brien, T., Ge, Z.: Privacy-preserving early detection of epileptic seizures in videos. In: International Conference on Medical Image Computing and Computer-Assisted Intervention, pp. 210–219. Springer (2023)
12. Park, S.H., Han, K.: Methodologic guide for evaluating clinical performance and effect of artificial intelligence technology for medical diagnosis and prediction. Radiology **286**(3), 800–809 (2018)
13. Pediaditis, M., Tsiknakis, M., Leitgeb, N.: Vision-based motion detection, analysis and recognition of epileptic seizures–systematic review. Comput. Methods Programs Biomed. **108**(3), 1133–1148 (2012)
14. Ryvlin, P., Nashef, L., Tomson, T.: Prevention of sudden unexpected death in epilepsy: a realistic goal? Epilepsia **54**, 23–28 (2013)
15. Saminu, S., et al.: A recent investigation on detection and classification of epileptic seizure techniques using EEG signal. Brain sci. **11**(5), 668 (2021)
16. Smith, L.N.: A disciplined approach to neural network hyper-parameters: part 1–learning rate, batch size, momentum, and weight decay. arXiv preprint arXiv:1803.09820 (2018)
17. Sultani, W., Chen, C., Shah, M.: Real-world anomaly detection in surveillance videos. In: Proceedings of the IEEE conference on computer vision and pattern recognition, pp. 6479–6488 (2018)
18. Sveinsson, O., Andersson, T., Carlsson, S., Tomson, T.: The incidence of sudep: a nationwide population-based cohort study. Neurology **89**(2), 170–177 (2017)
19. Tzallas, A.T., Tsipouras, M.G., Fotiadis, D.I.: Automatic seizure detection based on time-frequency analysis and artificial neural networks. Comput. Intell. Neurosci. **2007**(1), 080510 (2007)
20. Voigt, P., Von dem Bussche, A.: The EU general data protection regulation (GDPR). A practical guide, 1st ed., Cham: Springer International Publishing **10**(3152676), 10–5555 (2017)
21. Xu, Y., et al.: Vsvig: real-time video-based seizure detection via skeleton-based spatiotemporal VIG. In: European Conference on Computer Vision, pp. 228–245. Springer (2024)

An Explainable Graph Learning Framework for Severe Maternal Morbidity Prediction

Haohui Lu[1(✉)], Ye Lin[1,2], Rongcheng Wu[1,2], Pingyang Sun[1], Yu Gao[1], and Zhidong Li[2]

[1] Molly Wardaguga Institute for First Nations Birth Rights, Faculty of Health, Charles Darwin University, Casuarina, NT 0810, Australia
{haohui.lu,rongcheng.wu,pingyang.sun,yu.gao}@cdu.edu.au

[2] Faculty of Engineering and Information Technology, University of Technology Sydney, Ultimo, NSW 2007, Australia
ye.lin@student.uts.edu.au, zhidong.li@uts.edu.au

Abstract. Severe maternal morbidity (SMM) is a critical challenge in obstetric care, where early identification of high-risk patients is essential for preventing life-threatening outcomes. In this study, we present a novel framework, named Graph-based eXplainable Prediction of Severe Maternal Morbidity (SMMGraphX), that constructs a heterogeneous graph from electronic health records to improve the early prediction of SMM. Unlike conventional approaches that treat clinical data as flat tables, our framework explicitly models the rich relational structure connecting patients, diagnosis codes, and laboratory results. This heterogeneous graph captures the multifaceted nature of clinical interactions, preserving both node and relation-type information essential for understanding complex care pathways. The graph-based model learns to prioritise critical information embedded in this network, capturing which types of clinical relationships and entities contribute most to risk prediction. Importantly, the framework offers interpretability through analysis of learned importance weights, allowing post hoc identification of influential clinical factors such as abnormal laboratory values and key diagnostic patterns. This provides clinicians with transparent insights into the drivers of individual risk predictions, supporting more informed and trustworthy decision-making. Evaluated on the MIMIC-IV dataset, the framework achieves an accuracy of 91.78%, precision of 95.40%, recall of 84.50%, and F1-score of 89.53%, outperforming classical machine learning models and state-of-the-art homogeneous graph neural networks. This work offers a robust and transparent solution for risk stratification in maternal health, leveraging heterogeneous graph construction and interpretable learning to inform clinical decision-making and improve outcomes for at-risk mothers.

Keywords: Heterogeneous graph learning · Explainable AI · Risk prediction · Severe maternal morbidity · Health informatics

M. Liu et al. (Eds.): AI 2025, LNAI 16371, pp. 335–346, 2026.
https://doi.org/10.1007/978-981-95-4972-6_26

1 Introduction

Severe maternal morbidity (SMM) refers to life-threatening complications that occur during pregnancy, childbirth, or the postpartum period [10]. These include conditions such as major haemorrhage, hypertensive emergencies, and organ failure, often considered "near misses" for maternal death [10]. Each case of SMM presents both a critical clinical event and an opportunity to intervene before progression to maternal mortality [2]. Traditionally, clinicians assess maternal risk using screening criteria based on known comorbidities and pregnancy history. For example, the American College of Obstetricians and Gynecologists (ACOG) recommends identifying high-risk pregnancies based on pre-existing conditions including preeclampsia, chronic hypertension, and diabetes [12]. However, such rule-based methods show limited sensitivity. In one study, only 42.4% of women who experienced pregnancy-related hypertensive complications were flagged as high-risk by ACOG criteria [12]. This underlines the need for improved, data-driven approaches to SMM prediction.

Electronic health records (EHRs) contain clinical information including vital signs, laboratory test results, diagnosis codes, and treatment data, all of which may contribute to better prediction of adverse outcomes. Prior research has demonstrated the potential of machine learning (ML) models to leverage such data for predicting maternal conditions. For example, Sufriyana et al. [15] developed ML models for preeclampsia, while Hu et al. [6] applied tree-based models to predict postpartum hemorrhage. However, most of these methods treat each patient independently, without accounting for the relational structure among clinical features. Graph-based modeling provides an alternative by representing patients and clinical entities as interconnected nodes, allowing models to capture complex patterns. Graph neural networks (GNNs) have shown strong performance in predicting medical outcomes and supporting treatment decisions [11,16]. More recent approaches, such as Re-weighted Adversarial Graph Convolutional Network, Receptive Field Aware Graph Convolutional Network, and Self-attention equipped graph convolutions for disease prediction, incorporate medical ontologies and temporal dynamics for prediction [4,8,9]. However, these models rely on homogeneous graph structures, treating all nodes and relationships as identical. In contrast, EHR data is inherently heterogeneous, encompassing different types of nodes such as diagnoses, laboratory results, and patients.

Despite advances in GNN applications to healthcare, limited work has addressed their use in SMM prediction. This study aims to fill this gap by developing a heterogeneous graph learning framework. Patients are connected to diagnoses and lab results to form a structured graph that reflects real-world clinical complexity. In addition, to support clinical implementation, we include interpretability through GNNExplainer [18], which identifies key clinical variables influencing predictions. This assist clinicians in understanding why a patient is considered high risk, fostering greater confidence in decision-making.

Our contributions are as follows:

- We propose SMMGraphX, a novel framework that constructs a graph from the EHR database, linking patients, diagnoses, procedures, and lab results to capture complex clinical interactions.

- The framework allows comparison between heterogeneous and homogeneous graph learning, leveraging node and relation types to identify the most informative pathways and interactions for SMM risk prediction.
- To enhance clinical relevance, we integrate explanation methods that reveal key entities and connections driving each prediction, improving interpretability and supporting clinician trust.

2 Materials and Methods

2.1 Dataset and Cohort

We used data from the Medical Information Mart for Intensive Care IV (MIMIC-IV) [7], a publicly available database containing de-identified EHR from ICU and emergency admissions at a tertiary academic hospital. The dataset includes vitals, laboratory tests, diagnoses, procedures, and medication records.

2.2 Study Population and Outcomes

A retrospective cohort of obstetric patients was identified by selecting hospital admissions with International Classification of Diseases (ICD)-9 and 10 codes related to pregnancy, limited to females of reproductive age. The primary outcome of this study is the prediction of SMM among ICU-admitted patients. SMM is defined according to CDC guidelines [1], based on the presence of qualifying ICD-9 or ICD-10 diagnosis codes. The specific conditions and corresponding codes used for classification are listed in Table 1.

Table 1. Representative ICD-10 and ICD-9 codes for defining SMM conditions.

Condition	ICD-10	ICD-9
Sepsis	A41.9	038.9, 995.91
Shock	R57.9	785.59
Disseminated intravascular coagulation	D65	286.6
Transfusion	30233N1	99.04
Acute renal failure	N17.9	584.9
Respiratory distress	J80, R06.00	518.5, 799.1
Eclampsia	O15.9	642.6
Heart failure	I50.9	428.0
Pulmonary edema	J81.0	518.4
Hysterectomy	0UT90ZZ	68.3
Mechanical ventilation	5A1935Z	96.70

2.3 Heterogeneous Graph Construction

To represent the multi-modal EHR data, we constructed a heterogeneous graph $G = (V, E)$ with three node types: patients, diagnoses, and laboratory tests. Specifically, $V = V_{\text{patient}} \cup V_{\text{diagnosis}} \cup V_{\text{lab}}$.

- **Patient nodes (V_{patient}):** Each patient is a node with features including age and initial vital signs (e.g., blood pressure, temperature).
- **Diagnosis nodes ($V_{\text{diagnosis}}$):** Each unique ICD-9/10 code corresponds to a diagnosis node, representing medical conditions.
- **Lab test nodes (V_{lab}):** Each lab test type is a node with features such as normal ranges or test identifiers.

Edges in the heterogeneous graph represent relations between these entities:

- **Patient–Diagnosis edges:** If patient p_i has diagnosis d_j, an edge (p_i, d_j) is created, representing a "has_diagnosis" relation. These are unweighted.
- **Patient–Lab edges:** If lab test l_k is performed for patient p_i, we add edge (p_i, l_k) with the test value as a numerical edge feature.

This bipartite structure forms star-shaped subgraphs centered on patients, linking to associated diagnoses and labs. Indirect patient–patient connections emerge through shared diagnosis or lab nodes.

We define a patient's neighborhood as: $\mathcal{N}(p) = \{d_j \in V_{\text{diagnosis}} : (p, d_j) \in E\} \cup \{l_k \in V_{\text{lab}} : (p, l_k) \in E\}$.

Diagnosis and lab nodes are connected to all patients sharing those entities.

2.4 Graph-Based Models

Heterogeneous Graph Attention Network. We employ a Heterogeneous Graph Attention Network (HAN) to perform node classification on the patient-centered heterogeneous graph. HAN is specifically designed for heterogeneous graphs [17], allowing us to predict a binary outcome for each patient node.

Type-Specific Transformation: To project features from different node types (patients, diagnoses, labs) into a shared space, we apply type-specific linear transformations. For example, a patient node with initial features $h_p^{(0)}$ and a diagnosis node with $h_d^{(0)}$ are transformed as $x_p = W^{(p)} h_p^{(0)}$ and $x_d = W^{(d)} h_d^{(0)}$, respectively. Similarly, $x_l = W^{(l)} h_l^{(0)}$ for lab nodes.

Node-Level Attention: To aggregate messages from neighbors, HAN applies attention mechanisms to learn weights $\alpha_{p,v}$ for each neighbor v of patient p, using:

$$\alpha_{p,v} = \frac{\exp(e_{p,v})}{\sum_{u \in \mathcal{N}(p)} \exp(e_{p,u})} \tag{1}$$

where $e_{p,v}$ is computed from the transformed features of p and v.

Relation-Level Attention: To integrate information across relation types (Patient–Diagnosis and Patient–Lab), HAN computes separate embeddings $z_p^{(D)}$ and $z_p^{(L)}$, then combines them via:

$$h_p^{\text{new}} = \tanh\left(\beta^{(D)} z_p^{(D)} + \beta^{(L)} z_p^{(L)}\right) \tag{2}$$

where $\beta^{(D)}$ and $\beta^{(L)}$ are learned attention weights over relation types.

Prediction Layer: The final output is computed as $\hat{y}_p = \text{sigmoid}(w_o^\top h_p^{\text{new}})$ representing the predicted probability of SMM. The model is trained end-to-end using binary cross-entropy loss.

Homogeneous GNN Models. The heterogeneous graph can be converted to a homogeneous graph using the Pytorch Geometric (PyG) package [11]. We also compare the HAN model with homogeneous GNN models to evaluate it effectiveness. These include Graph Convolutional Network (GCN) [11], GraphSAGE [5], and Graph Attention Network (GAT) [16]. In a GCN, each node's representation is updated by a weighted average of its neighbours' representations and the node's own, using a shared weight matrix. We implemented a GCN using the same graph structure but treating it as undirected and single-typed. To apply GCN on our data, we treated every edge (i.e., patient–diagnosis or patient–lab) as an untyped edge in a single graph. We also had to create initial features for all nodes, for patient nodes we used the same features as in HAN, and for diagnosis/lab nodes we gave them identity embeddings, which are learnable vectors of the same dimension as patient features. In this case, GCN can propagate information from diagnosis and lab nodes to patient nodes, but without specialised attention for each type. In addition, GraphSAGE [5] is an inductive GNN that samples a fixed number of neighbours and aggregates them. We used a mean aggregator GraphSAGE, which essentially is similar to GCN but without the normalisation by node degree. Furthermore, GAT [16] is used to weight the influence of neighbouring nodes through attention mechanisms; however, it is originally designed for homogeneous graphs and does not natively handle heterogeneity in node or edge types.

2.5 Interpretability and Explainability

After training the graph-based models, we applied GNNExplainer [18] to interpret its predictions. For a given test patient node that labelled as SMM, either correctly or as a false positive, GNNExplainer was used to identify a small subgraph of the patient's neighbourhood that most influenced the prediction. This subgraph typically includes a subset of the patient's connections and highlights which features on those nodes or edges were important. Concretely, GNNExplainer optimises a mask on the adjacency matrix and node feature matrix to maximise the mutual information with the model's output [18]. For each explained patient, the output is a ranking of the features by importance and a visualisation of the subgraph.

2.6 Summary of the Proposed Framework

As shown in Fig. 1, the proposed framework SMMGraphX integrates EHR into a graph-based structure to enable early prediction of SMM. A graph is first constructed from EHR, in which nodes represent patients, diagnoses, and laboratory test results, and edges capture the relationships between them. This graph serves as input to a range of graph-based models, including both heterogeneous (e.g. HAN) and homogeneous (e.g. GCN, GAT) graph neural networks. To support interpretability, the framework includes a model explanation component, which highlights the important features. This approach enables both high performance and transparency in modelling maternal health risk.

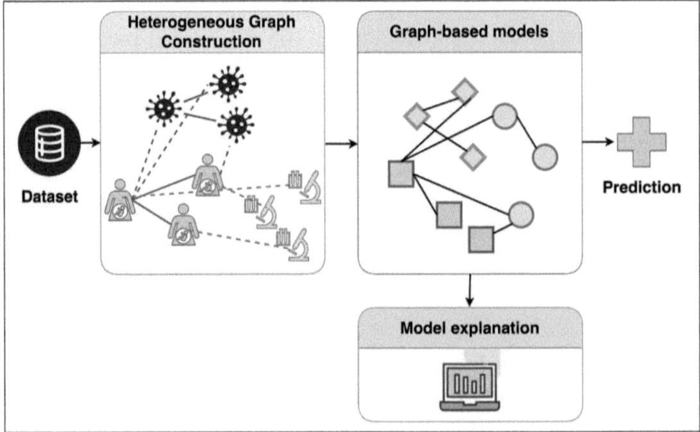

Fig. 1. SMMGraphX Framework for graph-based prediction of severe maternal morbidity (SMM). A heterogeneous graph is constructed from EHR data linking patients to diagnoses and lab results.

3 Experiments and Results

3.1 Patient Cohorts

A total of 835 ICU-admitted obstetric patients were included in the final cohort, with 535 (64.1%) experiencing SMM and 300 (35.9%) without SMM. All admissions with missing values in the selected features were excluded from the final analysis. No outlier removal or adjustment was performed in this study. Information on patient ethnicity was not consistently available in the MIMIC-IV dataset, and comorbidity conditions were not explicitly extracted or analysed as part of this study.

3.2 Experimental Settings

We develop risk prediction models based on a graph-structured dataset derived from the MIMIC-IV database, where patients are linked to recorded diagnosis codes and laboratory results during hospitalisation. The heterogeneous graph comprises three types of nodes (patients, diagnoses, labs) and multiple edge types reflecting clinical interactions. To predict SMM, we train both heterogeneous graph models using an HAN and homogeneous graph models (GCN, GraphSAGE, and GAT) after converting the graph to a homogeneous structure. The dataset is randomly split into 70% training and 30% testing sets based on patient nodes. Models are trained using Python, with classical models implemented in Scikit-learn (sklearn) [14] and graph models implemented using the PyTorch Geometric (PyG) library [3].

For the HAN model, we consider several hyperparameters during development, including the number of layers, hidden layer size, learning rate, optimiser type, dropout rate, and activation function. We apply dropout with a rate of 0.3 at all layers to mitigate overfitting. Models are trained for a maximum of 300 epochs using the Adam optimizer with an initial learning rate of 0.01 and early stopping based on validation loss with a patience of 30 epochs. We use two-layer architecture with hidden size of 64 units, ReLU activation functions after hidden layers, and sigmoid activation for output classification. We also implement 4 attention heads for the HAN model.

For the homogeneous graph models, the GCN model uses a two-layer architecture with a hidden size of 64 units, ReLU activation functions after hidden layers, and sigmoid activation for output classification. For the GAT model, we implement a two-layer architecture where the first layer consists of 4 attention heads, each computing 4 features, followed by an Exponential Linear Unit activation. The second output layer includes a single attention head for binary classification using a sigmoid activation. Hyperparameter tuning and model architecture designs are selected empirically.

3.3 Baselines

To evaluate the effectiveness of our proposed SMMGraphX framework for predicting SMM, we compared its performance against several state-of-the-art models that have demonstrated strong results in disease prediction tasks:

- **Re-weighted Adversarial Graph Convolutional Network (RA-GCN):** RA-GCN is designed to address the challenges of imbalanced medical datasets by incorporating adversarial training. It emphasises minority class samples during learning, which enhances disease prediction accuracy in skewed datasets, a common issue in clinical applications [4].
- **InceptionGCN:** InceptionGCN introduces inception modules into GNN architectures to capture multi-scale neighborhood information. It allows the model to consider both local and broader graph structures, improving disease prediction performance by aggregating features across varying receptive fields [8].

– **Self-Attention Equipped Graph Convolutions (SEGC):** SEGC integrates self-attention mechanisms into graph convolutional layers, enabling the network to dynamically weigh the importance of different neighboring nodes and features. This fine-grained attention leads to improved representation learning and predictive performance in healthcare settings [9].

These baselines were selected to benchmark our framework not only against conventional homogeneous and heterogeneous GNNs but also against advanced architectures specifically developed for biomedical and clinical graph data. All baseline models were implemented and tuned using published hyperparameter settings where available, and evaluated under the same training and testing splits to ensure fair comparison.

3.4 Result Analysis

We evaluate the predictive performance of the proposed SMMGraphX framework against several baseline models. Table 2 summarises the results. We ran each experiment five times to compute the average metrics and report the standard deviations. Among all models, SMMGraphX$_{\text{HAN}}$ achieves the highest overall performance, with an accuracy of 91.78%, precision of 95.40%, recall of 84.50%, and an F1-score of 89.53%. This surpasses both homogeneous graph-based models and alternative heterogeneous architectures. SMMGraphX$_{\text{GAT}}$ achieves slightly lower performance, with an accuracy of 91.22% and an F1-score of 88.67%, followed by SMMGraphX$_{\text{GraphSAGE}}$ at 90.07% accuracy and 87.27% F1-score.

Among the baseline heterogeneous graph models, SEGC achieves the best performance (accuracy 78.20%, F1-score 70.50%), followed by InceptionGCN (accuracy 76.80%, F1-score 68.90%) and RA-GCN (accuracy 74.50%, F1-score 66.80%). In contrast, the SMMGraphX$_{\text{GCN}}$ variant outperforms all these baselines with 81.56% accuracy and a 73.47% F1-score.

These findings highlight the value of explicitly modelling heterogeneous clinical relationships and incorporating attention mechanisms for SMM prediction. The performance gains in recall and F1-score demonstrate that SMMGraphX, particularly the HAN variant, effectively identifies true SMM cases, which is crucial for early clinical intervention. Overall, the results validate the advantage of heterogeneous graph learning combined with explainability for advancing maternal health risk stratification.

3.5 Feature Importance and Explainability Results

A key advantage of our framework is model interpretability. While heterogeneous graphs capture rich clinical relationships, their complexity makes direct interpretation difficult. To address this, we convert the graph to a homogeneous form and use the best-performing SMMGraphX$_{\text{GAT}}$, to extract feature importance. We conduct a case study by randomly selecting a patient node from the test set, whose features are as follows: age 30 years, systolic blood pressure 106.9 mmHg, diastolic blood pressure 60.9 mmHg, body temperature 98.7°F, heart rate

Table 2. Performance comparison of baseline models and SMMGraphX variants on SMM prediction. Values are reported as mean ± standard deviation (%). Best results are highlighted in bold.

Model	Accuracy	Precision	Recall	F1-score
RA-GCN	74.50 ± 0.9	85.30 ± 1.1	55.20 ± 2.1	66.80 ± 1.3
InceptionGCN	76.80 ± 1.0	87.10 ± 1.2	57.60 ± 2.0	68.90 ± 1.7
SEGC	78.20 ± 1.0	88.50 ± 1.2	59.10 ± 0.9	70.50 ± 1.2
$\text{SMMGraphX}_{\text{GCN}}$	81.56 ± 0.9	92.31 ± 1.0	61.02 ± 1.8	73.47 ± 1.4
$\text{SMMGraphX}_{\text{GraphSAGE}}$	90.07 ± 0.7	94.12 ± 0.9	81.36 ± 1.5	87.27 ± 1.1
$\text{SMMGraphX}_{\text{GAT}}$	91.22 ± 0.6	95.10 ± 0.8	83.05 ± 1.3	88.67 ± 1.0
$\text{SMMGraphX}_{\text{HAN}}$	**91.78 ± 0.5**	**95.40 ± 0.7**	**84.50 ± 1.2**	**89.53 ± 0.9**

81.0 bpm, respiratory rate 15.8 bpm, blood glucose 122.3 mg/dL, white blood cell count $11.03 \times 10^3/\mu L$, creatinine 11.5 mg/dL, platelet count $78.1 \times 10^3/\mu L$, hemoglobin 9.08 g/dL, and blood urea nitrogen 113.3 mg/dL.

Feature analysis using GNNExplainer identifies key features contributing to the model's prediction. As shown in Fig. 2, the most influential features include coagulation/thrombocytopenia risk, diastolic blood pressure, white blood cell count, blood urea nitrogen, and age. These findings are clinically consistent, as thrombocytopenia, abnormal blood pressure, impaired renal function, and elevated inflammatory markers are known risk factors for maternal morbidity [13]. The explainability analysis suggests that the GAT model prioritises meaningful clinical indicators, reinforcing the clinical relevance of the model's predictions.

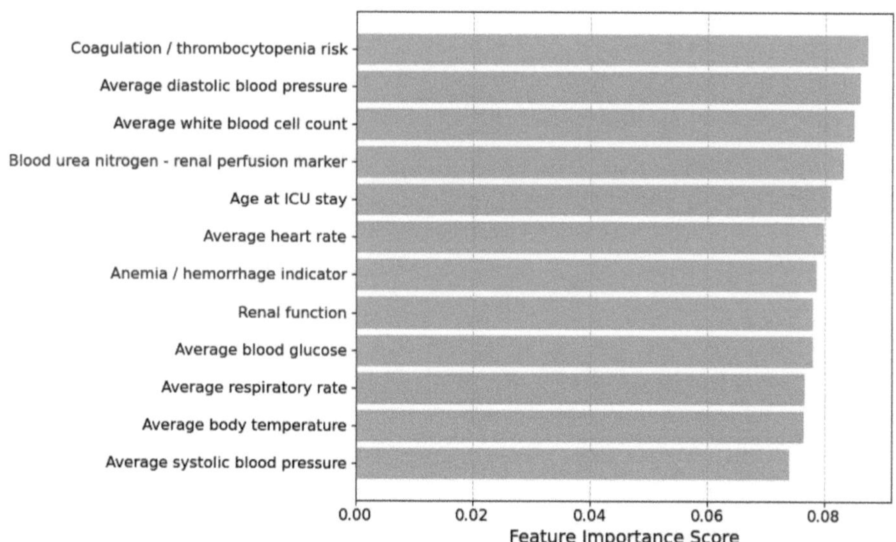

Fig. 2. Feature importance for a patient.

In addition to case-level explanations, we assess global feature importance across the test cohort. For each patient node, we rank features according to their contribution to the model's prediction and aggregate the results. As shown in Table 3, haemoglobin level (anaemia and haemorrhage indicator) is most frequently ranked as the top feature, appearing as the most important feature in 15 patients out of 141 patients. Other commonly top-ranked features include white blood cell count (14 patients), body temperature (14 patients), blood glucose (13 patients), and blood urea nitrogen (12 patients).

When examining mean importance scores across the test set, blood glucose, white blood cell count, and haemoglobin show the highest average contributions. These findings align with clinical understanding, where markers of hematologic status, inflammatory response, and metabolic stability are critical predictors of maternal morbidity [13].

Table 3. Global feature importance across the test cohort. The table reports the number of times each feature was ranked first and the mean importance score.

Feature Description	No. of Times Ranked 1st	Mean Importance Score
Anaemia/haemorrhage indicator	15	0.0827
Average white blood cell count	14	0.0826
Average body temperature	14	0.0823
Average blood glucose	13	0.0830
Blood urea nitrogen	12	0.0826
Average heart rate	12	0.0822
Average diastolic blood pressure	12	0.0822
Average respiratory rate	11	0.0825
Age at ICU stay	10	0.0822
Renal function (creatinine)	10	0.0822
Coagulation risk	10	0.0822
Average systolic blood pressure	8	0.0826

4 Discussion

This study proposes a graph-based framework for the early prediction of SMM from EHRs. Our framework preserves the rich relational structure of clinical data by representing patients, diagnoses, and laboratory tests as distinct nodes within a heterogeneous graph, moving beyond conventional flat feature vectors. The use of a $\text{SMMGraphX}_{\text{HAN}}$ allows the model to focus on the most relevant clinical factors through attention mechanisms, leading to superior predictive performance. The significant improvement in sensitivity and F1-score, demonstrates that explicitly modelling clinical heterogeneity is beneficial for maternal risk prediction.

The framework further includes explainability modules. We highlight key clinical features, such as thrombocytopenia and inflammatory markers, that contribute to the prediction of SMM. This interpretability is crucial for clinical adoption, offering transparent insights into why a patient is flagged as high-risk. Moreover, from the global feature importance across the cohort, we find that features such as haemoglobin, white blood cell count, and blood glucose consistently rank among the most influential predictors.

In addition, our proposed framework can flexibly accommodate additional clinical entities (e.g., medications, procedures) to further enhance predictive power. The modular design allows comparison between heterogeneous and homogeneous graph models, providing a clear evaluation of the value added by modelling heterogeneity. Importantly, while this study focuses on retrospective ICU data from MIMIC-IV, the framework is generalisable to other EHR datasets.

Limitations of this study include reliance on coded diagnosis and procedure data for defining SMM outcomes, which may miss uncoded clinical events or lead to misclassification. Validation using external datasets from diverse healthcare settings is also lacking, which limits confidence in the generalisability of the findings. In addition, the exclusion of patients with missing data and the absence of important demographic and clinical characteristics such as ethnicity and comorbidities may introduce bias and reduce the applicability of the results across different populations. Future work should prioritise external validation and the use of more comprehensive data sources to address these gaps. Nevertheless, this study demonstrates a promising approach for advancing risk stratification and clinical decision support in maternal health by integrating graph-based learning, attention mechanisms, and explainability into a unified framework.

5 Conclusion

This study introduces a graph-based framework for early prediction of SMM using electronic health records, capturing the structured relationships among patients, diagnoses, and laboratory results. The framework demonstrates that modelling the complex interactions in clinical data with a heterogeneous graph and applying GNN leads to improved predictive performance over both traditional machine learning and homogeneous graph-based methods. The integration of attention mechanisms in SMMGraphX$_{\text{HAN}}$ enables the model to prioritise clinically meaningful features, while explainability techniques enhance transparency and support clinical interpretation. Our findings highlight the potential of this approach for maternal health risk prediction, combining high sensitivity with interpretable, feature-level insights. Overall, the framework advances precision and interpretability in risk stratification for obstetric care.

References

1. Callaghan, W.M., Creanga, A.A., Kuklina, E.V.: Severe maternal morbidity among delivery and postpartum hospitalizations in the United States. Obstet. Gynecol. **120**(5), 1029–1036 (2012)
2. Dzakpasu, S., et al.: Severe maternal morbidity surveillance: monitoring pregnant women at high risk for prolonged hospitalisation and death. Paediatr. Perinat. Epidemiol. **34**(4), 427–439 (2020)
3. Fey, M., Lenssen, J.E.: Fast graph representation learning with pytorch geometric. arXiv preprint arXiv:1903.02428 (2019)
4. Ghorbani, M., Kazi, A., Baghshah, M.S., Rabiee, H.R., Navab, N.: RA-GCN: graph convolutional network for disease prediction problems with imbalanced data. Med. Image Anal. **75**, 102272 (2022)
5. Hamilton, W., Ying, Z., Leskovec, J.: Inductive representation learning on large graphs. In: Advances in Neural Information Processing Systems, vol. 30 (2017)
6. Hu, P., et al.: A comparison of lasso regression and tree-based models for delayed cerebral ischemia in elderly patients with subarachnoid hemorrhage. Front. Neurol. **13**, 791547 (2022)
7. Johnson, A.E., et al.: MIMIC-IV, a freely accessible electronic health record dataset. Sci. Data **10**(1), 1 (2023)
8. Kazi, A., et al.: InceptionGCN: receptive field aware graph convolutional network for disease prediction. In: Chung, A.C.S., Gee, J.C., Yushkevich, P.A., Bao, S. (eds.) IPMI 2019. LNCS, vol. 11492, pp. 73–85. Springer, Cham (2019). https://doi.org/10.1007/978-3-030-20351-1_6
9. Kazi, A., Shekarforoush, S., Kortuem, K., Albarqouni, S., Navab, N., et al.: Self-attention equipped graph convolutions for disease prediction. In: 2019 IEEE 16th International Symposium on Biomedical Imaging (ISBI 2019), pp. 1896–1899. IEEE (2019)
10. Kilpatrick, S.K., et al.: Severe maternal morbidity: screening and review. Am. J. Obstet. Gynecol. **215**(3), B17–B22 (2016)
11. Kipf, T.N., Welling, M.: Semi-supervised classification with graph convolutional networks. arXiv preprint arXiv:1609.02907 (2016)
12. Lee, S.M., et al.: Development of early prediction model for pregnancy-associated hypertension with graph-based semi-supervised learning. Sci. Rep. **12**(1), 15793 (2022)
13. Mattar, F., Sibai, B.M.: Eclampsiaviii. Risk factors for maternal morbidity. Am. J. Obstet. Gynecol. **182**(2), 307–312 (2000)
14. Pedregosa, F., et al.: Scikit-learn: machine learning in python. J. Mach. Learn. Res. **12**, 2825–2830 (2011)
15. Sufriyana, H., Wu, Y.W., Su, E.C.Y., et al.: Prediction of preeclampsia and intrauterine growth restriction: development of machine learning models on a prospective cohort. JMIR Med. Inform. **8**(5), e15411 (2020)
16. Velickovic, P., et al.: Graph attention networks. Stat **1050**(20), 10–48550 (2017)
17. Wang, X., et al.: Heterogeneous graph attention network. In: The World Wide Web Conference, pp. 2022–2032 (2019)
18. Ying, Z., Bourgeois, D., You, J., Zitnik, M., Leskovec, J.: Gnnexplainer: generating explanations for graph neural networks. In: Advances in Neural Information Processing Systems, vol. 32 (2019)

Personalized Federated Graph Learning for Heterogeneous Incomplete EHRs

Liping Chen[1], Guangqing Bai[2], Zhuoyang Han[2], Jing Ren[1], Mujie Liu[3], Tao Tang[4](✉), Shuo Yu[2], and Ivan Lee[4]

[1] RMIT University, Melbourne, Australia
{lp.chen,jing.ren}@ieee.org
[2] Dalian University of Technology, Dalian, China
baigq@mail.dlut.edu.cn, {zhuoyang.han,shuo.yu}@ieee.org
[3] Federation University Australia, Ballarat, Australia
mujie.liu@ieee.org
[4] University of South Australia, Adelaide, Australia
tao.tang@ieee.org, Ivan.Lee@unisa.edu.au

Abstract. Real-world electronic health records (EHRs) are decentralized and typically exhibit incomplete information across both modalities and features. This heterogeneous missingness poses a major challenge for federated graph learning, which aims to collaboratively model distributed medical data while preserving privacy. Existing methods often assume uniform data completeness or apply imputation strategies, which struggle to maintain consistent predictive performance under non-uniform and severe data incompleteness. To address this issue, we propose MissPCL, a personalized federated graph contrastive learning framework tailored to heterogeneous client-level missingness. Each client estimates its missingness rate and constructs a bipartite patient-modality graph from observed features. Clients with low missingness participate in global training using contrastive learning on local graphs, while high-missingness clients are excluded from aggregation and instead perform personalized fine-tuning. Experiments on two real-world EHR datasets under diverse missingness scenarios demonstrate that MissPCL consistently outperforms state-of-the-art baselines in classification accuracy and robustness. Notably, our framework achieves stable performance across varying degrees of data incompleteness, showing its practical utility in realistic federated clinical settings. Code implementations and the supplementary materials are available at https://github.com/TutaResearch/MISSPCL.

Keywords: Personalized Federated Learning · Graph Contrastive Learning · Heterogeneous Missingness

1 Introduction

Graph learning offers a powerful paradigm for capturing complex relationships across data entities and has demonstrated strong performance in domains such as

M. Liu et al. (Eds.): AI 2025, LNAI 16371, pp. 347–358, 2026.
https://doi.org/10.1007/978-981-95-4972-6_27

healthcare, transportation, and social networks [5,12]. In the healthcare domain, multimodal graph learning integrates heterogeneous information from structured electronic health records (EHRs), imaging, and lab tests, improving patient representation learning for tasks such as disease prediction and biomarker discovery [14,22,29]. However, multimodal clinical data are typically fragmented across institutions due to non-standardized collection protocols and strict privacy regulations [3], motivating the use of federated learning as a privacy-preserving paradigm for collaborative modeling without centralized data sharing. Recently, federated multimodal graph learning has gained attention as a promising approach to leverage distributed, complementary medical data across clients [7].

Beyond data decentralization, a major challenge lies in pervasive and non-random data incompleteness, which occurs at both the modality and feature levels [11]. Such fine-grained missingness arises from clinical heterogeneity, varying resource availability, and differences in patient characteristics, and is further amplified in federated settings where each client may exhibit distinct and sparse patterns [15]. For example, while one hospital may record full lab and imaging data, another may only collect basic vitals due to operational constraints. This heterogeneity severely limits global model optimization, particularly when clients with richer data dominate training [13].

To address missing data, prior studies have proposed modality imputation [17] and modality-invariant representation learning [30]. However, these methods often assume consistent modality availability or rely on reconstruction, which may be infeasible or privacy-sensitive in federated environments [4]. In contrast, graph contrastive learning (GCL) has recently emerged as a powerful alternative [18], offering robust representation learning without requiring full data reconstruction. Recent studies demonstrate its ability to effectively mitigate modality collapse and extract discriminative patient features even under severe missingness [24,27]. Current GCL-based methods are typically designed for centralized settings and do not account for federated heterogeneity or client-level missingness variation.

To bridge this gap, we propose MissPCL, a missingness-aware personalized federated graph contrastive learning framework for decentralized healthcare. Each client estimates its feature-level missingness and builds a bipartite graph connecting patients and available modalities. Clients with low missingness collaborate in global model training using a mutual-consistent contrastive learning objective that combines label supervision and graph augmentation. High-missingness clients are excluded from global updates and instead fine-tune the global model locally. This personalized framework jointly addresses client heterogeneity and missingness variation, yielding robust, consistent performance across diverse real-world scenarios. We summarize our key contributions as follows:

1) We propose MissPCL, a novel personalized federated graph contrastive learning framework tailored to heterogeneous missingness in decentralized multimodal healthcare data.

2) MissPCL estimates client-specific missingness rates and dynamically selects participants for global training, enabling personalized model adaptation for high-missingness clients.
3) Extensive experiments on real-world EHR datasets demonstrate that MissPCL achieves superior generalization and personalization under diverse missingness patterns.

2 Related Work

2.1 Multimodal Graph Learning with Missing Modalities

Multimodal graph learning enables integration of structured healthcare data such as imaging, clinical notes, and lab tests. Existing methods addressing missing modalities include imputation-based models, which reconstruct missing data via latent or clustering strategies, and representation learning approaches that aim for modality-invariant embeddings [17,28]. For instance, M3Care [28] uses modality-specific graphs for latent reconstruction, while Cafe [17] clusters clients by missingness for feature completion. Recent fusion-based models (e.g., ViLT [10], HGMF [1]) bypass reconstruction by learning from observed modalities, though they may overlook modality-specific cues. Contrastive frameworks like MUSE [24] and FuseMoE [6] improve flexibility but often ignore feature-level informativeness. However, these methods often treat modalities as atomic units and do not explicitly quantify feature-level informativeness under heterogeneous missingness. In contrast, our approach uses graph-based contrastive learning to selectively propagate informative signals, improving robustness to both modality and feature-level incompleteness.

2.2 Personalized Federated Learning on Heterogeneous Data

Personalized federated learning (PFL) mitigates Non-IID issues by adapting models to client-specific distributions. Existing methods include global model personalization (e.g., fine-tuning [21], reweighted aggregation [2,25]) and client-specific models using generative adversarial networks [23], prompt tuning [26], or adapters [19]. While effective for data heterogeneity, most existing methods assume uniform modality availability across clients-an unrealistic assumption in federated healthcare due to inconsistent data acquisition protocols. In contrast, our method explicitly models heterogeneous missingness and adopts selective client participation and fine-tuning, enhancing global model robustness and local adaptation in the presence of incomplete and non-uniform data.

3 Proposed Method

In this paper, we propose MissPCL, a missingness-aware personalized federated graph contrastive learning framework, designed to address heterogeneous data incompleteness in decentralized healthcare environments. The framework

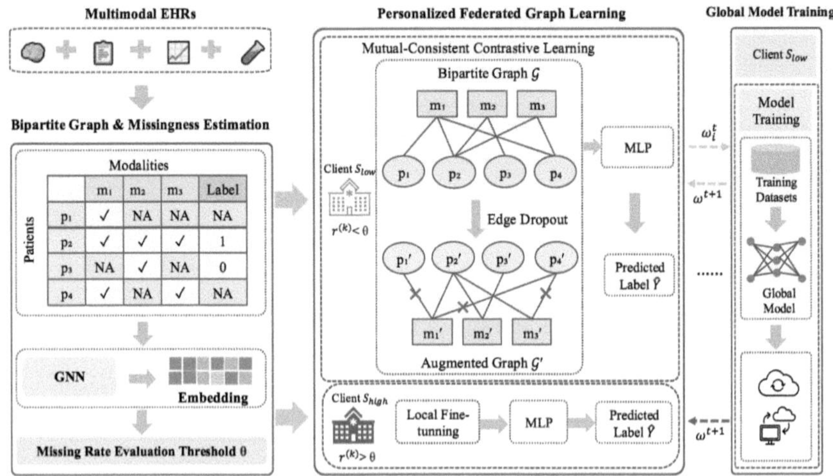

Fig. 1. Overview of the proposed MissPCL framework.

operates in three key stages: (1) each client quantifies its local feature-level missingness to guide personalized participation; (2) low-missingness clients collaboratively train a global model via federated optimization on locally constructed bipartite patient-modality graphs using contrastive learning; and (3) high-missingness clients personalize the global model via fine-tuning on their incomplete data. The method integrates mutual-consistent contrastive learning to enhance robustness and a personalized client selection mechanism to improve adaptability under diverse missingness conditions. This design enables more reliable disease prediction across clients with varying levels of data quality. The overall framework of the proposed MissPCL is shown in Fig. 1.

3.1 Missingness Estimation and Bipartite Graph Construction

Missingness Estimation. To quantify the degree of data incompleteness on each client, we compute a missingness rate based on the proportion of missing feature values. Specifically, for client $k \in (0, \dots, K)$, let $N^{(k)}$ be the number of patients and $\mathcal{F}$ the set of features. We define the missingness rate $r^{(k)}$ as the number of missing entries in its local dataset, divided by the total number of entries on client k, denoted as $N^{(k)}|\mathcal{F}|$. The number of missing entries is counted by scanning all feature values for all patients and checking whether a value is missing (e.g., equals NA or null). This count is then divided by the total number of entries.

Bipartite Graph Construction. After quantifying missingness rates, each client constructs an undirected bipartite patient-modality graph $\mathcal{G}^{(k)}$, following [24], to explicitly model the relationship between patients and modalities under the federated learning setting. For clarity, we omit the superscript k in subsequent descriptions.

Formally, each bipartite graph is defined as $\mathcal{G} = (\mathcal{V}, \mathcal{E})$, where $\mathcal{V}$ denotes the set of nodes and $\mathcal{E}$ the set of edges. The node set $\mathcal{V}$ consists of two disjoint subsets: the set of patient nodes $\mathcal{V}_P = \{v_1, v_2, \ldots, v_N\}$ and the set of modality nodes $\mathcal{V}_M = \{v_{N+1}, v_{N+2}, \ldots, v_{N+M}\}$, where N and M denote the number of patients and modalities, respectively. The edge set $\mathcal{E}$ is determined by the modality presence matrix $\mathbf{M} \in \{0,1\}^{N \times M}$, where an edge e_{v_i,v_j} exists if patient node $v_i \in \mathcal{V}_P$ has records corresponding to modality node $v_j \in \mathcal{V}_M$.

To capture semantic relationships between patients and modalities, we initialize edge embeddings based on raw modality features.

Specifically, for an edge e_{v_i,v_j}, its embedding is computed by applying a modality-specific encoder to the corresponding raw input by Eq. (1).

$$\mathbf{e}_{v_i,v_j} = \text{Encoder}^{(j)}(\mathbf{x}_{i,j}), \tag{1}$$

where $\mathbf{x}_{i,j}$ denotes the raw feature vector of modality j for patient i, and $\text{Encoder}^{(j)}(\cdot)$ is a learnable encoder dedicated to modality j.

3.2 Personalized Federated Graph Learning

Representation Learning on Bipartite Graphs. After constructing the bipartite patient-modality graph $\mathcal{G}$ locally at each client, we utilize a multi-layer Graph Neural Network (GNN) to iteratively aggregate and update node embeddings.

Specifically, we adopt distinct initialization strategies depending on node type. Each modality node v_j is initialized as $\mathbf{h}_{v_j}^{(0)} = \text{Embedding}(j)$, while each patient node v_i is initialized with a constant vector of ones, i.e., $\mathbf{h}_{v_i}^{(0)} = 1 \in \mathbb{R}^d$, where the superscript (0) denotes the initial layer, and d represents the embedding dimension.

Therewith, the message from modality node v_j to patient node v_i at the l-th propagation layer is computed by Eq. (2).

$$\mathbf{m}_{v_j \to v_i}^{(l)} = \text{GNN}^{(l)}\left(\left[\mathbf{h}_{v_j}^{(l-1)} \| \mathbf{h}_{v_i}^{(l-1)} \| \mathbf{e}_{v_j \to v_i}^{(l-1)}\right]\right), \tag{2}$$

where $\mathbf{h}_{v_j}^{(l-1)}$ and $\mathbf{h}_{v_i}^{(l-1)}$ represent embeddings of the source and target nodes at layer $(l-1)$ respectively, $\mathbf{e}_{v_j \to v_i}^{(l-1)}$ denotes the embedding of the edge from v_j to v_i, and $\|$ denotes vector concatenation.

Subsequently, each patient node v_i updates its embedding by aggregating messages from its neighboring modality nodes using mean pooling:

$$\mathbf{h}_{v_i}^{(l)} = \text{Aggregate}\left(\left\{\mathbf{m}_{v_j \to v_i}^{(l-1)} \mid v_j \in \mathcal{N}(v_i)\right\}\right), \tag{3}$$

where $\mathcal{N}(v_i) \subseteq \mathcal{V}_M$ denotes the set of modality nodes connected to patient node v_i.

After L propagation layers, we obtain the final patient node embeddings $\mathbf{h}_{v_i}^{(L)}$, which are then used for the contrastive learning.

Personalized Client Selection. To mitigate performance degradation caused by clients with high missingness rates, we introduce a personalized federated framework that dynamically selects participating clients based on their local missingness scores.

Given a tunable threshold θ, the client set is partitioned into:

- **Low-missingness clients** $\mathcal{S}_{\text{low}} = \{i \mid r^{(i)} < \theta\}$, who participate in global model training.
- **High-missingness clients** $\mathcal{S}_{\text{high}} = \{i \mid r^{(i)} \geq \theta\}$, who receive the trained global model and perform local fine-tuning.

The global training objective over $\mathcal{S}_{\text{low}}$ is defined as:

$$\min_{\mathbf{w}} \mathcal{L}_{\text{global}}(\mathbf{w}) = \sum_{i \in \mathcal{S}_{\text{low}}} \frac{n_i}{n_{\text{low}}} \mathcal{L}_i(\mathbf{w}), \tag{4}$$

where n_i is the number of samples on client S_i, $n_{\text{low}} = \sum_{i \in \mathcal{S}_{\text{low}}} n_i$, and $\mathcal{L}_i(\mathbf{w})$ is the local loss computed on client i.

Mutual-Consistent Contrastive Learning. To enhance robustness and representation consistency, we adopt Mutual-Consistent Contrastive Learning (MCCL) [24]. Specifically, we generate an augmented bipartite graph $\mathcal{G}'$ by randomly dropping edges from the original graph $\mathcal{G}$. We then construct contrastive pairs based on patient identity, encouraging embeddings of different modalities from the same patient (positive pairs) to be close, while pushing apart embeddings from different patients (negative pairs).

Local Training Objectives. On each selected client $S_i \in \mathcal{S}_{\text{low}}$, the local training objective integrates three components: unsupervised contrastive loss, supervised contrastive loss, and classification loss. These losses are computed based on patient node embeddings obtained from the local bipartite graph. We adopt NT-Xent loss [27] for contrastive objective and standard cross-entropy loss for local supervision. It is to be noted that the details of the constrastive loss, the supervised loss, and classification loss are introduced in the supplementary materials.

Personalized Fine-Tuning for High-Missingness Clients. After global training, the server distributes the optimized global model $\mathbf{w}^*$ to high-missingness clients $S_i \in \mathcal{S}_{\text{high}}$. Each client then performs local fine-tuning to better adapt to its heterogeneous and incomplete data distribution. The fine-tuning objective is formulated in Eq. (5).

$$\min_{\mathbf{w}_i} \mathcal{L}_i(\mathbf{w}_i) + \frac{\alpha}{2}\|\mathbf{w}_i - \mathbf{w}^*\|^2, \tag{5}$$

where $\mathcal{L}_i(\mathbf{w}_i)$ denotes the local training loss on client S_i, and α controls the strength of regularization towards the global model.

4 Experiments

4.1 Experimental Setup

Datasets. We evaluate our method on two real-world EHR datasets: Alzheimer's Disease Neuroimaging Initiative (ADNI) [8] and MIMIC-IV [9]. For ADNI, we use the curated TADPOLE challenge dataset [16], which includes structural MRI, FDG-PET, diffusion tensor imaging (DTI), CSF biomarkers, and clinical assessments, covering 1,737 patients and 12,741 visits. For MIMIC-IV, we extract structured data modalities including diagnoses (ICD codes), procedures, prescriptions, and laboratory tests, following prior categorization [20]. Each dataset is split into training, validation, and test sets with a 70%:10%:20% ratio.

Task Settings. We consider three prediction tasks: (i) mortality prediction and (ii) hospital readmission prediction, both based on MIMIC-IV, and (iii) Alzheimer's disease progression prediction based on ADNI.

Evaluation Metrics. For the mortality prediction task, we report the Area Under the ROC Curve (AUC-ROC) and the Area Under the Precision–Recall Curve (AUC-PRC). For the Alzheimer's disease progression (ADP) task, we report Accuracy and AUC-ROC.

Baselines. We compare our model with sota methods including **Cafe** [17], **PEARL** [21], and **M3Care** [28]. Additionally, we examine the effect of differential privacy (DP) by applying DP-SGD to global updates to assess robustness under stricter privacy constraints.

Data Pre-processing. Since ADNI and MIMIC-IV are centralized datasets, we simulate a federated setting by partitioning each into 8 clients with roughly equal patient counts. To capture real-world heterogeneity, we design three synthetic missingness configurations by assigning different missing rates to clients. Empirically, both datasets exhibit ~30% feature-level missingness. Based on this, we simulate client-wise missingness ranging from 0.3 to 1.0, where 1.0 represents fully missing input and is used to test robustness under extreme incompleteness.

Specifically, we define three settings:

- **Case 1 (increasing missingness):** (0.3, 0.4, 0.5, 0.6, 0.7, 0.8, 0.9, 1.0).
- **Case 2 (mixed missingness):** (0.3, 0.3, 0.5, 0.5, 0.7, 0.7, 0.9, 0.9).
- **Case 3 (grouped high missingness):** (0.6, 0.6, 0.7, 0.7, 0.8, 0.8, 0.9, 0.9).

For each client, missingness is simulated by randomly masking features to match the assigned rate. This setup enables a systematic evaluation of robustness under varying degrees and patterns of modality incompleteness. Further details of the experimental setup are provided in the supplementary material.

4.2 Results Analysis

As shown in Table 1, our method **MissPCL** achieves the best overall performance across all three tasks under increasing missingness (Case 1). On ADNI (ADP task), it obtains the highest AUC-ROC of 0.6427, outperforming all baselines. On MIMIC-IV, MissPCL also delivers strong results, reaching the highest accuracy for both mortality and readmission prediction tasks. While **Cafe** shows competitive AUC-ROC for the mortality task, it underperforms on the ADP task. **M3Care** attains good accuracy but degrades with severe missingness, and **PEARL** consistently lags due to limited ability to handle incomplete modalities.

MissPCL's advantage stems from its missingness-aware client selection. We set $\theta = 0.8$ for ADNI and $\theta = 0.9$ for MIMIC-IV. For example, in ADNI Case 1, clients with missingness below 0.8 (i.e., 0.3–0.7) join global training, while higher-missingness clients skip and fine-tune locally. This design lets the global model learn from more complete data while still enabling adaptation for sparse clients.

Table 1. Baseline results with Case 1 client settings. AUC represents AUROC, ACC represents Accuracy. Best values are in bold.

Baseline	ADP (AUC)	Mortality AUC	Mortality ACC	Readmission AUC	Readmission ACC
Cafe	0.5509	**0.8923**	0.8880	**0.6834**	0.7634
PEARL	0.4683	0.6026	0.4918	0.5330	0.6105
M3Care	0.5963	0.6318	0.9127	0.5665	0.7615
MissPCL (ours)	**0.6427**	0.8210	**0.9132**	0.6491	**0.7675**

Table 2. Baseline results with Case 2 client settings.

Baseline	ADP (AUC)	Mortality AUC	Mortality ACC	Readmission AUC	Readmission ACC
Cafe	0.5489	**0.8914**	0.8906	**0.6882**	0.7637
PEARL	0.6166	0.6922	0.8724	0.5261	0.5814
M3Care	0.6018	0.6388	0.9127	0.5714	0.7615
MissPCL (ours)	**0.6621**	0.8252	**0.9137**	0.6485	**0.7670**

As shown in Table 2, **MissPCL** delivers strong results under mixed missingness (Case 2) across all tasks. On ADNI-ADP, it achieves the highest AUC-ROC of 0.6621, surpassing all baselines. For MIMIC-IV, it attains the best accuracy on both mortality (91.37%) and readmission (76.70%) prediction, reflecting superior generalization. Although **Cafe** yields the top AUC-ROC on MIMIC-IV tasks, its ADP performance is poor, likely due to reliance on imputation that benefits binary outcomes but falters under complex missingness. **PEARL** and **M3Care** also underperform, struggling with heterogeneous missingness. In this case, we set $\theta = 0.6$ for ADNI and $\theta = 0.8$ for MIMIC-IV, enabling clients below the threshold to join collaborative training while others fine-tune locally. This strategy lets **MissPCL** exploit reliable data globally while adapting to sparse clients, enhancing robustness.

Table 3. MissPCL performance under different θ thresholds across three client cases. PRC represents AUPRC.

(a) Case 1

θ	ADP (AUC)	Mortality-AUC	Mortality-PRC	Readmission-AUC	Readmission-PRC
0.4	0.6224	0.7956	0.3001	0.6323	0.3678
0.5	0.6244	0.8087	0.3167	0.6417	0.3776
0.6	0.6216	0.8141	0.3262	0.6448	0.3831
0.7	0.6304	0.8175	0.3429	0.6459	0.3844
0.8	**0.6427**	0.8194	0.3494	0.6470	0.3861
0.9	0.6390	**0.8210**	0.3513	0.6491	**0.3874**
1.0	0.6362	0.8195	**0.3543**	**0.6506**	0.3868

(b) Case 2

θ	ADP (AUC)	Mortality-AUC	Mortality-PRC	Readmission-AUC	Readmission-PRC
0.4	0.6531	0.8129	0.3228	0.6395	0.3769
0.6	**0.6621**	0.8199	0.3499	0.6452	0.3845
0.8	0.6586	**0.8252**	**0.3657**	**0.6485**	**0.3877**
1.0	0.6547	0.8185	0.3594	0.6431	0.3862

(c) Case 3

θ	ADP (AUC)	Mortality-AUC	Mortality-PRC	Readmission-AUC	Readmission-PRC
0.7	0.6123	0.8056	0.3110	0.6364	0.3737
0.8	0.6155	0.8145	**0.3351**	0.6120	0.3815
0.9	**0.6202**	0.8191	0.3464	**0.6457**	**0.3848**
1.0	0.6176	**0.8215**	**0.3551**	0.6419	0.3814

4.3 Hyperparameter Analysis

Effect of Threshold θ. We analyze the sensitivity of the threshold parameter θ across three client configurations to evaluate its impact on performance. As shown in Table 3, higher θ generally improves **AUC-ROC** and **AUC-PRC**. The best results occur at $\theta = 0.8$ for ADNI and $\theta = 0.9$ for MIMIC-IV, suggesting that filtering out clients with severe missingness improves generalization by reducing noisy updates.

Beyond these points (e.g., $\theta = 1.0$), performance plateaus or slightly declines, indicating that including highly sparse clients may degrade the global model. Accordingly, we set $\theta = 0.8$ for ADNI and $\theta = 0.9$ for MIMIC-IV.

Case 2 (mixed missingness) shows MissPCL performing best on ADNI at $\theta = 0.6$ and on MIMIC-IV at $\theta = 0.8$, underscoring the trade-off between data quality and participation. Case 3 (high missingness) yields lower but stable performance, with optimal results again at $\theta = 0.9$, indicating robustness under challenging conditions.

Structured Modality-level Missingness. To assess robustness under more realistic conditions, we simulate missingness patterns based on empirical modality availability observed in the original datasets. In ADNI, cognitive assessments, MRI, and PET/CSF biomarkers are available in roughly 50%, 35%, and 15% of cases, respectively. In MIMIC-IV, discharge notes and lab vectors follow a 0.72-to-0.28 distribution. These ratios guide modality-wise masking, yielding client data distributions that reflect real-world clinical heterogeneity—such as limited access to PET scans or missing labs in smaller hospitals.

Under this setting (aligned with Case 1), **MissPCL** achieves 0.7836 AUC-ROC on ADNI, and 0.6233 AUC-ROC with 91.26% accuracy on MIMIC-IV mortality prediction. For readmission, it obtains 0.5767 AUC-ROC and 76.20% accuracy. These results highlight MissPCL's robustness under structured, non-uniform missingness.

Finally, we also evaluate MissPCL under varying values of θ and missingness rates r for Case 2 clients, and report detailed **AUC-PRC** results in the supplementary materials. To address privacy concerns in federated learning, we additionally extend MissPCL with differential privacy mechanisms; results and analysis of this extension are likewise included in the supplementary materials.

5 Conclusion

We propose MissPCL, a missingness-aware personalized federated graph contrastive learning framework designed for decentralized healthcare data with heterogeneous incompleteness. By estimating client-level missingness and dynamically selecting participants, MissPCL balances data quality and participation, while enabling personalized adaptation for highly incomplete clients. Leveraging bipartite graph modeling and mutual-consistent contrastive learning, it effectively captures patient-disease relationships under varying missingness conditions. Extensive experiments on real-world EHR datasets show that MissPCL consistently outperforms state-of-the-art baselines across diverse missingness patterns.

In future work, we aim to explore more fine-grained missingness modeling beyond scalar rates, integrate uncertainty-aware representations, and deploy the framework in real-world clinical federated settings to further evaluate its scalability and practical utility.

References

1. Chen, J., Zhang, A.: HGMF: heterogeneous graph-based fusion for multimodal data with incompleteness. In: Proceedings of the 26th ACM SIGKDD International Conference on Knowledge Discovery and Data Mining, pp. 1295–1305 (2020)
2. Chen, Y., Lu, W., Qin, X., Wang, J., Xie, X.: Metafed: federated learning among federations with cyclic knowledge distillation for personalized healthcare. IEEE Trans. Neural Netw. Learn. Syst. **35**(11), 16671–16682 (2024)
3. Esmaeilzadeh, P.: Challenges and strategies for wide-scale artificial intelligence (AI) deployment in healthcare practices: a perspective for healthcare organizations. Artif. Intell. Med. **151**, 102861 (2024)
4. Guan, H., Yap, P.T., Bozoki, A., Liu, M.: Federated learning for medical image analysis: a survey. Pattern Recogn. 110424 (2024)
5. Han, H., et al.: Openhgnn: an open source toolkit for heterogeneous graph neural network. In: Proceedings of the 31st ACM International Conference on Information & Knowledge Management, pp. 3993–3997 (2022)
6. Han, X., Nguyen, H., Harris, C.W., Ho, N., Saria, S.: Fusemoe: mixture-of-experts transformers for fleximodal fusion. In: The Thirty-eighth Annual Conference on Neural Information Processing Systems (2024)
7. Huang, W., Wang, D., Ouyang, X., Wan, J., Liu, J., Li, T.: Multimodal federated learning: concept, methods, applications and future directions. Inf. Fusion **112**, 102576 (2024)
8. Jack, C.R., Jr., et al.: The Alzheimer's disease neuroimaging initiative (ADNI): MRI methods. J. Magn. Resonan. Imaging: Official J. Int. Soc. Magn. Resonan. Med. **27**(4), 685–691 (2008)
9. Johnson, A.E., et al.: MIMIC-IV, a freely accessible electronic health record dataset. Sci. Data **10**(1) (2023)
10. Kim, W., Son, B., Kim, I.: Vilt: vision-and-language transformer without convolution or region supervision. In: International Conference on Machine Learning, pp. 5583–5594. PMLR (2021)
11. Koo, H.: Overcoming uncertain incompleteness for robust multimodal sequential diagnosis prediction via curriculum data erasing guided knowledge distillation. In: 2025 IEEE International Conference on Acoustics, Speech and Signal Processing (2025)
12. Lablack, M., Yu, S., Xu, S., Shen, Y.: Long-sequence model for traffic forecasting in suboptimal situation. In: Proceedings of the 18th Workshop on Mobility in the Evolving Internet Architecture, pp. 25–30 (2023)
13. Li, Y., et al.: A vertical federated multi-view fuzzy clustering method for incomplete data. IEEE Trans. Fuzzy Syst. (2025)
14. Liu, Q., et al.: Multimodal recommender systems: a survey. ACM Comput. Surv. **57**(2) (2024)
15. Liu, Y., Wang, C., Yuan, X.: Fedmobile: enabling knowledge contribution-aware multi-modal federated learning with incomplete modalities. In: Proceedings of the ACM on Web Conference 2025, pp. 2775–2786 (2025)
16. Marinescu, R.V., et al.: Tadpole challenge: accurate Alzheimer's disease prediction through crowdsourced forecasting of future data. In: Predictive Intelligence in Medicine: Second International Workshop, pp. 1–10. Springer, Cham (2019)
17. Min, S., Asif, H., Wang, X., Vaidya, J.: Cafe: improved federated data imputation by leveraging missing data heterogeneity. IEEE Trans. Knowl. Data Eng. **37**(5), 2266–2281 (2025)

18. Ren, J., Hou, M., Liu, Z., Bai, X.: Eagle: contrastive learning for efficient graph anomaly detection. IEEE Intell. Syst. **38**(2), 55–63 (2022)
19. Saha, P., Mishra, D., Wagner, F., Kamnitsas, K., Noble, J.A.: Fedpia – permuting and integrating adapters leveraging Wasserstein barycenters for finetuning foundation models in multi-modal federated learning. In: Proceedings of the AAAI Conference on Artificial Intelligence, vol. 39, no. 19 (2025)
20. Sung, M., et al.: Event prediction model considering time and input error using electronic medical records in the intensive care unit: retrospective study. JMIR Med. Inform. **9**(11), e26426 (2021)
21. Tang, T., et al.: Personalized federated graph learning on non-IID electronic health records. IEEE Trans. Neural Netw. Learn. Syst. **35**(9), 11843–11856 (2024)
22. Tang, T., Han, Z., Yu, S., Bagirov, A., Zhang, Q.: Fuzzy multiview graph learning on sparse electronic health records. IEEE Trans. Fuzzy Syst. **32**(10), 5520–5532 (2024)
23. Wijesinghe, A., Zhang, S., Ding, Z.: Ps-fedgan: an efficient federated learning framework with strong data privacy. IEEE Internet Things J. **11**(16), 27584–27596 (2024)
24. Wu, Z., et al.: Multimodal patient representation learning with missing modalities and labels. In: The Twelfth International Conference on Learning Representations (2024)
25. Wu, Z., et al.: Fedcache: a knowledge cache-driven federated learning architecture for personalized edge intelligence. IEEE Trans. Mob. Comput. **23**(10), 9368–9382 (2024)
26. Yang, F.E., Wang, C.Y., Wang, Y.C.F.: Efficient model personalization in federated learning via client-specific prompt generation. In: Proceedings of the IEEE/CVF International Conference on Computer Vision, pp. 19159–19168 (2023)
27. You, Y., Chen, T., Sui, Y., Chen, T., Wang, Z., Shen, Y.: Graph contrastive learning with augmentations. In: Advances in Neural Information Processing Systems, vol. 33, pp. 5812–5823 (2020)
28. Zhang, C., et al.: M3care: learning with missing modalities in multimodal healthcare data. In: Proceedings of the 28th ACM SIGKDD Conference on Knowledge Discovery and Data Mining, pp. 2418–2428 (2022)
29. Zheng, S., et al.: Multi-modal graph learning for disease prediction. IEEE Trans. Med. Imaging **41**(9), 2207–2216 (2022)
30. Zuo, H., Liu, R., Zhao, J., Gao, G., Li, H.: Exploiting modality-invariant feature for robust multimodal emotion recognition with missing modalities. In: 2023 IEEE International Conference on Acoustics, Speech and Signal Processing (2023)

Structure Matters: Brain Graph Augmentation via Learnable Edge Masking for Data-Efficient Psychiatric Diagnosis

Mujie Liu[1], Chenze Wang[2], Liping Chen[3], Nguyen Linh Dan Le[3], Niharika Tewari[3], Ting Dang[4], Jiangang Ma[1], and Feng Xia[3(✉)]

[1] Institute of Innovation, Science and Sustainability, Federation University Australia, Ballarat 3353, Australia
mujie.liu@ieee.org, j.ma@federation.edu.au

[2] School of Information and Electronic Engineering, Zhejiang Gongshang University, Hangzhou 315104, China
23020090112@pop.zjgsu.edu.cn

[3] School of Computing Technologies, RMIT University, Melbourne 3000, Australia
{lp.chen,f.xia}@ieee.org, {s4135772,S4044597}@student.rmit.edu.au

[4] School of Computing and Information Systems, The University of Melbourne, Melbourne 3052, Australia
ting.dang@unimelb.edu.au

Abstract. The limited availability of labeled brain network data makes it challenging to achieve accurate and interpretable psychiatric diagnoses. While self-supervised learning (SSL) offers a promising solution, existing methods often rely on augmentation strategies that can disrupt crucial structural semantics in brain graphs. To address this, we propose SAM-BG, a two-stage framework for learning brain graph representations with structural semantic preservation. In the pre-training stage, an edge masker is trained on a small labeled subset to capture key structural semantics. In the SSL stage, the extracted structural priors guide a structure-aware augmentation process, enabling the model to learn more semantically meaningful and robust representations. Experiments on two real-world psychiatric datasets demonstrate that SAM-BG outperforms state-of-the-art methods, particularly in small-labeled data settings, and uncovers clinically relevant connectivity patterns that enhance interpretability. Our code is available at https://github.com/mjliu99/SAM-BG.

Keywords: Psychiatric Diagnosis · Brain Graph Representation · Self-supervised Learning · Structure Semantics · fMRI · Graph Learning

1 Introduction

Functional magnetic resonance imaging (fMRI) is a widely adopted neuroimaging technique in neuroscience research. By capturing blood-oxygen-level-dependent

M. Liu et al. (Eds.): AI 2025, LNAI 16371, pp. 359–372, 2026.
https://doi.org/10.1007/978-981-95-4972-6_28

(BOLD) signals, fMRI enables functional connectivity measurement, allowing researchers to monitor brain activity and assess neurological development [27]. Increasing evidence shows that disruptions in brain network connectivity are closely linked to the onset and progression of various neurological and psychiatric disorders [11]. These findings underscore the importance of developing advanced computational methods tailored to fMRI data to better support clinical diagnostics and mental health care through physiologically grounded brain network analysis [2].

Graph learning [25] has emerged as a powerful tool in this context, representing fMRI data as brain graphs where nodes correspond to brain regions and edges represent functional connections. Leveraging graph neural networks (GNNs), these methods effectively capture complex connectivity patterns and have shown strong potential in tasks such as disease classification and biomarker discovery [18,23]. However, their success often depends on large-scale labeled datasets, which are scarce in psychiatric research due to privacy concerns and the high cost of expert annotations. For example, labeling patients with autism spectrum disorder (ASD) typically requires long-term behavioral observation and developmental assessments conducted by trained clinicians [9].

To alleviate this challenge, self-supervised learning (SSL) has emerged as a promising paradigm for learning meaningful representations from unlabeled data [14,26]. SSL-based methods typically rely on generic graph augmentation strategies such as random edge perturbation, node masking, or subgraph sampling to generate multiple views for contrastive or predictive learning [15,20,24,32]. While effective in general graph domains, these augmentation methods are task-agnostic and domain-independent, often neglecting domain-specific structural priors. In brain network analysis, such priors are particularly crucial, as they encode biologically meaningful structural semantics, often manifested as connectivity patterns (substructures), which are closely associated with psychiatric disorders [7]. Thus, indiscriminate perturbations of brain connections in current augmentation strategies can distort essential structural semantics in the resulting graph views, leading to unrealistic and less explainable representations. This distortion ultimately undermines the effectiveness of SSL-based methods in downstream clinical and diagnostic applications.

To address this issue, we propose SAM-BG, a framework designed for low-supervision settings that uses structure-aware augmentation via edge masking to learn robust, biologically meaningful brain graph representations from large-scale, unlabeled fMRI data. It follows a two-stage learning strategy: (i) *Structure Semantic Extraction*, which pre-trains an edge masker using a small set of labeled data to identify the most discriminative substructures, guided by the information bottleneck (IB) principle [29]. (ii) *SSL with Structure-semantic Preservation*, where the trained edge masker is applied to unlabeled brain graphs to retain key substructures while introducing controlled perturbations to the remaining graph. This ensures that augmented views preserve critical semantic information while adding sufficient variability. Through this design, SAM-BG improves both representation robustness and interpretability, offering insights into disorder-relevant connectivity patterns. Our main contributions are:

- We propose a novel two-stage learning framework, SAM-BG, which improves data efficiency for psychiatric disorder diagnosis by leveraging limited supervision to learn structurally meaningful representations from large-scale, unlabeled brain networks.
- We introduce a learnable masker that enhances the learning of biologically meaningful structural semantics, which in turn improves data augmentation for more effective representation learning in psychological applications.
- Extensive experiments on two real-world fMRI datasets demonstrate that SAM-BG outperforms state-of-the-art methods in low-data scenarios and reveals disorder-specific brain connectivity patterns.

2 Related Work

2.1 Brain Network Analysis on fMRI Data

The fMRI data captures brain connectivity through BOLD signal fluctuations and is widely used in brain network analysis. Traditional methods [17,21] typically flatten this data into feature vectors, ignoring the brain's topological structure, an essential aspect for detecting connectivity disruptions in neurological disorders. Graph learning addresses this limitation by representing fMRI data as graphs. BrainGB [4] provides a unified benchmark framework for evaluating graph learning methods on brain networks. Other models, such as BrainGNN [12], employ region-aware pooling to identify key regions of interest (ROIs), thereby enhancing disease-related interpretability. In contrast, IBGNN [5] highlights important ROIs and connections through a learnable global mask. However, these methods heavily depend on labeled data, which is often limited in psychiatric research. To address this challenge, we propose SAM-BG that improves representation learning from unlabeled fMRI data by preserving structural-semantic information during SSL.

2.2 Self-supervised Brain Network Representation Learning

SSL has recently emerged as a powerful approach for learning representations without relying on labeled data, effectively addressing data scarcity [14]. Through data augmentation strategies, SSL enables the extraction of discriminative features. In graph-based SSL, methods like GraphCL [28] generate positive pairs through edge perturbation, node dropping, and subgraph sampling. For brain networks, A-GCL [32] and SF-GCL [19] adopt Bernoulli masking for edge-level perturbations, while BraGCL [15] improves interpretability by incorporating node and edge importance. GCDA [24] further introduces a diffusion-based pretext task to enhance semantic augmentation. However, many existing methods neglect structural semantics during augmentation, leading to biased representations. To overcome this, we propose a structure-aware graph augmentation strategy that generates biologically meaningful views for learning more robust and explainable representations.

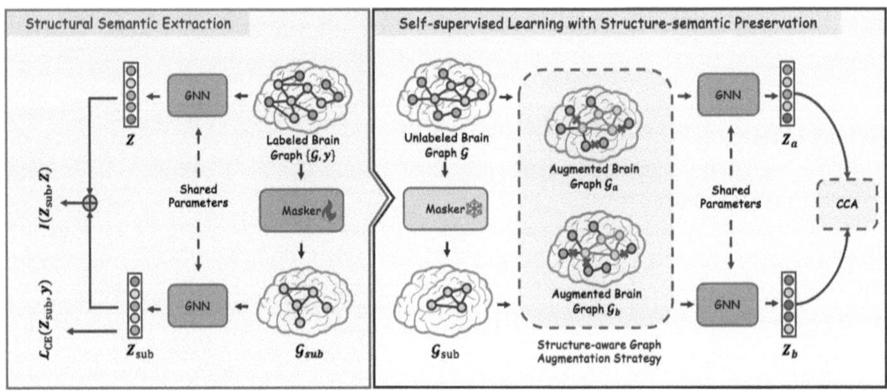

Fig. 1. Overview of the SAM-BG model consisting of structural semantic extraction and SSL with structure-semantic preservation phases.

3 The Proposed Method

The overall framework of SAM-BG is shown in Fig. 1. It consists of two main phases: (i) Structural semantic extraction, where an edge masker is trained to identify substructures containing task-relevant structural semantics; (ii) SSL with structure-sematic preservation, where augmentations are applied only to non-salient edges while retaining the extracted substructures, thus enhancing brain network representations by maintaining critical structural semantics.

3.1 Problem Definition

We represent a brain network as a graph $\mathcal{G} = (\mathcal{V}, \mathcal{E}, \mathbf{A}, \mathbf{X})$, where $\mathcal{V} = \{v_1, v_2, \ldots, v_N\}$ denotes the set of N ROIs, $\mathbf{A} \in \mathbb{R}^{N \times N}$ is the adjacency matrix computed via Pearson correlation [22], and $\mathbf{X} \in \mathbb{R}^{N \times D}$ is the node feature matrix with each row $\mathbf{x}_i \in \mathbb{R}^D$ representing the feature vector of ROI v_i. An edge $e_{i,j} \in \mathcal{E}$ exists if $a_{i,j} \in \mathbf{A}$ indicates significant functional connectivity. Given a small labeled dataset $\{\mathcal{G}^i, y^i\}_{i=1}^S$ and a large unlabeled set $\{\mathcal{G}^j\}_{j=1}^U$, where $U \gg S$ and $y \in \{0, 1\}$, our goal is to learn biologically meaningful brain graph representations $\mathbf{Z} \in \mathbb{R}^{N \times D}$, which are further used for downstream tasks.

3.2 Structural Semantic Extraction

Since brain connections (edges) are pivotal for diagnosing and interpreting the neural mechanisms of psychiatric disorders [7,10], we introduce an edge masker to distill discriminative substructures as structural priors (Fig. 2). For a given labeled brain network $\mathcal{G}$, the edge masker operates directly on $\mathbf{A}$, encoding it via a two-layer multi-layer perceptron (MLP) network followed by a sigmoid activation function to generate edge selection probabilities $\mathbf{P} \in \mathbb{R}^{N \times N}$. Each element $p_{i,j} \in [0, 1]$ denotes the selection probability of edge $e_{i,j}$. To impose sparsity while

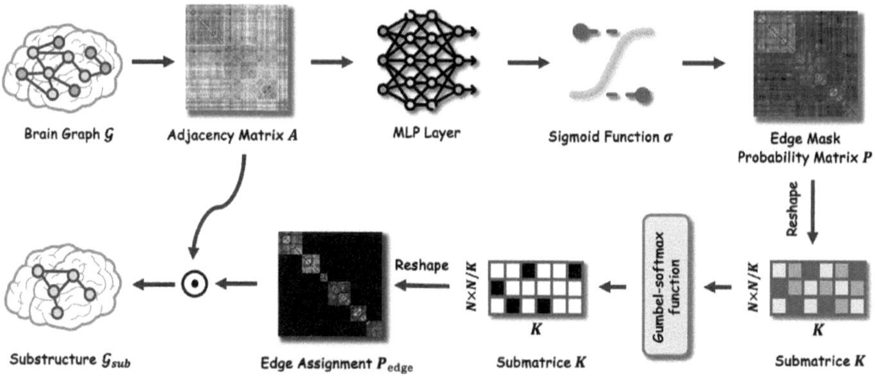

Fig. 2. Framework of the subgraph generation process. The input graph $\mathcal{G}$ is processed by the masker to sample a subset of edges and construct a subgraph $\mathcal{G}_{\text{sub}}$ that preserves key structural information.

avoiding over-pruning, we divide the probability matrix $\mathbf{P}$ into K submatrices, each containing $N \times N/K$ edges. Within each group, the Gumbel–Softmax reparameterization [16] selects one edge among K consecutive candidates, yielding a differentiable, structurally coherent mask. The grouped masks are reshaped back to $N \times N$ size and binarized to form the final edge selection mask $\mathbf{P}_{\text{edge}}$. The substructure of brain network $\mathcal{G}_{\text{sub}}$ is obtained by element-wise masking of the original graph: $\mathbf{A}_{\text{sub}} = \mathbf{A} \odot \mathbf{P}_{\text{edge}}$.

Then, the masker is trained under the Information Bottleneck (IB) principle [29], as illustrated in Fig. 1 (left), aiming to retain label-relevant semantics while discarding redundant topology. Specifically, the objective is to minimize the mutual information (MI) between $\mathcal{G}_{\text{sub}}$ and $\mathcal{G}$ to encourage compression, while maximizing the MI between $\mathcal{G}_{\text{sub}}$ and the label y to ensure discriminability. The objective is formally expressed as:

$$\min_{\mathcal{G}_{\text{sub}}} \beta I(\mathcal{G}_{\text{sub}}, \mathcal{G}) - I(\mathcal{G}_{\text{sub}}, y), \tag{1}$$

where β is the hyperparameters, and $I(\cdot,\cdot)$ is the MI estimation. To compute the first term, we encode the embeddings of $\mathcal{G}_{\text{sub}}$ and $\mathcal{G}$ using a shared GNN encoder as $\mathbf{Z}_{\text{sub}}$ and $\mathbf{Z}$. We then use the matrix-based Rényi's α-order MI estimator [30] to obtain their MI as $I(\mathcal{G}_{\text{sub}}, \mathcal{G}) := I(\mathbf{Z}_{\text{sub}}, \mathbf{Z})$. For the second term, we follow previous work [29] that minimizes the cross-entropy loss between the $\mathbf{Z}_{\text{sub}}$ and its ground truth label y as $-I(\mathcal{G}_{\text{sub}}, y) := \mathcal{L}_{\text{CE}}(\mathbf{Z}_{\text{sub}}, y)$. Thus, the final objective becomes:

$$\mathcal{L}_{\text{pre}} = \beta I(\mathbf{Z}_{\text{sub}}, \mathbf{Z}) + \mathcal{L}_{\text{CE}}(\mathbf{Z}_{\text{sub}}, y). \tag{2}$$

This formulation encourages the edge masker to extract compact, label-relevant substructures $\mathcal{G}_{\text{sub}}$ that preserve essential structural semantics, performing as the structural prior knowledge.

3.3 Self-supervised Learning with Structure-Sematic Preservation

As shown in Fig. 1 (right), we propose an SSL-based representation learning module, which incorporates structural priors to mitigate the information loss commonly introduced by random perturbations in existing methods. The framework comprises a structure-aware graph augmentation strategy with a tailored learning objective to learn semantically meaningful representations for data-efficient diagnosis of psychiatric disorders.

Structure-Aware Graph Augmentation Strategy. An effective augmentation should increase diversity without compromising semantic integrity. To this end, we integrate a trained edge masker to transfer structural priors from limited labeled data to large-scale unlabeled data, thereby generating biologically plausible augmented brain graphs that support meaningful representation learning. Specifically, the masker first extracts salient substructures $\mathcal{G}_{\text{sub}}$ from unlabeled brain graphs. To introduce variability, we perturb the complementary non-salient part $\Delta\mathcal{G} = \mathcal{G} \setminus \mathcal{G}_{\text{sub}}$ by applying uniform edge dropout to its edge set $\mathcal{E}_\Delta$. For each edge $e_{i,j} \in \mathcal{E}_\Delta$, an independent mask is sampled as $m_{i,j} \sim \text{Bernoulli}(1-\epsilon)$, where ϵ is the drop rate; edges in the salient subgraph are always preserved, i.e., $m_{i,j} = 1$ for $e_{i,j} \in \mathcal{G}_{\text{sub}}$. For brain networks, symmetry is enforced by setting $m_{i,j} = m_{j,i}$. The augmented adjacency is then updated element-wise as $\mathbf{A}' = \mathbf{A} \odot \mathbf{M}$ (equivalently, $a'_{i,j} = a_{i,j} m_{i,j}$). Resampling the masks yields two complementary yet semantically aligned views,$\mathcal{G}_a = (\mathcal{V}, \mathcal{E}_a, \mathbf{A}_a, \mathbf{X})$ and $\mathcal{G}_b = (\mathcal{V}, \mathcal{E}_b, \mathbf{A}_b, \mathbf{X})$, both preserving core semantic structure while providing sufficient diversity.

Structure-Aware Representation Learning Objective. After generating structurally meaningful augmented brain graphs, we adopt an SSL framework to learn graph-level representations. To capture shared structural semantics across views, SAM-BG employs canonical correlation analysis (CCA) [3] as its core learning objective. CCA encourages the model to learn perturbation-invariant yet semantically aligned embeddings by maximizing the correlation between the augmented view representations. Specifically, the two augmented brain graphs, $\mathcal{G}_a$ and $\mathcal{G}_b$, are fed into a shared GNN encoder, which is initialized using the encoder pre-trained during the structure masker phase. This encoder generates view-specific embeddings $\mathbf{Z}_a$ and $\mathbf{Z}_b$, which are then optimized via a CCA-based loss to align the representations in the latent space. The learning objective is defined as:

$$\mathcal{L}_{\text{CCA}} = \|\mathbf{Z}_a - \mathbf{Z}_b\|_{\text{F}}^2 + \lambda \left(\left\|\mathbf{Z}_a^\top \mathbf{Z}_a - \mathbf{I}\right\|_{\text{F}}^2 + \left\|\mathbf{Z}_b^\top \mathbf{Z}_b - \mathbf{I}\right\|_{\text{F}}^2 \right), \tag{3}$$

where the λ is the trade-off coefficient, and $\mathbf{I}$ is an identity matrix. The first term of the objective aligns the embeddings of the two augmented graphs in a shared latent space, promoting perturbation-invariant representations. The second term enforces decorrelation across embedding dimensions, preserving view-specific features and preventing representation collapse. By jointly optimizing these two terms, the trained GNN encoder can effectively learn meaningful and discriminative representations $\mathbf{Z}$ from any given brain network $\mathcal{G}$. These representations can then be applied to various downstream tasks, such as psychiatric disorder classification or biomarker discovery. The overall training procedure of SAM-BG is summarized in Algorithm 1.

Algorithm 1. Training Procedure of the SAM-BG Model

Input: Labeled brain network dataset $\{\mathcal{G}^i, y^i\}_{i=1}^S$, Unlabeled brain network dataset $\{\mathcal{G}^j\}_{j=1}^U$, Training epochs T
Output: Optimized model parameters $\hat{\theta}_{\text{GNN}}, \hat{\theta}_{\text{Masker}}$
Initialize: Model parameters $\theta_{\text{GNN}}, \theta_{\text{Masker}}$
Phase I: Structural Semantic Extraction Phase (supervised)
for epoch $= 1$ to T **do**
  for each labeled sample $i = 1$ to S **do**
    Compute edge mask probability $\mathbf{P}^i = f_{\theta_{\text{Masker}}}(\mathcal{G}^i)$
    Binarize $\mathbf{P}^i$ to obtain substructure $\mathcal{G}^i_{\text{sub}}$
    Obtain representations $\mathbf{Z}^i = f_{\theta_{\text{GNN}}}(\mathcal{G}^i)$ and $\mathbf{Z}^i_{\text{sub}} = f_{\theta_{\text{GNN}}}(\mathcal{G}^i_{\text{sub}})$
    Compute loss $\mathcal{L}_{\text{pre}}$ (Equation 2)
    Update θ_{GNN}, θ_{Masker} via gradient descent
  end for
end for
Return: Trained $\hat{\theta}_{\text{Masker}}$ and warm-start θ_{GNN}
Phase II: Self-supervised Representation Learning Phase
for epoch $= 1$ to T **do**
  for each unlabeled sample $j = 1$ to U **do**
    Extract substructure $\mathcal{G}^j_{\text{sub}}$ using trained masker $f_{\hat{\theta}_{\text{Masker}}}(\mathcal{G}^j)$
    Construct augmented graphs $\mathcal{G}^j_a$ and $\mathcal{G}^j_b$ by perturbing $\Delta\mathcal{G}^j = \mathcal{G}^j \backslash \mathcal{G}^j_{\text{sub}}$
    Obtain representations $\mathbf{Z}^j_a = f_{\theta_{\text{GNN}}}(\mathcal{G}^j_a)$ and $\mathbf{Z}^j_b = f_{\theta_{\text{GNN}}}(\mathcal{G}^j_b)$
    Compute loss $\mathcal{L}_{\text{CCA}}$ (Equation 3)
    Update θ_{GNN} via gradient descent
  end for
end for
Return: $\hat{\theta}_{\text{GNN}}, \hat{\theta}_{\text{Masker}}$

4 Experiment

4.1 Experiment Setup

Datasets. We conducted experiments using two real-world medical datasets: Autism Brain Imaging Data Exchange (ABIDE)[1] and Attention Deficit Hyperactivity Disorder (ADHD-200)[2] datasets. Both contain resting-state fMRI data commonly used in psychiatric research. For ABIDE, we selected a subset of 743 participants, comprising 364 ASD patients (ages 7–24) and 379 HC individuals (ages 6–22), with balanced age and gender distribution. For ADHD-200, we utilized a sample of 459 participants from this dataset, consisting of 229 typically developing (TD) individuals and 230 children and adolescents diagnosed with ADHD (ages 7–21).

[1] https://fcon_1000.projects.nitrc.org/indi/abide/.
[2] https://fcon_1000.projects.nitrc.org/indi/adhd200/.

Data Preprocessing. We preprocess resting-state fMRI into graph-ready brain networks. The pipeline includes slice-timing correction, head-motion correction, spatial normalization, and Gaussian smoothing, implemented with the Graph Theoretical Network Analysis (GRETNA) toolbox[3] and SPM12[4]. The brain is then parcellated using the Automated Anatomical Labeling (AAL) atlas, which defines 116 ROIs as nodes. For each ROI, we extract the BOLD time series and compute pairwise Pearson correlations [22] between ROI signals to weight their functional connectivity, resulting in a weighted functional brain network for each subject.

Baselines. We compare SAM-BG with several representative baselines, including supervised and SSL methods. For supervised learning, we adopt BrainGNN [12] and BrainGB [4], where BrainGNN employs an ROI-aware pooling mechanism, and BrainGB provides a standardized benchmark for brain graph analysis. For SSL, we include four competitive methods: GraphCL [28], GCA [33], MA-GCL [8], and CCA-SSG [31]. GraphCL explores several graph augmentation strategies; GCA improves this by adapting to salient graph structure; MA-GCL generates views based on learned embeddings; and CCA-SSG uses CCA to capture perturbation-invariant features without contrastive pairs. For a fair comparison, we adopt the official settings for BrainGNN and BrainGB. All SSL baselines use their original implementations with edge perturbation and 20% labeled data for fine-tuning, simulating low-supervision scenarios.

Implementation Details. Our model is implemented using PyTorch and trained on an NVIDIA Tesla P100 GPU. During pre-training, we leverage 20% of the labeled data to guide masker training. For downstream tasks, the dataset is split into 80% for fine-tuning, 10% for validation, and 10% for testing, with the test set strictly excluded from any supervised training during pre-training. The hyperparameters are configured as follows: the MI trade-off coefficient β is set to 0.01 for ABIDE and 0.001 for ADHD, while the weight of the decorrelation loss term λ is set to $1\mathrm{e}^{-4}$ for ABIDE and 0.001 for ADHD. To balance sparsity and structural completeness, the number of submatrices is set to $K = N/2$. The GNN encoder adopts graph isomorphism networks with bilinear second-order pooling to enhance node interaction and reduce parameters. We conduct 5-fold cross-validation and repeat experiments five times with different random seeds. Results are reported as mean and standard deviation of accuracy (ACC), the area under the ROC curve (AUC), recall, and F1-score.

4.2 Experimental Results and Discussion

Comparison Results. Table 1 presents the complete comparative results between our method and the baseline methods on the ABIDE and ADHD datasets. Overall, supervised learning methods tend to outperform most SSL

[3] https://www.nitrc.org/projects/gretna/.
[4] https://www.fil.ion.ucl.ac.uk/spm/software/spm12/.

Table 1. Performance (%) on the ABIDE and ADHD datasets. SL represents supervised learning, and SSL represents self-supervised learning. The best results are bold, and the second results are underlined.

	Methods	ABIDE				ADHD			
		ACC	AUC	Recall	F1-score	ACC	AUC	Recall	F1-score
SL	BrainGNN	54.9 ± 8.7	57.1 ± 8.3	55.0 ± 6.3	53.4 ± 6.0	55.8 ± 1.2	58.0 ± 4.9	46.3 ± 4.2	55.9 ± 4.2
	BrainGB	55.6 ± 4.3	56.3 ± 3.6	55.4 ± 5.2	53.6 ± 10.2	56.2 ± 7.1	59.6 ± 5.2	56.2 ± 4.6	61.6 ± 6.3
SSL	MA-GCL	49.1 ± 7.1	44.8 ± 9.2	52.5 ± 7.6	42.5 ± 11.8	50.1 ± 5.6	48.5 ± 6.2	46.5 ± 3.7	45.2 ± 6.4
	GraphCL	51.3 ± 4.6	50.4 ± 5.7	52.4 ± 7.5	50.3 ± 6.4	51.5 ± 6.8	47.6 ± 5.9	47.8 ± 7.5	46.5 ± 5.6
	GCA	52.8 ± 3.5	50.6 ± 2.4	53.2 ± 4.4	52.1 ± 2.3	50.8 ± 2.4	52.4 ± 5.1	59.4 ± 3.6	53.6 ± 4.4
	CCA-SSG	55.0 ± 6.1	57.2 ± 6.0	55.2 ± 7.7	53.3 ± 6.7	56.6 ± 5.8	60.3 ± 2.6	60.3 ± 4.4	58.0 ± 8.6
	BraGCL	53.5 ± 4.3	55.1 ± 5.2	53.4 ± 4.4	51.8 ± 5.0	63.8 ± 5.6	67.3 ± 3.7	63.8 ± 5.6	63.7 ± 5.5
	A-GCL	59.0 ± 1.6	63.3 ± 1.8	**71.6 ± 12.9**	63.0 ± 4.1	64.4 ± 2.5	**69.0 ± 3.2**	68.5 ± 17.2	64.7 ± 3.9
	SAM-BG (ours)	**63.1 ± 3.4**	**66.7 ± 4.3**	65.9 ± 4.2	**66.7 ± 5.2**	**67.7 ± 3.7**	66.5 ± 4.6	**68.8 ± 3.8**	**68.4 ± 5.4**

Table 2. Performance (%) of the ablation study for substructure evaluation on the ABIDE and ADHD datasets.

Methods	ABIDE				ADHD			
	ACC	AUC	Recall	F1-score	ACC	AUC	Recall	F1-score
SAM-BG (sub)	50.1 ± 4.3	51.4 ± 5.2	55.6 ± 3.9	51.8 ± 5.5	51.3 ± 3.4	52.5 ± 6.2	53.3 ± 5.6	52.9 ± 5.1
SAM-BG (rest)	46.4 ± 4.5	44.7 ± 4.3	48.1 ± 5.1	44.7 ± 4.8	48.3 ± 4.8	53.7 ± 5.1	49.5 ± 3.2	50.9 ± 6.2
SAM-BG (only)	58.3 ± 5.6	61.3 ± 4.6	57.8 ± 5.7	58.5 ± 7.3	61.3 ± 4.3	60.2 ± 4.2	61.5 ± 7.4	59.5 ± 6.5

methods in terms of accuracy, highlighting the importance of label information. Among them, BrainGB attains the highest F1-score, indicating strong robustness under full supervision. Despite this, CCA-SSG achieves competitive results across several metrics, benefiting from its non-contrastive objective, which facilitates better preservation of structural-semantic information. In particular, our proposed method, SAM-BG, consistently outperforms both state-of-the-art supervised and SSL methods across the two datasets. On ABIDE, it improves ACC by 4.1%, AUC by 3.4%, and F1-score by 3.7% compared to the best baseline, although its recall is slightly lower than A-GCL. On ADHD, it achieves 3.3% higher ACC, 0.3% higher recall, and 3.7% higher F1-score, with a marginally lower AUC than A-GCL. Notably, SAM-BG demonstrates more substantial improvements on the ABIDE dataset, suggesting that disease-related connectivity patterns may be more discriminative in autism brain networks. We attribute the superiority of SAM-BG to its two-stage training strategy, which uses limited labeled samples to guide structurally meaningful representation learning on large-scale, unlabeled brain networks.

Ablation Study. In this section, we evaluate the effectiveness of the structure-aware strategy and the semantic quality of extracted substructures, demonstrating our model's overall robustness and reliability.

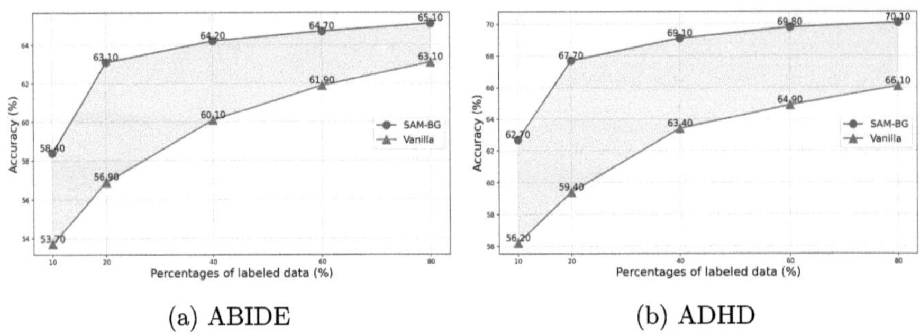

(a) ABIDE (b) ADHD

Fig. 3. Comparison between SAM-BG and Vanilla model with the increasing number of labeled data.

The Effectiveness of Extracted Substructure From Masker. To evaluate the role of the identified substructures, we compare three variants: (i) perturbing only the selected substructures (SAM-BG(sub)), (ii) removing them and perturbing the rest of the graph (SAM-BG(rest)), and (iii) using only the substructures for classification (SAM-BG(only)). As shown in Table 2, SAM-BG(rest) performs worst when the substructures are removed, while SAM-BG(only) is closest to the original model. This observation indicates that the identified substructures encode the brain network's most discriminative and label-relevant semantic features. Disrupting these substructures shifts the model's attention toward semantically irrelevant structural information, degrading learned representations and impairing downstream performance. These findings empirically validate the effectiveness of our structural masker in isolating non-essential components and preserving core topological semantics. Moreover, they highlight the potential of our method to guide the model toward learning biologically meaningful and task-relevant representations.

The Effectiveness of Structure-aware Graph Augmentation Strategy in Data-efficient Settings. To evaluate the impact of the learnable edge masking strategy in our augmentation process, we compared our model to a vanilla baseline that uses random structural perturbations under varying proportions of labeled data. As shown in Fig. 3, our approach consistently outperforms the baseline on both datasets, with accuracy improving as the amount of labeled data increases. Notably, with 20% labeled data, our model achieves the highest gains over the baseline, with improvements of 6.2% and 8.3%, respectively, and delivers performance comparable to the vanilla model trained with 80% labeled data. Furthermore, the results highlight the superior performance of our method in low-label scenarios, where it outperforms most existing SSL approaches even with only 10% labeled data. These findings validate the effectiveness of incorporating structural-semantic priors into the representation learning process. They also highlight the model's ability to deliver reliable performance even under limited supervision.

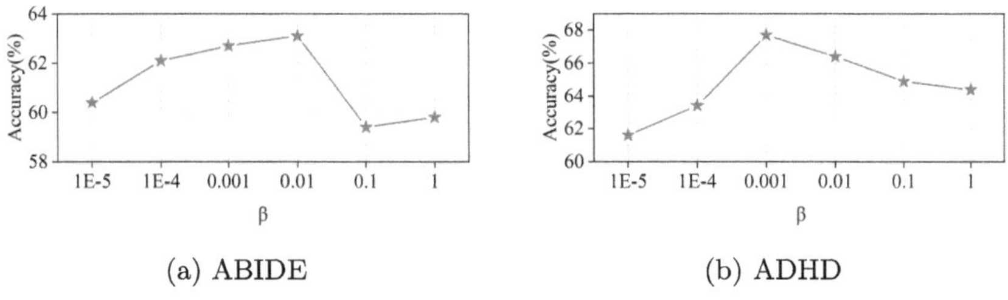

Fig. 4. Accuracy (%) of SAM-BG with different values of β in Eq. 2.

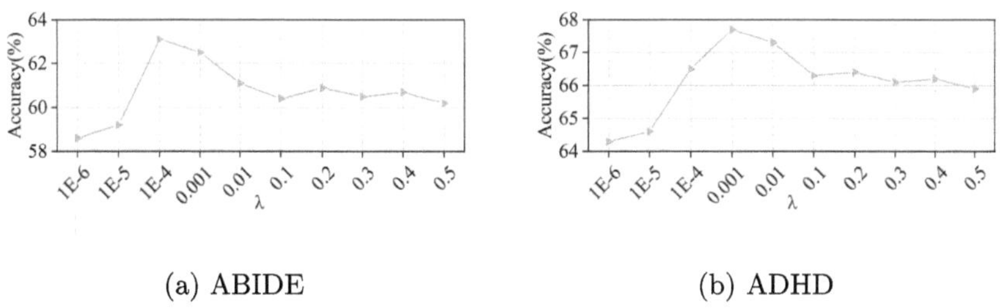

Fig. 5. Accuracy (%) of SAM-BG with different values of λ in Eq. 3.

Hyperparameter Tuning Study. We conduct a sensitivity analysis on two key hyperparameters: β and λ. First, β controls the sparsity regularization during the masker pre-training phase. We vary β in the range $[1\mathrm{e}^{-5}, 1.0]$ and evaluate its impact on classification performance for ABIDE and ADHD. As shown in Fig. 4, small values of β result in weak regularization, leading to substructures with redundant or noisy edges. Conversely, large β values over-prune the graph, potentially removing important connections. A balanced choice of $\beta = 0.01$ for ABIDE and $\beta = 0.001$ for ADHD achieves stable and high accuracy. Second, we examine λ, which weights the decorrelation constraint during representation learning to promote semantic diversity between augmented views. As illustrated in Fig. 5, a small λ fails to prevent feature collapse, while a large λ over-emphasizes diversity, impairing invariance learning. Optimal performance is achieved with $\lambda = 1\mathrm{e}^{-4}$ for ABIDE and $\lambda = 0.001$ for ADHD, balancing diversity and consistency across views.

Model Explainability Analysis. The edge masker in SAM-BG improves representation learning and interpretability by identifying discriminative substructures that reflect abnormal brain connectivity patterns in psychiatric disorders. To demonstrate this, Fig. 6 visualizes the top 20 connections from patients' adjacency matrices, with nodes color-coded by nine functional networks and edge thickness indicating connection strength. In the ABIDE dataset, our method identified increased internal connectivity within the VN and enhanced connectivity between the LIN and other networks, including the DMN and FPN, reflecting heightened sensory sensitivity and over-processing of emotional inputs in autism

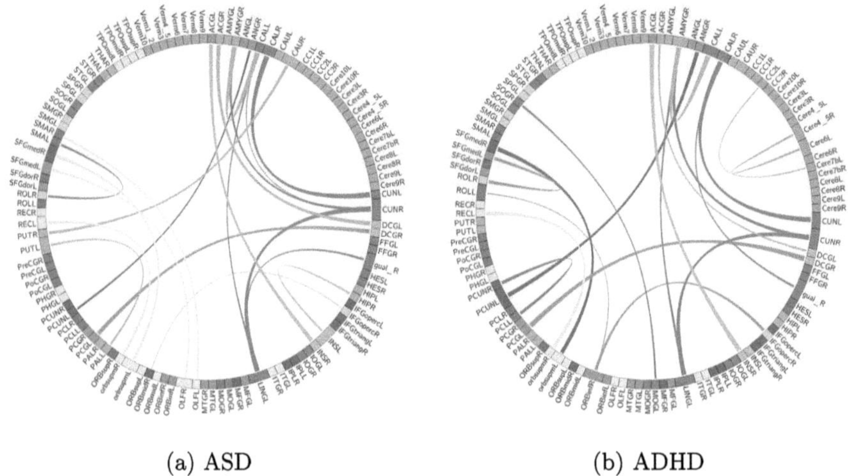

(a) ASD (b) ADHD

Fig. 6. Explainable graph connections in brain networks of patients with ASD and ADHD. The colors of brain networks are described as visual network (VN), somatomotor network (SMN), dorsal attention network (DAN), ventral attention network (VAN), limbic network (LIN), frontoparietal network (**FPN**), default mode network (DMN), cerebellum (CBL) and subcortical network (SBN), respectively. (Color figure online)

[1,6]. In the ADHD dataset, the model highlights hyperconnectivity in the VAN, particularly in the right hemisphere, along with elevated internal connectivity within the CBL. These patterns are consistent with evidence linking the cerebellum to attention control and the VAN to stimulus-driven attention, supporting their relevance in ADHD [13]. Interestingly, both datasets show increased VN connectivity, suggesting a shared neural signature that may underlie their frequent comorbidity. These results demonstrate that SAM-BG can uncover meaningful, explainable connectivity patterns, offering improved diagnostic insights into psychiatric disorders.

5 Conclusion

We propose SAM-BG, a novel framework that leverages structural priors to enable biologically meaningful brain graph representation learning under limited supervision. SAM-BG first trains an edge masker on a small labeled subset to extract salient connections, capturing key structural semantics. It then generates structure-aware augmentations by perturbing only non-essential edges, preserving semantic integrity for robust SSL. Experimental results demonstrate that SAM-BG substantially improves psychiatric disorder detection in low-data scenarios. Moreover, it provides explainable insights into brain connectivity patterns, underscoring its strong potential for real-world neuroscience applications.

References

1. Abbott, A.E., et al.: Repetitive behaviors in autism are linked to imbalance of corticostriatal connectivity: a functional connectivity MRI study. Soc. Cogn. Affect. Neurosci. **13**(1), 32–42 (2018)
2. Amemiya, S., Takao, H., Abe, O.: Resting-state fMRI: emerging concepts for future clinical application. J. Magn. Reson. Imaging **59**(4), 1135–1148 (2024)
3. Andrew, G., Arora, R., Bilmes, J., Livescu, K.: Deep canonical correlation analysis. In: International Conference on Machine Learning, pp. 1247–1255. PMLR (2013)
4. Cui, H., et al.: Braingb: a benchmark for brain network analysis with graph neural networks. IEEE Trans. Med. Imaging **42**(2), 493–506 (2022)
5. Cui, H., Dai, W., Zhu, Y., Li, X., He, L., Yang, C.: Interpretable graph neural networks for connectome-based brain disorder analysis. In: International Conference on Medical Image Computing and Computer-Assisted Intervention, pp. 375–385. Springer, Cham (2022)
6. Eggebrecht, A.T., et al.: Joint attention and brain functional connectivity in infants and toddlers. Cereb. Cortex **27**(3), 1709–1720 (2017)
7. Gallo, S., et al.: Functional connectivity signatures of major depressive disorder: machine learning analysis of two multicenter neuroimaging studies. Mol. Psychiatry **28**(7), 3013–3022 (2023)
8. Gong, X., Yang, C., Shi, C.: MA-GCL: model augmentation tricks for graph contrastive learning. In: Proceedings of the AAAI Conference on Artificial Intelligence, vol. 37, pp. 4284–4292 (2023)
9. Hyman, S.L., et al.: Executive summary: identification, evaluation, and management of children with autism spectrum disorder. Pediatrics **145**(1) (2020)
10. Li, A., et al.: A neuroimaging biomarker for striatal dysfunction in schizophrenia. Nat. Med. **26**(4), 558–565 (2020)
11. Li, Q., Seraji, M., Calhoun, V.D., Iraji, A.: Complexity measures of psychotic brain activity in the fMRI signal. In: 2024 IEEE Southwest Symposium on Image Analysis and Interpretation (SSIAI), pp. 9–12. IEEE (2024)
12. Li, X., et al.: BrainGNN: interpretable brain graph neural network for fMRI analysis. Med. Image Anal. **74**, 102233 (2021)
13. Lin, H., et al.: Functional connectivity of attention-related networks in drug-naive children with ADHD. J. Atten. Disord. **25**(3), 377–388 (2021)
14. Liu, Y., et al.: Graph self-supervised learning: a survey. IEEE Trans. Knowl. Data Eng. **35**(6), 5879–5900 (2022)
15. Luo, X., Dong, G., Wu, J., Beheshti, A., Yang, J., Xue, S.: An interpretable brain graph contrastive learning framework for brain disorder analysis. In: Conference on Web Search and Data Mining (WSDM 2024) (2024)
16. Maddison, C., Mnih, A., Teh, Y.: The concrete distribution: a continuous relaxation of discrete random variables. In: Proceedings of the International Conference on Learning Representations. In: International Conference on Learning Representations (2017)
17. Pan, X., Xu, Y.: A novel and safe two-stage screening method for support vector machine. IEEE Trans. Neural Netw. Learn. Syst. **30**(8), 2263–2274 (2018)
18. Peng, C., He, J., Xia, F.: Learning on multimodal graphs: a survey. arXiv Preprint arXiv:2402.05322 (2024)
19. Peng, C., Liu, M., Meng, C., Xue, S., Keogh, K., Xia, F.: Stage-aware brain graph learning for Alzheimer' s disease. MedRxiv, pp. 2024–04 (2024)

20. Peng, C., Liu, M., Meng, C., Yu, S., Xia, F.: Adaptive brain network augmentation based on group-aware graph learning. In: The Second Tiny Papers Track at ICLR 2024 (2024)
21. Plitt, M., Barnes, K.A., Martin, A.: Functional connectivity classification of autism identifies highly predictive brain features but falls short of biomarker standards. NeuroImage: Clin. **7**, 359–366 (2015)
22. Schober, P., Boer, C., Schwarte, L.A.: Correlation coefficients: appropriate use and interpretation. Anesthesia Analgesia **126**(5), 1763–1768 (2018)
23. Sun, K., Peng, C., Yu, S., Han, Z., Xia, F.: From EEG data to brain networks: graph learning based brain disease diagnosis. IEEE Intell. Syst. 1–9 (2024). https://doi.org/10.1109/MIS.2024.3352972
24. Wang, X., Fang, Y., Wang, Q., Yap, P.T., Zhu, H., Liu, M.: Self-supervised graph contrastive learning with diffusion augmentation for functional MRI analysis and brain disorder detection. Med. Image Anal. **101**, 103403 (2025)
25. Xia, F., et al.: Graph learning. arXiv preprint arXiv:2507.05636 (2025)
26. Xie, Y., Xu, Z., Zhang, J., Wang, Z., Ji, S.: Self-supervised learning of graph neural networks: a unified review. IEEE Trans. Pattern Anal. Mach. Intell. **45**(2), 2412–2429 (2022)
27. Yang, J., Gohel, S., Vachha, B.: Current methods and new directions in resting state fMRI. Clin. Imaging **65**, 47–53 (2020)
28. You, Y., Chen, T., Sui, Y., Chen, T., Wang, Z., Shen, Y.: Graph contrastive learning with augmentations. Adv. Neural. Inf. Process. Syst. **33**, 5812–5823 (2020)
29. Yu, J., Xu, T., Rong, Y., Bian, Y., Huang, J., He, R.: Recognizing predictive substructures with subgraph information bottleneck. IEEE Trans. Pattern Anal. Mach. Intell. **46**(3), 1650–1663 (2021)
30. Yu, S., Giraldo, L.G.S., Jenssen, R., Principe, J.C.: Multivariate extension of matrix-based rényi's α-order entropy functional. IEEE Trans. Pattern Anal. Mach. Intell. **42**(11), 2960–2966 (2019)
31. Zhang, H., Wu, Q., Yan, J., Wipf, D., Yu, P.S.: From canonical correlation analysis to self-supervised graph neural networks. Adv. Neural. Inf. Process. Syst. **34**, 76–89 (2021)
32. Zhang, S., et al.: A-GCL: adversarial graph contrastive learning for fMRI analysis to diagnose neurodevelopmental disorders. Med. Image Anal. **90**, 102932 (2023)
33. Zhu, Y., Xu, Y., Yu, F., Liu, Q., Wu, S., Wang, L.: Graph contrastive learning with adaptive augmentation. In: Proceedings of the Web Conference 2021, pp. 2069–2080 (2021)

Reinforcement Learning and Robotics

Evolving MCTS Macro-actions in Real-Time Domains

Andrew Gourley[1(✉)], Michael Dann[2], Xiaodong Li[1], and Fabio Zambetta[1]

[1] RMIT University, Melbourne, Australia
s4037984@student.rmit.edu.au
[2] Viewa Ltd., Melbourne, Australia

Abstract. Forward Search techniques such as Monte Carlo Tree Search (MCTS) take advantage of the ability to query a simulator and see the results of taking certain actions. While they are very effective in many board games they remain relatively ineffective in real-time environments. This is often due to the prohibitive expense of building a sufficiently deep search tree for effective play in these environments. We address this by introducing a new Evolutionary Algorithm which evolves macro-actions for a particular domain. A macro-action is a sequence of actions treated as a single action by MCTS; domain appropriate macro-actions increase search tree depth and avoid unproductive search. Our technique evolves effective macro-actions that significantly improve MCTS performance in multiple real-time domains.

Keywords: Macro-Actions · Evolutionary Algorithms · MCTS · Video Games

1 Introduction

Forward or look-ahead search methods such as Monte Carlo Tree Search (MCTS) [5,9] are a class of decision making algorithms that have had incredible success across many domains, most notably board games such as Chess and Go [19]. Unlike other techniques, these algorithms take advantage of the ability, where available, to query a simulator of the environment and investigate the results of future actions. Despite this, they remain relatively ineffective on real-time decision making tasks, including many standard benchmarks such as the Arcade Learning Environment (ALE) [2]. These environments' constraints and huge state sizes mean that forward search agents are often unable to make long term decisions, find sparse rewards, and are susceptible to dithering: moving back and forth without progressing.

To address these issues we present a new technique for evolving effective sequences of actions for a particular domain. These sequences of actions, called macro-actions, can significantly improve performance on real-time tasks when used by an MCTS agent. Our evolutionary technique finds high performing sets

M. Liu et al. (Eds.): AI 2025, LNAI 16371, pp. 375–386, 2026.
https://doi.org/10.1007/978-981-95-4972-6_29

of macro-actions in what is a complex optimisation task. We demonstrate this in video game environments, where there is often access to forward simulation.

Macro-actions have previously been shown to alleviate many of the issues that forward search algorithms face in real-time environments. These issues are largely caused by an inability to build a sufficiently deep forward search tree for effective decision making. Simulations in real-time environments, such as video games or robotics, are often extremely computationally expensive and additionally must be performed within very small available computational budgets: in some video game competitions as little as 40ms of computation time per action [17]. Exacerbating this, in real-time environments a single action usually has a very small effect on the state, such as moving a tiny distance in one direction. These 'fine-grained' action spaces allow for fine control but necessitate a significantly deeper search tree for medium and long term decision making. Macro-actions address these issues by significantly increasing the depth of a forward search tree without requiring additional computation. A tree searching over macro-actions of length n has a 'true' tree depth n times deeper than a similar tree searching over discrete actions. Additionally, macro-actions can exclude unproductive sequences of actions, such as dithering, from the search process. These effects have been demonstrated in relevant domains [8,16], however, to be effective a set of macro-actions must be domain appropriate. Where macro-actions are used in Planning and Reinforcement Learning (RL) there are a number of techniques that automatically generate appropriate macro-actions for a specific domain [3,4,11,14,18,24]. While these approaches are effective for their respective problem classes, the particular dynamics of forward search make it difficult to transfer their use into generating domain specific macro-actions for, for example, MCTS. Furthermore, there has been very little research towards this. Forward search macro-action approaches have instead used simple and hand-crafted macro-actions such as action repeats: repeating each action a set number of times. Addressing this gap we present a new Evolutionary Algorithm (EA) which evolves sets of macro-actions that considerably improve MCTS performance in a domain. The contributions of this work are as follows.

1. The design of a novel and specialised Evolutionary Algorithm which evolves sets of macro-actions to improve MCTS performance in a domain.
2. Results and insights demonstrating the effectiveness of this approach in relevant domains.
3. Analysis of limitations and promising future work.

2 Background and Related Work

2.1 Monte Carlo Tree Search

Monte Carlo Tree Search [5] is a forward search algorithm which selects actions in sequential decision-making domains by building and searching an unbalanced tree of future states. Most commonly used to guide the traversal and expansion of this tree (and often seen as a fundamental part of MCTS) is UCT [9] which aims to balance exploration and exploitation. Traditionally, the leaf nodes of this tree are evaluated by simulating a random series of actions from that state (also known as a rollout). The reward achieved from this is propagated back towards the root of the tree.

In its unmodified form MCTS does not require any kind of heuristic, due to the random rollouts requiring no specialised knowledge about a game besides its available actions. It also handles domains with a large branching factor well by prioritising high performing areas of its game tree. MCTS implementations are usually 'anytime' algorithms, meaning computation can be stopped at any point and the algorithm will return its current best solution.

MCTS has been widely used and adapted over the years in a huge variety of domains [22] including real-time environments such as video games [6,20]. However, video games pose a number of challenges, primarily due to extremely large state spaces and long time scales. Largely, this has led to MCTS being heavily modified in video game domains, and usually adapted to a specific game [15]. These modifications can be very effective but lack generality and require expert knowledge in their implementation.

2.2 Macro-actions in Forward Search

Macro-actions have been utilised by several video game agents to address the issues that forward search algorithms face in these domains. An implementation of macro-actions was found to be very effective on the *Physical Travelling Salesman Problem*, a combination of the Travelling Salesman Problem and a racing game [16]. The challenge of the game involves both long-term and short-term planning, and while the agent has a relatively small action space the game state is pseudo-continuous. Here, simple macro-actions (action repeats) considerably improved performance.

While investigating the effectiveness of MCTS in the game *Super Mario Bros.* action repeat macro-actions were also implemented [8]. However, the resulting coarseness of movement caused issues when the agent was near pits and enemies and therefore the agent reverted to using discrete actions in these situations. The use of a set containing both coarser and finer macro-actions could have helped to achieve a similar result. This coarseness of action caused the macro-actions to be fairly unsuccessful when compared to other modifications.

These uses demonstrate that an appropriate set of macro-actions can be very effective at addressing forward search issues that arise in video game domains. However, they also demonstrate that the effectiveness of a set of macro-actions is domain specific and that simple macro-actions are often unsuitable for a particular domain. By generating appropriate macro-actions designed for a specific environment a system could improve performance across many domains.

2.3 Generating Domain Appropriate Macro-actions

Planning. In Classical Planning tasks where there is a compact representation of the problem many solvers will directly extract macro-actions from it [4].

This is an effective approach where access to the problem dynamics is available, however, the macro-actions explored in this way can be limited to being fragments of generated plans. Due to this, other approaches have been investigated for designing macro-actions including Actor-Critic [11] and Evolutionary [14] methods.

Reinforcement Learning. In Reinforcement Learning, problems are modelled as Markov Decision Processes where an agent must interact with the environment to discover its dynamics. There has been considerable research into macro-actions and their effects in RL [12] where they are found to improve both exploratory behaviour and learning rates if they are appropriate to the task. Macro-actions have been designed in a variety of ways for RL agents including Classical Planning style extraction of subsequences from plans [24], statistical manipulation of Q functions [18], and temporally extended actions [3].

A framework that shares similarities to macro-actions is that of Hierarchical Reinforcement Learning [1], a paradigm that exploits hierarchical structure in problem spaces. Rather than acting over discrete actions, Hierarchical RL agents invoke 'options' [21], sub-policies which themselves may invoke discrete actions or further options. This can result in planning with higher order actions such as 'jump over an obstacle' or 'avoid an enemy' while the specifics of those activities are delegated. Macro-actions can be viewed as closely related to Hierarchical RL as they are similar to very simple options, policies that perform a series of primitive actions and then terminate, regardless of state or reward. The options framework offers greater flexibility, however, options are usually implemented using Deep Neural Networks which require significant computation to run a forward pass. This makes extending macro-actions into a full options based approach very difficult where computation budget is limited.

Forward Search. There has been very little work on generating macro-actions for forward search algorithms. Most prominently, a Genetic Algorithm for generating macro-actions for arbitrary video game domains was proposed [7]. Using a fairly standard binary representation and Genetic Algorithm this system achieves significant performance improvements across a number of video games when compared to using default actions. However, the approach is limited by its simplicity. We discuss this technique and how we have built upon it in our methodology.

3 Methodology

We introduce a new evolutionary technique which optimises domain specific sets of macro-actions for MCTS agents. This operates by generating a population of prospective sets of macro-actions using evolutionary operators, then evaluating each set's fitness by testing them in the domain. Once each set of macro-actions has an associated fitness the EA selects high performing sets and uses them to generate a new population. This process is detailed in Algorithm 1. High performing sets of macro-actions can be extracted from the optimisation process and applied to an MCTS agent to create an improved agent for the domain.

Algorithm 1. The Evolutionary Optimisation Process

```
population ← initialise()
for i = 1,...,MaxGenerations do
    for c in population do
        sumScores ← 0
        for n = 1,...,NumTrials do
            sumScores ← sumScores + MCTS(Domain, c.MacroActions)
        end for
        c.fitness ← sumScores/NumTrials
    end for
    population ← generateNewPopulation(population)
end for
```

Our work builds upon previous research which proposed similar evolutionary generation of macro-actions [7], however, this work was limited by a simple representation and operators. A binary string representation was used which identified which of a predetermined list of macro-actions was present in each chromosome. This limited the macro-actions that could be explored to those in the predetermined set (in their experiments this limited the maximum macro-action length to 5). The representation also did not account for the advantageousness of smaller macro-action sets, as they allows search algorithms to build deeper trees. Thus, binary strings including many macro-actions, which made up much of the search space, were almost always low quality solutions. Finally, the binary representation was very opaque, with the GA not knowing which discrete actions made up a particular macro-action. This considerably limited the power and flexibility of operators such as crossover and mutation, which were only able to blindly swap macro-actions in and out of chromosomes. The GA also had issues with rapid convergence towards a single high performing solution which was very likely a local optimum. The design of the algorithm contributed to this issue due to high selection pressure and operators which struggled to effectively explore a sparse search space.

Our work builds upon this research and successfully addresses many of these limitations. Our major contributions are a new method of representing macro-actions, and a new collection of crossover and mutation operators suitable for this representation.

3.1 Representation

Macro-action sets in our EA are represented very directly: each chromosome stores a set of sequences of discrete actions. For example:

$$\text{Set}_1 = (< up, up, left >, < down, right >, < jump >)$$

$$\text{Set}_2 = (< up, up, up, jump >, < up, right >, < down, down, left >)$$

$$\text{Set}_{\text{default}} = (< up >, < down >, < left >, < right >, < jump >)$$

This allows for the representation of an arbitrary amount of macro-actions of any length, including different lengths. While this means that the search space is of infinite size, the representation remains memory efficient for sets with relatively few macro-actions, which we again find to be the most promising areas of the search space, as they allow for deeper search trees. The representation also scales easily to domains with large actions spaces. Finally, it allows for far more specialised and flexible crossover and mutation operators as they have direct access to the contents of each macro-action. We detail our operators below.

There is a natural trade-off with this representation: with the ability to represent arbitrary macro-action sets we lose control over the properties of the search space. For example, if chromosomes are represented as a selection from a predetermined collection of macro-actions it is easy to limit the maximum length of macro-actions. This can be achieved by simply removing long macro-actions from the predetermined collection. Our representation has no intrinsic limits on macro-action length or the number of macro-actions in a set, as we discuss later we manage these properties using operator selection.

3.2 Initialisation

With the capability to represent infinitely many unique macro-action sets, randomly generating an initial set of macro-actions raises the difficult issue of how to design a distribution to sample from. We instead take inspiration from macro-action usage in Planning where macro-actions are often sections of a plan generated using discrete actions. We initialise our EA with a population made up of the default, discrete actions for a domain and allow the evolutionary process to optimise from there. This has the added benefit of the first generation's fitness being the fitness of an agent using the default actions, allowing us to easily examine the system's progress.

3.3 Fitness Evaluation

Fitness evaluation is the most computationally expensive component of our evolutionary system. We calculate fitness directly: an instance of the domain is created and an MCTS agent, equipped with the set of macro-actions, returns the score it achieves within a maximum number of actions. Where compute allows we average the score over $NumTrials$ runs to mitigate the stochasticity of both the domain and MCTS itself. We use a standard and unmodified implementation of MCTS [22] with a roughly tuned UCT exploration term.

3.4 Operators

Our new representation necessitates the design of a new collection of crossover and mutation operators. With full access to the make-up of each macro-action there is a very wide range of potential modifications and combinations of different chromosomes. The representation also simplifies the process of using sequences of actions from game-play (during chromosome evaluation) when modifying chromosomes. Using this kind of information from evaluations allows for more guided and intelligent operators. The process of generating a new population is summarised in Algorithm 2.

Algorithm 2 The Generation of a New Population

```
newPopulation[0...ElitismSize − 1] ← getBest(population, ElitismSize)
for i = ElitismSize, ..., PopulationSize do
    chromosome ← tournamentSelect(population)
    if p < CrossoverRate then
        chromosome2 ← tournamentSelect(population)
        chromosome ← crossover(chromosome, chromosome2)
    end if
    chromosome ← mutate(chromosome)
    newPopulation[i] ← chromosome
end for
```

Selection. We use *Tournament Selection* to select high performing chromosomes. This works by randomly choosing $n_{tournament}$ chromosomes and selecting the one with highest fitness. We also include a small amount of elitism. This means that the top $n_{elitism}$ highest performing chromosomes from the previous generation are copied directly into a new generation.

Mutation. We propose a large set of mutation operators. Many of these are based on a set of operators devised for generating macro-actions in Planning [13], which have been modified for our system. Separated by function they are:

- ***Adding a new macro-action to a set***
  - **Lift:** a new macro-action is added to a set by randomly lifting a sequence of actions used in game-play.
  - **Construct:** a new macro-action is added to a set by choosing a sequence of individual actions randomly.
- ***Modifying existing macro-actions in a set***
  - **Extend:** an action appearing immediately after a macro-action in game-play is appended to its end.
  - **Insert:** an action is inserted into a random position in a macro-action.
  - **Delete:** an action is deleted from a random position in a macro-action.
  - **Alter:** an action replaces one random action of a macro-action.
  - **Split:** a macro-action is split at one random position into two macro-actions.
  - **Merge:** two macro-actions are concatenated into one macro-action.

  - **Concatenate:** a random prefix of one macro-action is concatenated with a random suffix of another macro-action.

Additionally, we implemented an operator of our own design:

- ***Removing a macro-action from a set***
  - **Remove:** a macro-action that was not used in game-play is removed from the set.

When mutating we randomly select one of these operators until a unique macro-action set is created.

Crossover We crossover two distinct chromosomes by calculating the average number of macro-actions across their two sets. We then randomly choose that many macro-actions from all of the macro-actions contained in those two sets. This can be seen as taking the union of the two sets of macro-actions and then randomly choosing a subset of this union, such that the subset is the average size of the two initial sets.

4 Experiments

Our experiments are implemented in Gymnasium [23], parameters are listed in Table 1.

Table 1. System Parameters

Parameter	Value
Tournament Pool Size	2
Elitism Size	3
Crossover Rate	0.3
MCTS Rollouts Per Action	20
Maximum Rollout Depth	50
Exploration Weight	1
Max Evaluation Actions	100,000

4.1 Flappy Bird

Following previous work [7] we tested our new system using an implementation of the 2D side-scolling game *Flappy Bird* [10], allowing us to directly compare results. In this game the player controls a bird character which can either flap its wings to increase its height or fall downward. The aim is to survive by dodging obstacles in the form of pipes. The score increases when staying alive and passing obstacles and is penalised for dying and hitting the top of the screen. The game is conceptually simple but difficult enough that naive algorithms, including vanilla MCTS, struggle to achieve more than a small score.

In Flappy Bird our new system is able to evolve macro-actions that significantly outperform both default actions and previous evolutionary methods. It is in part able to do this by generating sets with fewer, longer macro-actions, highlighting the benefit of the new, more flexible representation. Although macro-actions with similar properties could have been generated using a different representation it would have required considerable hand testing and tuning. Each run of the algorithm also found a number of high performing but different macro-action sets, a useful feature. As an example, two sets of effective macro-actions from the same evolutionary run, **Evolved Set 1** and **Evolved Set 2**, are shown below. Both of these sets may seem counter-intuitive in a number of ways, and therefore very difficult to hand design, but they are nevertheless extremely effective. Having macro-actions of differing lengths seems to be beneficial, along with a small number of very long macro-actions which allow for considerably increased forward look-ahead. Table 2 includes a comparison of average scores achieved by an MCTS agent using different action sets, including default actions, action repeats of length 5, a comparison to the binary string approach, and both of the evolved sets displayed below. These results are averages over 100 evaluations.

$$\text{Flappy Bird Evolved Set 1} = \\ (<0,0>,\ <0,0,0,0,0>,<0,0,0,0,0,0,0,0,1,0,0,0,0,0,0,0,0>, \\ <0,0,0,0,0,0,0,1,0,0,0,0,0,0>)$$

$$\text{Flappy Bird Evolved Set 2} = \\ (<0,0,0,0,0>,<0,0,0,0,0,0,0,0,0,0,0,0,0,0,0,0,1,0,0>, \\ <0,0,0,0,0,0,0,0,0,0,0,0,0,0,1,0,0,0>, \\ <0,0,0,0,0,0,0,0,0,0,0,0,1,0,0>)$$

4.2 Balancing Operators

While able to evolve very effective macro-actions, we found our evolutionary system to be very unstable in our initial experiments. Figure 1 shows this behaviour over an example evolutionary run. Investigating the cause of this instability we found that when randomly selecting operators there was an unbalanced pressure towards longer macro-actions and larger macro-action sets. As Fig. 1 demonstrates, collapse in performance corresponded with an increase in the number of macro-actions per set. As a simple method of addressing this we implemented a manual system which biasses operator selection to avoid this collapse. Where the number and average length of macro-actions in a set grows larger or smaller than a manually determined effective range we increase the chance of selecting

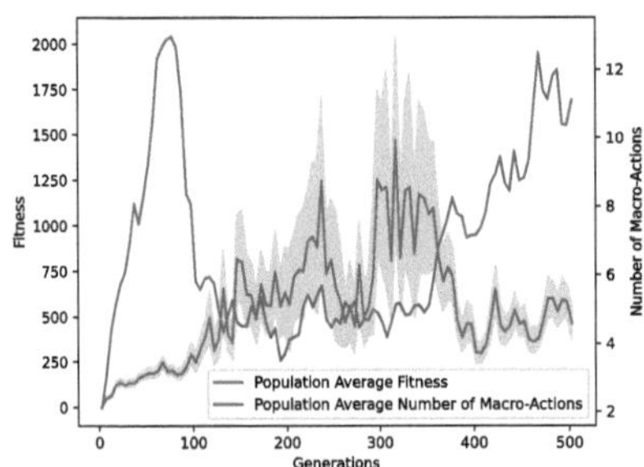

Fig. 1. An Example Evolutionary Run on Flappy Bird

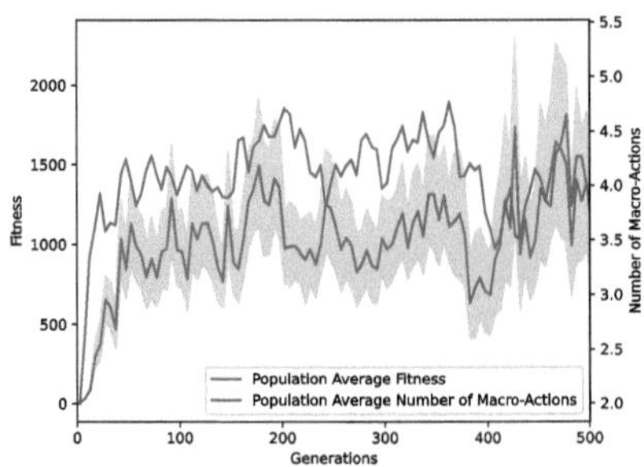

Fig. 2. An Example Run on Flappy Bird with Manual Operator Selection

operators which push these variables back into said range. This had the intended effect of maintaining the performance and stability of our evolutionary system, demonstrated in Fig. 2. As an added benefit this significantly accelerated the optimisation of macro-action sets in our system, generating significantly higher scores in only a few generations.

4.3 Breakout

We implemented our strategy in a more complex game: *Breakout*, from the suite of Atari 2600 games included in the Arcade Learning Environment (ALE) [2], these games are common benchmarks for real-time decision making agents. In this game the user controls a paddle to bounce a ball upward to break a wall of bricks. The score increases as bricks are destroyed and a life is lost if the ball is allowed to fall past the paddle. As bricks are cleared, the game becomes faster and more challenging. We limited lives to one to speed up evaluations and again used manual operator selection to maintain stability. Our system was once again successful at improving performance over default actions and previous methods and at evolving multiple high performing macro-action sets within the same evolutionary run (**Evolved Sets 1 and 2**).

We found that effective sets contained considerably more macro-actions than in Flappy Bird, for example **Evolved Set 2** had 18. Part of the reason for this is simply that Breakout has four default actions compared to Flappy Bird's two. This doubling of the number of default actions increases the complexity of the search problem significantly and implies that, while very promising, there may need to be further adjustments to the system to apply it to more complex environments including other ALE games. Table 3 includes a comparison of average scores achieved by an MCTS agent using different action sets, once again including a comparison to the binary string approach.

Table 2. Flappy Bird Performance

Action Set	Score, mean (sd)
Default	3.2 (4.0)
Action Repeats (len 5)	6.6 (2.4)
Binary String Evolved	113.4 (111.7)
Evolved 1	**2037.7 (1925.9)**
Evolved 2	**2729.9 (2616.4)**

Table 3. Breakout Performance

Action Set	Score, mean (sd)
Default	3.5 (6.0)
Action Repeats (len 5)	2.4 (1.2)
Binary String Evolved	50.2 (87.2)
Evolved 1	**65.5 (97.5)**
Evolved 2	**65.3 (91.6)**

5 Conclusion and Future Work

In this paper we detail an improved method for increasing MCTS performance in real-time environments by evolving domain specific macro-actions. We present a new evolutionary algorithm with a new, more powerful representation, and a collection of new operators which take advantage of this. Across the video games *Flappy Bird* and *Breakout* our method is able to generate very effective macro-action sets, demonstrating the promise of this new system. Areas for future work include: (1) Further optimising operator design to best take advantage of the new representation, (2) Introducing an adaptive operator selection system to avoid the need for hand-tuning, and (3) Further application to and analysis in new and varied domains.

References

1. Barto, A.G., Mahadevan, S.: Recent advances in hierarchical reinforcement learning. Discrete Event Dyn. Syst. **13**(4), 341–379 (2003)
2. Bellemare, M.G., Naddaf, Y., Veness, J., Bowling, M.: The arcade learning environment: an evaluation platform for general agents. J. Artif. Intell. Res. **47**, 253–279 (2013)
3. Biedenkapp, A., Rajan, R., Hutter, F., Lindauer, M.: TempoRL: learning when to act. In: Proceedings of the 38th International Conference on Machine Learning, pp. 914–924. PMLR (Jul 2021)
4. Botea, A., Enzenberger, M., Müller, M., Schaeffer, J.: Macro-FF: improving AI planning with automatically learned macro-operators. J. Artif. Intell. Res. **24**, 581–621 (2005)
5. Coulom, R.: Efficient selectivity and backup operators in Monte-Carlo tree search. In: Proceedings of the 5th International Conference on Computers and Games, pp. 72–83. Springer (2006)
6. Frydenberg, F., Andersen, K.R., Risi, S., Togelius, J.: Investigating MCTS modifications in general video game playing. In: 2015 IEEE Conference on Computational Intelligence and Games (CIG), pp. 107–113. IEEE (2015)
7. Gourley, A., Dann, M., Li, X., Zambetta, F.: Evolving macro-actions for MCTS video game agents. In: Proceedings of the Genetic and Evolutionary Computation Conference Companion, pp. 823–826. Association for Computing Machinery (2025)
8. Jacobsen, E.J., Greve, R., Togelius, J.: Monte mario: platforming with MCTS. In: Proceedings of the 2014 Annual Conference on Genetic and Evolutionary Computation, pp. 293–300. Association for Computing Machinery (2014)

9. Kocsis, L., Szepesvári, C.: Bandit based Monte-Carlo planning. In: Proceedings of the 17th European Conference on Machine Learning, pp. 282–293. Springer (2006)
10. Kubovčík, M.: Flappy Bird for Gymnasium (2024). https://github.com/markub3327/flappy-bird-gymnasium
11. Lee, Y., Cai, P., Hsu, D.: MAGIC: learning macro-actions for online POMDP planning using generator-critic. In: Proceedings of the Robotics: Science and Systems Conference (2021)
12. Mcgovern, A., Sutton, R.: Macro-actions in reinforcement learning: an empirical analysis. University of Massachusetts, Computer Science Department Faculty Publication Series, p. 15 (1998)
13. Newton, M.A.H., Levine, J.: Evolving macro-actions for planning. In: Proceedings of the Workshop on AI Planning and Learning held at ICAPS 07. AAAI (2007)
14. Newton, M.A.H., Levine, J., Fox, M.: Genetically evolved macro-actions in AI planning problems. In: Proceedings of the 24th Workshop of the UK Planning and Scheduling Special Interest Group (PlanSIG 2005), pp. 163–172 (2005)
15. Pepels, T., Winands, M.H.M.: Enhancements for Monte-Carlo tree search in Ms Pac-man. In: 2012 IEEE Conference on Computational Intelligence and Games (CIG), pp. 265–272. IEEE (2012)
16. Perez, D., et al.: Solving the physical traveling salesman problem: tree search and macro actions. IEEE Trans. Comput. Intell. AI Games **6**(1), 31–45 (2014)
17. Perez-Liebana, D., Samothrakis, S., Togelius, J., Schaul, T., Lucas, S.: General video game AI: competition, challenges and opportunities. In: Proceedings of the AAAI Conference on Artificial Intelligence, vol. 30, no. 1 (2016)
18. Randlov, J.: Learning macro-actions in reinforcement learning. In: Advances in Neural Information Processing Systems, vol. 11. MIT Press (1998)
19. Silver, D., et al.: A general reinforcement learning algorithm that masters chess, shogi, and go through self-play. Science **362**(6419), 1140–1144 (2018)
20. Soemers, D.J.N.J., Sironi, C.F., Schuster, T., Winands, M.H.M.: Enhancements for real-time monte-carlo tree search in general video game playing. In: 2016 IEEE Conference on Computational Intelligence and Games (CIG), pp. 436–443 (2016)
21. Sutton, R.S., Precup, D., Singh, S.: Between MDPs and semi-MDPs: a framework for temporal abstraction in reinforcement learning. Artif. Intell. **112**(1), 181–211 (1999)
22. Świechowski, M., Godlewski, K., Sawicki, B., Mańdziuk, J.: Monte Carlo tree search: a review of recent modifications and applications. Artif. Intell. Rev. **56**(3), 2497–2562 (2023)
23. Towers, M., et al.: Gymnasium: a standard interface for reinforcement learning environments (2024)
24. Vezhnevets, A.S., et al.: Strategic attentive writer for learning macro-actions. In: Proceedings of the 30th International Conference on Neural Information Processing Systems, pp. 3494–3502. Curran Associates Inc. (2016)

A Hybrid Multi-Agent Reinforcement Learning Framework for Decentralised Search-And-Interact Tasks Under Partial Observability

Anh Viet Do[1(✉)], Eranda Galhenage[2], Aneta Neumann[1], Frank Neumann[1], Anton V. Uzunov[2], and Claudia Szabo[1]

[1] The University of Adelaide, Adelaide, Australia
vietanh.do@adelaide.edu.au
[2] Defence Science and Technology Group, Canberra, Australia

Abstract. Using autonomous, decentralised control of unmanned aerial vehicles (UAVs) for critical scenarios such as bushfire firefighting provides clear safety and cost benefits. In this paper, we introduce a novel framework for autonomous, decentralised control based on multi-agent reinforcement learning (MARL) and a planning heuristic, designed to support learning a search-and-interact task in a multi-particle environment, which models a range of multi-UAV control applications. Our approach hybridises a planning heuristic with generic MARL methods under a discrete multi-action paradigm. The planning heuristic extracts only focused information from the environment, while also augmenting agent-centric observation data via exploiting observation-action symmetry to improve training. We employ this heuristic within a Centralised and Training Decentralised Execution (CTDE) paradigm to train autonomous control that is agnostic to the complexity of the environment, and effective under a range of action dynamics. The controller is encoded in a lean policy with reduced footprint, and can be deployed within resource-limited devices. Experimental results show that using even simple Q-learners within the framework provides highly effective agents for accurate firefighting in complex scenarios. Our approach is applicable to decision problems with similar search-and-interact characteristics.

Keywords: CTDE · Partial Observability · Search-And-Interact

1 Introduction

Wildfires are a significant cause of damage to the environment and the economy around the world, making firefighting an attractive application of intelligent systems. In particular, numerous works investigate the use of unmanned aerial vehicles (UAVs) in various problems related to wildfires [5,7,8,16]. Reinforcement learning [18] in general and multi-agent reinforcement learning (MARL) [2] in particular have been widely used in numerous real-world applied settings [12,23] such as traffic management, unmanned aerial vehicle control [6], swarm robotics, autonomous driving [24] and cellular base

M. Liu et al. (Eds.): AI 2025, LNAI 16371, pp. 387–401, 2026.
https://doi.org/10.1007/978-981-95-4972-6_30

station control [1], as they provide an effective way to tackle complex tasks through the collaboration of autonomous agents capable of independent decision-making.

In this paper, we consider an application of UAVs to extinguish fire zones via decentralised control. To this end, we study a simple abstraction of the fire-extinguishing task that can capture other applications of UAVs–the search-and-interact task. It involves autonomously acting agents that can move within an area *searching* for fire zones, and *interact* with these once located. Agents must interact with all fire zones quickly while avoiding collisions. Furthermore, we assume that each agent can only see objects within a radius, as restricted visual information is common in wildfire scenarios. This partial observability property necessitates the search component of the task, as agents may have to discover fire zones within the area via proximity.

To address the challenge of producing agent controllers for the search-and-interact task, we introduce a framework for hybrid MARL which allows classical MARL approaches to be enhanced through a hybridisation process. Specifically, our proposed framework allows agents to be trained in simplified environments in a way that they still perform well in complex environments due to the incorporation of heuristics, which give agents additional directions for tackling their given tasks. Our main contributions are as follows:

- We formulate a model of the search-and-interact task as a Decentralised Partially Observable Markov Decision Process (Dec-POMDP) in a multi-particle environment with a 2-dimensional action space. The task requires agents to search for (via movement) and interact with certain objects in the environment while avoiding collision. Each agent is allowed only to operate on information captured within a fixed radius around itself.
- We propose a simple framework that allows using MARL algorithms to train independent controllers to tackle decentralised search-and-interact tasks in a partially observable environment. The framework pessimistically assumes that the controller can only rely on presently observed information, and no inter-agent communication. It involves a simple planning heuristic that enables individualised exploration of the map, and a MARL-trained policy to handle low-level actions. The hybridisation is modular, allowing the use of any model-free Centralised and Training Decentralised Execution (CTDE) method.

2 Related Work

Reinforcement learning is well suited for optimising UAV while considering fire dynamics and resource constraints [2,3,23]. In particular, MARL approaches have been shown to enable UAV coordination for efficient wildfire tracking and response in firefighting [4,13,25]. To deal with the complexity of controlling multiple UAVs in wildfire monitoring, deep reinforcement learning approaches for training decentralised controllers were introduced in [8]. The approaches incorporate immediate observations of individual UAVs, a fire map, and a record of visited locations to enable collaboration. They facilitate accurate tracking of wildfire expansion and adapting to different team sizes and fire shapes. A cooperative reinforcement learning framework using deep Q-learning that allows multiple UAVs to learn how to track and monitor a wildfire front, was proposed in [21]. It uses CTDE approaches to address the typical sample complexity challenge in MARL, and the simulation results demonstrate superior sample efficiency over those of previous state-of-the-art approaches in the task of monitoring a wildfire front. In [5], the authors proposed a CTDE-based MARL framework for wildfire tracking using homogeneous UAV swarms. The framework accounts for the tracking and charging stages of tasks while maintaining network connectivity. It optimises UAV trajectories under energy and communication constraints, achieving over 90% coverage, with performance dependent on energy levels, altitude, and swarm size. A distributed deep reinforcement learning approach for coordinating UAVs in autonomous forest fire control were introduced in [7]. The approach implements CTDE by training a single policy via only local observations from all agents. It is demonstrated to outperform heuristic methods, and exhibit scalability to various forest sizes and UAV team configurations on real-world mobile robots benchmark.

3 Problem Statement

We model the problem as a Dec-POMDP within a continuous Multi Particle Environment (MPE). The problem involves homogeneous agents executing tasks of variable horizons: tasks require varying amount of time to complete.

3.1 Scenario Modelling

The environment is defined over a 2-dimensional map, containing objects whose shapes are abstracted as discs. The state of each object includes a coordinate, a radius, and a velocity vector. An episode is played out in discrete time steps, with a fixed interval Δt between consecutive steps. The object displacement logic in a step is identical to that in the Pettingzoo's Simple MPE [10, 11].

The environment consists of three types of objects: agent, obstacle[1], and fires (target objects). The agents are homogeneous. Each agent has a detection radius determining its detection area as a disc centred at itself. In a step, an agent directly receives information about the position and radius of an object if and only if that object overlaps with the agent's detection area at that step. Agents and obstacles are subject to collisions, while fires do not exert collision forces. Additionally, the state of each fire contains a lifetime variable, and the fire is "resolved" if this variable reaches 0. Finally, while objects' positions are initialised within a fixed boundary, their movements are not limited by it.

In this search-and-interact problem, the goal of the agents is to extinguish all fires on the map. The primary objective is to achieve this goal as quickly as possible. A secondary objective, often conflicting with the first one, is to minimise the total number of agent collisions. We do not specify any trade-off between objectives so as to keep our investigation open-ended.

[1] We include obstacles as collision avoidance is necessary simply by virtue of multiple agents being present. Obstacles provide a simple way to test avoidance behaviours.

We make several simplification assumptions. First, obstacles are static; agents can serve as each other's moving obstacles. Second, fires are static and their lifetimes do not increase over time. Third, no new fires appear over time. This basic model simplifies interpreting episodic metrics, allowing us to gain insight into the behaviours from the trained agent's policies. For illustration, we render agents as blue discs, fires as red discs, obstacles as black discs, and the map's bound as a rectangle. The detection range of each agent is denoted by a dash-line circle centred at that agent (Fig. 2).

3.2 Action Specification

The action space for each agent is $\mathcal{A} \times \mathcal{I}$, where. $\mathcal{A}$ is a finite set of movement actions and $\mathcal{I}$ is a finite set of target interaction actions. The movement actions trigger actuators that apply acceleration forces onto the agent. We define $\mathcal{A}$ for the square 8-directional movement regime with $\mathcal{A} := \{-1, 0, 1\}^2$. When a movement action is committed, the corresponding acceleration vector is multiplied with Δt and the acceleration coefficient, then added to the agent's current velocity to derive its next-step velocity. The no-op in $\mathcal{A}$ is $(0, 0)$, as is standard.

Each action in $\mathcal{I}$ represents a setting with which the agent interacts with targets. Committing an action in this set triggers an interaction with all targets that overlap with an annulus centred at the agent. That is, each action is characterised by a pair $(small\_radius, large\_radius)$. A target in contact with the triggered area of effect in a step has its lifetime decreased by 1, to a minimum of 0. Furthermore, this contact is only effective in the step after the interaction area becomes active, i.e. two steps after the corresponding action is committed. This delay can cause some wasteful actions that agents should minimise. Note that a target can touch areas of effect from multiple agents simultaneously; its lifetime is decreased by the number of active areas of effect it touches. We can denote the no-op in $\mathcal{I}$ with $(-\infty, -\infty)$, indicating no interaction.

4 Hybrid MARL Framework

We propose a framework for the optimisation of a controller for decentralised deployment, while allowing for independent agent behaviors. In our hybrid approach, the controller consists of an RL-trained policy that is modulated by a planning heuristic that is focused on distilling observations related to navigation and collision avoidance within a search-and-interact type of problem.

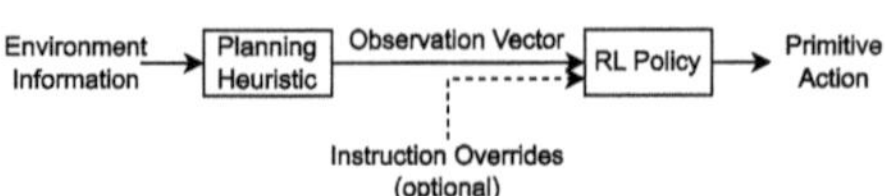

Fig. 1. Overview of the agent's controller in the proposed framework. In a multi-agent scenario, the controller is deployed in a decentralised manner, where the input environment information is agent-centric and reflects partial observability.

The framework is designed to accommodate model-free CTDE RL methods. The structure of the controller in our approach is illustrated in Fig. 1. While we use Q-learners in this work, any CTDE method is applicable.

The planning heuristic transforms the agent's observable environmental information into a fixed-size observation vector during training and deployment. This handles the dynamic information capture under partial observability in a way that facilitates the use of an action policy encoded in a neural-net. Moreover, by operating only on the observed information, the framework facilitates training under the same observation regime as in deployment environments of arbitrary complexity in terms of number of objects, which means the trained controllers are well-behaved in such environments. As the framework uses a CTDE method to train controllers, such agnosticism extends to agents as well–the learning and deployment of controllers make no assumption on the team dynamics or even the number of agents in the given scenario.

As will be described, the transformation of observation is simple and efficient, and thus can be used at high frequency in practice. Therefore, it is reasonable to assume that the planning heuristic can be invoked at every decision step. This transformation extracts only features that are relevant to the search-and-interact problem with a collision avoidance constraint. It is designed to guide the learning towards generalizable agent behaviours (e.g. across varying task horizons). Furthermore, the dimensionality reduction incurred speeds up learning while reducing the degree of expressiveness needed for the policy to capture the action dynamics, allowing the use of lean neural-net architectures with fast forward-pass; a helpful feature when controllers are run by resource-limited hardware.

5 Methods and Implementation Details

We propose a hybrid approach where the controller consists of a RL-trained policy modulated by a planning heuristic as shown in Fig. 1.

5.1 Observation Model and Definition

To facilitate CTDE methods, each observation vector is tailored to an individual agent. This means all linear features in the observation are encoded as vectors whose origin points are the agent's location. The observation is formulated to let RL train desired behaviours without violating the knowledge restrictions. Furthermore, the observation space is Markovian, which allows some agnosticism to uncertainties inherent to Dec-POMDP scenarios.

As mentioned, each agent runs a simple planner to process observations. This heuristic tells the agent where to go, relevant characteristics of fires if any, and which obstacles to prioritize. Each agent has a radius R for the purpose of collision check, and can observe objects within a certain radius around itself, assumed to be greater than R. The agent is given the following at each step:

- Map bounds in all dimensions,
- Its own coordinate, velocity vector and radius,
- For each observed object i: its radius r_i and relative positional vector pointing from the agent's coordinate to the observed object's coordinate.

To ease the training of the policy, the observation processor feeds the following information to the policy network (i and j are object indices, whose selection will be explained later):

1. The agent's radius R (static) and velocity vector v_t,
2. The destination vector d_t,
3. An obstacle vector $o_{t,i}$ paired with an obstruction factor $b_{t,i}$ correlating with the minimum deviation from the destination vector needed to avoid collision with obstacles (other agents count as obstacles for the purpose of collision avoidance),
4. An obstacle vector $o_{t,j}$ paired with a collision factor $c_{t,j}$ marking the most imminent collision,
5. The distance from the nearest target object, and its radius. If no target object is seen by the agent, these are set to its detection radius and 0, respectively.

We refer to the vectors, v_t, d_t, $o_{t,i}$, $o_{t,j}$, as linear features. For each observed obstacle i, $b_{t,i}$ and $c_{t,i}$ are calculated via ($c_{t,j}$ is calculated similarly)

$$c_{t,i} = \sqrt{1 - \left(\frac{R + r_i}{|o_{t,i}|}\right)^2}, \; b_{t,i} = \begin{cases} \frac{\langle d_t \cdot o_t \rangle}{|d_{t,i}||o_{t,i}|} - c_{t,i} & \text{if enlarged object obstructs } d_t, \\ 0 & \text{otherwise} \end{cases}.$$

Here, "enlarged object" refers to the object i with radius $R + r_i$, and smaller $c_{t,i}$ denotes being closer to collision with object i, with collision occurring at 0. On the other hand, smaller $b_{t,i}$, which is always non-negative, denotes less obstruction, with value 0 indicating the presence of a clear line of movement toward the destination. For input (4.), the chosen obstacle is $\text{argmin}_i \, c_{t,i}$, while for input (3.), the chosen obstacle is $\text{argmax}_i \, b_{t,i}$. If the agent does not observe an obstacle, $o_{t,j}$ and $o_{t,i}$ are set to $-d_t$, and $b_{t,i}$, $b_{t,j}$, $c_{t,j}$ are set to 0 to avoid misleading signals. Note that while including information from all observed objects can in principle lead to better decision making, partial-observability and the static-size constraint on the policy's inputs prevent this due to the number of observed objects being dynamic.[2]

We remark that the terminal states are exactly those where there are no target objects. This implies all observation vectors contain no target object. However, the inverse is not true due to partial observability, so the agents cannot discern the completion of the episode. As such, we do not use terminal rewards.

Destination Heuristic. To calculate d_t in the observation vector, the heuristic considers two cases. If a target object is detected by the agent, the destination is set to the nearest target's location, stopping at the edge of its radius. Otherwise, an exploration subroutine is used, which manages an uncertainty mapping over set of grid points P across the map. Each point $p \in P$ is assigned an uncertainty score $u_{t,p}$ which is set to the upper bound $\mathbf{u}$ at the start of the

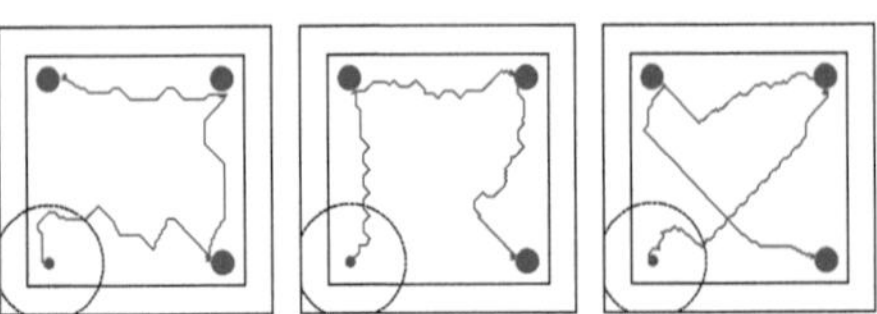

Fig. 2. Exploratory behaviours, illustrated via the agent's position tracing over the initial state's snapshot. Episodes are completed by a IQL-trained policy. Grid sizes left-to-right: 5×5, 11×11, 21×21. Grid points include the boundary corners. Upper bound $\mathbf{u}$ is 1000. Episodes' lengths left-to-right: 274, 266, 310. Momentum is zero.

[2] This can be circumvented by using a grid map encoding of sensory data similarly in [22]. We leave it as future work.

episode, and updated as follow: $u_{t+1,p} \leftarrow 0$ if p is not seen by the agent, otherwise $u_{t+1,p} \leftarrow \min\{\mathbf{u}, u_{t,p} + 1\}$. The destination is then the point in $\text{argmax}_p\, u_{t,p}/D_{t,p}$ where $D_{t,p}$ is the distance between p and the agent at step t. This heuristic tends to result in sweeping paths with infrequent backtracking. Some examples of induced navigation are shown in Fig. 2.

There are two edge cases in this approach. One is when $u_{t,p} = 0$ for all $p \in P$, in which case the destination is the nearest point. The other is when $D_{t,p'} = 0$ for some point p'. Firstly, this can occur with at most one point since points in P are distinct. Secondly, we can reasonably assume the agent has a positive detection radius, so it must be that $u_{t,p'} = 0$. As such, p' is not the destination, unless the first edge case also occurs.

An important feature of the exploration subroutine is that each agent maintains its own uncertainty mapping; this has three benefits. First, it allows agents to have different mapping settings, facilitating specialized exploratory behaviours if necessary. Second, no communication between agents is necessary; each agent has access to sufficient information in every step to update its own mapping. Third, it maintains high map coverage: assuming identical mapping settings among agents, if agents already occupy different regions in the map, they will see different regions as "least explored", leading to diverse future travel paths.

5.2 Action and Reward Models

We assume simple dynamics between the action types. The agent's task is to move toward targets' locations and interact with targets in the effective range. Under this assumption, appropriate movement actions and interaction actions can be derived from information extracted only from environmental states. Therefore, it suffices to treat a policy as consisting of two sub-policies, one for each action type and trained with its own reward function. This approach decomposes the action space, reducing its size and thus the amount of data needed to train the policy. Also, separating reward signals between action types prevents misleading situations where optimal choices in one action type yield lower rewards than suboptimal ones due to the choices in the other action type. While this is not necessary for learning optimal behaviours as long as the joint reward is decomposable (e.g. sum of individual rewards), it helps speed up the process.

In a multi-policy setup where each policy is dedicated to a specific task, the policies are typically trained separately. Our approach streamlines their training in a single process, and their respective updates are done simultaneously as in the single-policy paradigm. This facilitates integration of most existing RL methods with minimal modification. As sub-policies are updated via a single sum over individual losses, the rewards are normalised (bounded in $[-1, 1]$) so as to mitigate biases toward specific sub-policies in the updates.

Movement. The agents are to shorten the distance between themselves and their destination while avoiding collision. The seemingly simple task is challenging as the presence of obstacles frequently incurs multiple optimal paths requiring significantly different action sequences, making the return landscape multimodal in the policy space. This tends to confuse the learning trajectories induced by local-search-like policy optimisation routines employed by most RL methods, such as Stochastic Gradient Descent. As such, to keep the RL algorithms simple, we decide to smoothen this landscape via a dense reward function. We also make the reward a function of the observation, as this minimises noises in the rewards, which are already noisy in Dec-POMDPs.

Since the movement actions do not affect the agent's position directly, the reward is instead based on the agent's velocity. It is defined for two cases:

- When there is no collision, the agent is rewarded for aligning its velocity toward the destination, with deviation being encouraged when obstruction is present. Given the velocity vector v_t, the maximum velocity v_{max}, and the destination vector d_t, the function returns $\frac{\langle v_t \cdot d_t \rangle}{v_{max}|d_t|} \cdot (1 - b_{t,i}) - b_{t,i}$.
- When there is a collision with any other object, the reward is -1.

Interacting with targets The choices in this action type are straightforward, so we opt for a simple reward function. A target interaction action is successful if it is effective (e.g., reduces lifetime of a target object). The reward function returns 0 for a no-op. Otherwise, a success is rewarded $+1$ while failure is penalised -1. Note that since there is a 2-step delay between the point of committing and the effect on the target, the reward signal is also 2-step delayed.

5.3 Learning Methods

For the initial study of this particular problem, we use simple and effective multi-agent Q-learners to train a deterministic policy: IQL [19,20], VDN [17], QMIX [14]. We also implement IQL embedded in an ensemble learning framework, SUNRISE [9], which was proposed as a way to reduce learning's sensitivity to initial conditions. Note VDN and QMIX use team rewards, which we interpret as sums of individual agents' rewards. Details are left in the Appendix due to space limit.

6 Experimental Results

We investigate our hybrid approach with IQL, VDN, QMIX, and SUNRISE-IQL. IQL is a simple Q-learning method for MARL, and the rest are known to perform well in MARL problems. This selection of RL methods is conducive in revealing the effectiveness of the proposed hybrid approach, and isolates the effects of the RL method mechanisms on the observed performance metrics. Our implementation of the environment and algorithms is based on JaxMARL's [15]. We perform 10 runs per configuration, producing 10 corresponding policies.

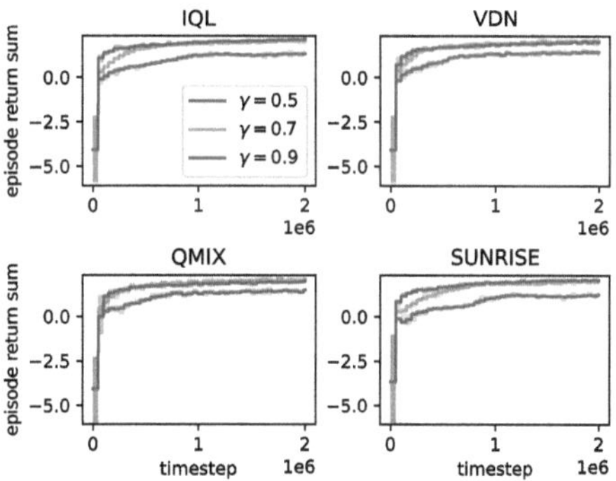

Fig. 3. Return curves (mean, min, max) as accumulated over undiscounted rewards, normalised by episodes' lengths and summed over agents. Each reward score is an average over 32 test episodes with 64 max steps and initial positions of objects sampled uniformly within the map's bound.

6.1 Training Performances

In our experiments, we focus on the effects of the discount factor γ on agents' behaviours. We train policies with the four algorithms and $\gamma \in \{0.5, 0.7, 0.9\}$ using parameters given in the Appendix. This setting densely populates the world map with agents, obstacles, and targets, which gives several benefits. Firstly, it helps increase the frequency of generating diverse observation vectors and, consequently, discerning reward signals. Secondly, it lets agents come into contact with targets frequently, regardless of movements. This means that the quality of the generated data for training the *interact* sub-policy is less dependent on the performance of the *search* sub-policy.

The learning curves are shown in Fig. 3. Here, we take the episodic per-step averages of rewards instead of reward sums, as shorter episodes indicate better performing agents. We observe similar training performances across algorithms, indicating that the dense reward functions ease the learning process.

Regarding the effect of γ, higher values seem to lead to faster convergence and higher final returns. This is expected as the learning target aligns more closely with the reward sums as γ approaches 1. However, magnifying long-term reward signals can cause the learning to over-fit the policy to the training setting. In Sect. 6.2, we will see that policies trained with larger γ are suboptimal when environmental parameters differ from those used during training.

6.2 Performances On Test Scenarios

To study in detail the agents' behaviours produced by the trained policies, we establish test scenarios with various characteristics. Completing these scenarios requires the agents to perform a wide range of maneuvers within environments that contain objects of varying sizes. The environmental parameters for these scenarios are listed in Table 3.

The key parameters among these are Max Velocity, Momentum Coefficient and Action Acceleration Coefficient, as these determine the dynamics of actions, which is what the RL methods aim to encode in the policies. The values of these parameters in the test scenarios differ from those used for the training. With this, we can see how well the learned correlations encoded in the trained policies generalise over different action dynamics. Note that the values of Max Velocity are higher than the effective maximum collision-free velocity, which is $\frac{\sqrt{2} \cdot action\_accel\_coeff \cdot \Delta t}{1 - momentum\_coeff}$.

An important feature of these scenarios is the long-horizon tasks. As described in Sect. 6.1, the policies were trained by observing independent trajectories spanning at most 64 steps. Meanwhile, each test scenario requires more than 200 steps to complete, as it takes longer for agents to move between targets, and involves a higher target-to-agent ratio. As such, the results in this section demonstrate the degree to which these policies can perform more complex tasks than those they were trained on.

In addition, we establish a baseline for each scenario while validating the effect of the reward formulation on inducing scenario-completing action sequences. To this end, we run a look-ahead heuristic on the test scenarios that, in each step, chooses the action maximising the reward sum over some steps ahead under the same reward formulation. It searches exhaustively within the joint action space by running $\sum_{j=1}^{l} 27^{m \cdot j}$ look-ahead environment steps per choice, where m is the number of agents and l is the number of steps to look ahead. Since the environment dynamics is deterministic, this heuristic is run once per scenario. Due to the high computational time cost on 2-agent scenarios, and the fact that the effects of the interaction actions have a 2-step delay, we set $l = 2$ (2-step search). We illustrate the behaviours generated by the heuristic, together with the rendering of the initial states, in the Appendix.

We run all policies trained by all algorithms as in Sect. 6.1 (120 policies in total) on each test scenario. We record four end-of-episode metrics: episode length, completion status, number of steps the agents spend in collision with a non-target, and number of steps the agents misfire. An agent misfires if it commits a non-no-op interaction action in the preceding step without interacting with a target object in the current step. The metrics are reported in Table 1, including those from the 2-step look-ahead runs where appropriate. We see that no training setting resulted in policies that produce the best measurements across all metrics. However, policies trained with $\gamma = 0.7$ seem to give the best trade-offs overall. Those trained with $\gamma = 0.9$ are effective in collision avoidance, but fail more often in avoiding misfires. Meanwhile, those trained with $\gamma = 0.5$ exhibit the opposite trend. Note that the 2-step look-ahead heuristic produced a similarly large number of collisions, suggesting that long-term reward signals are necessary to learn effective collision avoidance behaviours. The QMIX-trained and IQL-trained policies at $\gamma = 0.7$ produced the most successful runs overall.

We note that Scenario 8 is in theory the "hardest" as it requires a single agent to resolve the most number of targets while navigating through the most number of obstacles. Despite this, the trained policies completed this scenario more reliably than Scenario 7, which features two agents and the same arrangement of obstacles and targets. A key factor is the large momentum coefficient: the increased amount of velocity carried over the steps reduces the amount of work the agent needs to commit to move between points. This came at the cost of more frequent collisions, which require actions over

Table 1. Median metrics on test scenarios. Collision and misfiring statistics from each run are averaged over agents. A success is a completion of a scenario within the step limit. Collisions with target objects are not counted. Best values in each row are boldfaced.

	No.	iql-0.5	iql-0.7	iql-0.9	vdn-0.5	vdn-0.7	vdn-0.9	qmix-0.5	qmix-0.7	qmix-0.9	sunrise-0.5	sunrise-0.7	sunrise-0.9	look-ahead
Episode Lengths	1	302.0	282.5	356.0	295.5	349.0	285.0	387.0	259.0	285.5	**257.5**	285.5	284.0	364
	2	431.5	385.5	384.5	395.5	**357.5**	407.0	390.5	376.5	383.0	422.5	407.5	368.5	360
	3	239.5	233.0	345.5	219.0	226.5	259.0	230.0	232.5	430.5	237.5	215.5	334.5	**203**
	4	232.0	220.0	292.5	228.5	224.5	301.0	244.0	222.0	348.5	226.5	211.0	249.5	**201**
	5	325.5	306.5	**297.0**	353.0	300.0	317.5	317.5	320.5	344.0	323.5	301.5	380.5	302
	6	370.0	314.5	**297.5**	452.0	302.5	350.0	323.0	322.0	323.0	329.0	323.5	318.5	307
	7	295.0	**274.0**	497.5	476.0	313.0	500.0	468.0	348.0	500.0	294.0	276.0	500.0	281
	8	343.0	205.0	238.5	332.5	243.0	302.5	362.5	246.0	303.0	353.0	242.0	251.0	**202**
No. Successes	1	**10/10**	**10/10**	9/10	**10/10**	**10/10**	**10/10**	**10/10**	**10/10**	**10/10**	**10/10**	8/10	**10/10**	-
	2	**10/10**	**10/10**	**10/10**	**10/10**	**10/10**	**10/10**	**10/10**	**10/10**	**10/10**	**10/10**	**10/10**	**10/10**	-
	3	8/10	**10/10**	7/10	**10/10**	9/10	7/10	**10/10**	**10/10**	5/10	9/10	**10/10**	7/10	-
	4	8/10	**10/10**	9/10	**10/10**	**10/10**	**10/10**	9/10	**10/10**	7/10	**10/10**	**10/10**	8/10	-
	5	6/10	8/10	**10/10**	7/10	8/10	8/10	9/10	8/10	7/10	8/10	9/10	7/10	-
	6	7/10	9/10	**10/10**	5/10	9/10	7/10	9/10	9/10	8/10	9/10	8/10	8/10	-
	7	**9/10**	**9/10**	5/10	8/10	**9/10**	3/10	5/10	8/10	2/10	8/10	8/10	3/10	-
	8	9/10	9/10	9/10	**10/10**	**10/10**	9/10	**10/10**	**10/10**	**10/10**	**10/10**	**10/10**	**10/10**	-
Collision Steps	1	**0.0**	**0.0**	**0.0**	**0.0**	**0.0**	**0.0**	**0.0**	**0.0**	**0.0**	**0.0**	**0.0**	**0.0**	**0.0**
	2	2.0	**0.0**	**0.0**	0.5	**0.0**	**0.0**	**0.0**	**0.0**	**0.0**	5.0	**0.0**	**0.0**	**0.0**
	3	3.0	**0.0**	**0.0**	6.0	**0.0**	**0.0**	1.0	**0.0**	**0.0**	10.0	**0.0**	**0.0**	4.0
	4	6.0	0.25	**0.0**	7.25	0.5	**0.0**	4.25	**0.0**	**0.0**	10.5	**0.0**	**0.0**	1.5
	5	8.5	**0.0**	**0.0**	18.5	**0.0**	**0.0**	7.0	**0.0**	**0.0**	20.5	**0.0**	**0.0**	12.0
	6	10.0	1.0	**0.0**	18.0	1.0	**0.0**	7.5	2.0	0.5	27.0	5.0	0.5	17.0
	7	13.5	**0.75**	3.75	37.0	7.5	1.0	25.75	1.5	3.25	19.25	3.25	3.0	21.5
	8	31.5	10.0	**6.5**	30.5	14.0	15.5	36.0	19.0	10.5	43.0	12.0	9.5	9.0
Misfiring Steps	1	**4.0**	5.0	7.0	**4.0**	**4.0**	8.0	**4.0**	**4.0**	6.0	**4.0**	4.5	7.0	**4.0**
	2	5.25	5.5	11.75	**4.5**	5.0	8.0	5.0	5.0	7.0	5.25	5.0	7.0	5.5
	3	**5.0**	**5.0**	7.5	**5.0**	6.5	7.5	**5.0**	6.0	6.0	**5.0**	**5.0**	8.0	**5.0**
	4	**5.0**	**5.0**	20.0	**5.0**	5.75	8.5	**5.0**	5.5	6.5	**5.0**	6.0	8.5	**5.0**
	5	**4.0**	**4.0**	14.0	**4.0**	**4.0**	5.0	**4.0**	**4.0**	**4.0**	**4.0**	**4.0**	5.0	**4.0**
	6	4.0	4.0	15.0	**3.0**	4.0	5.0	4.0	4.0	4.0	4.0	4.0	7.0	4.0
	7	4.25	4.5	25.25	**4.0**	5.25	5.25	**4.0**	4.5	4.25	**4.0**	4.75	7.75	4.5
	8	7.0	9.0	10.0	9.0	9.0	10.5	8.0	10.0	10.0	**6.0**	9.0	9.0	**6.0**

more preceding steps to avoid. However, the 2-step look-ahead heuristic gave the fastest completion, which was achieved by maintaining low velocities. This is followed by the IQL-trained policies at $\gamma = 0.7$, which gave reasonably low collision rates.

7 Conclusion

Our proposed approach an effective solution for navigation and collision avoidance in search-and-interact tasks within multi-agent environments. By selectively focusing on relevant environmental data and augmenting agent-specific observations, our approach enhances training efficiency via increased action coverage. Within a CTDE framework, our approach trains autonomous control that is both adaptable and efficient, regardless of environmental complexity. Experimental results demonstrate that even simple Q-learners can produce controllers that effectively complete tasks such as accurate fire-fighting in intricate scenarios. We show that controllers trained with our approach can complete scenarios which exhibit roughly four times longer task horizons than, and significantly different action dynamics from, those seen by the learners in the training scenarios. This approach is versatile and can be applied to other complex decision-making problems with similar search-and-interact dynamics.

Acknowledgments. This study was funded by the SA Defence Innovation Partnership (funding number DIP00004CRF7). We would like to acknowledge the extensive support provided by DEWC Services in deploying these results within the DEWC visualisation tool.

Disclosure of Interests. The authors have no competing interests to declare that are relevant to the content of this article.

A Training Details

Here, we detail our training implementations, including the policy network architecture, and the training parameters.

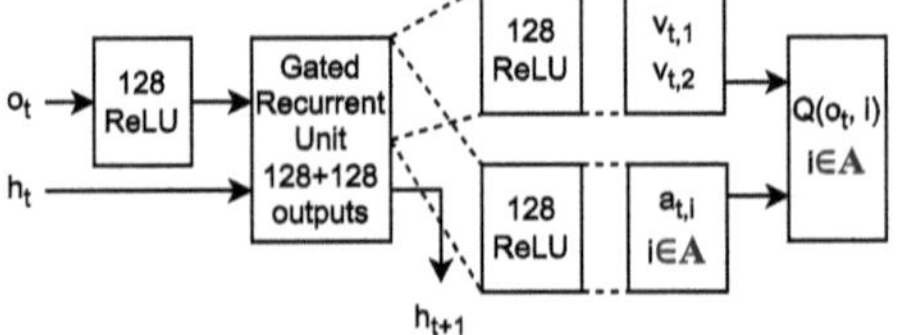

Fig. 4. The policy network architecture. $v_{t,i}$ and $a_{t,i}$ refer to the value and advantage from the sub-policy i at step t, respectively. The outputs cover the union between action types. The recurrent state h_t is reinitialized if t is the start of an episode. For policy ensemble, the sizes of outputs ($v_{t,i}$ and $a_{t,i}$) are multiplied by the ensemble size.

A.1 Policy Network Architecture

We include a diagram of the policy network architecture in Fig. 4. We choose the dueling architecture [26], which computes Q-values by averaging advantage estimates. We keep sub-policies densely connected to allow maximum learning flexibility. Policies in the ensemble (in SUNRISE) are likewise densely connected.

A.2 Parameter Setting

The environmental parameters and learning hyperparameters used in the experiment are listed in Table 2.

Table 2. Training parameter setting. Radii in (multi-)sets are distributed equally over objects of corresponding types.

Environment parameter	Setting	Hyperparameter	Setting
Trajectory length	64	No. new samples per update	8
No. agents/obstacles/targets	$4/6/12$	Replay buffer capacity	10000
Agent detection range	0.5	Mini-batch size	32
Agent radii	$\{0.02, 0.02, 0.04, 0.08\}$	Max total steps	2000000
Obstacle radii	$\{0.1, 0.2, 0.3\}$	No. epochs	2
Target radii	$\{0.05, 0.1, 0.2\}$	Learning rate	0.005
Step interval Δt	0.1	Gradient norm clip	10
Max velocity	0.8	Target policy update interval	200 online policy updates
Momentum coefficient	0.5		
Action acceleration coefficient	4		
Uncertainty grid	11×11		
Uncertainty upper bound	100		
Interaction set $\mathcal{I}$	$\{\text{no-op}, (0, 0.1), (0.1, 0.2)\}$		

B Additional Results on Test Scenarios

The specification detail for the test scenarios is given in Table 3. The agents' behaviors in the test scenarios generated by the 2-step look-ahead are illustrated in Fig. 5. The agent exhibits efficient path-finding when its velocity is small relative to its action acceleration.

Table 3. Test scenarios parameters. No. objects are shown in agents/obstacles/targets. Radii are distributed equally over objects of corresponding types.

Parameter	Scenario 1	Scenario 2	Scenario 3	Scenario 4	Scenario 5	Scenario 6	Scenario 7	Scenario 8
Max step	500	500	500	500	500	500	500	500
No. objects	$1/0/4$	$2/0/8$	$1/2/5$	$2/4/10$	$1/4/4$	$1/4/4$	$2/8/8$	$1/8/8$
Detection range	0.7	0.7	1.5	0.8	1	1	1	1
Agent radii	$\{0.02\}$	$\{0.02, 0.08\}$	$\{0.08\}$	$\{0.02, 0.08\}$	$\{0.04\}$	$\{0.02\}$	$\{0.02, 0.04\}$	$\{0.02\}$
Obstacle radii	-	-	$\{0.1, 0.2\}$	$\{0.1, 0.2\}$	$\{0.15\}$	$\{0.18\}$	$\{0.15, 0.18\}$	$\{0.15, 0.18\}$
Target radii	$\{0.1\}$	$\{0.1\}$	$\{0.1\}$	$\{0.1\}$	$\{0.1\}$	$\{0.1\}$	$\{0.1\}$	$\{0.1\}$
Max velocity	1	1	1	1	1	1	1	2
Momentum coeff.	0	0	0	0	0	0	0	0.9
Act. accel. coeff.	1	1	1	1	1	1	1	1

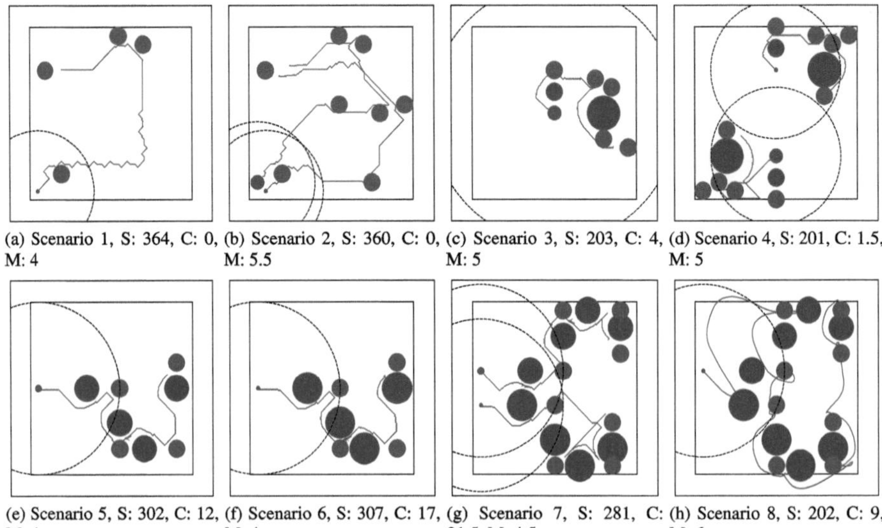

(a) Scenario 1, S: 364, C: 0, M: 4 (b) Scenario 2, S: 360, C: 0, M: 5.5 (c) Scenario 3, S: 203, C: 4, M: 5 (d) Scenario 4, S: 201, C: 1.5, M: 5

(e) Scenario 5, S: 302, C: 12, M: 4 (f) Scenario 6, S: 307, C: 17, M: 4 (g) Scenario 7, S: 281, C: 21.5, M: 4.5 (h) Scenario 8, S: 202, C: 9, M: 6

Fig. 5. Results and agents' position tracing on test scenarios from 2-step look-ahead heuristic, imposed on corresponding initial states. Metrics (S=Steps, C=Collision, M=Misfires) are averaged over agents. All episodes are terminated upon task completion.

References

1. Al-Nima, R.R.O., Han, T., Al-Sumaidaee, S.A.M., Chen, T., Woo, W.L.: Robustness and performance of deep reinforcement learning. Appl. Soft Comput. **105**, 107295 (2021)
2. Albrecht, S.V., Christianos, F., Schäfer, L.: Multi-agent reinforcement learning: foundations and modern approaches. MIT Press (2024)
3. Canese, L., et al.: Multi-agent reinforcement learning: a review of challenges and applications. Appl. Sci. **11**(11), 4948 (2021)
4. Cui, J., Liu, Y., Nallanathan, A.: Multi-agent reinforcement learning-based resource allocation for UAV networks. IEEE Trans. Wireless Commun. **19**(2), 729–743 (2019)
5. Diaz-Vilor, C., Lozano, A., Jafarkhani, H.: A reinforcement learning approach for wildfire tracking with UAV swarms. IEEE Trans. Wireless Commun. (2025)
6. Guo, J., Chen, Y., Hao, Y., Yin, Z., Yu, Y., Li, S.: Towards comprehensive testing on the robustness of cooperative multi-agent reinforcement learning. In: Proceedings of the IEEE/CVF Conference on Computer Vision and Pattern Recognition, pp. 115–122 (2022)
7. Haksar, R.N., Schwager, M.: Distributed deep reinforcement learning for fighting forest fires with a network of aerial robots. In: 2018 IEEE/RSJ International Conference on Intelligent Robots and Systems (IROS), pp. 1067–1074. IEEE (2018)
8. Julian, K.D., Kochenderfer, M.J.: Distributed wildfire surveillance with autonomous aircraft using deep reinforcement learning. J. Guid. Control. Dyn. **42**(8), 1768–1778 (2019)
9. Lee, K., Laskin, M., Srinivas, A., Abbeel, P.: SUNRISE: a simple unified framework for ensemble learning in deep reinforcement learning. In: Meila, M., Zhang, T. (eds.) Proceedings of the 38th International Conference on Machine Learning. Proceedings of Machine Learning Research, vol. 139, pp. 6131–6141. PMLR (2021)
10. Lowe, R., Wu, Y., Tamar, A., Harb, J., Abbeel, P., Mordatch, I.: Multi-agent actor-critic for mixed cooperative-competitive environments. Neural Inf. Process. Syst. (NIPS) (2017)

11. Mordatch, I., Abbeel, P.: Emergence of grounded compositional language in multi-agent populations. arXiv preprint arXiv:1703.04908 (2017)
12. Nguyen, T.T., Nguyen, N.D., Nahavandi, S.: Deep reinforcement learning for multiagent systems: a review of challenges, solutions, and applications. IEEE Trans. Cybern. **50**(9), 3826–3839 (2020)
13. Peng, H., Shen, X.: Multi-agent reinforcement learning based resource management in MEC- and UAV-assisted vehicular networks. IEEE J. Sel. Areas Commun. **39**(1), 131–141 (2020)
14. Rashid, T., Samvelyan, M., de Witt, C.S., Farquhar, G., Foerster, J., Whiteson, S.: Monotonic value function factorisation for deep multi-agent reinforcement learning. J. Mach. Learn. Res. **21**(178), 1–51 (2020)
15. Rutherford, A., et al.: JaxMARL: multi-agent RL environments and algorithms in JAX. In: Proceedings of the 23rd International Conference on Autonomous Agents and Multiagent Systems, pp. 2444–2446. AAMAS '24, International Foundation for Autonomous Agents and Multiagent Systems, Richland, SC (2024)
16. Seraj, E., Silva, A., Gombolay, M.: Multi-UAV planning for cooperative wildfire coverage and tracking with quality-of-service guarantees. Autonomous Agents Multi-Agent Syst. **36**(2) (2022). https://doi.org/10.1007/s10458-022-09566-6
17. Sunehag, P., et al.: Value-decomposition networks for cooperative multi-agent learning based on team reward. In: Proceedings of the 17th International Conference on Autonomous Agents and MultiAgent Systems, pp. 2085–2087. AAMAS '18, International Foundation for Autonomous Agents and Multiagent Systems, Richland, SC (2018)
18. Sutton, R.S., Barto, A.G.: Reinforcement learning: an introduction. The MIT Press, 2nd edn. (2018)
19. Tampuu, A., et al.: Multiagent cooperation and competition with deep reinforcement learning. PLoS ONE **12**(4), e0172395 (2017). https://doi.org/10.1371/journal.pone.0172395
20. Tan, M.: Multi-agent reinforcement learning: independent vs. cooperative agents, pp. 330–337. Elsevier (1993). https://doi.org/10.1016/b978-1-55860-307-3.50049-6
21. Viseras, A., Meissner, M., Marchal, J.: Wildfire front monitoring with multiple UAVS using deep Q-learning. IEEE Access (2021)
22. Wang, J., Wang, Y., Zhang, D., Yang, Y., Xiong, R.: Learning hierarchical behavior and motion planning for autonomous driving. In: 2020 IEEE/RSJ International Conference on Intelligent Robots and Systems (IROS). IEEE (2020). https://doi.org/10.1109/iros45743.2020.9341647
23. Zhang, K., Yang, Z., Başar, T.: Multi-agent reinforcement learning: a selective overview of theories and algorithms. Handbook of Reinforcement Learning and Control, 321–384 (2021)
24. Zhang, S.Q., Zhang, Q., Lin, J.: Succinct and robust multi-agent communication with temporal message control. Adv. Neural. Inf. Process. Syst. **33**, 17271–17282 (2020)
25. Zhang, Y., Mou, Z., Gao, F., Jiang, J., Ding, R., Han, Z.: UAV-enabled secure communications by multi-agent deep reinforcement learning. IEEE Trans. Veh. Technol. **69**(10), 11599–11611 (2020)
26. Wang, Z., Schaul, T., Hessel, M., Hasselt, H., Lanctot, M., Freitas, N.: Dueling network architectures for deep reinforcement learning. In: Proceedings of The 33rd International Conference on Machine Learning. Proceedings of Machine Learning Research, vol. 48, pp. 1995–2003. PMLR (2016)

Synergistic MARL: Unifying Physics-Informed, Meta Learning, and Hybrid Learning to Outperform Communication-Based Coordination

YiYao Zhang[1(✉)], YiFei Dong[2], Kun Yu[2], Wei Liu[2], Jianlong Zhou[2], Bin Liang[2], and Fang Chen[2]

[1] School of Computer Science and Mathematics, Adelaide University, Adelaide, Australia
yiyao.zhang@student.adelaide.edu.au
[2] Data Science Institute, University of Technology Sydney, Sydney, Australia
{Yifei.dong,Kun.Yu,Wei.Liu,Jianlong.Zhou,Bin.Liang,Fang.Chen}@uts.edu.au

Abstract. Multi-Agent Reinforcement Learning (MARL) has evolved from independent learning to sophisticated communication-based coordination systems like Adaptive Topology Communication (ATC). However, current communication-based approaches face limitations in domain knowledge integration, adaptation speed, and learning paradigm flexibility that hinder real-world deployment. This study proposes three synergistic MARL framework optimization techniques—Physics-Informed MARL, Meta-Learning MARL, and Hybrid Learning MARL—that extend beyond communication-based coordination to address these fundamental limitations. These methods integrate domain-specific knowledge, enable rapid adaptation to new scenarios, and provide flexible learning paradigms that dynamically balance centralized and decentralized approaches. Experimental evaluation in warehouse coordination tasks demonstrates statistically significant reward improvements of 2.4%–31.3% over state-of-the-art baselines (ATC), with Hybrid Learning MARL achieving the highest performance. The results validate that proposed techniques can substantially enhance communication-based MARL systems while maintaining safety and coordination effectiveness for practical deployment.

Keywords: Multi-Agent Reinforcement Learning · Communication-Based Coordination · Physics-Informed Learning · Meta-Learning · Hybrid Learning · Warehouse Automation

1 Introduction

The evolution of Multi-Agent Reinforcement Learning (MARL) has been a journey from isolated agents to interconnected, intelligent teams. Early independent learning approaches have given way to sophisticated communication-based coordination systems, where agents actively share information to achieve collective

M. Liu et al. (Eds.): AI 2025, LNAI 16371, pp. 402–413, 2026.
https://doi.org/10.1007/978-981-95-4972-6_31

goals. State-of-the-art methods like Multi-Agent Actor-Critic (MADDPG) and communication protocols exemplify this progress, enabling agents to dynamically optimize their communication networks [7,15]. For instance, in a multi-robot warehouse, these systems allow agents to learn who to "talk" to, perhaps forming a tight communication cluster to navigate a congested aisle or prioritizing messages from a robot that has located a high-priority package.

Despite their sophistication, these communication-centric methods face a performance ceiling due to three fundamental limitations. First, they are largely **domain-agnostic**, treating warehouse robots and financial trading bots as interchangeable black boxes, thus ignoring readily available domain knowledge about physics, energy consumption, or safety constraints. Second, they exhibit **brittle adaptation**; a minor change in the environment, like adding a new shelf or altering a robot's battery capacity, often requires extensive and costly retraining from scratch. Third, they commit to a **rigid learning paradigm**, locking agents into either a fully decentralized approach (fast but potentially uncoordinated) or a centralized one (coordinated but computationally expensive), without the flexibility to switch based on the task's dynamic needs [6].

This research argues that the next leap in MARL performance requires moving beyond communication-based coordination alone. We propose a framework of three advanced MARL techniques that address these bottlenecks directly:

1. **Physics-Informed MARL (PI-MARL):** To inject domain knowledge, this method integrates physical laws and operational constraints (e.g., energy efficiency, collision avoidance) directly into the reward structure and agent architecture.
2. **Meta-Learning MARL (ML-MARL):** To enable rapid adaptation, this approach trains agents on a wide distribution of tasks, allowing them to "learn to learn" and quickly fine-tune their policies for new, unseen scenarios with minimal data.
3. **Hybrid Learning MARL (HL-MARL):** To achieve paradigm flexibility, this technique introduces an intelligent controller that dynamically switches between centralized and decentralized learning, optimizing the trade-off between coordination and computational cost in real-time.

2 Related Work

Recent advances in Multi-Agent Reinforcement Learning (MARL) have focused on improving coordination mechanisms beyond traditional independent learning approaches. The field has progressed through several key paradigms, each addressing specific limitations of previous approaches.

Independent Learning Approaches: Early MARL research focused on independent learning where agents treat other agents as part of the environment [11]. While simple to implement, these approaches suffer from non-stationarity issues and poor coordination in complex scenarios. Independent Proximal Policy Optimization (IPPO) [13] represents a strong baseline in this category, demonstrating that simple independent learning can be surprisingly effective when properly implemented.

Communication-Based Coordination: Zhu et al. [15] developed a comprehensive survey of communication-based coordination in multi-agent deep reinforcement learning, categorizing approaches across nine dimensions of communication design. Lowe et al. [7] introduced Multi-Agent Deep Deterministic Policy Gradient (MADDPG), which uses centralized training with decentralized execution to address coordination challenges. These approaches improve coordination but remain domain-agnostic and inflexible to changing scenarios.

Domain Knowledge Integration: Banerjee et al. [2] provided a comprehensive survey of physics-informed reinforcement learning, demonstrating how domain knowledge can enhance sample efficiency and performance. Zhao and Liu [14] introduced physics-informed constraints in aircraft conflict resolution, showing enhanced performance in safety-critical applications. However, these approaches lack the communication mechanisms necessary for complex multi-agent coordination scenarios.

Meta-learning in MARL: Mao et al. [8] established theoretical results for meta-learning in fundamental MARL settings, showing sharper convergence rates with task similarity. Beck et al. [3] provided a comprehensive tutorial on meta-reinforcement learning, establishing foundations for rapid adaptation that can be extended to multi-agent settings. These approaches show promise for rapid adaptation but require extension to complex coordination scenarios.

Hybrid Learning Paradigms: Santos et al. [10] explored centralized training with hybrid execution, introducing paradigms that adapt communication levels at execution time. Amato [1] provided theoretical foundations for centralized training with decentralized execution in cooperative MARL. However, existing hybrid approaches use static paradigm selection rather than dynamic adaptation based on task requirements and resource constraints.

Table 1 presents a comprehensive comparison of existing approaches with our proposed methods across key parameters including domain knowledge integration, adaptation mechanisms, learning paradigm flexibility, communication efficiency, and scalability.

Table 1. Literature comparison across key parameters

Method	Domain Knowledge	Rapid Adaptation	Paradigm Flexibility	Communication	Scalability
IPPO [13]	×	×	×	×	✓
MADDPG [7]	×	×	×	✓	✓
Zhao & Liu [14]	✓	×	×	×	✓
Mao et al. [8]	×	✓	×	×	✓
Santos et al. [10]	×	×	✓	✓	×
Zhu et al. [15]	×	×	×	✓	✓
Physics-Informed MARL	✓	✓	×	✓	✓
Meta-Learning MARL	×	✓	✓	✓	✓
Hybrid Learning MARL	✓	✓	✓	✓	✓

Our proposed methods demonstrate comprehensive coverage across all key parameters, with Hybrid Learning MARL providing the most complete solution by integrating domain knowledge, enabling rapid adaptation, offering paradigm flexibility, maintaining communication efficiency, and ensuring scalability. This comprehensive approach represents a significant advancement over existing methods that typically excel in only one or two dimensions.

3 Methodology: Proposed MARL Optimization Techniques

The following sections detail three distinct yet complementary techniques designed to enhance a core MARL framework from different optimization perspectives. Each method is a targeted solution to a specific limitation in current MARL systems: Physics-Informed MARL (PI-MARL) optimizes for physical realism by integrating domain knowledge, Meta-Learning MARL (ML-MARL) optimizes for rapid adaptation to new tasks, and Hybrid-Learning MARL (HL-MARL) optimizes the learning process itself by dynamically selecting the most effective training paradigm.

3.1 Physics-Informed MARL (PI-MARL)

Physics-Informed MARL (PI-MARL) integrates domain-specific physical constraints, such as energy considerations, velocity limits, and collision avoidance, directly into the learning process.

Network Architecture

- **Shared State Encoder:** A 2-layer MLP with ReLU activations processes the input state s into a shared, high-level feature vector, h_{feat}. This allows the model to learn a common representation of the system state.
- **Specialized Prediction Heads:** This feature vector h_{feat} is fed into three parallel, independent prediction heads. Each head is a single-layer MLP that specializes in predicting a specific scalar component of the total value:
  - V_{base} **Head:** Predicts the expected discounted task-related return.
  - V_{energy} **Head:** Predicts the expected discounted future energy costs.
  - V_{safety} **Head:** Predicts the expected discounted future safety constraint violations.

Mathematical Formulation. Let $\mathcal{S}$ be the state space, $\mathcal{A}$ the action space, and $\mathcal{P}$ the set of physical constraints. The core of PI-MARL is a composite reward function that balances task completion with physical considerations. This physics-informed reward is defined as:

$$R_{\text{PI}}(\boldsymbol{s}, \boldsymbol{a}) = R_{\text{task}}(\boldsymbol{s}, \boldsymbol{a}) - \lambda_E R_{\text{energy}}(\boldsymbol{s}, \boldsymbol{a}) - \lambda_S R_{\text{safety}}(\boldsymbol{s}, \boldsymbol{a}) \tag{1}$$

where R_{task} is the original task-completion reward, R_{energy} penalizes energy inefficiency, and R_{safety} enforces safety constraints. The hyperparameters λ_E and λ_S

are weighting parameters that balance these objectives. This formulation is supported by the following mechanisms.

Reward Shaping. To encourage smoother and more physically realistic motion, the reward signal is augmented to penalize energy-inefficient behaviors, excessive acceleration or high velocity. This is formulated as:

$$R_{\text{energy}}(\boldsymbol{s}, \boldsymbol{a}) = (1 + w_{load} \cdot L) \cdot (\alpha \cdot \|\boldsymbol{a} - \boldsymbol{a}_{\text{prev}}\|^2 + \beta \cdot \|\boldsymbol{v}\|^2) \tag{2}$$

where L is a binary variable indicating whether the agent is carrying a load, and w_{load} determines the impact of load on energy consumption.

Feasible Action Projection. To ensure that all executed commands are physically viable and safe, any raw action $\boldsymbol{a}_{\text{raw}}$ proposed by the policy is projected onto the nearest action that satisfies all system constraints $\mathcal{P}$, which is in $\mathcal{A}_{\text{feasible}}$. This projection is defined as the solution to the following optimization problem:

$$\boldsymbol{a}_{\text{proj}} = \underset{\boldsymbol{a} \in \mathcal{A}_{\text{feasible}}}{\arg\min} \|\boldsymbol{a} - \boldsymbol{a}_{\text{raw}}\|^2 \tag{3}$$

Structured, Domain-Aware Critic. Instead of a monolithic network, the critic architecture is explicitly structured to reason about different physical aspects of the environment. This leads to more accurate and disentangled value estimates. The final composite value function is a weighted sum from the specialized value heads:

$$V_{\text{PI}}(\boldsymbol{s}) = V_{\text{base}}(\boldsymbol{s}) + \gamma_E V_{\text{energy}}(\boldsymbol{s}) + \gamma_S V_{\text{safety}}(\boldsymbol{s}) \tag{4}$$

3.2 Meta-Learning MARL (ML-MARL)

Meta-Learning MARL treats each variation in layout, obstacle placement, package distribution or team makeup as a distinct "task". By training across a diverse set of such tasks, agents learn a policy initialization that can be quickly adapted to a new warehouse scenario with just a few gradient updates.

Mathematical Formulation: Given a distribution of tasks $p(\mathcal{T})$, the meta-learning objective is:

$$\theta^* = \arg\min_{\theta} \mathbb{E}_{\mathcal{T} \sim p(\mathcal{T})} \left[\mathcal{L}_{\mathcal{T}}(f_{\theta - \alpha \nabla_\theta \mathcal{L}_{\mathcal{T}}(f_\theta)})\right] \tag{5}$$

where the loss $\mathcal{L}_{\mathcal{T}}$ is computed on the test set of task $\mathcal{T}$ after adapting the model parameters θ using the support set of the same task. The inner gradient update adapts the policy to the specific task, while the outer loop optimizes for a policy that can be effectively adapted.

3.3 Hybrid-Learning MARL (HL-MARL)

Hybrid-Learning MARL dynamically balances centralized versus decentralized learning by monitoring how tightly coupled the agents' decisions are and by checking available computation and communication resources, as shown in Fig. 1.

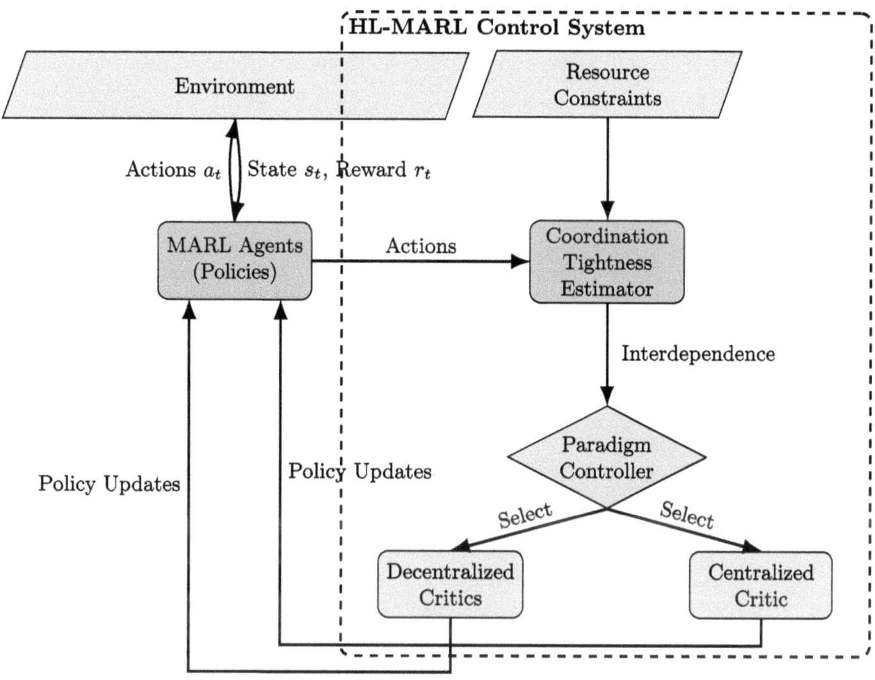

Fig. 1. System architecture of Hybrid-Learning MARL (HL-MARL). The HL-MARL system doesn't just learn what to do (the policies), but also learns how to learn. By dynamically estimating the need for coordination, the Paradigm Controller acts as a smart switch, ensuring that the system uses the powerful but expensive centralized approach only when necessary and defaults to the efficient decentralized approach otherwise.

Mathematical Formulation: The paradigm controller uses mutual information to assess coordination tightness:

$$I(A_i; R_{global}) = \sum_{a_i, r} p(a_i, r) \log \frac{p(a_i, r)}{p(a_i)p(r)} \tag{6}$$

where A_i represents agent i's actions and R_{global} is the global reward.

Paradigm Controller. A lightweight module evaluates current task demands (e.g. how much information each robot's action contributes to team performance) and resource constraints (compute budget, network bandwidth). It then selects between three modes: fully centralized training, fully decentralized training, or a hybrid mix. The controller selects paradigm π based on:

$$\pi^* = \arg \max_{\pi \in \{\text{cent,decent,hybrid}\}} \frac{I(A; R_{global}) \cdot \eta_\pi}{C_\pi} \tag{7}$$

where η_π is the paradigm efficiency and C_π is the computational cost.

Smooth Mode Switching. To avoid instability, HL-MARL incorporates hysteresis: the system only switches modes if the measured benefits persist over several episodes. Transitions themselves follow a gradual hand-off protocol, ensuring that neither learning performance nor system stability suffers when moving between paradigms. Paradigm transitions use exponential smoothing:

$$w_{cent}(t+1) = \rho \cdot w_{cent}(t) + (1-\rho) \cdot \mathbb{I}[\pi^*(t+1) = \text{cent}] \tag{8}$$

where ρ is the smoothing factor and $\mathbb{I}$ is the indicator function.

Hybrid Critic Architecture: The value function combines centralized and decentralized components:

$$V_{hybrid}(s) = w_{cent} \cdot V_{cent}(s, a_{-i}) + (1 - w_{cent}) \cdot V^i_{decent}(s_i) \tag{9}$$

4 Experimental Evaluation

4.1 Experimental Setup

We evaluate our methods in a configurable grid-based warehouse environment. The baseline scenario consists of a 15×15 grid with 3 homogeneous agents, 6 static obstacles, and a dynamic package generation process following a Poisson distribution ($\lambda = 0.3$), which is Fig. 2. To address the point on generalizability, we introduce a series of more complex scenarios:

- **Scalability Analysis:** We test our methods with increasing numbers of agents (3, 5, 7, and 10) in proportionally larger grids (15×15, 20×20, 25×25, and 30×30) to evaluate the scalability of each approach.
- **Dynamic Obstacles:** We introduce dynamic obstacles that move randomly within the environment, forcing agents to adapt to a more unpredictable environment.
- **Heterogeneous Agents:** We use a mix of agent types with different capabilities (e.g., some agents move faster but have less battery, while others are slower but can carry more packages).
- **Priority-Based Tasks:** We introduce a priority system for packages, where some packages have higher delivery priority and yield greater rewards, requiring agents to coordinate and prioritize tasks effectively.

In all scenarios, each complete robot has a 5-cell observation radius and a 3-cell communication range. To benchmark our proposed methods, we compare them against two baselines: IPPO [13], a fundamental independent learning approach, and MADDPG [7], a state-of-the-art actor-critic method. All approaches are trained for 200 episodes across five random seeds using shared hyperparameters (learning rate 3×10^{-4}, $\gamma = 0.99$) and a 256-unit MLP architecture, with our proposed methods utilizing additional specialized networks for their unique components.

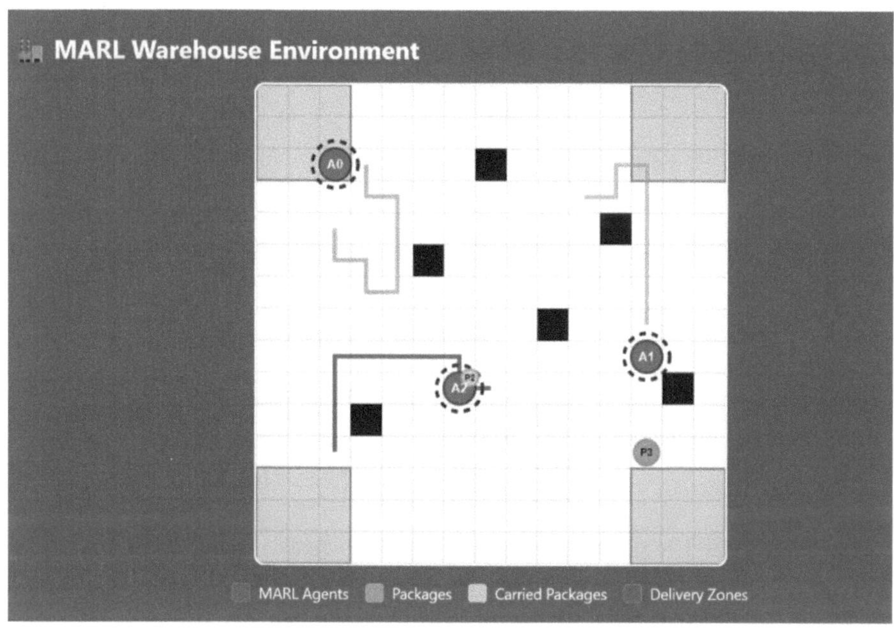

Fig. 2. The multi-agent warehouse coordination environment, shown here with 3 agents in a 15×15 grid.

4.2 Deepening PI-MARL's Domain Knowledge Integration

To address the point on deepening the domain knowledge integration of PI-MARL, we have made the following enhancements:

Load-Dependent Energy Penalty. The original energy penalty in Eq. 2 has been updated to include load dependence. The new energy penalty is defined as shown in Eq. 2, where L is a binary variable indicating whether the agent is carrying a load (1 if carrying, 0 otherwise), and w_{load} is a weight that determines the impact of the load on energy consumption. This updated formulation more accurately reflects the increased energy cost associated with transporting packages.

Constraint Weight Tuning and Heuristics. The weights λ_E and λ_S in Eq. 1 play a crucial role in balancing safety and efficiency. A higher λ_S value will encourage safer behavior at the cost of potentially slower task completion, while a higher λ_E will prioritize energy efficiency, which might lead to more conservative movements.

We propose a simple heuristic for setting these weights based on the operational priorities of the warehouse. For environments where safety is paramount (e.g., those with fragile goods or human co-workers), we suggest starting with a higher λ_S (e.g., 0.4) and a moderate λ_E (e.g., 0.1). Conversely, for environments where throughput is the primary concern, a lower λ_S (e.g., 0.1) and a higher

λ_E (e.g., 0.3) may be more appropriate. These values can then be fine-tuned based on simulation results to achieve the desired balance between safety and efficiency.

4.3 Ablation Studies

To isolate the contributions of the individual components of our proposed methods, we conducted a series of ablation studies. The results are summarized in Table 2.

Table 2. Ablation study results for our proposed methods.

Method	Ablated Component	Final Reward
PI-MARL	Energy Penalty	−34.15
	Safety Penalty	−35.02
ML-MARL	Meta-Learning	−33.89
HL-MARL	Paradigm Controller	−31.23
	Shared Policy	−29.56

PI-MARL Ablation. We tested PI-MARL without the energy penalty and without the safety penalty. In both cases, the performance degraded, indicating that both components are crucial for the method's success. The removal of the safety penalty had a slightly larger negative impact, highlighting the importance of this component in the warehouse environment.

ML-MARL Ablation. We ablated the meta-learning component from ML-MARL, effectively turning it into a standard MARL algorithm. The performance dropped significantly, demonstrating the importance of the meta-learning approach for rapid adaptation.

HL-MARL Ablation. We conducted two ablation studies for HL-MARL. First, we removed the paradigm controller, forcing the system to use a fixed hybrid approach. This resulted in a significant performance drop, confirming the importance of the dynamic switching mechanism. Second, we ablated the shared policy component (Eq. 9), forcing agents to rely solely on private adaptation. While the performance drop was less severe than with the removal of the paradigm controller, it was still significant, indicating that the shared policy component is important for maintaining coordination.

4.4 Computational Overhead Analysis

To provide a clear understanding of the computational costs associated with our proposed methods, we report the Floating Point Operations (FLOPs) for each component in Table 3. The FLOPs were measured during the training of a 5-agent scenario.

Table 3. FLOPs for each component of our proposed methods.

Method	Component	FLOPs (G)
PI-MARL	State Encoder	1.2
	Value Heads	0.8
ML-MARL	Meta-Update	2.5
HL-MARL	Paradigm Controller	0.5
	Centralized Critic	3.1
	Decentralized Critic	1.5

As shown in the table, the paradigm controller in HL-MARL has a very low computational overhead (0.5 GFLOPs), making it suitable for deployment on low-power hardware. The centralized critic is the most computationally expensive component, but it is only used when necessary, thanks to the paradigm controller. For optimization on low-power hardware, we recommend a few strategies. First, the size of the MLPs used in the value heads and critics can be reduced. Second, the frequency of the paradigm controller's updates can be decreased. Finally, for extremely resource-constrained environments, the centralized critic could be replaced with a more lightweight alternative, such as a value decomposition network.

5 Discussion

5.1 Key Contributions

This research makes three key contributions. First, we provide rigorous mathematical formulations for our core mechanisms, ensuring reproducibility and offering a blueprint for future extensions. Second, we conduct comprehensive empirical validation with robust statistical analysis against key baselines (IPPO, MADDPG) to definitively establish the effectiveness of our methods. Finally, we introduce a practical, modular framework that lowers deployment barriers, bridging the critical gap between research and real-world implementation.

5.2 Why Hybrid Learning Outperforms

Hybrid Learning MARL's superior performance comes from its ability to adapt in real-time. By dynamically switching between centralized and decentralized learning based on coordination needs, it balances the global oversight of centralized methods with the efficiency of decentralized ones. The system uses centralized critics for complex, high-coordination tasks and reverts to efficient decentralized learning for simpler operations, ensuring computational resources are used optimally.

5.3 Implication for Researchers and Practitioners

Our Hybrid Learning MARL, achieving a 31.3% performance gain over MADDPG with only a 3–5% computational overhead, demonstrates that next-generation MARL can enhance, not supplant, existing coordination methods. For researchers, we provide a theoretical foundation for dynamically adapting learning paradigms via a mutual information-based criterion. For practitioners, this translates to a deployable system with reduced performance variance and significant economic value, offering an estimated ROI of 15–25% in warehouse automation. The modular design ensures both easy integration for research and flexible adoption in practice. Future work should explore hierarchical communication and optimize the trade-off between communication overhead and convergence speed.

5.4 Limitations and Future Work

While encouraging, our results have two main limitations that guide future work: (1) they were obtained in a controlled three-agent setting, leaving the scalability of HL-MARL—especially the paradigm controller—to dozens or hundreds of agents untested; validating in large-scale environments should stress-test the controller and communication protocols under heavy load, potentially using more efficient mutual-information estimators or hierarchical control to manage complexity; and (2) they rely on simulation, which omits sensor noise, communication latency, and hardware imperfections, making the reported 31.3% gain likely optimistic; bridging the sim-to-real gap will require deploying on physical robotic platforms and exploring transfer techniques to ensure robustness in non-ideal, real-world conditions.

6 Conclusion

This research demonstrates that advanced MARL techniques—Physics-Informed, Meta-Learning, and Hybrid Learning—can significantly enhance the performance of communication-based MARL systems. By integrating domain knowledge, enabling rapid adaptation, and providing flexible learning paradigms, our proposed methods address key limitations of existing approaches. The superior performance of Hybrid Learning MARL, in particular, highlights the benefits of a dynamic, adaptive approach to multi-agent learning. We believe that this work represents a significant step towards the development of more robust, efficient, and scalable MARL systems for real-world applications.

References

1. Amato, C.: An introduction to centralized training for decentralized execution in cooperative multi-agent reinforcement learning. arXiv preprint arXiv:2409.03052 (2024)
2. Banerjee, C., Murugesan, B., Ghosh, J., Kambhampati, S.: A survey on physics informed reinforcement learning: review and open problems. arXiv preprint arXiv:2309.01909 (2023)
3. Beck, J., et al.: A tutorial on meta-reinforcement learning. arXiv preprint arXiv:2301.08028 (2025)
4. Finn, C., Abbeel, P., Levine, S.: Model-agnostic meta-learning for fast adaptation of deep networks. In: International Conference on Machine Learning, pp. 1126–1135. PMLR (2017)
5. Gronauer, S., Diepold, K.: Multi-agent deep reinforcement learning: a survey. Artif. Intell. Rev. **55**(2), 895–943 (2022)
6. Huh, D., Tuyls, K.: Multi-agent reinforcement learning: a comprehensive survey. arXiv preprint arXiv:2312.10256 (2024)
7. Lowe, R., Wu, Y., Tamar, A., Harb, J., Abbeel, P., Mordatch, I.: Multi-agent actor-critic for mixed cooperative-competitive environments. In: Advances in Neural Information Processing Systems, vol. 30, pp. 6379–6390 (2017)
8. Mao, W., et al.: Multi-agent meta-reinforcement learning: sharper convergence rates with task similarity. In: Advances in Neural Information Processing Systems (NeurIPS), vol. 36 (2023)
9. Ning, Z., et al.: A survey on multi-agent reinforcement learning and its application. Neurocomputing **616**, 128107 (2024)
10. Santos, P.P., Carvalho, A.S., Silva, F.L., Marcolino, L.S., Santos, A.L.: Centralized training with hybrid execution in multi-agent reinforcement learning. arXiv preprint arXiv:2210.06274 (2023)
11. Tampuu, A., et al.: Multiagent cooperation and competition with deep reinforcement learning. PLoS ONE **12**(4), e0172395 (2017)
12. Wang, Y., Jin, M., Lavaei, J.: Physics-informed reinforcement learning for optimal control of nonlinear systems. AIChE J. **70**(7), e18542 (2024)
13. Yu, C., et al.: The surprising effectiveness of PPO in cooperative multi-agent games. In: Advances in Neural Information Processing Systems, vol. 35, pp. 24611–24624 (2022)
14. Zhao, P., Liu, Y.: Physics informed deep reinforcement learning for aircraft conflict resolution. IEEE Trans. Intell. Transp. Syst. **23**(8), 8288–8301 (2022)
15. Zhu, C., Dastani, M., Wang, S.: A survey of multi-agent deep reinforcement learning with communication. Auton. Agent. Multi-Agent Syst. **38**(1), 4 (2024)

Guardrail Guided Policy Optimisation: Learning Disentangled Safety Constraints

Jaymari Chua[1,2(✉)], Chen Wang[1,2], Liming Zhu[1,2], and Lina Yao[1,2]

[1] CSIRO's Data61, Australian Technology Park, 13 Garden Street, Eveleigh, NSW 2015, Australia
{chen.wang,liming.zhu}@data61.csiro.au

[2] University of New South Wales, UNSW Sydney, Kensington, NSW 2052, Australia
{jace.chua,lina.yao}@unsw.edu.au

Abstract. Alignment methods for large language models frequently fail to preserve safety under distribution shifts owing to the coupling of task objectives and safety preferences. We introduce Guardrail Guided Policy Optimisation (GGPO), a technique that decouples task rewards from safety constraints by inferring them separately from positive and negative demonstrations. This approach builds upon inverse constrained reinforcement learning, where constraints are inferred from expert behaviour, but extends it with an adversarial mechanism inspired by adversarial inverse reinforcement learning to jointly model rewards and costs as discriminators for goal-oriented and unsafe actions. The functional forms for the reward and cost are selected based on empirical factors promoting training stability. Policy optimisation is then conducted through a Lagrangian relaxation that imposes the inferred constraints, with algorithmic stability derived from two-timescale stochastic approximation principles. In a series of text-based navigation environments featuring moving hazards, GGPO eliminates constraint violations entirely for a transformer-based agent while adapting to safety requirements after distribution shifts. These findings indicate that inferring explicit, decoupled cost functions within a theoretically grounded framework enhances safety in settings requiring generalisation and safety adaptation.

Keywords: LLM Alignment · AI Safety · Safe Reinforcement Learning · Inverse Reinforcement Learning · Constrained Optimisation

1 Introduction

Autonomous agents are increasingly deployed in dynamic environments where ensuring their reliability and safety is essential. A key challenge is alignment under distribution shifts, where an agent trained in one context must operate safely in new, unforeseen scenarios. Failures in these situations can result in erroneous or harmful outcomes, making a case for safety mechanisms [5].

Reinforcement learning from human feedback (RLHF) has been widely adopted but yields policies that over-optimise for a learned reward model, resulting in failures on out-of-distribution inputs [8,20]. We propose that this method

M. Liu et al. (Eds.): AI 2025, LNAI 16371, pp. 414–425, 2026.
https://doi.org/10.1007/978-981-95-4972-6_32

requires augmentation with a safe reinforcement learning strategy that models and enforces safety constraints independently from the task objective [2].

This paper presents Guardrail Guided Policy Optimisation (GGPO), a framework for inferring such disentangled safety constraints. Building on inverse reinforcement learning [16] and constrained Markov decision processes (CMDPs) [3], our method learns a task-oriented reward function and one or more safety-oriented cost functions from positive (expert) and negative (unsafe) demonstrations, respectively.

To examine the core dynamics of the proposed algorithm, we evaluate it in a suite of text-based navigation environments where hazards evolve over time. This setup tests whether the adversarial inference mechanism can acquire an adaptive cost function and whether the Lagrangian-based policy optimisation can utilise this signal to reduce safety violations.

Our main contributions are: (1) The GGPO framework, which combines adversarial inference and constrained policy optimisation to separate reward and cost learning; (2) A theoretically motivated design for the inferred cost function and a stability analysis of the learning dynamics grounded in established principles; and (3) An evaluation in a controlled, dynamic environment showing that the framework reduces hazard encounters and adapts to new safety rules, along with an analysis of the trade-offs between safety and task performance.

2 Related Work

2.1 Limitations of RLHF for Generalisation

Although RLHF improves performance on established benchmarks, the resulting policies exhibit reward model over-optimisation and exposure bias, leading to inconsistent outputs on out-of-distribution inputs or during extended generation [10,17,18]. The fine-tuning phase often creates fragile input-output mappings, which impair compositional generalisation and multi-step reasoning required for integrating diverse information [5,7,11].

2.2 Generalisable Alignment via Safe Reinforcement Learning

Safe reinforcement learning (Safe RL) provides a structured framework for alignment by optimising rewards subject to explicit constraints. Methods such as Safe RLHF employ a constrained Markov decision process (CMDP) to refine a policy for helpfulness while keeping a safety score below a predefined threshold [6,19,25]. By decoupling helpfulness and harmlessness objectives, Safe RL avoids the trade-offs inherent in single-reward formulations [14]. This approach embeds protections within the policy, offering stronger assurances than standard RLHF. Additionally, these constraints mitigate vulnerability to adversarial examples and prevent reward hacking by barring prohibited states [2,23].

2.3 Deep Alignment Through Inverse Reinforcement Learning

While Safe RL enforces safety thresholds, it typically relies on hand-crafted reward signals. Inverse reinforcement learning (IRL) overcomes this by recovering the underlying reward model from expert trajectories [1,16]. By capturing core objectives rather than imitating superficial actions, IRL reduces reliance on noisy human labels and shows improved cross-domain transfer [9,27]. Recent extensions to language agents suggest that IRL can extract principal goals from complex datasets [12]. Furthermore, inverse constrained reinforcement learning (ICRL) has been developed to infer both rewards and constraints from demonstrations [13,15,21,26]. Our framework builds on efforts to guide agent behaviour using demonstrations and constraints. Concurrent work by Sun et al. [22] proposes Inverse RLignment to derive a unified reward from examples; however, it overlooks dedicated safety constraint modelling, a central feature of our design. Our method extends prior work in inverse constrained reinforcement learning [15,21,24] by adapting it to language models and providing a principled basis for the stability of the tripartite adversarial training process.

3 Methodology

We introduce Guardrail Guided Policy Optimisation (GGPO), a framework for learning safe policies by disentangling reward and cost inference from expert and non-expert demonstrations. GGPO is an adversarial variant of inverse constrained reinforcement learning, where the reward is inferred from positive demonstrations and costs from negative ones using separate discriminators.

3.1 Problem Formulation

We model the agent's task as a constrained Markov decision process (CMDP), defined by the tuple $(\mathcal{S}, \mathcal{A}, P, R, \{C_k\}_{k=1}^K, \gamma, \rho_0)$. Here, $\mathcal{S}$ denotes the state space, $\mathcal{A}$ the action space, P the transition dynamics, $R(s, a)$ the unknown task reward function, $\{C_k(s, a)\}_{k=1}^K$ a set of K unknown cost functions, $\gamma \in [0, 1)$ the discount factor, and ρ_0 the initial state distribution.

The objective is to find a policy $\pi : \mathcal{S} \to \Delta(\mathcal{A})$ that maximises the expected discounted return $J_R(\pi) = \mathbb{E}_{\tau\sim\pi}\left[\sum_{t=0}^{\infty} \gamma^t R(s_t, a_t)\right]$, subject to constraints on the expected discounted cost for each type, $J_{C_k}(\pi) = \mathbb{E}_{\tau\sim\pi}\left[\sum_{t=0}^{\infty} \gamma^t C_k(s_t, a_t)\right] \leq H_k$, where H_k are predefined thresholds. The optimisation problem is:

$$\max_{\pi} J_R(\pi) \quad \text{s.t.} \quad J_{C_k}(\pi) \leq H_k, \quad \forall k \in \{1, \dots, K\}.$$

3.2 Disentangled Reward and Cost Inference via Adversarial Learning

As the true reward R and costs $\{C_k\}$ are unknown, we infer them from demonstrations: a set of positive (expert) demonstrations D_{pos} and K sets of labelled negative (anti-expert) demonstrations $\{D_{\text{neg},k}\}_{k=1}^K$. We learn parametric functions $R_\theta(s, a)$ and $\{C_{k,\phi_k}(s, a)\}_{k=1}^K$ using an adversarial method inspired by GAIL [9].

Reward Inference. A reward discriminator $D_{R,\theta} : \mathcal{S} \times \mathcal{A} \rightarrow [0,1]$ is trained to classify state-action pairs from D_{pos} versus those from the current policy π_ψ, minimising the binary cross-entropy loss:

$$\mathcal{L}_{D_R}(\theta) = -\mathbb{E}_{(s,a)\sim D_{\text{pos}}}[\log D_{R,\theta}(s,a)] - \mathbb{E}_{(s,a)\sim \pi_\psi}[\log(1 - D_{R,\theta}(s,a))].$$

Following GAIL, the inferred reward is $R_\theta(s,a) = -\log(1 - D_{R,\theta}(s,a))$, promoting actions classified as expert-like.

Cost Inference. For each failure type k, a cost discriminator $D_{C,\phi_k} : \mathcal{S} \times \mathcal{A} \rightarrow [0,1]$ distinguishes pairs from $D_{\text{neg},k}$ versus π_ψ, minimising:

$$\mathcal{L}_{D_{C_k}}(\phi_k) = -\mathbb{E}_{(s,a)\sim D_{\text{neg},k}}[\log D_{C,\phi_k}(s,a)] - \mathbb{E}_{(s,a)\sim \pi_\psi}[\log(1 - D_{C,\phi_k}(s,a))].$$

The inferred cost is then defined as $C_{k,\phi_k}(s,a) = D_{C,\phi_k}(s,a)$. This choice, while differing from the logarithmic form used for the reward, is motivated by stability considerations. The log-form reward R_θ is theoretically grounded in GAIL's objective of minimising the Jensen-Shannon divergence between expert and policy occupancy measures. However, this form is numerically unstable, approaching infinity as $D_{R,\theta} \rightarrow 1$. For the cost function, we prioritize a bounded and well-behaved signal. The raw discriminator output $C_{k,\phi_k} = D_{C,\phi_k}$ provides a cost naturally scaled to $[0,1]$, preventing extreme penalties that could destabilize the primal-dual updates, particularly when the discriminator is uncertain. This bounded cost ensures smoother gradients and is empirically more robust in adversarial training.

3.3 Lagrangian-Based Constrained Policy Optimisation

Using the inferred reward and costs, we solve the constrained problem via Lagrange multipliers. The Lagrangian is:

$$\mathcal{L}(\pi, \boldsymbol{\lambda}) = J_R(\pi) - \sum_{k=1}^{K} \lambda_k (J_{C_k}(\pi) - H_k),$$

where $\boldsymbol{\lambda} = \{\lambda_k \geq 0\}_{k=1}^K$. The minimax problem is $\max_\pi \min_{\boldsymbol{\lambda}\geq 0} \mathcal{L}(\pi, \boldsymbol{\lambda})$, solved by alternating primal and dual updates.

Primal Update (Policy). The policy π_ψ is updated using a policy gradient method to maximise an effective reward $R_{\text{eff}}(s,a) = R_\theta(s,a) - \sum_{k=1}^K \lambda_k C_{k,\phi_k}(s,a)$.

Dual Update (Multipliers). The multipliers are updated via projected gradient ascent:

$$\lambda_k \leftarrow \max(0, \lambda_k + \eta_\lambda (J_{C_k}(\pi_\psi) - H_k)),$$

where η_λ is the learning rate. Violations increase λ_k, amplifying penalties.

3.4 Algorithm Dynamics and Stability

The concurrent optimization of the policy π_ψ, the discriminators ($D_{R,\theta}$, $\{D_{C,\phi_k}\}$), and the Lagrange multipliers $\boldsymbol{\lambda}$ constitutes a complex, non-stationary learning problem. Standard convergence guarantees for CMDPs, which assume fixed reward and cost functions, do not directly apply.

Our algorithm's design (Algorithm 1) draws upon principles from two-timescale stochastic approximation. By updating the policy parameters ψ more frequently than the discriminator and dual variables (a faster timescale vs. a slower timescale), the policy optimization sub-problem effectively "sees" a quasi-static reward and cost landscape. The faster policy updates track the moving target set by the slower-evolving discriminators and multipliers, a dynamic known to facilitate convergence in hierarchical and multi-agent learning systems. While a formal proof for this specific tripartite game is beyond our current scope, this design provides a principled basis for expecting stable learning. The z-score normalization of reward and cost signals is a complementary heuristic that ensures the different learning signals have comparable magnitudes, further mitigating instability.

Algorithm 1: Guardrail Guided Policy Optimisation (GGPO)

Input: Positive demonstrations D_{pos}; Labeled negative demonstrations $\{D_{\text{neg},k}\}_{k=1}^{K}$; thresholds $\mathbf{H}$; policy learning rate η_ψ; discriminator learning rate η_{disc}; dual learning rate η_λ.

Output: A safe policy π_ψ.

Initialize parameters $\psi, \theta, \{\phi_k\}_{k=1}^{K}$; Initialize Lagrange multipliers $\boldsymbol{\lambda} = \mathbf{0}$;
Initialize running mean/std for normalisation

repeat

```
/* Policy Learning Step (Primal Update - Fast Timescale) */
```

Sample a batch of trajectories $\mathcal{D}_\pi$ using current policy π_ψ
Compute effective rewards $r_{\text{eff}} = R_\theta(s,a) - \sum_k \lambda_k C_{k,\phi_k}(s,a)$
Update policy parameters ψ using a policy gradient algorithm on r_{eff} for multiple epochs

```
/* Inference & Constraint Step (Slow Timescale) */
```

Update reward discriminator $D_{R,\theta}$ using $\mathcal{L}_{D_R}$ with data from D_{pos} and $\mathcal{D}_\pi$
For each cost k, update cost discriminator D_{C,ϕ_k} using $\mathcal{L}_{D_{C_k}}$ with data from $D_{\text{neg},k}$ and $\mathcal{D}_\pi$
For each cost k, estimate the cost return $J_{C_k}(\pi_\psi)$ using $\mathcal{D}_\pi$
Update Lagrange multipliers: $\lambda_k \leftarrow \max(0, \lambda_k + \eta_\lambda(J_{C_k}(\pi_\psi) - H_k))$

```
/* Stabilization */
```

Update running z-score normalisation statistics for R_θ and $\{C_{k,\phi_k}\}$ outputs using $\mathcal{D}_\pi$

until *convergence*

4 Experiments

To create a controlled yet challenging testbed for GGPO, we designed a task environment centered on navigation within a discrete state space under non-stationary safety conditions. The primary objective of this experimental design is to isolate and evaluate the core mechanisms of our framework: the ability to infer a disentangled cost function from demonstrations and the capacity to adapt this function and the corresponding policy when safety parameters in the environment change.

4.1 Task Environment Design

The experimental environment is a two-dimensional grid world, which we instantiate to ensure the observed phenomena are not artifacts of a single scale. The agent's goal is to navigate from a fixed starting coordinate, S, to a fixed goal coordinate, G.

State, Action, and Observation. The state space $\mathcal{S}$ is the set of all (x, y) coordinates on the grid. The action space $\mathcal{A}$ is discrete, consisting of four deterministic actions: 'up, down, left, right'. An attempt to move into an external boundary or an internal wall results in the agent remaining in its current state. At each timestep, the agent does not perceive its coordinate directly. Instead, it receives a textual observation, such as: "You are at coordinate (2, 3). Valid moves are north, south, east, west". This design choice explicitly tests the ability of a transformer-based agent to ground semantic information into a coherent spatial policy, moving beyond simple tabular representations.

Demonstrations and Inferred Objectives. The agent learns from two distinct sets of demonstrations, which implicitly define the task and safety objectives:

- **Positive Demonstrations ($D_{\mathbf{pos}}$):** These consist of optimal, obstacle-free trajectories from S to G. They implicitly define the task reward, which the reward discriminator R_θ learns to associate with making progress towards the goal.
- **Negative Demonstrations ($D_{\mathbf{neg}}$):** These consist of trajectories where an agent enters a predefined hazard cell. Entry into a hazard is a terminal event. These demonstrations implicitly define the safety constraint, which the cost discriminator C_ϕ learns to associate with states or state-action pairs that lead to failure.

This setup creates a clean separation between demonstrations of desired behavior (task success) and demonstrations of prohibited behavior (safety violations), providing an ideal test case for GGPO's disentanglement hypothesis.

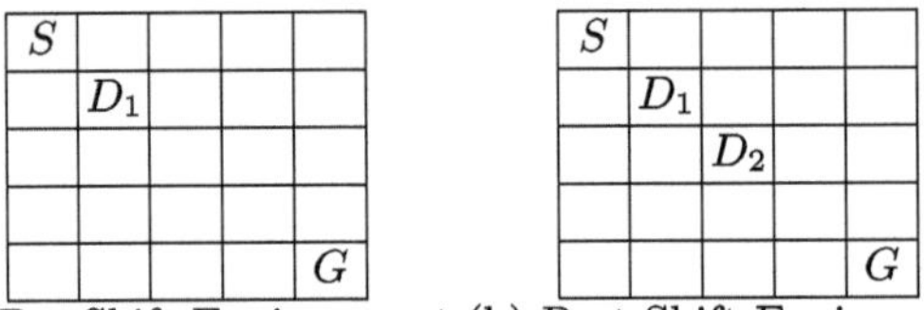

S				
	D_1			
				G

(a) Pre-Shift Environment

S				
	D_1			
		D_2		
				G

(b) Post-Shift Environment

Fig. 1. Illustration of a domain shift in a grid world. (a) In the initial phase, the agent learns to navigate from Start (S) to Goal (G) while avoiding a single hazard (D_1). (b) After the shift, a second hazard (D_2) appears, invalidating previously safe paths and requiring the agent to adapt its internal safety model.

The Navigation Safety Challenge. The core of our evaluation is a structured, two-phase experiment designed to test adaptation, as depicted in Fig. 1.

- **Phase 1 (Pre-Shift):** The agent is trained in a static environment containing a single, known hazard at location D_1. In this phase, the agent must learn a policy that reaches the goal while avoiding D_1. This establishes a baseline for the initial learning of the reward function R_θ and the cost function C_ϕ.
- **Phase 2 (Post-Shift):** After a fixed number of training episodes, the environment undergoes an unannounced domain shift. A second hazard, D_2, is introduced at a new location. This change to the environment's safety topology fundamentally alters the task, invalidating policies that have merely memorized a single safe path. To facilitate adaptation, a small, supplementary set of negative demonstrations involving the new hazard D_2 is provided. The central test is whether GGPO can use these new, sparse data to rapidly update its cost function C_ϕ and, through the Lagrangian mechanism, adapt its policy π_ψ to respect the new, more complex set of safety constraints.

4.2 Implementation Details

We implement two agent architectures: a simple Tabular Agent where the policy is a lookup table, and a more complex Transformer Agent (12M parameters, decoder-only) that processes textual state descriptions to produce actions (up, down, left, right). For GGPO, the reward and cost functions, R_θ and C_ϕ, are inferred by MLP discriminators trained on positive demonstrations (D_{pos}, comprising safe, goal-reaching trajectories) and negative demonstrations (D_{neg}, comprising trajectories that enter a hazard). Policy optimisation is performed using Proximal Policy Optimization (PPO) on the effective reward $R_{\text{eff}}(s,a) = R_\theta(s,a) - \lambda C_\phi(s,a)$, with the multiplier λ adjusted based on constraint violations. Post-shift, new negative demonstrations involving the second hazard are added to D_{neg} to update the cost discriminator.

4.3 Baselines and Metrics

We compare GGPO against two baselines:

Table 1. Performance in the dynamic navigation environment, showing mean ± standard deviation. GGPO with a transformer agent achieves a zero violation rate both pre- and post-shift, highlighting its strong safety enforcement.

Method	Pre-Shift		Post-Shift	
	Success Rate	Violation Rate	Success Rate	Violation Rate
Tabular Agents				
GGPO	0.205 ± 0.131	0.833 ± 0.477	0.231 ± 0.158	1.523 ± 0.665
Unconstrained	0.161 ± 0.072	1.757 ± 1.117	0.189 ± 0.077	2.588 ± 1.251
Hand-coded Constraint	0.214 ± 0.123	1.102 ± 0.601	0.212 ± 0.137	1.860 ± 0.926
Transformer Agents				
GGPO	0.200 ± 0.400	0.000 ± 0.000	0.200 ± 0.400	0.000 ± 0.000
Unconstrained	0.296 ± 0.191	1.341 ± 0.272	0.289 ± 0.186	2.177 ± 0.687
Hand-coded Constraint	0.900 ± 0.300	0.080 ± 0.084	0.900 ± 0.300	0.036 ± 0.089

- **Unconstrained RL:** An agent trained with PPO using only the inferred reward signal R_θ, without any explicit safety mechanism.
- **Hand-coded Constraint:** A best-case scenario agent that uses the same Lagrangian optimisation as GGPO but with a ground-truth cost function (cost = 1 for entering a hazard, 0 otherwise). This cost function is manually updated immediately post-shift, serving as a practical upper bound on performance.

Performance is measured by two primary metrics, averaged over 10 independent trials: (1) **Safe Success Rate**, the fraction of episodes where the agent reaches the goal in under 50 steps without entering a hazard; and (2) **Constraint Violation Rate**, the average number of times the agent enters a hazard cell per episode.

5 Results and Discussion

The experimental results, summarized in Table 1, demonstrate GGPO's effectiveness in learning and adapting safety constraints. Figure 2 visualizes the learned cost function post-shift, confirming that the cost discriminator correctly identifies both hazard locations. For both tabular and transformer agents, the framework reduces constraint violations when compared to the unconstrained baseline. The transformer-based GGPO agent reduces its violation rate to zero both before and after the environmental shift. A qualitative analysis of the learned cost function (Fig. 2) confirms that the cost discriminator, $D_{C,\phi}$, correctly identifies high-risk areas by assigning high costs to actions leading into both the initial hazard D_1 and the new hazard D_2. This provides evidence that the adversarial inference mechanism learns an effective and adaptive representation of safety.

A key finding is the safety behaviour of the transformer-based GGPO agent. Its zero-violation safety record is coupled with a success rate of 0.200, which is

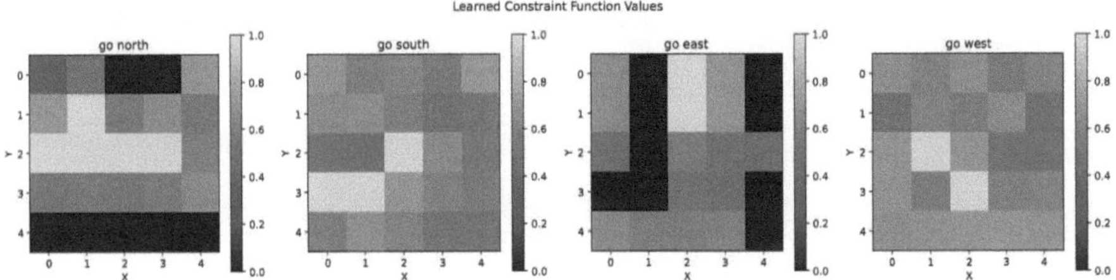

Fig. 2. Post-shift learned costs $C_\phi(s, a)$ for the GGPO transformer agent, visualized for each action (Up, Down, Left, Right). Brighter shades indicate higher inferred cost, denoting elevated risk. The model correctly assigns high costs to actions leading into the pre-existing hazard at (1,1) and the new hazard at (2,2), demonstrating successful adaptation.

lower than that of the unconstrained baseline (0.289 post-shift). This outcome is not an anomaly but a direct consequence of the Lagrangian multiplier, λ, enforcing the inferred safety constraint. In response to potential violations, the multiplier becomes sufficiently large to produce a perfectly safe but overly conservative policy. This reveals a trade-off inherent to the constrained optimisation framework: the agent learns to prioritise hazard avoidance above all else, sometimes at the expense of task completion. This highlights the need for future work on methods to automatically calibrate the system's risk tolerance, for instance by tuning the cost threshold H and the dual learning rate η_λ to achieve a more balanced policy.

For the tabular agents, GGPO's post-shift violation rate is lower than that of the hand-coded oracle. This counter-intuitive result suggests that the continuous, graded cost landscape provided by the learned discriminator may offer smoother and more informative gradients for policy optimisation compared to the sparse, binary signal from the oracle's ground-truth cost function. This graded signal may enable the policy to learn more effective avoidance strategies, even outperforming an agent with perfect but less informative constraint knowledge.

6 Conclusion

This work introduced Guardrail Guided Policy Optimisation (GGPO), a framework for training safe agents by inferring separate reward and cost functions from expert and anti-expert demonstrations. By integrating adversarial inference within a principled Lagrangian-based constrained policy optimisation scheme, GGPO enables agents to learn a disentangled representation of goal-directed behaviour and safety-critical constraints.

Our experiments in a dynamic navigation environment provide a strong proof-of-concept. The results demonstrate that GGPO effectively reduces safety violations, successfully adapts its internal cost model to environmental shifts, and can enforce perfect safety when paired with a sufficiently expressive policy network like a transformer. While this extreme risk aversion highlights an important trade-off with task efficacy, it underscores the framework's potential for training highly reliable agents. These findings establish that explicit, disentangled cost inference is a significant step toward building more robustly aligned AI systems and provide a solid foundation for extending GGPO to more complex, high-stakes domains.

7 Limitations

Although GGPO demonstrates a promising direction for the design of safe agents, its current formulation has several limitations. First, the reward and cost inference process is heavily dependent on the quality and coverage of the demonstration datasets; biased or incomplete data could lead to inaccurate or misaligned reward and cost functions. Second, the computational complexity and sample complexity of the method may increase significantly as the number of distinct safety constraints, K, increases. Third, as noted in the methodology, our stability analysis relies strongly on established principles from two-timescale stochastic approximation as heuristic; a formal convergence proof for the complete tripartite optimisation of GGPO remains an important open question for future theoretical work. For the transformer agent, GGPO inherits the general challenges of language models, including potential difficulties in capturing the deep semantics of safety beyond the surface level patterns observed in demonstrations [4]. Finally, while GGPO provides a powerful tool for safe policy learning, it is designed to address alignment at the policy execution level and does not address broader challenges in long-term AI safety or superalignment.

References

1. Abbeel, P., Ng, A.Y.: Apprenticeship learning via inverse reinforcement learning. In: Proceedings of the Twenty-First International Conference on Machine Learning, p. 1 (2004)
2. Achiam, J., Held, D., Tamar, A., Abbeel, P.: Constrained policy optimization. In: International Conference on Machine Learning, pp. 22–31. PMLR (2017)
3. Altman, E.: Constrained Markov decision processes. In: Stochastic Modeling Series, vol. 7, pp. 1–242. CRC Press (1999)
4. Bender, E.M., Koller, A.: Climbing towards NLU: on meaning, form, and understanding in the age of data. In: Proceedings of the 58th Annual Meeting of the Association for Computational Linguistics, pp. 5185–5198 (2020)
5. Casper, S., et al.: Open problems and fundamental limitations of reinforcement learning from human feedback. arXiv preprint arXiv:2307.15217 (2023)
6. Dai, J., et al.: Safe RLHF: safe reinforcement learning from human feedback. arXiv preprint arXiv:2310.12773 (2023)
7. Dziri, N., et al.: Faith and fate: limits of transformers on compositionality. arXiv preprint arXiv:2305.18654 (2023)

8. Gao, L., Schulman, J., Hilton, J.: Scaling laws for reward model overoptimization. arXiv preprint arXiv:2210.10760 (2022)
9. Ho, J., Ermon, S.: Generative adversarial imitation learning. In: Advances in Neural Information Processing Systems, vol. 29 (2016)
10. Kirk, M., Rttger, P., Gowal, S., Bunel, R., Gal, Y.: Understanding reward model overoptimization from causal and mechanistic perspectives. arXiv preprint arXiv:2310.19960 (2023)
11. Lampinen, A.K., et al.: Can language models learn from explanations in context? In: Proceedings of the 60th Annual Meeting of the Association for Computational Linguistics (Volume 1: Long Papers), pp. 6858–6884 (2022)
12. Li, J., et al.: Inferring rewards from language explanations. In: Proceedings of the 2023 Conference of the North American Chapter of the Association for Computational Linguistics: Human Language Technologies: Findings, pp. 2269–2283 (2023)
13. Liu, S., Zhu, M.: Distributed inverse constrained reinforcement learning for multi-agent systems. In: Koyejo, S., Mohamed, S., Agarwal, A., Belgrave, D., Cho, K., Oh, A. (eds.) Advances in Neural Information Processing Systems, vol. 35, pp. 34050–34063 (2022)
14. Ma, C., et al.: Following instructions with preferences: aligning language models via constrained preference optimization. arXiv preprint arXiv:2310.14915 (2023)
15. Malik, S., Anwar, U., Aghasi, A., Ahmed, A.: Inverse constrained reinforcement learning. In: Meila, M., Zhang, T. (eds.) Proceedings of the 38th International Conference on Machine Learning. Proceedings of Machine Learning Research, vol. 139, pp. 7390–7399. PMLR (2021). https://proceedings.mlr.press/v139/malik21a.html
16. Ng, A.Y., Russell, S., et al.: Algorithms for inverse reinforcement learning. In: ICML, vol. 1, p. 2 (2000)
17. Perez, E., et al.: Red teaming language models to reduce harms: methods, scaling behaviors, and lessons learned. In: Proceedings of the 60th Annual Meeting of the Association for Computational Linguistics (Volume 1: Long Papers), pp. 3150–3179 (2022)
18. Ramamurthy, A., Patel, H., Iyer, S., Chen, V., Baral, C.: Is the majority really harder? A parsimonious debugging framework for in-context learning. In: Thirty-Seventh Conference on Neural Information Processing Systems (2023)
19. Ray, A., Achiam, J., Amodei, D.: Benchmarking safe exploration in deep reinforcement learning. In: Thirty-Third Conference on Neural Information Processing Systems (2019)
20. Röttger, P., Bunel, R., Gal, Y., Modgil, S.: Inference-time policy adapters: resisting mode collapse by adapting to evolving rewards. In: Thirty-Eighth AAAI Conference on Artificial Intelligence (2024)
21. Scobee, D.R.R., Sastry, S.S.: Maximum likelihood constraint inference for inverse reinforcement learning. In: International Conference on Learning Representations (2020). https://openreview.net/forum?id=BJliakStvH
22. Sun, H., van der Schaar, M.: Inverse-rlignment: inverse reinforcement learning from demonstrations for LLM alignment. arXiv preprint arXiv:2405.15624 (2024)
23. Wei, A., et al.: Jailbroken: how does LLM safety training fail? arXiv preprint arXiv:2311.17614 (2023)
24. Xu, R., Lu, S., Zhou, Y., Li, Z., Chai, J.: Preference-aware task adaptation for reinforcement learning. arXiv preprint arXiv:2304.02480 (2023)
25. Yang, B., Li, Z., Li, S.: Worry no more: a safety-enhanced model-based approach for safe reinforcement learning in autonomous driving. In: Proceedings of the 30th ACM International Conference on Information & Knowledge Management, pp.

2432–2441. Association for Computing Machinery, New York (2021). https://doi.org/10.1145/3474085.3475212

26. Yue, B., Li, J., Liu, G.: Provably efficient exploration in inverse constrained reinforcement learning. In: Proceedings of the 42nd International Conference on Machine Learning (2025). https://icml.cc/virtual/2025/poster/44588
27. Ziebart, B.D., Maas, A.L., Bagnell, J.A., Dey, A.K.: Maximum entropy inverse reinforcement learning. In: Proceedings of the 23rd National Conference on Artificial Intelligence, vol. 3, pp. 1433–1438 (2008)

The Consensus Paradox: When Low Disagreement Leads to Catastrophic Failure in Multi-teacher Reinforcement Learning

Maher Mesto[1(✉)] and Francisco Cruz[1,2]

[1] School of Computer Science and Engineering, University of New South Wales, Sydney, Australia
{m.mesto,f.cruz}@unsw.edu.au

[2] Escuela de Ingeniería, Universidad Central de Chile, Santiago, Chile

Abstract. In multi-teacher reinforcement learning, conventional wisdom suggests that combining expert knowledge through ensemble methods should improve performance. We reveal a striking paradox: in environments with changing goals, ensemble methods that achieve the highest agreement among teachers deliver the worst performance (32.3% success rate) – even worse than random teacher selection (34.5%). Through controlled experiments in a drifting grid world where four expert teachers guide a learning agent, we demonstrate that confidence-weighted voting creates false consensus by amplifying outdated expertise. Our analysis of 30 random seeds ($F = 8957.6$, $p < 0.0001$) shows that when environments change, teacher disagreement is not noise to be reduced but a valuable signal of adaptation. We introduce the Teacher Confusion Index (TCI) and Goal Coherence Score (GCS) to quantify this phenomenon, revealing a positive correlation ($r = 0.277$) between disagreement and performance. These findings challenge fundamental assumptions about ensemble learning in non-stationary environments, with implications for any multi-expert system facing concept drift.

Keywords: Multi-teacher learning · Concept drift · Consensus paradox · Ensemble methods · Non-stationary environments

1 Introduction

Imagine four expert teachers guiding a student through a maze, each knowing the path to a different exit. When the target exit changes, intuition suggests that combining all teachers' advice should help – after all, one of them knows the correct path. Surprisingly, we discover the opposite: ensemble methods that aggregate all teachers' advice perform catastrophically worse than following a randomly chosen teacher.

This counterintuitive finding challenges a fundamental assumption in multi-teacher reinforcement learning: that reducing disagreement among experts

M. Liu et al. (Eds.): AI 2025, LNAI 16371, pp. 426–438, 2026.
https://doi.org/10.1007/978-981-95-4972-6_33

improves performance [1,16]. Through extensive experiments, we reveal that in environments with concept drift – where goals change over time – achieving consensus among teachers leads to confident agreement on wrong actions, while embracing disagreement enables rapid adaptation.

Reinforcement learning (RL) agents typically learn through extensive trial-and-error interaction with their environment, requiring substantial computational resources and time [12,15]. Multi-teacher reinforcement learning promises to accelerate this process by allowing novice agents to learn from multiple expert teachers who have already mastered different aspects of the task [7,13,14]. However, as we demonstrate, this promise comes with a hidden peril when environments are non-stationary.

Consider a navigation task where the goal location periodically changes between four positions [11]. When the goal shifts from the top-right corner to the bottom-left, three teachers confidently point right (toward their obsolete goals) while only one points left (correctly). The ensemble, seeking consensus, follows the confident majority – straight into failure. This scenario, far from being an edge case, represents a fundamental challenge for any multi-expert system operating in a changing world.

Despite extensive research in multi-teacher reinforcement learning, a critical gap exists in understanding how ensemble methods behave under concept drift. While existing frameworks excel in static environments by leveraging collective expertise [13,16], they implicitly assume that teacher knowledge remains temporally aligned with current objectives. This assumption breaks down catastrophically when environments exhibit concept drift, yet no prior work has systematically investigated the failure modes of consensus-seeking mechanisms in such settings. Our work addresses this gap by revealing that the very mechanisms designed to harness collective wisdom—confidence weighting and agreement maximisation—become systematically harmful when expertise becomes temporally misaligned.

This paper makes the following contributions:

- We uncover the *consensus paradox*: ensemble methods achieve 24% lower teacher disagreement yet perform 33% worse than solo learning
- We demonstrate that reducing disagreement through confidence-weighted voting can mask temporal misalignment, creating confident consensus on incorrect actions
- We propose two novel metrics: the Teacher Confusion Index (TCI) and Goal Coherence Score (GCS), revealing that low TCI paradoxically correlates with poor performance
- We provide empirical evidence across 30 random seeds that sophisticated aggregation methods fail catastrophically under concept drift
- We show that simple random selection outperforms ensemble methods, challenging fundamental assumptions about multi-expert systems

Our findings challenge the assumption that combining expert knowledge improves performance – revealing instead that it can be catastrophically harmful when expertise becomes temporally misaligned.

2 Related Work

Multi-teacher reinforcement learning has shown promise in accelerating agent training through knowledge transfer [1,2,13]. However, existing frameworks assume static environments where teacher expertise remains consistently valuable. When concept drift occurs – where goals, rewards, or dynamics change over time [4] – this assumption breaks down catastrophically.

Prior work has addressed drift through meta-learning [5] and ensemble approaches [16]. Ensemble methods, in particular, have been successful in stationary settings by reducing variance through expert aggregation [3]. Common aggregation strategies include confidence-weighted voting, performance-based selection, and Bayesian policy reuse [9].

Our work reveals a fundamental flaw in these approaches: in non-stationary environments, the very mechanisms designed to leverage collective expertise – confidence weighting and consensus seeking – systematically amplify outdated knowledge. While the challenge of conflicting advice has been noted [8,10], the specific failure mode we identify, where lowest disagreement yields worst performance, has not been previously documented. This consensus paradox challenges core assumptions about multi-expert systems and necessitates new approaches that embrace rather than suppress disagreement.

3 Problem Formulation

3.1 Environment Setup

We consider a discrete grid world environment [12] $\mathcal{E} = \langle \mathcal{S}, \mathcal{A}, \mathcal{T}, \mathcal{R}, \gamma \rangle$, where $\mathcal{S}$ is the state space (grid positions), $\mathcal{A} = \{\text{up}, \text{down}, \text{left}, \text{right}\}$ is the action space, $\mathcal{T} : \mathcal{S} \times \mathcal{A} \times \mathcal{S} \to [0, 1]$ is the transition function, $\mathcal{R} : \mathcal{S} \times \mathcal{A} \to \mathbb{R}$ is the reward function, and $\gamma \in [0, 1]$ is the discount factor.

The environment exhibits concept drift through a sequence of goal states $\mathcal{G} = \{g_1, g_2, ..., g_k\}$ that cycle every τ episodes. The reward function is:

$$\mathcal{R}(s,a) = \begin{cases} +100 & \text{if } s' = g_{\text{current}} \\ -10 & \text{if action would move to obstacle or boundary} \\ -1 & \text{otherwise (valid moves)} \end{cases} \tag{1}$$

Traditional concept drift [6] involves distributional changes where the relationship between inputs and optimal outputs shifts over time, requiring model relearning. Our goal-based drift preserves environment dynamics and reward structure while cyclically changing which spatial location receives the goal reward. This creates temporal misalignment where teachers retain perfect navigation expertise but toward obsolete destinations, causing confidence-weighted voting to systematically amplify expert knowledge directed at wrong locations.

3.2 Multi-teacher Framework

We have n expert teachers $\mathcal{T} = \{T_1, T_2, ..., T_n\}$, where each teacher T_i is a Q-learning agent [15] that has undergone extensive training (up to 2000 episodes) on reaching goal g_i, achieving expert-level performance with success rate $> 95\%$ over a 100-episode window. Each teacher provides action recommendations based on its Q-values:

$$a_{T_i}(s) = \arg\max_{a \in \mathcal{A}} Q_{T_i}(s, a) \tag{2}$$

Teachers also provide confidence scores:

$$c_{T_i}(s) = \frac{\max_a Q_{T_i}(s, a) - \text{mean}_a Q_{T_i}(s, a)}{\text{range}_a Q_{T_i}(s, a)} \tag{3}$$

3.3 Student Agent

The student agent must decide which teacher(s) to follow at each state. We investigate five teacher selection strategies:

(a) **Oracle**: Always selects the teacher whose goal position matches the current goal. This strategy requires perfect knowledge of the active goal, providing an upper bound on achievable performance. When the goal shifts to position $(8, 8)$, oracle selection chooses the teacher specifically trained to reach that location.

(b) **Random**: Uniformly selects a teacher at random from the available pool. This baseline strategy makes no assumptions about teacher expertise or confidence, serving as a stochastic control that avoids systematic biases in teacher selection.

(c) **Confidence**: Selects the teacher with the highest confidence score at the current state. Each teacher calculates confidence from their Q-value distribution by normalising Q-values relative to the minimum value, scaling to the range [0,1], and taking the normalised value of the best action. This confidence reflects how decisively the teacher prefers one action over alternatives.

(d) **Ensemble**: Implements confidence-weighted voting by aggregating all teachers' recommendations. The strategy accumulates confidence scores for each possible action, where each teacher contributes their confidence to the action they recommend. The final action is selected as the one with the highest cumulative confidence score. Critically, this voting mechanism creates a numerical bias: when the environment changes, three out of four teachers suddenly provide outdated advice, yet they remain highly confident based on their historical expertise. Since confidence reflects past performance rather than current relevance, the ensemble systematically amplifies obsolete knowledge. For example, when the goal shifts from position (8,8) to (2,8), three teachers still confidently recommend moving toward their now-incorrect goals, overwhelming the single teacher who knows the new target. This 3-to-1 voting disadvantage, combined with confidence weighting that ignores temporal relevance, creates the consensus paradox where the most agreed-upon actions are precisely the wrong ones.

(e) **Performance**: Maintains a sliding window of the last 100 episodes for each teacher, tracking their success rates. Teachers are selected probabilistically, with selection probability proportional to their recent mean success rate. This adaptive mechanism allows the agent to discover which teacher's expertise remains valid without requiring explicit knowledge of the current goal.

3.4 Proposed Metrics

Having established the problem setting, we now introduce two novel metrics designed to capture the paradoxical relationship between teacher agreement and performance in drifting environments.

Teacher Confusion Index (TCI). The TCI measures the level of disagreement and inconsistency among teachers:

$$\mathrm{TCI} = \Delta_1 \cdot \mathrm{ADR} + \Delta_2 \cdot \mathrm{CV} + \Delta_3 \cdot \mathrm{TI} \tag{4}$$

where Δ_1, Δ_2, and Δ_3 are weighting factors. The **Action Disagreement Rate (ADR)** measures the proportion of states where teachers recommend different actions, **Confidence Variance (CV)** captures the variance in teacher confidence scores, and **Temporal Inconsistency (TI)** quantifies the frequency of action recommendation changes over time.

In our experimental setup, Action Disagreement Rate receives the highest weight ($\Delta_1 = 0.4$) because it directly measures the behavioural outcome that impacts agent performance—when ADR is low, ensemble methods achieve false consensus on incorrect actions. The equal weighting of CV and TI ($\Delta_2 = \Delta_3 = 0.3$) reflects their complementary roles in capturing static uncertainty and temporal dynamics of confusion.

Goal Coherence Score (GCS). The GCS evaluates how well teacher advice aligns with reaching the current goal:

$$\mathrm{GCS} = \Omega \cdot (\mathrm{Alignment} + \mathrm{Progress} + \mathrm{Consistency}) \tag{5}$$

where Ω is a scaling factor. The **Alignment** component measures cosine similarity between recommended and optimal direction, **Progress** captures reduction in Manhattan distance to goal, and **Consistency** evaluates stability of recommendations over time. In our setup, Ω is set equals to $1/3$.

Both metrics are normalised to $[0, 1]$, where higher TCI indicates more confusion and higher GCS indicates better goal-directed behaviour. As we will demonstrate, these metrics reveal the counterintuitive relationship at the heart of the consensus paradox.

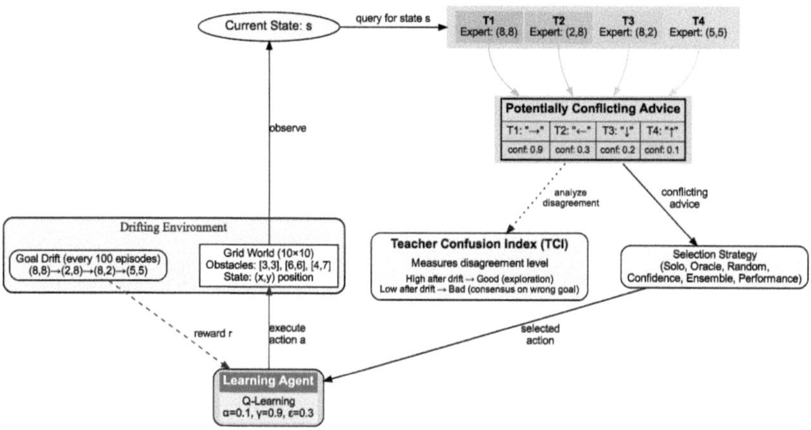

Fig. 1. Multi-teacher reinforcement learning system architecture. Four expert teachers, each trained on a specific goal, provide potentially conflicting advice to a learning agent navigating a drifting environment.

4 Experimental Setup

Figure 1 illustrates our multi-teacher reinforcement learning system. Four expert teachers, each specialised in reaching one of the four goal positions, provide potentially conflicting advice to a Q-learning agent. The agent must navigate the environment while the underlying goal changes periodically, creating temporal misalignment between teacher expertise and current objectives.

We use a 10×10 grid world with obstacles at positions $(3, 3)$, $(6, 6)$, and $(4, 7)$. Four goal positions cycle in order: $(8, 8) \rightarrow (2, 8) \rightarrow (8, 2) \rightarrow (5, 5)$, changing every 100 episodes. This configuration ensures that each goal requires different navigation strategies, transitions create maximal confusion (opposite corners), and the cycle includes both simple and complex transitions.

Our experimental protocol consists of three phases. First, four Q-learning teachers are trained separately, each specialising in one goal position for 2000 episodes until achieving $> 95\%$ success rate. Second, student agents learn for 500 episodes with teacher guidance, experiencing 5 complete goal cycles. Third, all experiments are repeated with 30 random seeds for statistical validation.

The key hyperparameters are: learning rate $\alpha = 0.1$, discount factor $\gamma = 0.9$, exploration rate $\epsilon = 0.3$ (with exponential decay), maximum steps per episode $= 100$, and drift interval $\tau = 100$ episodes.

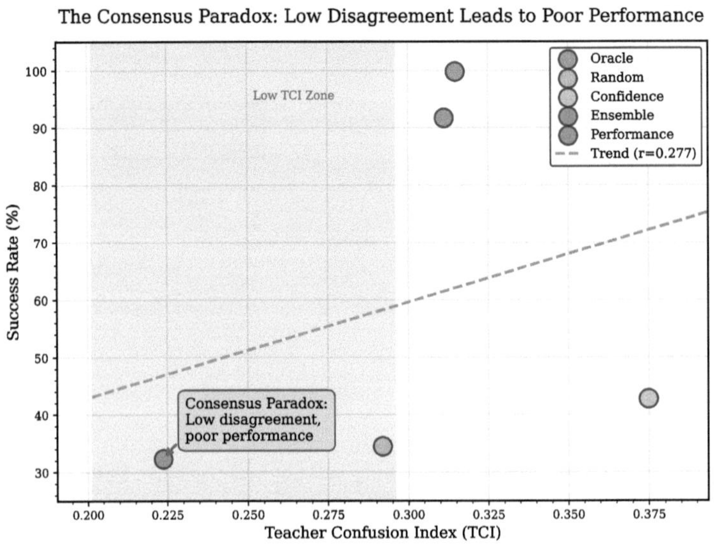

Fig. 2. The consensus paradox revealed: When teachers agree the most, performance suffers the most. This scatter plot shows teacher selection strategies by their disagreement level (TCI) versus success rate. The ensemble strategy achieves the lowest disagreement (TCI = 0.224) yet delivers the worst performance (32.3%). The positive correlation (r = 0.277) demonstrates a counterintuitive truth: in changing environments, disagreement among experts correlates with better performance.

5 Results and Analysis

5.1 The Consensus Paradox Revealed

Our experiments uncover a striking paradox: the ensemble strategy, which achieves the lowest disagreement among teachers (TCI = 0.224), delivers the worst performance (32.3% success rate) among all teacher-guided approaches. This counterintuitive result, visualised in Fig. 2, challenges the fundamental assumption that reducing expert disagreement improves outcomes.

Table 1 presents the complete results across all strategies, revealing striking patterns that illuminate the consensus paradox. The results demonstrate the oracle strategy's near-perfect performance (99.8% success rate), validating that appropriate expertise exists within the teacher pool. Most remarkably, the ensemble strategy's confidence-weighted voting delivers the worst performance among all teacher-guided approaches, achieving only 32.3% success rate worse than the solo baseline of 48.3%. This severe temporal goal confusion contrasts sharply with the performance-based strategy, which achieves 91.7% success by tracking recent outcomes, enabling effective teacher selection without requiring oracle knowledge of the current goal. Figure 3 further illustrates this paradoxical relationship between teacher agreement and performance:

Figure 4 reveals the fundamental trade-offs between teacher selection strategies when environments change.

Table 1. Performance comparison across teacher selection strategies (mean ± std over 30 seeds)

Strategy	Success Rate	Avg. Steps	TCI	GCS
Solo	48.3 ± 4.9%	68.4 ± 36.7	N/A	N/A
Oracle	**99.8 ± 0.2%**	**16.3 ± 10.1**	0.315 ± 0.062	0.448 ± 0.045
Random	34.5 ± 2.4%	82.4 ± 28.8	0.292 ± 0.096	0.381 ± 0.067
Confidence	42.7 ± 1.1%	66.8 ± 39.1	**0.375 ± 0.068**	**0.457 ± 0.072**
Ensemble	**32.3 ± 2.0%**	75.9 ± 37.1	0.224 ± 0.097	0.399 ± 0.078
Performance	91.7 ± 3.8%	24.2 ± 25.3	0.311 ± 0.067	0.448 ± 0.054

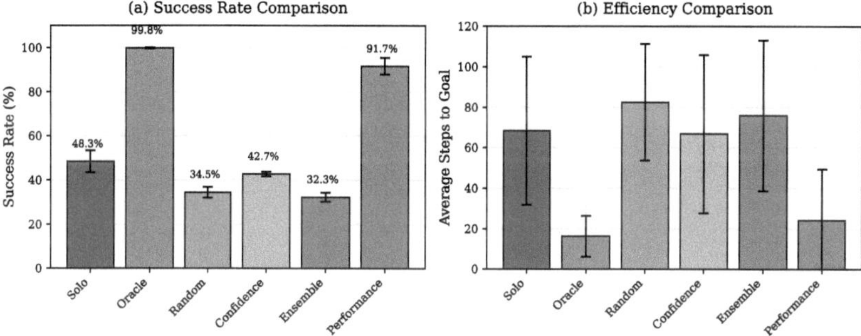

Fig. 3. Performance comparison across strategies. (a) Success rate comparison with error bars showing standard deviation across 30 random seeds. (b) Average steps to goal with standard deviation. The Oracle strategy achieves near-perfect performance while the Ensemble strategy, despite having access to all teachers, shows the poorest performance among teacher-guided approaches.

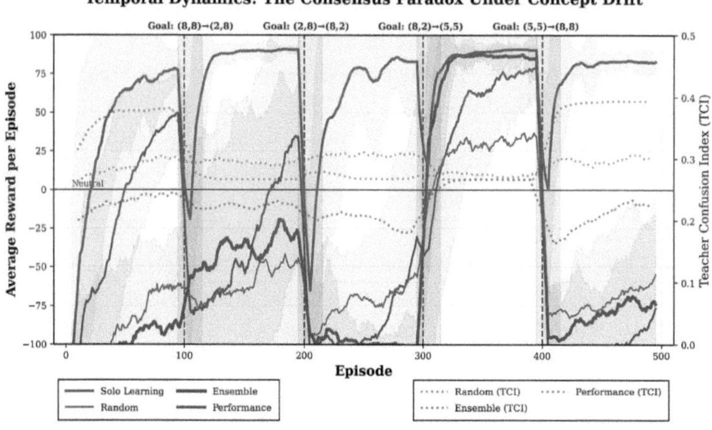

Fig. 4. Average reward during goal transitions. Shaded regions indicate drift points. Ensemble strategy shows persistent negative performance during transitions.

Ensemble Strategy Paradox. Confidence-weighted voting is designed to leverage collective expertise through aggregation. However, under concept drift, this mechanism becomes counterproductive: when goals change, three out of four teachers possess outdated expertise yet remain highly confident. The ensemble amplifies this false consensus, exhibiting catastrophic drops at each drift point with prolonged recovery (>20 episodes).

Performance-Based Adaptation. Performance-based selection achieves rapid adaptation (5–10 episodes) by tracking recent teacher success rather than confidence. This strategy identifies which teacher's expertise remains valid, achieving 91.7% success without requiring explicit goal knowledge.

Baseline Comparisons. Random selection (34.5% success) outperforms ensemble by avoiding systematic bias toward confident but outdated teachers. Oracle selection (99.8% success) validates that appropriate expertise exists, establishing the upper bound.

The temporal dynamics demonstrate that disagreement spikes at drift points signal valuable information about teacher relevance, not noise to be suppressed through consensus—revealing why seeking agreement becomes harmful when expertise becomes temporally misaligned.

5.2 Statistical Analysis and Drift Response

Our proposed metrics quantitatively confirm what the performance data suggests – a paradoxical relationship between consensus and success. We observe a positive correlation between TCI and success rate ($r = 0.277$, $p = 0.652$); while this correlation is not statistically significant, the pattern is consistent across all experiments and the effect sizes are substantial. More importantly, the ensemble strategy achieves the lowest TCI (0.224) yet delivers the worst teacher-guided performance (32.3%) – a 33% performance drop compared to solo learning. This stark contrast between lowest disagreement and worst performance empirically demonstrates the consensus paradox in action.

One-way ANOVA reveals significant differences between strategies ($F(5,174) = 8957.6$, $p < 0.0001$). Post-hoc Tukey HSD tests confirm that Oracle significantly outperforms all strategies ($p < 0.001$), Ensemble significantly underperforms the solo baseline ($p < 0.01$), and Performance-based selection significantly outperforms confidence-based selection ($p < 0.001$).

Analysis of strategy behaviour immediately after drift reveals distinct adaptation patterns. Oracle maintains consistent performance across drifts, as expected given its perfect knowledge. Performance-based selection shows rapid adaptation within 5–10 episodes, quickly identifying which teacher's expertise aligns with the new goal. In stark contrast, the ensemble strategy exhibits prolonged confusion, requiring over 20 episodes for even partial recovery as it struggles to overcome the collective momentum of outdated expertise. Confidence-based selection shows high variance in its response, often misled by teachers who remain overconfident despite their knowledge becoming obsolete.

6 Theoretical Analysis of the Consensus Paradox

Our empirical results reveal a stark pattern: ensemble methods achieve the lowest TCI (0.224) yet deliver the worst performance (32.3%), while performance-based selection achieves high success (91.7%) despite moderate TCI (0.311). This counterintuitive relationship demands theoretical explanation. Why does reducing disagreement lead to catastrophic failure? How can a simple performance-tracking strategy outperform sophisticated confidence-weighted voting by nearly 60% points?

The answer lies in understanding how ensemble methods interact with temporal misalignment. When environments are static, confidence-weighted voting successfully leverages collective expertise. However, when goals change, this same mechanism becomes a trap that systematically amplifies outdated knowledge.

To understand why ensemble methods fail so dramatically, consider what happens when the environment changes. Three out of four teachers suddenly possess obsolete knowledge, yet they remain highly confident in their recommendations. The ensemble mechanism, designed to find wisdom in collective agreement, instead amplifies this outdated consensus. The following theorem formalises this intuition:

Theorem 1 (Temporal Misalignment Amplification). *Let T_i be teacher i with Q-function $Q_i(s,a)$ trained on goal g_i. Under confidence-weighted ensemble voting with weights $w_i(s) = \frac{c_i(s)}{\sum_j c_j(s)}$ where $c_i(s)$ is teacher confidence, the ensemble action selection:*

$$a_{ensemble}(s) = \arg\max_a \sum_{i=1}^{n} w_i(s) \cdot Q_i(s,a)$$

converges to the incorrect action with probability approaching $\frac{n-1}{n}$ when only one teacher knows the current goal.

Proof Sketch: The key insight is that after environmental drift, outdated teachers maintain high confidence in their obsolete expertise. When goal g_j changes to g_k, only teacher T_k provides correct guidance, while the remaining $n-1$ teachers confidently recommend actions toward their (now incorrect) goals. Under confidence-weighted voting, the weight of the single correct teacher becomes $w_k = \frac{c_k}{c_k+(n-1)\bar{c}_{wrong}}$, where $\bar{c}_{wrong}$ is the average confidence of incorrect teachers. As n increases, this weight approaches $\frac{1}{n}$, making the ensemble select incorrect actions with probability $\frac{n-1}{n}$. Thus, paradoxically, more teachers lead to worse performance when expertise is temporally misaligned. For our 4-teacher scenario, this predicts $\frac{3}{4} = 75\%$ incorrect actions, yielding approximately 25% success rate—closely matching our observed 32.3% ensemble performance.

Lemma 1 (False Consensus Creation). *The measured disagreement $D = \frac{1}{n(n-1)} \sum_{i \neq j} \mathbf{1}(a_i \neq a_j)$ decreases while performance degrades when confidence weighting creates consensus among incorrect teachers, where $\mathbf{1}(a_i \neq a_j)$ equals 1 when teachers i and j recommend different actions, and 0 otherwise.*

Explanation: Under drift, incorrect teachers often agree on suboptimal actions (e.g., moving toward their respective obsolete goals). Confidence weighting amplifies this agreement, reducing measured disagreement while selecting incorrect actions with high collective confidence.

Corollary 1. *Low teacher disagreement is necessary but not sufficient for good performance in non-stationary environments. The relationship between disagreement and performance becomes inverted under concept drift.*

Three conditions create the perfect situation for consensus failure. First, persistent confidence occurs when teacher confidence remains high after environmental change, despite their knowledge becoming obsolete. Second, majority obsolescence arises when most teachers possess outdated knowledge, creating a numerical advantage for incorrect guidance. Third, aggregation bias emerges when the mechanism weights confidence over temporal relevance, systematically favouring assured but outdated expertise.

This theoretical framework directly explains our empirical observations. The ensemble strategy achieves TCI = 0.224 (lowest disagreement) precisely because confidence-weighted voting creates artificial consensus among outdated teachers – the $\frac{n-1}{n} = \frac{3}{4}$ probability of incorrect action selection manifests as the observed 32.3% success rate. In contrast, the performance-based strategy (91.7% success) succeeds because it implicitly tracks which teacher's knowledge remains temporally aligned, while random selection (34.5% success) avoids systematic bias by giving each teacher equal chance regardless of their obsolete confidence. The theorem mathematically confirms what our experiments demonstrate: when expertise becomes temporally misaligned, the mechanism designed to harness collective wisdom becomes the architect of collective failure.

7 Discussion and Future Work

The consensus paradox forces us to reconsider fundamental principles of ensemble learning. While combining expert opinions reduces variance in static environments, this same mechanism becomes a fatal flaw under concept drift: the pursuit of agreement systematically amplifies outdated expertise.

The implications extend beyond reinforcement learning. Large language model ensembles may reach consensus on outdated information; federated learning systems may amplify obsolete patterns when data distributions shift; financial trading algorithms could collectively reinforce strategies from expired market regimes.

Our theoretical analysis suggests that addressing the consensus paradox requires fundamental changes to how we aggregate expert knowledge in non-stationary environments. Future algorithms should: (1) weight recent performance over historical expertise, (2) use disagreement spikes as drift detection signals, and (3) abandon consensus-seeking in favour of tracking which expert's knowledge remains valid. Preliminary experiments show promise, achieving 87.3% success rate compared to ensemble's 32.3%.

While our findings robustly demonstrate the consensus paradox across 30 random seeds, several limitations exist. Our experiments used a discrete 10×10 grid world with predictable cyclic drift; real-world environments exhibit stochastic transitions and irregular drift patterns. We tested homogeneous Q-learning teachers; heterogeneous ensembles might show different failure modes. Future work should validate the paradox in continuous domains and develop drift-adaptive aggregation mechanisms.

8 Conclusion

We began with a puzzle: why do ensemble methods fail when they achieve the highest agreement among expert teachers? Our investigation revealed a fundamental paradox in multi-teacher reinforcement learning – reducing disagreement can be catastrophically harmful when the world changes.

The consensus paradox emerges from a cruel irony: confidence-weighted voting, designed to find truth in agreement, instead amplifies obsolete knowledge when most experts possess outdated expertise. In our experiments, ensemble methods achieved 24% lower disagreement yet performed 33% worse than learning alone, and even worse than random teacher selection. With probability approaching $\frac{n-1}{n}$, the ensemble will choose incorrectly when only one teacher knows the current truth.

This finding has profound implications for any multi-expert system facing change – from AI ensembles to human committees. When large language models trained on different time periods reach consensus, they may be agreeing on yesterday's facts. When trading algorithms converge on a strategy, they may be collectively racing toward an obsolete optimum. The pursuit of consensus, far from ensuring correctness, can become the very mechanism of collective failure.

Our work suggests a new principle for adaptive systems: in dynamic environments, disagreement is not noise to be eliminated but a signal to be interpreted. The challenge is not achieving consensus but recognising when consensus itself has become the problem. As we build increasingly complex multi-agent systems, we must remember that when the ground truth shifts, the expert who disagrees might be the only one who's right.

References

1. Bignold, A., Cruz, F., Dazeley, R., Vamplew, P., Foale, C.: Human engagement providing evaluative and informative advice for interactive reinforcement learning. Neural Comput. Appl. **35**(25), 18215–18230 (2023)
2. Christiano, P.F., Leike, J., Brown, T., Martic, M., Legg, S., Amodei, D.: Deep reinforcement learning from human preferences. In: Advances in Neural Information Processing Systems, vol. 30 (2017)
3. Faußer, S., Schwenker, F.: Neural network ensembles in reinforcement learning. Neural Process. Lett. **41**(1), 55–69 (2015)

4. Feng, F., Huang, B., Zhang, K., Magliacane, S.: Factored adaptation for non-stationary reinforcement learning. Adv. Neural. Inf. Process. Syst. **35**, 31957–31971 (2022)
5. Finn, C., Abbeel, P., Levine, S.: Model-agnostic meta-learning for fast adaptation of deep networks. In: International Conference on Machine Learning, pp. 1126–1135. PMLR (2017)
6. Gama, J., Žliobaitė, I., Bifet, A., Pechenizkiy, M., Bouchachia, A.: A survey on concept drift adaptation. ACM Comput. Surv. (CSUR) **46**(4), 1–37 (2014)
7. Griffith, S., Subramanian, K., Scholz, J., Isbell, C.L., Thomaz, A.L.: Policy shaping: integrating human feedback with reinforcement learning, vol. 26 (2013)
8. Littman, M.L.: Markov games as a framework for multi-agent reinforcement learning, pp. 157–163 (1994)
9. Rosman, B., Hawasly, M., Ramamoorthy, S.: Bayesian policy reuse. Mach. Learn. **104**(1), 99–127 (2016). https://doi.org/10.1007/s10994-016-5547-y
10. Shoham, Y., Leyton-Brown, K.: Multiagent Systems: Algorithmic, Game-Theoretic, and Logical Foundations. Cambridge University Press (2008)
11. Sridhar, A., Shah, D., Glossop, C., Levine, S.: Nomad: goal masked diffusion policies for navigation and exploration. In: 2024 IEEE International Conference on Robotics and Automation (ICRA), pp. 63–70. IEEE (2024)
12. Sutton, R.S., Barto, A.G., et al.: Reinforcement Learning: An Introduction, vol. 1. MIT Press, Cambridge (1998)
13. Taylor, M.E., Stone, P.: Transfer learning for reinforcement learning domains: a survey. J. Mach. Learn. Res. **10**(7) (2009)
14. Torrey, L., Taylor, M.: Teaching on a budget: agents advising agents in reinforcement learning, pp. 1053–1060 (2013)
15. Watkins, C.J., Dayan, P.: Q-learning. Mach. Learn. **8**(3), 279–292 (1992)
16. Wiering, M.A., Van Hasselt, H.: Ensemble algorithms in reinforcement learning, vol. 38, pp. 930–936. IEEE (2008)

Towards Search Node-Specific HTN Heuristics

Lijia Yuan(✉) and Pascal Bercher

School of Computing, The Australian National University, Canberra, Australia
lijia.yuan@alumni.anu.edu.au, pascal.bercher@anu.edu.au

Abstract. While effective in hierarchical task network (HTN) planning, heuristic search currently relies on few pre-selected heuristics. However, during progression-based search, many search nodes exhibit specific properties, e.g., they may become totally ordered or acyclic, allowing for the application of specialized heuristics. For these search nodes, we conducted an experimental evaluation, employing reachability analysis to examine the special cases encountered during search. Measuring how often various special cases occur informs us which heuristics developed for special cases – selected on a per-search node basis – are most promising. As a side contribution, we propose a slight modification of a technical definition used to define properties of the task hierarchy.

Keywords: HTN planning · Progression search · Heuristics

1 Introduction

Hierarchical Task Network (HTN) planning [4,11,12] is an *AI planning* framework where tasks are organized hierarchically into directly executable primitive tasks and decomposable abstract tasks. *Progression-based search* [17] is a forward-operating technique for solving HTN problems, adhering to specific orderings from left to right. It uses heuristics to refine the search trajectory, minimizing exhaustive exploration and guiding the path to the goal. The effectiveness of heuristic search in HTN planning depends on general-purpose heuristics (*TDG Heuristic* [5], *Relax Composition Heuristic (RC Model)* [16,17], *ILP-based Heuristics* [15,19], and *Landmark Heuristics* [14,20]). While there is one heuristic that targets a specific case – namely, for totally ordered HTN problems [19] – most are designed for the general case and do not account for specific properties that emerge during search, such as problems becoming totally ordered, or recursive parts becoming non-recursive after certain decisions. Designing general-case heuristics is complex, suggesting that specialized heuristics for specific properties hold greater potential. For instance, a recent pruning technique for totally ordered HTN problems [18] and a heuristic for totally ordered HTN problems [19] demonstrate this benefit. We demonstrate that special cases arise during search, enabling the dynamic selection of specialized heuristics to enhance performance.

M. Liu et al. (Eds.): AI 2025, LNAI 16371, pp. 439–452, 2026.
https://doi.org/10.1007/978-981-95-4972-6_34

To assess the potential of choosing specialized heuristics during search, we conducted an experimental analysis of the frequency of specific cases (focusing on totally ordered problems). As an additional contribution, we introduce a minor refinement to a technical definition that facilitates the determination of properties of the task hierarchy.

2 HTN Planning Formalism

Our work builds upon the HTN planning formalization initially introduced by Geier and Bercher [12] and further developed by Bercher, Alford, and Höller [4].

A *task network* tn, represented as a tuple $(T, \prec, \alpha)$, consists of a finite set of *task id symbols* T, a strict partial order on T denoted by $\prec \subseteq T \times T$ (which is irreflexive, asymmetric, and transitive), and a mapping α that assigns each task id in T to either a *primitive task name* in N_p or *abstract task name* in N_a. An *HTN domain* D is a tuple (F, N_p, N_a, δ, M), consisting of a finite set of *facts* F, a finite set of primitive task names N_p, a finite set of abstract task names N_a, a mapping $\delta : N_p \rightarrow 2^F \times 2^F \times 2^F$ that assigns each primitive task (also called an *action*) to its *preconditions*, *add effects*, and *delete effects*, and a finite set of *(decomposition) methods* M, where each method $m \in M$ is a tuple (c, tn) pairing an abstract task c with a task network tn. An *HTN problem* is a tuple $\mathcal{P} = (D, s_I, \text{tn}_\text{I}, g)$, comprising an HTN domain D, an *initial state* $s_I \in 2^F$, an *initial task network* tn_I, and a goal description $g \subseteq F$.

A task network $\text{tn}_\text{a} = (T_a, \prec_a, \alpha_a)$ can be decomposed by a method $m = (c, \text{tn}_\text{m})$ into a new task network $\text{tn}_\text{b} = (T_b, \prec_b, \alpha_b)$ if and only if there exists a task identifier $t \in T_a$ such that $\alpha_a(t) = c$ is replaced by subtasks in tn_m, and all ordering constraints from t will be inherited, written as $\text{tn}_\text{a} \xrightarrow{t,m} \text{tn}_\text{b}$ Formally, let the task network $\text{tn}' = (T', \prec', \alpha')$ be equivalent to tn_m, such that $T' \cap T_a = \emptyset$, i.e., the only difference between tn' and tn_m are task identifiers. The application of m to tn_a results into the task network tn_b given as follows[1].

$$
\begin{aligned}
tn_b &= \left((T_a \setminus \{t\}) \cup T', (\prec_a|_{(T_a\setminus\{t\})\times(T_a\setminus\{t\})}) \cup \prec' \cup \prec_x, (\alpha_a|_{T_a\setminus\{t\}}) \cup \alpha'\right) \\
\prec_x &= \{(t_a, t_b) \in T_a \times T' \mid (t_a, t) \in \prec_a\} \;\cup\; \{(t_a, t_b) \in T' \times T_a \mid (t, t_b) \in \prec_a\}
\end{aligned}
$$

A task network is *executable* if it has an *executable linearization* of its tasks, with primitive task $p \in N_p$ linked to action a, where $\delta(p) = (pre(a), add(a), del(a))$ is executable in state s if and only if $pre(a) \subseteq s$, and its execution modifies s to the resulting state $(s \setminus del(a)) \cup add(a)$. An executable linearization for task network $\text{tn} = (T, \prec, \alpha)$ is a sequence $(t_1, t_2, ..., t_n)$ where each $t_i \in T$ with $\alpha(t_i) \in N_p$ can be executed sequentially. A task network $\text{tn}_\text{s} = (T_s, \prec_s, \alpha_s)$ is called a solution of an HTN problem $\mathcal{P} = (D, s_I, \text{tn}_\text{I}, g)$ if and only if tn_I can be decomposed into tn_s through a series of decompositions, tn_s solely comprises primitive tasks $(\forall t \in T_s : \alpha(t) \in N_p)$, and tn_s has an executable linearization. The notation $\text{tn} \xrightarrow{*} \text{tn}'$ indicates tn can be decomposed into tn' by using a (possibly empty) sequence of methods.

[1] Here, we use $R|_X$ to denote the restriction of relation R to the domain X.

3 Known Special Cases

In this section, we provide the definitions for known problem classes, which are *primitive HTN problems*, *totally ordered problems*, *regular problems*, *acyclic problems* [11], and *tail-recursive problems* [1,2]. We refine the *stratification* by Alford et al. [2] and propose a more tight *HTN stratification* to assist in defining acyclic and tail-recursive problems. While the existing definition was not wrong, our "tighter version" allows us to differentiate more classes based on the stratification alone, without having to consult the underlying HTN problem.

HTN planning is undecidable [11,12] but various special cases can make the plan existence problem easier, such as totally ordered ones.

Primitive HTN problems (when all tasks in the initial task network are primitive) are the base case, appearing at the end of the search if a solution exists.

Definition 1 (Totally Ordered Problem). *An HTN problem $\mathcal{P}$ is called* totally ordered *iff the ordering of its initial task network* $\mathrm{tn_I}$ *is totally ordered and all decomposition methods are totally ordered, i.e., for each $m \in M$ with $m = (c, \mathrm{tn_m})$, $\mathrm{tn_m}$ is a totally ordered task network.*

Totally ordered HTN planning is EXPTIME-complete [1,11].

Definition 2 (Regular Problem). *An HTN problem $\mathcal{P}$ is* regular *iff its initial task network* $\mathrm{tn_I}$ *and* $\mathrm{tn_m}$ *in all its methods $(c, \mathrm{tn_m}) \in M$ are regular. A task network $tn = (T, \prec, \alpha)$ is regular if there is at most one task in T that is abstract; and if $t \in T$ and $\alpha(t) \in N_a$, it is the last task in tn, i.e., for all $t' \in T$ with $t' \neq t$, we have $(t', t) \in \prec$.*

Regular problems were shown to be PSPACE-complete [11].

For the more complex problem restrictions like tail-recursive ones [1], we provide stratifications, as introduced by Alford et al. [2].

Definition 3 (Stratification). *A set $R \subseteq N_a \times N_a$ is called a* stratification *iff it is a total preorder (i.e., reflexive, transitive, and total). A set $S \subseteq N_a$ is called* stratum *iff it is inclusion-maximal and for all $x, y \in S$, $(x, y) \in R$ holds.*

According to this definition, stratifications are a concept independent of the underlying HTN problem; the connection to the HTN problem is imposed by the respective problem definitions. While the original definition of tail-recursive problems [1,2] requires a specific stratification to exist, it implicitly allows for *multiple* such stratifications adhering to the restrictions. Thus, even for tail-recursive problems (defined later), the set of possible stratifications defined over the respective abstract tasks is not unique. In fact, totality requires that any two abstract tasks are "artificially" put into some relationship, even if neither of tasks can be decomposed into another. We propose a stricter definition that removes the requirement of totality and imposes an *exact* relationship between the decomposition hierarchy and the underlying stratification – thus making the stratification a formal one-to-one mapping of the underlying task hierarchy. We believe that this also prevents errors, as we argue that this definition is less artificial and hence more natural.

Beyond achieving a *unique* stratification per domain, *clear, intuitive semantics*, and simplified tail-recursive problem definitions, our proposed definition allows differentiating whether singleton strata represent recursive tasks or not. The original definition's requirement for reflexivity ($(c, c) \in R$ for every abstract task) makes it impossible to identify actual recursion solely from stratification, as reflexivity is demanded by definition, not by reachability. In our tighter definition, we do not demand reflexivity, so this property will only hold if a task can indeed reach itself via decomposition. To remain consistent with the way methods are written (e.g., (c, tn) decomposes c into tn), we define stratifications to "decrease towards the right", e.g., $(A, B) \in R$ will mean that B is on a lower stratum than A (not necessarily strictly).

Definition 4 (HTN Stratification). *Given an HTN domain $D = (F, N_p, N_a, \delta, M)$, a set $R_{HTN} \subseteq N_a \times N_a$ is called an* HTN stratification *of D iff it is transitive and it holds $(c, c') \in R_{HTN}$ if and only if c' is reachable from c via decomposition. A set $S \subseteq N_a$ is called a* (HTN) stratum *iff it is inclusion-maximal, and $|S| = 1$ or for all $c, c' \in S$, $(c, c') \in R_{HTN}$ holds.*

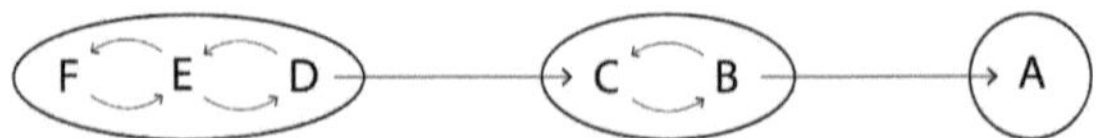

Fig. 1. HTN Stratification R_{HTN} (height and size 3) with abstract tasks A to F. Strata: $S_1 = \{A\}$, $S_2 = \{B, C\}$, $S_3 = \{D, E, F\}$. Arrows (e.g., $(B, A) \in R_{HTN}$) denote the direct decomposition; transitive relations are omitted.

For simplicity, we omit the "HTN" prefix from "stratum" when the meaning is clear from context. The height of a stratum equals the highest number of strata reachable rooted in it, including itself. The height of an HTN stratification equals its highest stratum height. Its size equals its number of strata.

HTN stratifications can be depicted as directed graphs as shown, e.g., in Fig. 1. There, task A cannot be decomposed into any other abstract task and his hence on the lowest stratum. In contrast, tasks B and C are reachable from each other and thus form another stratum, as do D, E, and F. Note that primitive tasks are not depicted, which we assign a stratum height of 0. Also notice that this HTN stratification is total (and therefore compatible with Definition 3), which is not always the case. In contrast, Fig. 2 shows an example for a non-total HTN stratification (which is still compatible with Definition 4). In this example, task F has a stratum height of 2, as do tasks B and C. However, task F is not in the same stratum as tasks B, C, since tasks B, C and task F cannot be reached from each other. A stratum with a single abstract task c implies that no other task c' exists for which c can reach c' while c' can also reach c via decomposition. However, c may be decomposed into itself, indicating a self-loop if $(c, c) \in R_{HTN}$. So, in our example, there are two strata of size 1, $S_1 = \{A\}$ and $S_3 = \{F\}$, yet only F can decompose into itself as now visible from the HTN stratification, i.e., $(F, F) \in R$, but $(A, A) \notin R$. A and F are in their own stratum. Therefore, R_{HTN} effectively differentiates whether an abstract task c allows a self-loop.

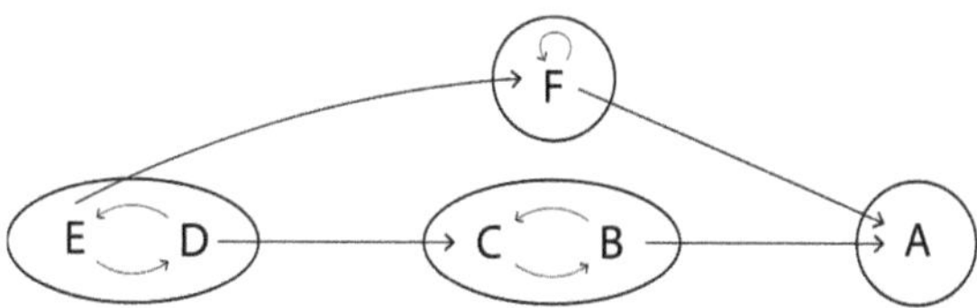

Fig. 2. Partially ordered HTN stratification R_{HTN} (height 3 and size 4) for tasks A to F. Arrows denote directed decomposition (the transitive relations are not shown). Strata: $S_1 = \{A\}$, $S_2 = \{B, C\}$, $S_3 = \{F\}$, $S_4 = \{D, E\}$.

Although HTN stratifications were initially introduced to define tail-recursive problems, we first use them to define acyclic problems.

Definition 5 (Acyclic Problem). *An HTN problem is acyclic iff for its HTN stratification $R_{HTN} \subseteq N_a \times N_a$ holds: if $(c, c') \in R_{HTN}$, then $(c', c) \notin R_{HTN}$.*

The HTN stratification R_{HTN} for the acyclic problem is asymmetric – so demanding that R_{HTN} is asymmetric is an alternative definition.

Acyclicity will bring the computational complexity of an HTN problem down to NEXPTIME-complete [1]. If an HTN problem is acyclic and totally ordered, it will be PSPACE-complete [1].

The definition of tail-recursiveness requires the concept of (non-)last tasks. A task is called *last* in a task network if and only if all other tasks in that task network are ordered to occur before it; otherwise it is a *non-last task*. Notably, except for task networks of size 0 or 1, all task networks have non-last tasks (potentially all), but not every task network has a last task.

Definition 6 (Tail-Recursive Problem). *An HTN problem $\mathcal{P}$ is* tail-recursive *iff for its HTN stratification R_{HTN} and for all methods $(c, \mathrm{tn_m}) \in M$ it holds that for any non-last abstract task c_m in $\mathrm{tn_m}$ with $c_m \in N_a$, $(c_m, c) \notin R_{HTN}$.*

In a tail-recursive problem, every method $(c, \mathrm{tn_m})$ allows its abstract tasks c_m in $\mathrm{tn_m}$ to decompose back into c only if c_m is the last task in $\mathrm{tn_m}$. In Fig. 3a, the task with a stratum height 4 in $\mathrm{tn_1}$ decomposes into $\mathrm{tn_a}$, which contains only tasks that are on a (strictly) lower stratum, so no issues may arise. The only other abstract task in $\mathrm{tn_1}$ can be decomposed into a task network $\mathrm{tn_b}$ which contains a task with the same stratum height as the decomposing task, but this is the last one on $\mathrm{tn_b}$, so no issue arises. Note that only positions within $\mathrm{tn_a}$ and

tn_b are relavant, not in tn_1. In Fig. 3b, we see a non-tail-recursive case. Method m_2 poses no issue, but method $m_3 = (c, \text{tn}_\text{c})$ does. tn_c contains a task of the same stratum height (4) as c, which should therefore be its last task—but is not. Even without the ordering constraint, it would not be tail-recursive, because the abstract task would still not be the last.

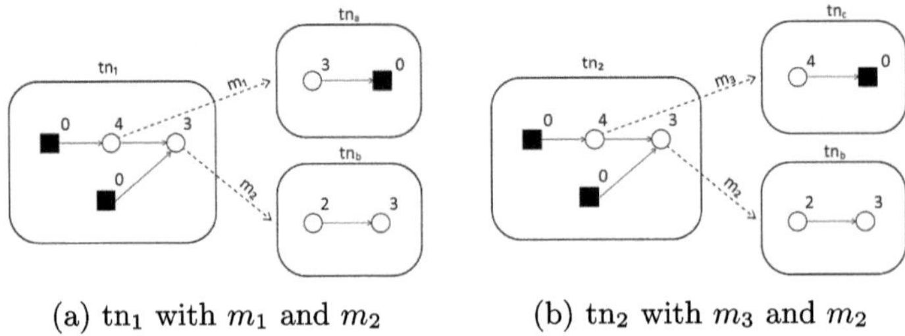

(a) tn_1 with m_1 and m_2 (b) tn_2 with m_3 and m_2

Fig. 3. Comparison of tail-recursiveness (left) and non-tail-recursiveness (right). Filled squares denote primitive tasks; circles represent abstract tasks; numbers beside indicate the height of the stratum. Example borrowed from an ICAPS tutorial on HTN planning [6].

The computational complexity of tail-recursive problems is EXPSPACE-complete and becomes PSPACE-complete when also totally ordered [1].

Testing these special cases can be done in polynomial time (relative to the size of the model), as HTN stratifications are structurally equivalent to Task Decomposition Graphs (see next section), which can be computed in polynomial time [5], reducing the the problem to checking simple graph properties. While we omit implementation details, the order of tests significantly affects efficiency: some are dependent or more expensive than others. Due to the way problem classes relate (e.g., any regular problem is tail-recursive), tests can be performed in a specific order, as laid out in Fig. 4.

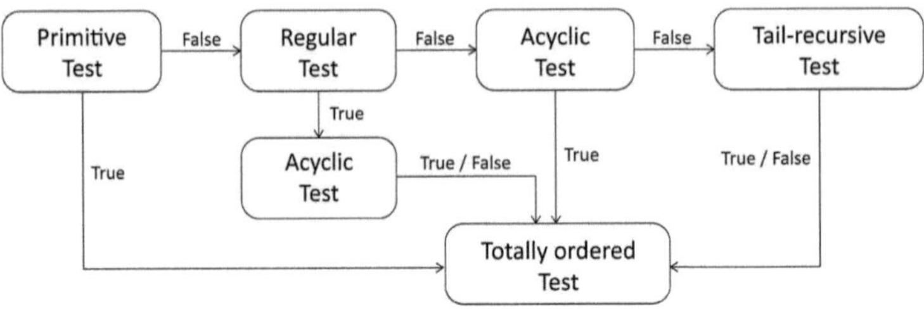

Fig. 4. Control flow graph: Tests ordered to minimize costs and handle property dependencies. Arrows show paths based on outcomes.

In progression search, if a task network within a search node possesses a particular property, all task networks within its child nodes will inherit this property but gain further ones due to the decisions made. Figure 5 illustrates this in an example search space.

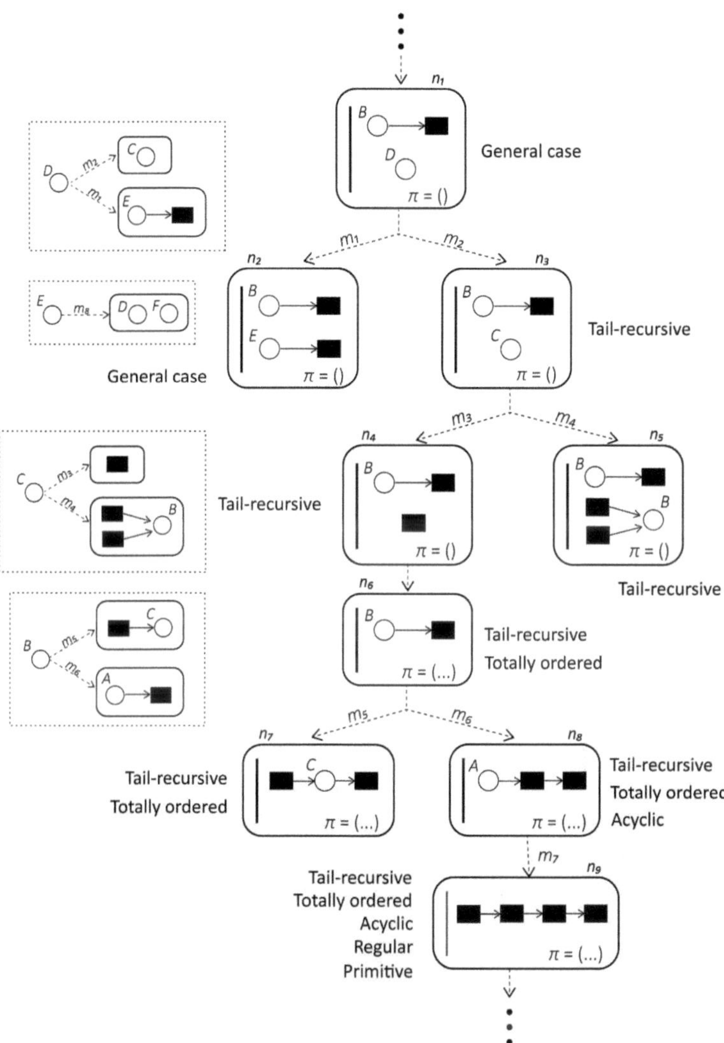

Fig. 5. Special cases during search: Filled squares are primitive tasks, unfilled circles are abstract tasks. π is the generated plan prefix. Vertical lines (in node n_1 to n_9) show search states (no changes assumed, as primitive tasks have no preconditions/effects). Task network properties and methods for B, C, D, E are shown next to each node. The HTN Stratification matches both Fig. 1 and Fig. 2.

4 Obtaining Reachability Information

In a task network, its properties like total ordering, regularity, acyclicity, or tail-recursiveness depend on both the initial task network and all its reachable methods. The *task decomposition graph (TDG)* is the foundational data structure for hierarchical reachability analysis. It was first introduced by Elkawkagy et al. [9] and refined by Bercher et al. [5].

The TDG can represent the hierarchical reachability of an HTN planning problem, i.e., tasks and methods that can be reached via decomposition. According to the simplistic definition provided, it is only based on the reachability from decomposition without considering the states. However, Elkawkagy, Schattenberg, and Biundo [10] and Bercher et al. [5] suggested that method nodes containing unreachable primitive tasks based on the state reachability analysis can be removed. (For a more formal definition, consult Def. 4 by Behnke et al. [3], where this idea is incorporated for grounding lifted HTN planning problems.)

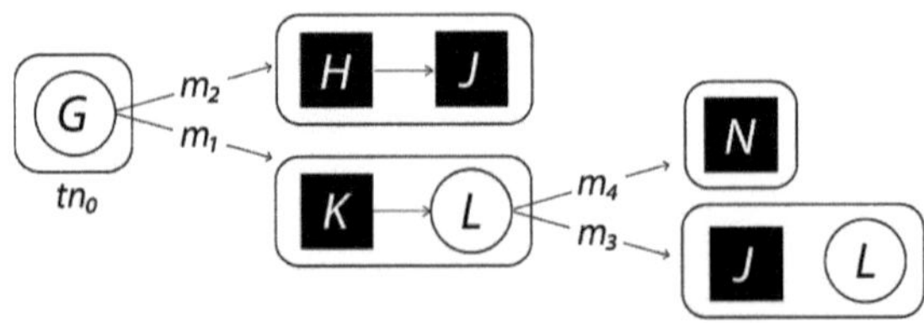

Fig. 6. A fragment of a TDG [5]: task networks are denoted by surrounding boxes, abstract tasks by circles, and primitive tasks by filled square boxes.

It can be observed that the reachable methods per search node change, as illustrated in Fig. 6. Here, the reachable method set from the initial task network is initially m_1 to m_4. However, after m_2 was applied, the reachable methods even became empty, since the new current task network is primitive. If alternatively m_1 would have been chosen, at least m_2 will not be reachable anymore, but potentially even one of m_3 and m_4 as it might be that those actions from m_2's task network were used to (inaccurately) determine the reachability of m_3 or m_4. Hence, recomputing the TDG after some decisions were made can lead to a reduced reachable method set [5]. This is highly relevant for our endeavor of determining search space properties since these properties directly depend on the reachable methods. Going back to our example (Fig. 6), if task K cannot be executed, method m_1, and consequently m_3 and m_4 become unreachable. This leaves only m_2 as reachable and thus the sub-TDG makes tn_0 acyclic and totally ordered – properties that the initial TDG did not show. Thus, when computing the properties of a search node, one has two possibilities: either use the initial TDG or recompute the TDG from the current search node (which is more expensive but leads to potentially fewer reachable methods), and hence a tighter special case analysis.

In our empirical evaluation, we do both. For recomputing the TDG, we use the Relaxed Composition Heuristic [16,17] as a basis. It transforms an HTN search node into a classical problem, which can represent a superset of all reachable decomposition methods (expressed as classical actions). We used this transformation as a basis, estimating all reachable methods by computing a fixed point in a relaxed planning graph (RPG) computed from the classical problem.

5 Evaluation

We analyzed the entire total-order (TO) benchmark set from the IPC 2023. We restrict to TO domains both due to space restrictions and also since the IPC showed that in most instances, the partial order can be compiled away in advance without making the respective problem unsolvable [21,22].

Experiments were conducted in a virtualized environment. The underlying hardware was powered by an Intel(R) Core(TM) i9-8950HK CPU, clocking at 2.90GHz, which was allocated a single CPU core and provisioned with 8GB of RAM for the experiments, and 30 min limited for each instance (memory and time limit of IPC 2023). We run the latest progression-based version of the PANDA$_\pi$ system [13,16,17]. We opted for the currently best-performing configurations, i.e., Greedy Best First Search (GBFS) combined with the Relax Composition (RC) heuristic [16], which utilizes the classical Add heuristic [7] that measures the distance to the solution by accounting for both action applications and method decompositions.

We analyze the search nodes in terms of the problem classes outlined in Section *Known Special Cases*. However, since several classes overlap (e.g., any regular problem and any acyclic problem are also tail-recursive), we account for those overlappings. We define the following classes:

- *Primitive*: If all tasks in the current task network are primitive.
- *Regular & (A)cyclic*: If a problem is regular (and acyclic), but not primitive.
- *Acyclic*: If a problem is acyclic but not regular.
- *Tail-recursive*: If a problem is tail-recursive but not acyclic.
- *Unrestricted*: If a problem is not tail-recursive.

By adhering to these definitions, all classes are exclusive (and so values should add up to 100% per line for the same test in Table 1).

5.1 Results

A summary of our results (per domain) is provided in Table 1. Note that we only include instances that were solved by both runs: those that don't recompute the TDG (referred to as "Simple" in the table) and those that do (referred to as "Reachable"). This is to make the results comparable because only then are the explored search spaces the same. This is also why we only report on 22 domains, although the IPC TO benchmark set encompasses 23. This discrepancy arises because the "Freecell" domain was excluded due to the absence of problem instances that were solved in both runs. In the table, we report per domain over all solved instances the minimum ($\downarrow$), maximum ($\uparrow$), and average (μ) percentage of the respective problem class.

Computation Time. Ideally, class identification of problem classes should exert negligible influence on the computational time. In particular, when our ultimate goal is to choose the most informed heuristic based on the respective

Table 1. The minimum (↓), maximum (↑), and average (μ) percentages (%) of unrestriction, tail-recursion, acyclicity, regularity, and primitiveness in the search space without and with TDG-recomputation ("Simple" vs. "Reachable"). The computational complexity of properties decreases progressively from left to right. Next to the name of each domain is the number of (solved) problem instances that were used as a basis. "*A*" (short for "all") represents 100%. Values highlighted in bold show where improvements were achieved due to TDG-recomputation.

Domain		Unrestricted						Tail-recursive						Acyclic					
		Simple			Reachable			Simple			Reachable			Simple			Reachable		
		↓	↑	μ	↓	↑	μ	↓	↑	μ	↓	↑	μ	↓	↑	μ	↓	↑	μ
Assembly	(30)	0	0	0	0	0	0	82	~*A*	97	82	~*A*	97	0	0	0	0	~0	~0
Barman-BDI	(15)	0	0	0	0	0	0	0	0	0	0	0	0	96	~*A*	~*A*	96	~*A*	~*A*
Blocksw.-GTOHP	(29)	0	~*A*	**94**	0	~*A*	**93**	0	0	0	0	0	0	0	99	**5**	0	99	**6**
Blocksw.-HPDDL	(28)	0	0	0	0	0	0	98	~*A*	~*A*	96	~*A*	~*A*	0	0	**0**	0	~0	**~0**
Depots	(22)	0	~*A*	**35**	0	~*A*	**32**	0	0	**0**	0	24	**1**	0	~*A*	**63**	~0	~*A*	**65**
Factories	(8)	0	0	0	0	0	0	96	~*A*	99	96	~*A*	99	0	0	0	0	0	0
Hiking	(24)	87	97	92	87	97	92	3	12	7	3	10	7	0	0	**0**	~0	1	**1**
Lamps	(16)	0	0	0	0	0	0	80	~*A*	95	80	~*A*	95	0	0	0	0	0	0
Logistics-Learned	(44)	0	0	0	0	0	0	98	~*A*	~*A*	98	~*A*	~*A*	0	0	0	0	0	0
Minecraft Pl.	(1)	0	0	0	0	0	0	84	84	84	84	84	84	15	15	15	15	15	15
Minecraft Reg.	(42)	0	0	0	0	0	0	0	0	0	0	0	0	97	~*A*	99	97	~*A*	99
Monroe FO	(17)	74	~*A*	95	74	~*A*	95	0	0	0	0	0	0	0	17	2	0	17	2
Monroe PO	(8)	25	~*A*	80	25	~*A*	80	0	0	0	0	0	0	0	58	14	0	58	14
Multiarm-Blocksw.	(74)	0	0	0	0	0	0	98	~*A*	**~*A***	6	96	**50**	0	0	**0**	0	94	**48**
Robot	(20)	0	0	0	0	0	0	67	~*A*	**97**	4	83	**31**	0	0	**0**	0	~0	**~0**
Rover	(21)	0	~*A*	**89**	0	~*A*	**88**	0	0	0	0	0	0	0	97	10	~0	97	10
Satellite	(19)	83	99	95	83	99	95	0	0	0	0	0	0	~0	8	2	~0	8	2
SharpSAT	(10)	0	0	0	0	0	0	0	0	0	0	0	0	98	~*A*	99	98	~*A*	99
Snake	(2)	0	0	0	0	0	0	~*A*	~*A*	~*A*	99	~*A*	~*A*	0	0	0	0	0	0
Towers	(13)	0	0	0	0	0	0	75	~*A*	97	75	~*A*	97	0	0	0	0	0	0
Transport	(25)	88	~*A*	97	88	~*A*	97	0	0	0	0	0	0	0	2	~0	0	2	~0
Woodworking	(19)	0	0	0	0	0	0	0	0	0	0	0	0	0	~*A*	85	0	~*A*	85

Domain		Regular & Cyclic						Regular & Acyclic						Primitive					
		Simple			Reachable			Simple			Reachable			Simple			Reachable		
		↓	↑	μ	↓	↑	μ	↓	↑	μ	↓	↑	μ	↓	↑	μ	↓	↑	μ
Assembly	(30)	0	7	**2**	0	2	**~0**	0	9	**~0**	~0	9	**2**	~0	9	1	~0	9	1
Barman-BDI	(15)	0	0	0	0	0	0	~0	2	~0	~0	2	~0	~0	2	~0	~0	2	~0
Blocksw.-GTOHP	(29)	0	0	0	0	0	0	0	3	~0	0	3	~0	~0	5	1	~0	5	1
Blocksw.-HPDDL	(28)	0	0	0	0	0	0	0	0	0	~0	2	~0	~0	2	~0	~0	2	~0
Depots	(22)	0	0	0	0	0	0	0	0	0	0	0	0	~0	8	2	~0	8	2
Factories	(8)	0	0	0	0	0	0	0	0	0	0	0	0	~0	4	1	~0	4	1
Hiking	(24)	0	0	0	0	0	0	0	0	0	0	0	0	~0	1	1	~0	1	1
Lamps	(16)	0	0	0	0	0	0	0	0	0	0	0	0	~0	20	5	~0	20	5
Logistics-Learned	(44)	0	0	0	0	0	0	0	0	0	0	0	0	~0	2	~0	~0	2	~0
Minecraft Pl.	(1)	0	0	0	0	0	0	0	0	0	0	0	0	1	1	1	1	1	1
Minecraft Reg.	(42)	0	0	0	0	0	0	~0	1	~0	~0	1	~0	~0	1	~0	~0	1	~0
Monroe FO	(17)	0	0	0	0	0	0	0	4	~0	0	4	~0	~0	14	3	~0	14	3
Monroe PO	(8)	0	0	0	0	0	0	0	8	2	0	8	2	~0	11	5	~0	11	5
Multiarm-Blocksw.	(74)	0	0	0	0	0	0	0	0	**0**	~0	60	**2**	~0	2	~0	~0	2	~0
Robot	(20)	0	0	0	0	0	0	0	0	**0**	0	96	**67**	~0	33	3	~0	33	3
Rover	(21)	0	0	0	0	0	0	0	0	0	0	0	0	~0	7	2	~0	7	2
Satellite	(19)	0	0	0	0	0	0	0	5	1	0	5	1	~0	8	2	~0	8	2
SharpSAT	(10)	0	0	0	0	0	0	0	0	0	0	0	0	~0	2	1	~0	2	1
Snake	(2)	0	0	0	0	0	0	0	0	**0**	~0	~0	**~0**	~0	~0	~0	~0	~0	~0
Towers	(13)	0	0	0	0	0	0	0	0	0	0	0	0	~0	25	3	~0	25	3
Transport	(25)	0	0	0	0	0	0	0	5	1	0	5	1	~0	12	2	~0	12	2
Woodworking	(19)	0	0	0	0	0	0	0	75	8	0	75	8	~0	25	6	~0	25	6

search node property, it is essential that the overhead incurred from identifying this special case pays off.

In the "simple" experiment, the total computational time for problem class identification varied across domains, ranging from 0.64% to 42% of total computation time (with an average of 11%), compared to the heuristic's average computation time of 34%. This aligns with expectations, suggesting class identification times are acceptable. However, runtimes can still be reduced by optimizing the way classes are identified based on reachable methods (see below).

Conversely, in the "reachable" experiment, the average total computation time for class identification surged to 64%, much overshadowing the heuristic computation time, which averaged 14%. This indicates that class identification significantly impacts the overall computation time in this context, underlining a need for optimization in re-computing the TDG. That said, we did not put any effort into recomputing the TDG effectively, so there are still lots of possible optimizations that could be done. For example, since we use the RC heuristic, we could have used the RPG computed by it when extracting the reachable methods. However, for the sake of simplicity, we re-computed this anyway. Also, we hypothesize that our TDG construction (based on these reachable methods) *and* the way we identify special cases based on the reachable methods could still be optimized (which is the reason why we don't report individual runtimes).

Thus, our reported results can only be used to show which special cases occur and how frequently, but not to draw conclusions about the feasibility of their computation. They do show, however, that it will be beneficial to optimize TDG recomputation and the identification of special cases.

Difference Between Experiments. Before we report on the main findings of our experiments, namely the frequency of certain problem classes, we report on the impact of TDG recomputation. Table 1 reveals that only a select few domains exhibit notable differences between the "simple" and "reachable" experiments.

The domain exhibiting the most striking substantial disparity is "Multiarm-Blocksworld", where the average percentage of tail-recursion in the "simple" experiment approaches 100%. However, in the reachable experiment, this average drops to just half. In the remaining 50%, 48% become acyclic (and non-regular), and 2% become regular & acyclic. "Robot" is another domain showing significant differences. In the "simple" experiment, it only has tail-recursive and primitive search spaces. However, in the "reachable" experiment average percentage for tail-recursion decreased dramatically from 97 to 31 in favor of 67% regular and acyclic search nodes.

All in all, it seems that TDG recomputation only pays off in few domains, so unless recomputation cost can be reduced significantly, our results suggest that, on average, this might not pay off for the sake of deploying specialized heuristics.

Frequency of Special Cases. We can see that only 5 out of the 22 domains have 90% or more of undecidable search nodes, with 2 further domains having 80% and 89%. Those are the domains where heuristics dealing with the general case (in TO HTN planning) would be required. So in the huge majority of domains, at least tail-recursiveness could be exploited. Domains such as "Blocksworld-GTOHP" and "Hiking" exhibit very high ratios of "undecidability", which are over 90%. The number of solved instances of these domains is large, indicating that the existing general heuristic is efficient.

Three domains – "Barman-BDI", "Minecraft Regular" and "SharpSAT' show nearly 100% or exactly 100% acyclicity (in Table 1). Since acyclic counts exclude nodes that are also regular (cyclic) or primitive, these domains are fully acyclic upon closer inspection. However, acyclicity patterns vary: for example, "Blocksworld-GTOHP" ranges wildly in acyclicity, from being virtually cyclic to fully acyclic, as evidenced by its minimum of approximately 0% and a maximum of 99%. Conversely, "Monroe-FO" and "Satellite-GTOHP" show more restrained ranges, peaking at 17% and 8% respectively. It indicates that there are only relatively few nodes in the domain that are acyclic. While the distribution of acyclicity across domains is varied, certain domains inherently exhibit acyclicity, whereas others only manifest it as the problem simplifies during the search. Nonetheless, acyclicity seems to occur often enough within the search space that it seems beneficial to develop specialized heuristics. It is particularly noteworthy that only one single *domain* seems to be acyclic. For all others, acyclicity only occurs within the search space, but not for the initial search node.

Another *clear* observation is that regularity seems to only occur extremely rarely. In only one domain, the ratio of search nodes is about 67%, in one it is 8%, in most others, however, it is 0%, and in just a few 1% or 2%. This however shows a possible optimization as it means that we can stop the TDG recomputation and special case identification once acyclic search nodes are discovered, thus speeding up computation time.

6 Conclusion

Based on the total-order IPC 2023 benchmark set, we analyzed the explored search space of a progression-based planner for the frequency of certain special cases within the problem classes (although some more novel special cases [8,23] were not yet tested). Our motivation for this investigation is the deployment of node-specific heuristics, tailored towards the respective special case. Our empirical findings indicate a high potential for such heuristics aimed at tail-recursiveness and acyclicity due to the high number of such search nodes. However, computation time for the identification of these classes is still relatively high and thus requires consideration as well.

Acknowledgements. Pascal Bercher is the recipient of an Australian Research Council (ARC) Discovery Early Career Researcher Award (DECRA), project number DE240101245, funded by the Australian Government.

References

1. Alford, R., Bercher, P., Aha, D.W.: Tight bounds for HTN planning. In: ICAPS 2015, pp. 7–15. AAAI Press (2015)
2. Alford, R., Shivashankar, V., Kuter, U., Nau, D.S.: HTN problem spaces: structure, algorithms, termination. In: SOCS 2012, pp. 2–9. AAAI Press (2012)
3. Behnke, G., Höller, D., Schmid, A., Bercher, P., Biundo, S.: On succinct groundings of HTN planning problems. In: AAAI 2020, pp. 9775–9784. AAAI Press (2020)
4. Bercher, P., Alford, R., Höller, D.: A survey on hierarchical planning – one abstract idea, many concrete realizations. In: IJCAI 2019, pp. 6267–6275. IJCAI (2019)
5. Bercher, P., Behnke, G., Höller, D., Biundo, S.: An admissible HTN planning heuristic. In: IJCAI 2017, pp. 480–488. IJCAI (2017)
6. Bercher, P., Höller, D.: Tutorial: introduction to hierarchical task network (HTN) planning. ICAPS (2018). Accessed 15 July 2025
7. Bonet, B., Geffner, H.: Planning as heuristic search. AIJ 2001 **129**(1-2), 5–33 (2001)
8. Dekker, M., Behnke, G.: Barely decidable fragments of planning. In: ECAI 2024, pp. 4198–4206. IOS Press (2024)
9. Elkawkagy, M., Bercher, P., Schattenberg, B., Biundo, S.: Improving hierarchical planning performance by the use of landmarks. In: AAAI 2012, pp. 1763–1769. AAAI Press (2012)
10. Elkawkagy, M., Schattenberg, B., Biundo, S.: Landmarks in hierarchical planning. In: ECAI 2010, pp. 229–234. IOS Press (2010)
11. Erol, K., Hendler, J.A., Nau, D.S.: Complexity results for HTN planning. Ann. Math. AI (AMAI) **18**(1), 69–93 (1996)
12. Geier, T., Bercher, P.: On the decidability of HTN planning with task insertion. In: IJCAI 2011, pp. 1955–1961. AAAI Press (2011)
13. Höller, D., Behnke, G.: Loop detection in the PANDA planning system. In: ICAPS 2021, pp. 168–173. AAAI Press (2021)
14. Höller, D., Bercher, P.: Landmark generation in HTN planning. In: AAAI 2021, pp. 11826–11834. AAAI Press (2021)
15. Höller, D., Bercher, P., Behnke, G.: Delete- and ordering-relaxation heuristics for HTN planning. In: IJCAI 2020, pp. 4076–4083. IJCAI (2020)
16. Höller, D., Bercher, P., Behnke, G., Biundo, S.: A generic method to guide HTN progression search with classical heuristics. In: ICAPS 2018, pp. 114–122. AAAI Press (2018)
17. Höller, D., Bercher, P., Behnke, G., Biundo, S.: HTN planning as heuristic progression search. JAIR **2020**(67), 835–880 (2020)
18. Olz, C., Bercher, P.: A look-ahead technique for search-based HTN planning: reducing the branching factor by identifying inevitable task refinements. In: SoCS 2023, pp. 65–73. AAAI Press (2023)
19. Olz, C., Lodemann, A., Bercher, P.: A heuristic for optimal total-order HTN planning based on integer linear programming. In: ECAI 2024, pp. 4303–4310. IOS Press (2024)
20. Putrich, V.S., Meneguzzi, F., Pereira, A.G.: Landmark generation in HTN planning, revisited. In: ICAPS 2025. AAAI Press (2025)

21. Wu, Y.X., Lin, S., Behnke, G., Bercher, P.: Finding solution preserving linearizations for partially ordered hierarchical planning problems. In: PuK 2022 "Planen, Scheduling und Konfigurieren, Entwerfen" (2022)
22. Wu, Y.X., Olz, C., Lin, S., Bercher, P.: Grounded (lifted) linearizer at the IPC 2023: solving partial order HTN problems by linearizing them. In: IPC 2023 (2023)
23. Zhang, Y., Bercher, P.: Computational complexity of planning for recursive primitive task networks: selective action nullification with state preservation. In: IJCAI 2025. IJCAI (2025)

Planner-Independent Extraction of Goals and Constraints from Natural Language for Open-World Mobile Robot Missions

Björn Döschl(✉) and Jane Jean Kiam

University of the Bundeswehr Munich, 85579 Neubiberg, Germany
{bjoern.doeschl,jane.kiam}@unibw.de

Abstract. In real-world applications, the demand for unmanned, autonomous platforms rises; yet these robot platforms are often deployed in highly dynamic environments and are operated by personnel with little to no expertise in formal planning, for whom plan validation must be simple and follow a human-in-the-loop ("four-eyes") principle. To address this challenge, we propose a two-stage pipeline for automatic goal and constraint extraction for problem instance acquisition, where extracted goals complement the problem instance and constraints guide execution in open-world mobile robot mission planning. Our Large Language Model (LLM)-based goal and constraint extraction (GCE) module parses the operator-defined natural language mission instruction into a structured planner-independent goal and constraint representation while being robust to slightly misphrased mission instructions. In a second stage, a planner-specific projection layer maps these general goals and constraints to the syntax of any downstream planner. In a first validation step, on a fixed-domain benchmark of operator-authored mission instructions, our LLM-based extractor achieves higher exact-match goal identification and constraint coverage than a strong rule-based baseline, and we situate these results within current work. In a second step, we integrate the full pipeline into AUSPEX and showcase the planner-independent to planner-specific projection functionality based on different planners for a sample Search and Rescue (SAR) scenario with UAVs.

Keywords: Automated planning · Problem instance acquisition · Natural language processing

1 Introduction

Mobile robots are increasingly used in time-critical operations such as Search and Rescue (SAR) and surveillance. In order to improve the effective operation area, scaling up the fleet size appears to be a reasonable next disruptive step in their deployment. However, this can exponentially increase the operator's workload, since each additional robot introduces another stream of tasks to assign, control and monitor. To reduce this workload, which is particularly essential in

M. Liu et al. (Eds.): AI 2025, LNAI 16371, pp. 453–465, 2026.
https://doi.org/10.1007/978-981-95-4972-6_35

applications with real-time planning and acting, current research [12,23] trends towards issuing high-level commands in Natural Language (NL), interpreted by Large Language Models (LLMs). Directly mapping NL input to robot actions via chatbot-like methods is efficient [22], but can suffer from a lack of robustness and verifiability [14]. In many applications, especially safety- or time-critical ones, learning the domain is infeasible as it requires online verification. Rather, an approach is required that considers the domain as given, while leveraging the LLM's world knowledge, via Natural Language Processing (NLP), to acquire situation-specific details. This includes acquiring the problem instance, e.g., the extraction of goal conditions that make up a goal state, as well as the acquisition of constraints to be adhered to during execution.

Some recent works studied the possibility of interpreting NL for acquiring initial states and/or goals [19,21,23]. However, these methods typically assume that human operators are rational and are able to specify full logical and perfect natural language problem specifications. Moreover, previous work on problem instance extraction [4,6] is tailored to specific modelling syntaxes such as the Planning Domain Definition Language (PDDL). Clearly, decoupling the NL input layer from the underlying planner and its syntax, by introducing an intermediate representation, broadens the set of supported planners. This abstraction enables compatibility with diverse planning back-ends and allows a generalised monitoring of the execution process.

In this work, we propose an LLM-based pipeline that converts natural language mission descriptions into planner-independent, structured goals and constraints for open-world mobile robots. Our approach explicitly accounts for the logical inconsistencies in operator-provided mission instructions that are inevitable in real-world missions and are essential for assessing goal completion during execution. The extracted abstract goal and constraint representations can be further utilised for any downstream planner. The detailed approach can be summarised as follows:

1. By leveraging Natural Language mission instructions as input to a one-shot NLP module powered by LLMs grounded in domain-specific goal knowledge, we can bypass the need for operators to manually encode mission goals and eliminate the need for logically correct mission instructions.
2. By exploring a two-stage modular pipeline that separates goal and constraint extraction from planning, thereby generating a simpler intermediate space, more reliable producible by LLMs, that supports domain-specific error correction.

In this paper, we start by outlining the relevant related works, followed by a problem formulation of our two-stage goal and constraint acquisition pipeline, and a detailed method description. In the validation section, we show that our LLM-based NLP outperforms classic NLP for 100 SAR related NL mission samples. Moreover, we validate the effectiveness of our approach by integrating the pipeline into an aerial robotics framework, AUSPEX [10], and demonstrate, using its built-in planners, a range of SAR use-case scenarios.

2 Related Works

Open-world planning complicates problem instance acquisition due to underspecified or only partially observable state spaces. Jiang et al. allow agents to act on hypothesised objects but require a narrow hypothesis space, thereby struggling in more dynamic environments [13]. The Unified Planning [18] supports planning in more dynamic environments yet still needs manual problem specification, limiting its effectiveness in open-world settings. Stern et al. update internal models online, while reacting to detected changes in the environment [20]. Continual Open-World Planning (COWP) [8] fuses symbolic planning with LLMs but it inherits the need for validating an LLM's output. Recent research leverages LLMs and their embedded world knowledge for full planning problem learning [4,19,21]. However, in open-world scenarios, where domain modelling is incomplete, training LLMs on dynamic, open-world domains is not feasible, due to the vast amount of data the operator has to provide.

Considering a domain-independent planner or a predefined planning domain, only the problem instance must be generated or complemented. Current approaches to extract goals and constraints from NL instructions rely on NLP-based methods that span *classical, rule-based* [11,24], *hybrid* (e.g., BERT [7] encoders) [1], or *LLM-only* approaches [2,19,21,23]. Tantakoun et al. provide a detailed survey in their work [21], showing recent approaches regarding translating NL to goals or full problem instances [5,12,15,17,23,25]. In contrast to these previous works, for our LLM-based NLP, we neither assume that the human operator understands planning logic nor can formulate perfectly rational mission instructions. *We explicitly consider mission instructions that may be logically incorrect.*

3 Problem Statement of an NLP-Based Goal and Constraint Extraction

Our problem can be defined as extracting the mission goals and constraints from operator-provided natural language instruction l, with $l \in \mathcal{L}$, and $\mathcal{L}$ being the space of all possible natural language instructions. Our specific problem can be divided into the acquisition of planner-independent goals and constraints Γ and the translation of the planner-specific goals and constraints γ for a planner P. Based on this foundation, Γ can be defined as:

$$\Gamma = \langle G, C \rangle, \quad \text{where } G \subseteq \mathcal{G},\ C \subseteq \mathcal{C}, \tag{1}$$

with $\mathcal{G}$ and $\mathcal{C}$ as the space of all possible intermediate goals and constraints. To obtain the set of extracted goals G and constraints C packed as Γ, a function $\Phi : \mathcal{L} \rightarrow \mathcal{G} \times \mathcal{C}$ needs to be defined, which maps a natural language instruction $l \in \mathcal{L}$ to a subset $G \subseteq \mathcal{G}$ and a subset $C \subseteq \mathcal{C}$.

Moreover, to utilise Γ for planning, a planner-specific projection function, $\Pi : \mathcal{G} \times \mathcal{C} \rightarrow \gamma$, is required to map the planner-independent goals G and constraints C to a planner-specific representation γ[1] for a selected planner P.

4 Goal and Constraint Extraction for Open-World Scenarios

Our system follows a modular pipeline architecture that bridges high-level natural language input with planner-specific syntax (see Fig. 1).

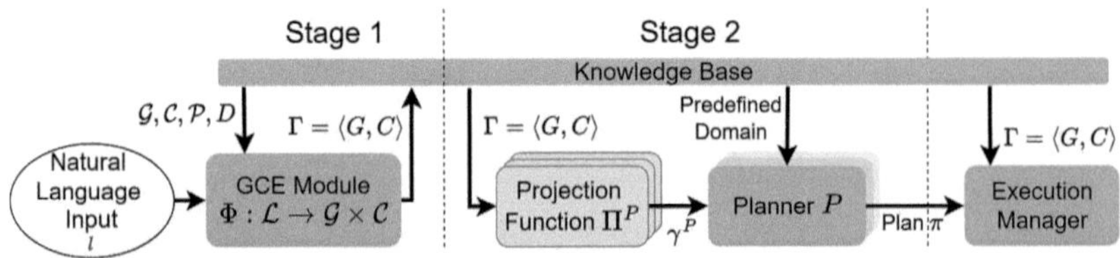

Fig. 1. Two-stage pipeline that extracts goals and constraints from natural language, then projects them for domain-specific planning and execution.

This pipeline can be divided into two stages: (1) the extraction of planner-independent goals and constraints from natural language and (2) the projection of these to planner-specific syntax, which are explained in the following.

4.1 Stage 1: Planner-Independent Goal and Constraint Extraction

In the first stage, an NL input l is provided (see Fig. 1) by the operator to specify mission goals and constraints. This input is processed by the *goal and constraint extraction (GCE)* module, implementing extraction function $\Gamma = \Phi(l)$, which extracts semantic content and maps the NL input l to a planner-independent, formal goal G and constraint C representation $\Gamma = \langle G, C \rangle$. This structured tuple is decoupled from any specific planning formalism.

To extract goals and constraints, our GCE module employs NLP to effectively handle open-world scenarios, while relying on LLMs for their extensive world knowledge. The detailed process is shown in Algorithm 1. Before querying an LLM, it is necessary to first specify the following components:

1. the goal space $\mathcal{G}$,
2. the constraint space $\mathcal{C}$,
3. the parameter space $\mathcal{P}$.

[1] Here, γ can be PDDL goal conditions, a goal formulated as Unified Planning (UP) goal, a goal in natural language, or any planner-specific goal representation.

Algorithm 1 Goal and constraint extraction (GCE) Module $\Phi : \mathcal{L} \rightarrow \mathcal{G} \times \mathcal{C}$

Given: Goal, constraint and parameter spaces: $\mathcal{G}$, $\mathcal{C}$, $\mathcal{P}$.
Given: Domain prompt p
Input: Natural language input $l \in \mathcal{L}$.
1: Initialise extracted goal set $G \leftarrow \emptyset$
2: **while** $G = \emptyset$ **do**
3: $\quad m \leftarrow \texttt{concat}(p, \mathcal{G}, \mathcal{C}, \mathcal{P}, l)$ ▷ Generate Prompt
4: $\quad r \leftarrow \text{LLM}(m)$ ▷ Query LLM
5: $\quad \langle G, C \rangle = \texttt{parse}(r)$ ▷ Parse Result to JSON
6: $\quad$ **if** $G = \emptyset$ **then**
7: $\qquad l \leftarrow \texttt{user\_refine}(l)$
8: $\quad$ **end if**
9: **end while**
10: **return** $\langle G, C \rangle$

The above components collectively define the valid goal, constraint, and parameter spaces. The parameter space $\mathcal{P}$ entails all possible parameters with which a goal $g \in G$ and constraint $c \in C$ can be parameterised. Goal, constraint, and parameter spaces, $\mathcal{G}$, $\mathcal{C}$, and $\mathcal{P}$, are dynamically acquired for each evaluation of the goal and constraint extraction (GCE) module Φ, by querying a knowledge base for feasible goals $\mathcal{G}$ (e.g., `searched`, `found`, `scanned`, `deployed`), constraints $\mathcal{C}$ (e.g., `WITHIN_BOUNDS`, `ABOVE_MIN_ALTITUDE`, `WITHIN_ TIME_LIMIT`) and parameters $\mathcal{P}$, such as potential search areas for a considered platform or possible boundaries for a mission constraint. The input to Φ is a natural language instruction $l \in \mathcal{L}$, such as "*Search the forest and find a person.*", specified by the platform operator and the domain prompt p, which provides detailed instructions to the LLM on how to process the input and what output format is expected. To illustrate the format, an instruction always begins with: "*You are a mission-planning assistant. Your job is to parse user instructions and return ONLY valid JSON. ...*", which directs the LLM to the required output format. Moreover, p already includes an example input-output pair to prevent the LLM from hallucinating output formats. While the extracted goal set G is empty, the function Φ generates a prompt message m by concatenating the domain prompt p with the goals, constraint, and parameter spaces $\mathcal{G}$, $\mathcal{C}$, and $\mathcal{P}$, along with the natural language input l (see Algorithm 1, Lines 3 and 4). This prompt is then evaluated by an LLM, yielding response r in the JSON format, specified in the domain prompt p. By loading the JSON r in Python, we can parse it to goals and constraints $\Gamma = \langle G, C \rangle$. If the goal set G is empty, the user is prompted to refine the input l, and the process is repeated.

To be robust to slightly misphrased mission instructions, since humans are not perfectly rational, in a second step, we exploit domain-specific goal knowledge to link the individual extracted goals with logical connectives `AND` and `OR`. A goal condition can either be classified as a `task` goal or a `mission` goal. A task goal is achieved by executing a specific action (e.g., `search(area1)`), whereas a mission goal is a condition that "must" be satisfied (e.g., `found(person)`). A function then joins goals of the same class with conjunctions (AND) and combines goals with different classes with a disjunction (OR), yielding a first-order logic tree that evaluates to `true` only if all of its leaf nodes do, providing a robust criterion for verifying goal completion during execution.

Our goal and constraint representation adheres to a predefined ROS 2 message format[2], populating fields such as `goal_id`, `goal_type`, and `parameters_json`.

4.2 Stage 2: Projecting Γ to Planner-Specific Goals and Constraints

Our approach extracts goals and constraints in a planner-independent format, without tailoring them to any specific modelling syntax. Therefore, each planner needs its own planning front-end capable of understanding the syntax used for modelling planning problems[3]. This front-end is responsible for projecting the goals G and constraints C to the syntax-specific tuple γ.

This is realised in the second stage, where a projection function Π^P (see Fig. 1) of a planner P maps the planner-independent pair $\langle G, C\rangle$ to planner-specific goals and constraints $\gamma = \langle G^P, C^P\rangle$. Hence, each Π^P has to implement a respective conversion tailored to its planning back-end P. Constraints are only projected when the planner can process them; otherwise, they are omitted and only used during execution for monitoring the validity of the plan.

To construct γ^P, Π^P iterates over all goals $g \in G$ and constraints $c \in C$, processing a translation function that realises the semantics and syntax required by planner P. This function projects each element to its planner-specific representation g^P and c^P, yielding the tuple $\gamma^P = \langle G^P, C^P\rangle$. Note that $G^P \not\subseteq G$ and $C^P \not\subseteq C$, as planner-specific projections, introduce structural or semantic transformations. Since most planners come with their own syntax and semantic conventions, this translation function is highly planner- and implementation-dependent.

Subsequently, the planner P computes with γ^P and the predefined planning domain as input, a corresponding plan trajectory $\pi = P(\gamma^P)$.

As an illustrative example, we implemented the projection function for three representative planners, as shown in the following:"

1. **PDDL-based Planner Library P^{UP}:**
   UP [18] includes a range of PDDL-based planners using one corresponding projection function Π^{UP} which we can define by implementing the conversion from Γ to a PDDL-specific syntax $\gamma^{UP} = \Pi^{UP}(\Gamma)$. For this, Π^{UP} matches the given goals to its predefined PDDL fluents and adds the respective fluent states as goals to the planning problem.
2. **LLM-modulo Planner $P^{Say'n'Fly}$** [9]:
   $P^{Say'n'Fly}$ expects a natural language mission description as input. To provide input to $P^{Say'n'Fly}$, the projection function $\Pi^{Say'n'Fly}$ converts all goals g

[2] AUSPEX-MSGS GitHub.

[3] For example, AI planners mostly rely on PDDL for problem definition.

to natural language, including the corresponding parameters, spelling out the logical connectives between the goals and stripping away unnecessary filler text.

3. **Pattern Planner $P^{Pattern}$:**
The pattern planner $P^{Pattern}$ is only able to plan for `scanned` or `searched` goals. For these goal types, the projection function $\Pi^{Pattern}$ extracts the goal parameters, such as search area, desired altitude, turning radius and field-of-view of the respective platform. Based on these, $\Pi^{Pattern}$ generates a low-level mission description variable, which is used as input to $P^{Pattern}$. The planner, thereupon, is able to plan a search pattern (e.g., lawn mower pattern) for the given search area.

4.3 Execution Monitoring

In real-world applications, robots can always deviate from plans, e.g., due to hardware failures, or violate constraints not considered during planning; therefore, continuous execution monitoring is essential for short-term intervention. Since $\Gamma = \langle G, C \rangle$ is decoupled from any specific planning back-end, it can be leveraged as a high-level reference for goal monitoring and constraint compliance checks, beyond its usage for planning. For this, we introduce an Execution Manager (see Fig. 1), which is responsible for evaluating the abstract goals G and constraints C against the dynamic world state by computing a nominal/actual value comparison. By checking if any goals are completed or constraints are violated, the *Execution Manager* can notify the respective planning engine P about states like *completed*, *cancelled*, or *incomplete*.

- *completed*: all goals in G evaluate to true, implying a successful execution.
- *cancelled*: some constraint $c \in C$ is violated, the operator cancelled the plan, or the execution platform does not adhere to the plan.
- *incomplete*: the current plan π is insufficient to complete all goals G, i.e., all actions from a plan are executed, but some goals are still not completed.

For example, if the flag `WITHIN_TIME_LIMITS` is set and the execution time, entailed in the world state, exceeds the specified duration, the Execution Manager cancels the plan and notifies the planner about the new *cancelled* state.

5 Evaluation and Results

This section presents the experimental validation of the proposed NLP-based goal and constraint extraction pipeline, along with its integration into the AUSPEX [10] framework.

5.1 Validation Stage 1: The Abstract Goal and Constraint Extraction

To evaluate the accuracy of Stage 1, we consider two configurations of our goal and constraint extraction module: (i) one leveraging an LLM for NLP and

(ii) another using a classic, rule-based, keyword-driven approach. We benchmark these configurations against the L2P model [21] with the DomainBuilder pipeline. L2P is an offline toolkit that uses LLMs to translate natural language task and domain descriptions into PDDL domain + problem files, making the resulting specification solvable by any classical planner. In this evaluation, we use OpenAI's Chat-GPT model `gpt-4o-mini`(https://openai.com) as the underlying LLM, and the NLP library `nltk` [3] as a rule-based variant. Since we focus on zero- to one-shot extraction without previous fine-tuning, we omitted NLP models such as BERT [7] and RoBERTa [16].

For the comparison, we construct a test dataset $\mathcal{L}$ consisting of 100 natural language instructions for SAR scenarios with Unmanned Aerial Vehicles (UAVs). The set includes 25 simple, single-goal instructions, 50 instructions with multiple goals and constraints formulated, and 25 complex instructions with variability in the number and types of goals. Each instruction $l \in \mathcal{L}$ intentionally varies in phrasing and keyword usage, with some instructions referencing parameters that cannot be matched to the predefined domain space, in order to reflect open-world conditions. For example, considering the natural language instruction l_1 =*Search the open areas and find a person with a red t-shirt. Please stay within bounds and above 20 m altitude.*, we can compute our planner-independent goals by evaluating the GCE module with input l_1 to $G = \{g_1, g_2\}$ and $C = \{c_1, c_2\}$:

$$\langle \{g_1, g_2\}, \{c_1, c_2\} \rangle = \Phi(l_1) \tag{2}$$

yielding the following abstract goal and constraint representation Γ:

$$g_1 = \{\texttt{type} : \texttt{search},\ \texttt{areas} : \texttt{openareas}\}, \tag{3}$$

$$g_2 = \{\texttt{type} : \texttt{find},\ \texttt{obj} : \texttt{person},\ \texttt{clr} : \texttt{red}\} \tag{4}$$

$$c_1 = \{\texttt{type} : \texttt{WITHIN\_BOUNDS}\} \tag{5}$$

$$c_2 = \{\texttt{type} : \texttt{MIN\_ALTITUDE}\}, \tag{6}$$

In the second step of Stage 1, SAR domain knowledge is used to create the respective logical connectives. Given the insight that a SAR mission succeeds when either the search area is scanned or the target object is located, we can combine the two goals into a logic tree rooted at an `OR` node with leaves g_1 and g_2, yielding expression $g_1 \vee g_2$. By leveraging this domain-specific goal knowledge, the GCE module corrects slightly misphrased mission instructions. The task is to "*search and find*", but the module recognises the logical goal disjunction and replaces the written *and* with a logical `OR`.

The evaluation results for input set $\mathcal{L}$ are presented in Table 1. We compared the outputs generated by the classic NLP pipeline, the LLM-based approach, and the L2P against an expert-generated ground truth. Specifically, we assessed the accuracy by counting the correctly extracted goal names, goal details (i.e., parameters), and constraints.

All methods achieve ≈ 90 % on goal names. However, rule-based approaches fall short due to the infeasibility of defining all possible keyword variations for each goal name. Our LLM-based approach and L2P achieved higher accuracy

Table 1. Accuracy comparison of goal and constraint extraction methods: rule-based, our LLM-based approach, and L2P.

	Rule-based	L2P	Our Approach
Goal Name	0.890	0.910	0.926
Goal Details	0.627	0.8142	0.853
Constraints	0.807	0.9167	0.951
Overall	**0.77**	**0.88**	**0.91**

in identifying goal details (62.7 % vs. ≈ 83 %) and constraints (80.7 % vs. ≈ 93 %), indicating the LLM's superior ability to leverage world knowledge. Our LLM-based approach performs slightly better than L2P due to the embedded domain-specific knowledge in the goal-extraction step (see Section Stage 1). This comparison highlights that in open-world scenarios, where instructions can vary in phrasing and rely on implicit knowledge, the extensive world knowledge embedded in LLMs provides a substantial advantage over rule-based NLP systems.

We further investigate the performance impact of the chosen LLM of the GCE module. For this, we compare the previously evaluated Chat-GPT (`gpt-4o-mini`) model with two alternative state-of-the-art models: DeepSeek V3(DSv3) (`deep- seek-v3-0324`) and Claude 3 Opus (`claude-3-opus-20240229`) by Anthropic. The comparative results of this evaluation are summarised in Table 2.

Table 2. Comparing the accuracy of different LLMs for LLM-based NLP goal and constraint extraction.

	ChatGPT-4o	Claude 3	DSv3
Goal Name	0.926	0.906	0.917
Goal Details	0.853	0.854	0.801
Constraints	0.951	0.877	0.9717
Overall	**0.91**	**0.879**	**0.897**

These results demonstrate that, as long as the model is sufficiently recent and capable, the exact choice of the LLM does not significantly affect the extraction quality. The considered LLMs have all been trained on large-scale knowledge, making them more effective than keyword- or rule-based methods.

5.2 Validating Stage 2: The Planner-Specific Goal and Constraint Extraction

To validate the projection function Π^P, the GCE module Φ is integrated into AUSPEX [10], a decision-making framework designed for UAVs in rescue missions. AUSPEX supports the integration of various planners, ranging from symbolic approaches such as classical PDDL-based planners, to more data-driven back-ends, including language-driven (Say'n'Fly [9]) and path planners. To enable seamless integration with the GCE module, a dedicated projection function Π^P for each planner P was implemented within AUSPEX. Since Π^P is implementation-specific, we showcase the functionality using an example rather than scoring it.

For this validation, we pass the projection function the abstract goals and constraints $\Gamma = \langle \{g_1, g_2\}, \{c_1, c_2\} \rangle$ (see Eqs. 3, 4, 5, and 6) and test it with the following planners. For Say'n'Fly (P^{LLM}), an LLM-modulo planner, its projection function $\Pi^{Say'n'Fly}$ rewrites Γ as a concise NL prompt, removing any filler text: $\gamma^{Say'n'Fly}$ = *Search openareas and find a person in red. Stay within bounds and above min altitude.*

For any classical planner (P^{UP}) accessed via UP expecting a PDDL-like syntax for the planner-specific goal and constraint representation γ^{UP}. Because UP has a standardised input syntax across its supported planners, a single projection function Π^{UP} is sufficient. The respective Π^{UP} processes the goals $G = \{g_1, g_2\}$ and returns a UP-specific goal representation γ^{UP}:

$$\begin{aligned} \gamma^{UP} = \ & (\texttt{searched(openarea1)} \wedge \texttt{searched(openarea2)}) \\ & \vee \texttt{found(person)}. \end{aligned} \tag{7}$$

Here, Π^{UP} queries the knowledge base for all areas tagged as `openareas` and creates sub-goals `searched(openarea1)` and `searched(openarea2)`. Because these sub-goals are derived from a single task goal, they are combined with a logical `AND`. The goals `searched` (task goal) and `found` (mission goal) are combined with a logical `OR` (see Section Stage 1), since the execution is successful if either the searched-for object is found or all areas are searched.

In both cases, the constraints are used in the execution monitoring step to determine possible violations of the computed plan trajectory.

5.3 Discussion and Verification

In safety-critical missions, such as Search and Rescue scenarios, a false goal extraction could be fatal. Therefore, the pipeline is designed to prevent false goals from propagating into the planning layer. Our GCE module entails a closed, domain-specific goal space, inhibiting the hallucination of new goal conditions. Moreover, during Stage 1, if no goals are extracted, a `user_refine` function is called, prompting the operator to refine the language input to provide more details regarding the goals of the mission. Once planned, the plan trajectory is saved to the knowledge base and can be displayed for human-in-the-loop verification; execution is initiated only after the operator finally approves the plan. Considering these sanity checks, the accuracy on our 100-instruction test cases from Sect. 5.1 will increase to 100 %, meaning that every intended high-level goal was captured correctly and no unintended goals were introduced. This ensures that downstream planners receive only validated mission goals, eliminating major safety risks.

6 Conclusion

In this work, we introduced a two-stage pipeline that couples an LLM-based NLP module for planner-independent goal and constraint extraction, given a mission description in natural language, with a planner-specific projection layer that can be implemented for any downstream planner. By parsing goals and constraints in a planner-independent representation, we are decoupling them from the planning formalism, leveraging a more general interface to provide support for various planners. Validation results on 100 SAR-related mission descriptions in natural language showed that LLM-based NLP modules are superior to a classic rule-based extraction of goal and constraint instances with fixed domains. Furthermore, in a UAV SAR scenario, our goal and constraint extraction (GCE) module successfully parsed goals and constraints, by using domain-specific goal knowledge, injecting logical goal connectives, thereby making it robust against slightly misphrased mission instructions. Furthermore, we evaluated the effectiveness of our sample projection functions, based on the syntax of two AUSPEX-integrated planners. Each function converted the planner-independent goals and constraints into their planner-specific format for downstream planning.

As future work, we plan to strengthen the integration of the knowledge base and add ontology support. These enhancements will enable the problem acquisition to handle more precise knowledge solicitation, such as querying only areas near a given location instead of all areas of the same descriptor. Moreover, we plan to expand the possible cardinality of the goal, constraint, and parameter space the GCE module can reliably interpret to improve downstream quality, e.g., details for object detection.

As future work, we plan to strengthen the integration of the knowledge base and add ontology support. These enhancements will enable the problem acquisition to handle more precise requests, such as querying only areas near a specific location instead of all areas. Moreover, we plan to expand the number of parameters the GCE module can reliably interpret to improve downstream quality, e.g., details for object detection.

Acknowledgments and Disclosure of Funding. This work is funded by the Bundesministerium für Forschung, Technologie und Raumfahrt (BMFTR) (Project MENTHON) and by the Wehrwissenschaftliches Institut für Werk- und Betriebsstoffe (WIWeB) (Project EMUE) which we gratefully acknowledge. The modelling of the validation scenario benefited from valuable input provided by domain experts of the Penzberg Mountain Rescuers (Bergwacht Penzberg).

Disclosure of Interests. The authors have no competing interests to declare that are relevant to the content of this article.

References

1. Asai, M., Fukunaga, A.: Classical Planning in Deep Latent Space: Bridging the Subsymbolic-Symbolic Boundary (2017). arXiv:1705.00154

2. Bai, D., Singh, I., Traum, D., Thomason, J.: TwoStep: Multi-agent Task Planning using Classical Planners and Large Language Models (2025). arXiv:2403.17246
3. Bird, S., Loper, E., Klein, E.: Natural Language Processing with Python. O'Reilly Media, Inc., Sebastopol (2009). http://www.nltk.org/book
4. Capitanelli, A., Mastrogiovanni, F.: A framework for neurosymbolic robot action planning using large language models. Front. Neurorobot. **18** (2024). https://doi.org/10.3389/fnbot.2024.1342786. https://www.frontiersin.org/journals/neurorobotics/articles/10.3389/fnbot.2024.1342786
5. Collins, K.M., Wong, C., Feng, J., Wei, M., Tenenbaum, J.B.: Structured, Flexible, and Robust: Benchmarking and Improving Large Language Models Towards More Human-Like Behavior in Out-of-Distribution Reasoning Tasks (2022). arXiv:2205.05718
6. Dang, X., Kudláčková, L., Edelkamp, S.: Planning with Vision-Language Models and a Use Case in Robot-Assisted Teaching (2025). arXiv:2501.17665
7. Devlin, J., Chang, M.W., Lee, K., Toutanova, K.: BERT: pre-training of deep bidirectional transformers for language understanding. In: Burstein, J., Doran, C., Solorio, T. (eds.) Proceedings of the 2019 Conference of the North American Chapter of the Association for Computational Linguistics: Human Language Technologies, Volume 1 (Long and Short Papers), pp. 4171–4186. Association for Computational Linguistics, Minneapolis, Minnesota (2019). https://doi.org/10.18653/v1/N19-1423. https://aclanthology.org/N19-1423/
8. Ding, Y., et al.: Robot Task Planning and Situation Handling in Open Worlds (2024). arXiv:2210.01287
9. Döschl, B., Kiam, J.J.: Say'n'Fly: an LLM-modulo online planning framework to automate UAV command and control. In: Proceedings of the 2025 IEEE International Conference on Robot and Human Interactive Communication (RO-MAN). IEEE (2025, accepted for publication)
10. Döschl, B., Sommer, K., Kiam, J.J.: AUSPEX: An Integrated Open-Source Decision-Making Framework for UAVs in Rescue Missions (2025). https://doi.org/10.36227/techrxiv.174123265.55724570/v1. https://www.techrxiv.org/doi/full/10.36227/techrxiv.174123265.55724570/v1
11. Funkquist, M., Ståhlberg, S., Geffner, H.: Learning to Ground Existentially Quantified Goals (2024). arXiv:2409.20259
12. Gestrin, E., Kuhlmann, M., Seipp, J.: NL2Plan: Robust LLM-Driven Planning from Minimal Text Descriptions (2024). arXiv:2405.04215
13. Jiang, Y., Walker, N., Hart, J., Stone, P.: Open-world reasoning for service robots. In: Proceedings of the International Conference on Automated Planning and Scheduling, vol. 29, no. 1, pp. 725–733 (2019). https://doi.org/10.1609/icaps.v29i1.3541
14. Kambhampati, S., et al.: LLMs Can't Plan, But Can Help Planning in LLM-Modulo Frameworks (2024)
15. Liu, B., et al.: LLM+P: Empowering Large Language Models with Optimal Planning Proficiency (2023). arXiv:2304.11477
16. Liu, Y., et al.: RoBERTa: A Robustly Optimized BERT Pretraining Approach. CoRR abs/1907.11692 (2019). arXiv:1907.11692
17. Mahdavi, S., Aoki, R., Tang, K., Cao, Y.: Leveraging Environment Interaction for Automated PDDL Translation and Planning with Large Language Models (2024). https://openreview.net/forum?id=RzlCqnncQv
18. Micheli, A., et al.: Unified planning: modeling, manipulating and solving AI planning problems in python. SoftwareX **29**, 102012 (2025). https://doi.org/10.1016/j.softx.2024.102012

19. Miglani, S., Yorke-Smith, N.: NLtoPDDL: One-Shot Learning of PDDL Models from Natural Language Process Manuals (2020)
20. Stern, R., Piotrowski, W., Klenk, M.: Model-based adaptation to novelty for open-world AI. In: Proceedings of the Workshop on Planning and Robotics (PRL) at ICAPS (2022). https://prl-theworkshop.github.io/prl2022-icaps/papers/PRL2022_paper_3.pdf
21. Tantakoun, M., Zhu, X., Muise, C.: LLMs as Planning Modelers: A Survey for Leveraging Large Language Models to Construct Automated Planning Models (2025). arXiv:2503.18971
22. Vemprala, S.H., Bonatti, R., Bucker, A., Kapoor, A.: ChatGPT for robotics: design principles and model abilities. IEEE Access **12**, 55682–55696 (2024)
23. Xie, Y., Yu, C., Zhu, T., Bai, J., Gong, Z., Soh, H.: Translating Natural Language to Planning Goals with Large-Language Models (2023). arXiv:2302.05128
24. Xu, J., Cui, J., Li, J., Rong, W., Matsuda, N.: Entity Aware Syntax Tree Based Data Augmentation for Natural Language Understanding (2022). arXiv:2209.02267
25. Zhang, T., et al.: PROC2PDDL: Open-Domain Planning Representations from Texts (2024). arXiv:2403.00092

PERCY: Personal Emotional Robotic Conversational System

Zhijin Meng[1], Mohammed Althubyani[2], Shengyuan Xie[1], Imran Razzak[3], Eduardo B. Sandoval[1,4], Mahdi Bamdad[1], and Francisco Cruz[1,5(✉)]

[1] School of Computer Science, University of New South Wales, Sydney, Australia
[2] University of Technology Sydney, Sydney, Australia
[3] Mohamed bin Zayed University of Artificial Intelligence, Masdar City, Abu Dhabi, United Arab Emirates
[4] School of Art and Design, University of New South Wales, Sydney, Australia
[5] Escuela de Ingeniería, Universidad Central de Chile, Santiago, Chile

Abstract. Traditional rule-based conversational robots, constrained by fixed scripts and static response mappings, fundamentally lack adaptability for sustained personalized human interaction. Although large language models (LLMs) such as GPT-4 enable open-domain dialogue capabilities, most existing social robot approaches remain deficient in emotional awareness and longitudinal personalization continuity. To address this critical gap, we present PERCY (Personal Emotional Robotic Conversational sYstem) – an innovative framework that dynamically integrates: (1) real-time affective signals through facial expression recognition, (2) semantic content of user utterances, and (3) contextual profile data, synthesizing these multimodal inputs into emotion-aware prompt engineering for GPT-4. This integration drives both contextually appropriate verbal responses and synchronized non-verbal robot behaviors. PERCY utilizes GPT-4 to dynamically model the robot's internal affective state, with non-verbal feedback primarily expressed through facial expressions. The system architecture leverages ROS-based multimodal processing: visual emotion recognition via fine-tuned MobileNetV2, textual sentiment analysis using NLTK's VADER, decision-level sensor fusion, and GPT-4 prompt conditioning to orchestrate ARI robot behaviors. Empirical evaluation with 30 human participants demonstrated statistically significant improvements in dialogue coherence, contextual relevance, and response diversity compared to baseline systems. PERCY highlights the potential of integrating advanced multimodal perception and personalization to build a scalable foundation for next-generation emotionally intelligent human-robot interaction systems, rooted in contextually conditioned, multimodal affective computing.

Keywords: Cognitive modelling and computer-human interaction · Social Robotics · Human-Robot Interaction

1 Introduction

The evolution of social robotics has transformed robots from task-specific tools into collaborative partners capable of operating in human-centric environments,

M. Liu et al. (Eds.): AI 2025, LNAI 16371, pp. 466–478, 2026.
https://doi.org/10.1007/978-981-95-4972-6_36

where emotion and personalization play key roles, such as healthcare, hospitality, and education [24]. Consequently, social robots must be able to cognize human emotional states and support natural, profile-aware conversations, leveraging both verbal and non-verbal reactions back to humans. Although these systems excel at closed-domain tasks, traditional conversational systems struggle to adapt to the dynamic and open-ended human interaction due to their rule-based architectures [7].

The emergence of Large Language Models (LLMs), such as GPT-4, has marked a milestone in conversational AI. These models, trained on vast amounts of text data, demonstrate remarkable capabilities in understanding and generating human-like language. Unlike traditional rule-based systems, LLMs can handle a wide range of topics, adapt to diverse conversational contexts, and generate contextually relevant responses [1]. This makes them particularly well-suited for social robotics. However, large language models (LLMs) were not originally designed for social robotics contexts because their inputs are predominantly unidimensional – text only [10]. Relying solely on textual information is insufficient for social robots to infer users' emotional states. Integrating multimodal user data directly into the LLM prompt represents a promising direction to elicit more empathetic and contextually appropriate responses.

In this paper, we present PERCY, a novel architecture that bridges this gap by embedding user multimodal signals, such as persona, facial expressions, into the LLM prompt. PERCY's core innovation lies in its bidirectional affective grounding mechanism: (i) a real-time multimodal emotion recognition algorithm to provide user emotion state for LLM prompt, (ii) a GPT-4 core enhanced via prompt engineering, fed with facial and vocal signals, to dynamically modulate its response generation and (iii) a ROS-based behavior planner translates emotional semantics into synchronized non-verbal actions. The contributions of this work are threefold:

- We introduce a multimodal conversational framework that combines LLMs with real-time emotion recognition and personalized dialogue management, enabling social robots to engage in open-domain, emotionally aware interactions.
- We demonstrate the effectiveness of PERCY through automatic and user evaluations, showing that the system achieves high levels of naturalness, engagement, fluency, relevance, empathy, and consistency in its conversations.
- We provide insights into the potential of integrating LLMs into social robots, paving the way for future advancements in emotionally intelligent human-robot interaction.

2 Related Work

The design of PERCY integrates advances from three interrelated research strands: (1) large language models (LLMs) in social robotics, (2) multimodal affective computing, and (3) embodied interaction design.

2.1 Language Models in Social Robotics

Early conversational systems in robotics relied on rule-based or finite-state architectures [3], which achieved high precision in narrow domains but lacked the flexibility for open-ended dialogue. The advent of transformer-based LLMs has dramatically expanded the capabilities of dialogue systems through contextual awareness, few-shot learning, and fluent generation. Recent work has explored integrating off-the-shelf LLMs such as GPT-3 into social robots [9]; however, these approaches remain purely text-based and fail to perceive or respond to users' affective states.

To address this limitation, recent efforts have focused on endowing LLMs with empathetic capabilities. Qian et al. [20] conduct a systematic evaluation of LLMs (e.g., GPT-3.5) on empathetic dialogue benchmarks and propose improvements through semantically guided in-context learning, two-stage interactive generation, and augmentation with external knowledge bases. Their methods yield strong gains in both automatic and human evaluations, yet they operate exclusively in the textual modality and do not incorporate real-time user modeling, multimodal perception, or physical embodiment.

Further advancing this direction, Chen et al. [6] introduce a chain-of-thought fine-tuning framework that explicitly reasons about the causes behind user emotions, integrates commonsense knowledge from COMET, and enhances listener role awareness via structured prompting. Their approach achieves state-of-the-art performance on empathy, coherence, and response diversity. Nevertheless, it remains a text-only, non-embodied system requiring model fine-tuning. In contrast, PERCY builds on these LLM-based empathetic dialogue strategies but avoids fine-tuning altogether. Instead, it combines prompt engineering with real-time multimodal perception and persona modeling to ground responses affectively within an embodied robotic platform.

2.2 Multimodal Affective Computing

Robust affect recognition in naturalistic interaction requires the synergistic integration of verbal and nonverbal cues–mirroring how humans interpret emotions through facial expressions, vocal prosody, and body language [15]. While benchmark datasets like IEMOCAP [5] have enabled progress in controlled, dyadic settings, they poorly reflect the dynamic, open-world conditions of humanrobot interaction (HRI). Contemporary systems often fuse facial expression analysis with paralinguistic speech features (e.g., pitch, rhythm, pauses) to improve emotion detection accuracy [23], yet few support continuous personalization or adaptive user modeling in live interactions.

A notable recent effort is STICKERCONV [25], which enriches empathetic dialogue with expressive non-textual elements (stickers). The authors collect 12.9K dialogue sessions across 2K scenarios and propose the PEGS model to jointly generate text and sticker responses. While this expands the expressive modality of dialogue systems, STICKERCONV remains disembodied and does not integrate real-time facial emotion recognition, user profiles, or robotic non-verbal behaviors. PERCY addresses this gap by unifying multimodal perception–facial expression, voice, and user history–within a physically embodied agent capable of coordinated verbal and nonverbal output.

2.3 Embodied Interaction Design

Effective social robots must tightly coordinate linguistic content with congruent nonverbal behavior [4]. Foundational work by Mutlu et al. [14] established design principles for expressive robot body language, while systems like MIT's Huggable [8] demonstrated the therapeutic potential of embodied companions–albeit with scripted, domain-limited interactions. PERCY advances this line through a ROS-based behavior planner that couples dialogue generation with expressive output. Specifically, before verbalizing a response, PERCY internally evaluates the emotional valence of its utterance and assigns an affective label to the robot's state. This label is then mapped–via a predefined lookup table–to a facial expression, which the Emotion Control Module renders in real time. This architecture enables context-sensitive, affectively congruent nonverbal feedback that dynamically responds to both the semantic content of the conversation and the user's emotional state, thereby enhancing social presence and rapport.

3 Methodology

The PERCY framework, illustrated in Fig. 1, comprises three core components: (1) Speech Processing System, (2) Real-time Emotion Recognition and Non-Verbal Planner, and (3) GPT-4 Reasoning Engine.

3.1 Speech Processing System

Our speech processing system contains automatic speech recognition (ASR) and text-to-speech (TTS) synthesis to facilitate vocal human-robot interaction.

The ASR subsystem leverages the OpenAI Whisper [21], specifically the large-v2 model variant with an impressive 2.73% word error rate (WER). This high accuracy ensures that the robot can reliably understand and process users' spoken commands, even in noisy environments.

PERCY leverages the ARI robot's integrated TTS module [16] to articulate the system's verbal responses to users. While it offers limited emotional expressiveness, its simple encoder-decoder architecture ensures low GPU memory usage and rapid response times, facilitating effective communication with users.

3.2 Real-Time Emotion Recognition and Non-verbal Planner

The proposed system integrates a multi-modal emotion recognition framework that combines computer vision and sentiment analysis to enable empathetic human-robot interaction through perceiving users' emotional status (happy, sad, angry, confused, fearful, disgusted, neutral), with architecture shown in Fig. 2.

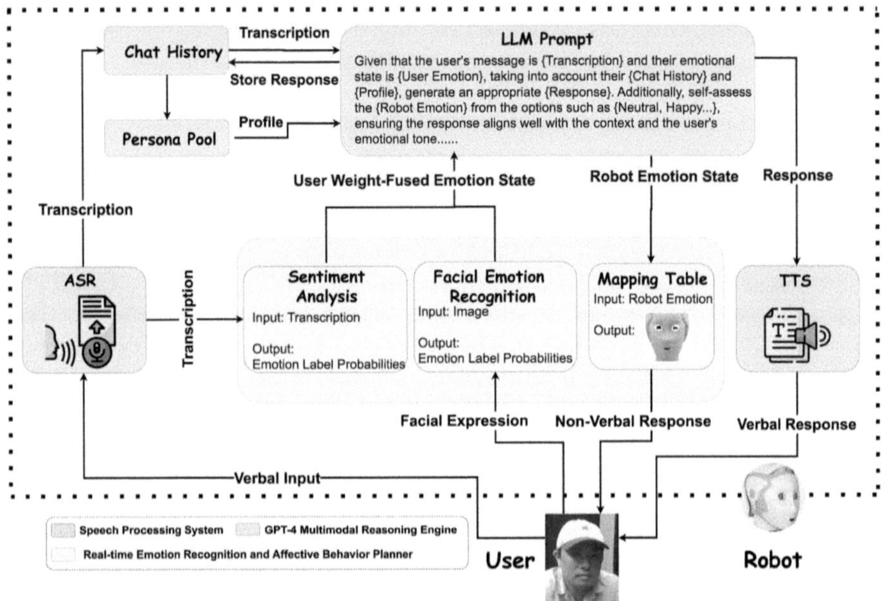

Fig. 1. PERCY's architecture assesses the user's real-time emotional states by analyzing their speech and facial expressions. By combining this information with short-term chat history and long-term user profile data, PERCY generates both verbal and non-verbal responses. These responses are customized based on the system's own evaluation of its emotional states, ensuring that interactions are coherent and contextually appropriate.

Computer Vision. For visual emotion recognition, a MobileNetV2-based backbone network extracts multi-scale feature maps from RGB-D facial data captured by the Intel RealSense D435i depth camera mounted on the ARI robot. These features are processed through dual prediction headers: a classification header for emotion scoring and a regression header for bounding-box detection. The Single Shot Detection (SSD) model refines predictions using prior boxes and applies Non-maximum Suppression (NMS) to eliminate low confidence detections, achieving robust real-time emotion classification.

Sentiment Analysis. Simultaneously, the sentiment analysis component utilizes the NLTK package with the VADER lexicon to analyze sentiment from speech-to-text outputs. This textual emotion analysis complements the visual data, providing a more holistic view of the user's emotional state. By combining these light-weight computer vision and natural language processing techniques, our emotion recognition system can accurately and rapidly interpret and respond to both visual and verbal cues, enhancing the robot's ability to engage with users in a timely, meaningful, and empathetic manner.

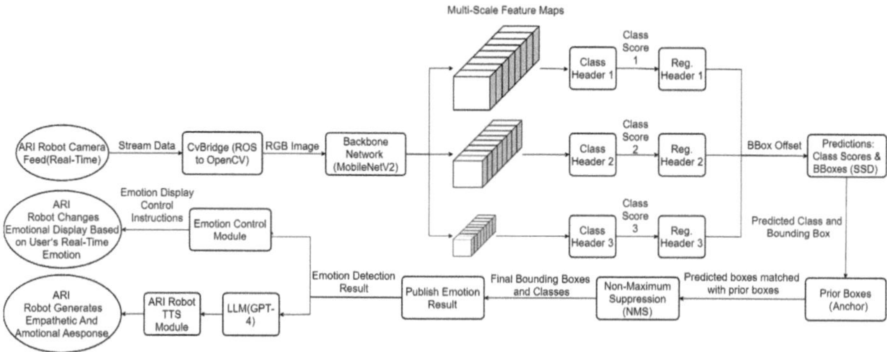

Fig. 2. PERCY captures RGB data with a depth camera. The data is processed using MobileNetV2 with an SSD detector for emotion classification and bounding-box prediction, while user speech is analyzed by NLTK's VADER. Both visual and textual emotion probabilities are combined using a weighted approach–0.6 for visual and 0.4 for textual sentiment–to determine the final emotion state. This multimodal input is used in the GPT-4 prompt to generate context-aware responses and adjust the robot's facial expressions via an Emotion Control Module.

Non-verbal Response Planner. The robot's emotional state, assessed by GPT-4 from its generated responses, is translated into non-verbal actions, specifically eye expressions, using predefined patterns on the ARI robot [22]. When PERCY generates a response, it selects an emotion from a list of 25 predefined eye expressions, which include happiness, sadness, surprise, and others. This selection is then matched to corresponding non-verbal reactions via a fixed mapping table, as illustrated in Fig. 3.

3.3 GPT-4 Reasoning Engine

We designed the GPT-4 Reasoning Engine with prompt engineering[1] based on the principle shown in Eq. (1). PERCY customizes responses by creating a long-term persona from the summary of all chat history, enhanced with the last 10 sentences annotated with detected emotions, to balance long-term personalization with short-term user patterns. As shown in Fig. 4, the system integrates LLM capabilities with affective reasoning through four operational layers.

$$R_t = \Phi_\theta(P_u \oplus E_t \oplus H_{t-1}), \quad (1)$$

where R_t is the generated response, Φ_θ represents GPT-4's parameters, P_u the user persona, E_t the decision-level fused emotional state, and H_{t-1} the 10 recent sentences history.

[1] The PERCY prompt and system code are accessible via our GitHub repository: https://github.com/zhijinMeng/PERCY.

```
{
"role": "system",
"content": "After generating your reply,
immediately append exactly **one** emotion
label from this fixed list of **25** emotions—
**netural, angry, sad, happy, surprised,
disgusted, scared, pleading, vulnerable,
desparied, guilty, disappointed, embarrassed,
horrified, skepical, annoyed, furious, suspicious,
rejected, bored, tired, asleep, confused, amazed,
excited**—prefixed by `| ` (pipe + space). For
example:\n\n"Here's your empathic reply... |
joy"\n\nUse this delimiter format so the emotion
can be easily parsed and passed along."
}
```

(a) GPT-4 prompt for non-verbal response mapping table

(b) Examples of ARI robot eye expressions aligned with the robot's emotional states

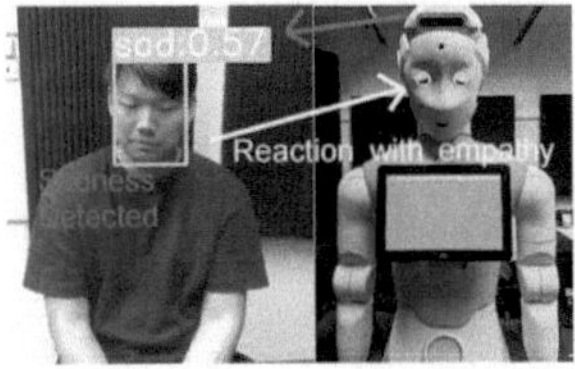

(c) PERCY generated a sad emotion to show empathy to a sad user

Fig. 3. PERCY evaluates its generated responses using prompts, assigns an emotional state, and translates this into non-verbal actions like eye expressions. As shown in Figure (a), PERCY first assesses its response to determine an emotion. It then looks up the corresponding eye expression from a predefined set of 25 emotions [17] in Figure (b). Finally, as illustrated in Figure (c), PERCY displays the selected expression to the user, effectively conveying the emotion through the ARI robot's expressive eyes.

4 Experimental Setup

4.1 Dataset

To comprehensively evaluate the multimodal capabilities of PERCY, we designed a dual-aspect evaluation framework focusing on its emotion-aware intelligence and generative competence. The experiments were conducted using the MERCI Dataset [2].

This custom-collected dataset contains 30 participants' open-domain conversations with an ARI robot, capturing multimodal interaction patterns (textual, vocal, and visual modalities). We leveraged this dataset to assess PERCY's contextual response generation quality in real-world human-robot interaction scenarios.

4.2 Models

Our evaluation framework compared PERCY's performance with two strategically selected baselines representing divergent approaches to affective human-AI interaction:

GPT-4. The standard GPT-4 implementation served as our text-centric baseline, maintaining its original 175B-parameter transformer architecture without multimodal augmentation.

EmpGPT-3. As the current state-of-the-art in empathic dialogue systems, EmpGPT-3 [11] employs a hybrid architecture combining a 175B-parameter GPT-3 backbone with:

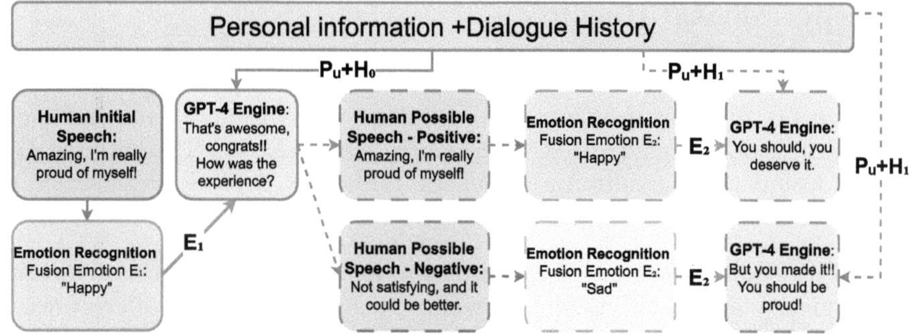

Fig. 4. GPT-4 Multimodal Reasoning Flow: PERCY fuses visual affect (Sect. 3.2) and textual sentiment (Sect. 3.2) via Eq. (1), integrating user personal information, emotional state, and dialogue history data to generate responses. The algorithm integrates emotional context with long-term user persona memory and short-term recent dialogue (last 10 sentences) to produce responses that align with both emotion and personalization. It combines LLM capabilities with ethical constraints and affective alignment for context-aware interaction.

- Emotion Context Database: Curated repository of 1.2M emotion-cause pairs
- Affective Alignment Module: Supervised attention layers for emotional lexicon weighting
- Rule-based Response Post-processing: Safety filters and empathy templates

While achieving notable performance in single-modal (text-only) emotional resonance. EmpGPT-3's late-fusion approach to affective computing (applying emotion rules post-generation) contrasts with PERCY's embedded multimodal affect perception architecture.

4.3 Evaluation Protocol

To evaluate the performance of PERCY, we conducted an automatic evaluation, we used metrics such as dialogue quality, emotion recognition accuracy, and personalization effectiveness to demonstrate that PERCY outperforms the baseline GPT-4 by involving multimodal inputs. In addition, we compared the performance of PERCY's embedded empathy framework and that of the EmpGPT-3's rule-based emotional adjustment framework.

Contrastive Analysis: The baseline selection strategically decouples two key factors:

- **Multimodal vs. Unimodal Input**: Text-only GPT-4 vs. PERCY integrating text, audio, and visual modalities.
- **Superficial vs. Embedded Empathy**: EmpGPT-3's rule-based emotional adjustment vs. PERCY's deep multimodal affect modeling.

5 Experimental Results

5.1 Automated Evaluation

Dialogue Quality Evaluation. We evaluated the coherence, relevance, and diversity of responses using BERT [12], BLEU [18], MAUVE [19], and Perplexity [13]. Comparison results are shown in Table 1. A detailed explanation of PERCY performance is as below:

- **BERT = 0.32**. The score of 0.32 indicates moderate alignment with human-generated responses. While the model produces fluent and relevant responses, it may miss deeper emotional or contextual nuances expected in empathetic dialogue.
- **BLEU = 0.51**. This score indicates PERCY's responses are relatively varied (a perfect score of 1 would mean identical responses). A score of 0.51 suggests a healthy balance between diversity and coherence, crucial for engaging in conversations.
- **MAUVE = 0.98**. This score suggests that the generated responses are diverse and closely mimic the statistical properties of human responses. The high MAUVE score confirms that the model produces varied and fluent responses that contribute positively to the flow of conversation.
- **Perplexity = 24.53**. Indicating PERCY's performance in predicting the next word. Lower perplexity is better, showing the model's confidence. A score of 24.53 is moderate, suggesting the model is reasonably good at generating meaningful responses but occasionally struggles with complex or emotionally nuanced dialogue due to the variability in conversational data.

Table 1. Model comparison showing PERCY's superior empathy (BERT) and diversity (BLEU/MAUVE), with GPT-4 leading in fluency.

Model	Empathy	Diversity		Fluency
	BERT	BLEU	MAUVE	PPL
PERCY	**0.32**	**0.51**	**0.98**	24.53
EmpGPT-3	0.28	0.48	0.92	28.75
GPT-4	0.25	0.45	0.95	**18.20**

Table 2. Comparison of PERCY with state-of-the-art models and PERCY achieves the highest accuracy in emotion recognition.

Model	Detection Type	Accuracy (%)
PERCY	Facial + Sentiment	**92.0**
YOLOv8n	Facial + Sentiment	91.3
YOLOv5n	Facial + Sentiment	90.5
FERPlus (ResNet-50)	Sentiment Only	91.4
OpenFace	Sentiment Only	89.8

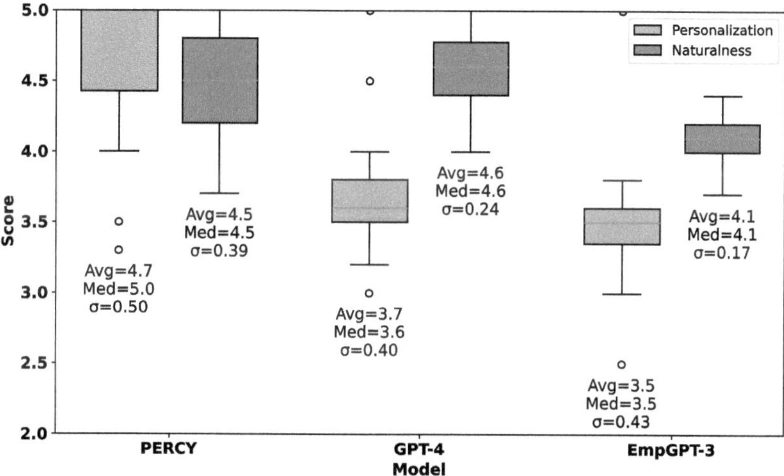

Fig. 5. Human Evaluation Result Box-plot: While PERCY, GPT-4, and EmpGPT-3 exhibited comparable naturalness scores, PERCY outperformed both in personalization, with average scores of 4.7, 3.7, and 3.5, respectively. This suggests that PERCY's integration of multimodal emotional cues and personalized user profiles contributes to more contextually relevant interactions.

Emotion Recognition Accuracy. Our MobileNetV2 + SSD model achieves an impressive **92.0%** emotion accuracy, outperforming OpenFace (89.8%) and matching FERPlus (91.4%). It also surpasses the YOLO variants (91.3% and 90.5%), confirming the advantage of MobileNetV2 + SSD for challenging facial analysis tasks. These comparisons are detailed in Table 2.

5.2 Human Evaluation

Despite the insights from automatic evaluations, human assessment remains vital as it captures nuanced aspects like tone and personalization, which are essential for understanding user experience. For this evaluation, 30 participants interacted with PERCY, GPT-4, and EmpGPT-3 through chat sessions and then rated each model's Personalization and Naturalness on a 1–5 scale with one decimal precision, providing detailed feedback based on their experiences.

The results of the human evaluation are summarized in Fig. 5. Thirty participants assessed PERCY, GPT-4, and EmpGPT-3 on personalization and naturalness using a 15 scale. While naturalness scores were similar across models, PERCY excelled in personalization with a score of 4.7/5, compared to GPT-4's 3.7 and EmpGPT-3's 3.5. This suggests that PERCY's integration of multimodal emotional cues and personalized user profiles contributes to more contextually relevant interactions.

6 Conclusion

This work presented PERCY, a multimodal conversational system framework combining GPT-4 with real-time emotion recognition and adaptive dialogue management to enhance human-robot interaction. Our contributions include (i) Integrating LLMs with real-time emotion recognition and adaptive dialogue management enables robots to engage in emotionally sensitive, open-domain conversations, bridging the gap between technical functionality and social robot requirements. (ii) PERCY outperforms GPT-4 and EmpGPT-3 in empathy and response diversity, maintaining competitive fluency, proving the potential of the use of multimodal information in HRI. (iii) Our deployment experience provides guidelines for balancing computational efficiency with social nuance, informing future work on scalable, ethically grounded HRI systems.

These findings suggest PERCY's potential as a foundation for future social robots. Due to the functional limitations of the ARI robot, no functional nonverbal responses could be explored. Future endeavors will focus on employing more advanced robotic platforms and action planners to construct more intricate interaction frameworks. These results highlight PERCY's potential in laying the groundwork for next-generation social robots. Further investigations should aim at exploring cross-cultural adaptations as well as the development of empathy coordination among multiple robots. Additionally, by offering an open-source implementation, our goal is to provide a reference standard for the community, emphasizing the critical importance of combining language generation capabilities with multimodal perception for enhancing human-robot interaction experiences.

References

1. Alawida, M., Mejri, S., Mehmood, A., Chikhaoui, B., Isaac Abiodun, O.: A comprehensive study of chatgpt: advancements, limitations, and ethical considerations in natural language processing and cybersecurity. Information **14**(8), 462 (2023)
2. Althubyani, M., et al.: Merci: multimodal emotional and personal conversational interactions dataset (2024). arXiv:2412.04908. Accessed 17 Sept 2025
3. Bohus, D., Horvitz, E.: Interactive visual dialog with robots. In: Proceedings of the 2014 ACM/IEEE International Conference on Human-Robot Interaction, pp. 65–72 (2014)

4. Breazeal, C.: Role of expressive behaviour for robots that learn from people. Philos. Trans. Roy. Soc. B Biol. Sci. **364**(1535), 3527–3538 (2009)
5. Busso, C., Bulut, M., Lee, C.C., et al.: Iemocap: interactive emotional dyadic motion capture database. Lang. Resour. Eval. **42**, 335–359 (2008). https://doi.org/10.1007/s10579-008-9076-6
6. Chen, X., et al.: Cause-aware empathetic response generation via chain-of-thought fine-tuning. arXiv preprint arXiv:2408.11599 (2024)
7. Cherakara, N., et al.: Furchat: an embodied conversational agent using LLMs, combining open and closed-domain dialogue with facial expressions. arXiv preprint arXiv:2308.15214 (2023)
8. Jeong, S., Breazeal, C., Logan, D., Weinstock, P.: Huggable: the impact of embodiment on promoting socio-emotional interactions for young pediatric inpatients. In: Proceedings of the 2018 CHI Conference on Human Factors in Computing Systems, pp. 1–13 (2018)
9. Kim, T.L., Lu, K., Havoutis, I.: Inteliplan: interactive lightweight LLM-based planner for domestic robot autonomy. arXiv preprint arXiv:2409.14506 (2024)
10. Kong, K., et al.: Opentab: advancing large language models as open-domain table reasoners. arXiv preprint arXiv:2402.14361 (2024)
11. Lee, Y.J., Lim, C.G., Choi, H.J.: Does GPT-3 generate empathetic dialogues? A novel in-context example selection method and automatic evaluation metric for empathetic dialogue generation. In: Proceedings of the 29th International Conference on Computational Linguistics, pp. 669–683 (2022)
12. Li, B., Han, L.: Distance weighted cosine similarity measure for text classification. In: Yin, H., et al. (eds.) IDEAL 2013. LNCS, vol. 8206, pp. 611–618. Springer, Heidelberg (2013). https://doi.org/10.1007/978-3-642-41278-3_74
13. Meister, C., Cotterell, R.: Language model evaluation beyond perplexity. arXiv preprint arXiv:2106.00085 (2021)
14. Mutlu, B., Shiwa, T., Kanda, T., Ishiguro, H., Hagita, N.: Footing in human-robot conversations: how robots might shape participant roles using gaze cues. In: Proceedings of the 7th Annual ACM/IEEE International Conference on Human-Robot Interaction, pp. 313–320 (2012)
15. Ngiam, J., Khosla, A., Kim, M., et al.: Multimodal deep learning. In: ICML (2011)
16. PAL Robotics: Ari SDK speech documentation (2023). https://docs.pal-robotics.com/ari/sdk/23.1.15/speech.html. Accessed 31 Jan 2025
17. PAL Robotics: Expressive eyes (2025). https://docs.pal-robotics.com/ari/sdk/23.1.13/expressiveness/expressive-eyes.html. Accessed 31 July 2025
18. Papineni, K., Roukos, S., Ward, T., Zhu, W.J.: Bleu: a method for automatic evaluation of machine translation. In: Proceedings of the 40th Annual Meeting of the Association for Computational Linguistics, pp. 311–318 (2002)
19. Pillutla, K., et al.: Mauve: measuring the gap between neural text and human text using divergence frontiers. Adv. Neural. Inf. Process. Syst. **34**, 4816–4828 (2021)
20. Qian, Y., Zhang, W.N., Liu, T.: Harnessing the power of large language models for empathetic response generation: empirical investigations and improvements. arXiv preprint arXiv:2310.05140 (2023)
21. Radford, A., Kim, J.W., Xu, T., Brockman, G., McLeavey, C., Sutskever, I.: Robust speech recognition via large-scale weak supervision. In: International Conference on Machine Learning, pp. 28492–28518. PMLR (2023)
22. Robotics, P.: Expressive eyes: SDK documentation (2023). https://docs.pal-robotics.com/ari/sdk/23.1.13/expressiveness/expressive-eyes.html. Accessed 24 Feb 2025

23. Schuller, B., et al.: The interspeech 2018 computational paralinguistics challenge: atypical & self-assessed affect, crying & heart beats. In: INTERSPEECH, pp. 122–126 (2018)
24. Shimoda, S., et al.: What is the role of the next generation of cognitive robotics? Adv. Robot. **36**(1–2), 3–16 (2022)
25. Zhang, Y., Kong, F., Wang, P., Sun, S., Wang, L., Feng, S., et al.: Sticker-conv: generating multimodal empathetic responses from scratch. arXiv preprint arXiv:2402.01679 (2024)

Understanding Human Situation Awareness in One-to-Many Human-Robot Interaction Scenarios

Wasura D. Wattearachchi(✉), Erandi Lakshika, Kathryn Kasmarik, and Michael Barlow

University of New South Wales (UNSW), Canberra, Australia
w.wattearachchi@unsw.edu.au

Abstract. With evolving technology, enabling a single operator to control multiple robots or a robotic swarm is becoming technically feasible. However, policy and regulatory frameworks do not yet address these scenarios, and we do not yet have a strong understanding of the impact of controlling large numbers of robots on human situation awareness. This paper addresses this gap through an empirical study that systematically degrades interface information to assess its impact on operator situation awareness and task performance in a human-swarm interaction scenario. Using a tablet-based interface, participants guide a robotic swarm through a simulated spreading fire task, and situation awareness and task performance are monitored. Results reveal that degrading information leads operators to focus on a single robot, often neglecting the rest of the swarm, a narrowing of attention that raises concerns for single-operator-multi-robot operations. Additionally, only certain information types were found to significantly affect situation awareness and performance, suggesting that some elements are critical while others are not. These findings underscore the need for regulatory support mechanisms such as secondary operators and call for flexible, task-specific policies that prioritise the most critical information elements in multi-robot operations.

Keywords: Human-computer interaction (HCI) · human-swarm interaction (HSI) · robotics · drones · situation awareness (SA) · user interface design

1 Introduction

This paper contributes to ongoing discussions around regulatory standards on single-operator, multi-robot control. Currently, aviation and robotics regulations often limit one operator to a single robot [1]. However, with growing interest in enabling a single operator to control multiple robots or a Robotic Swarm (RS), it becomes critical to understand how their ability to manage the swarm effectively with awareness.

RSs have increasingly been deployed in high-risk environments such as disaster response, where Human-Swarm Interaction (HSI) must be carefully managed [2]. A key factor in successful HSI is the operator's Situation Awareness (SA). This paper presents an empirical study where participants interact with a robotic swarm through a tablet

M. Liu et al. (Eds.): AI 2025, LNAI 16371, pp. 479–490, 2026.
https://doi.org/10.1007/978-981-95-4972-6_37

interface in a simulated fire scenario. We extend an existing framework [3] to examine how specific types of degraded information influence both SA, measured using the Situation Awareness Global Assessment Technique (SAGAT) and Situation Awareness Rating Technique (SART), and task performance (TP) assessed through robot distance-to-target metrics. By systematically degrading information, the study investigates the relationship between SA and TP under degraded interface conditions, while also revealing insights into the challenges of single-operator-multi-robot control. These insights may be valuable for informing future regulatory frameworks. By further examining SA, the study explores whether all types of information equally affect SA and whether identifying the most critical information dimensions is essential for guiding future policy and regulation in single-operator-multi-robot systems.

This study provides two key contributions:

- Operational insight for multi-robot control under degraded information: It reveals that under fully degraded information, operators tend to focus on a single robot—neglecting the rest of the swarm—highlighting limitations in single-operator-multi-robot control and the potential need for regulatory support mechanisms such as secondary operators.
- Identify high-impact information elements: It shows that not all information impacts SA equally, emphasising the need for flexible, task-informed regulations based on empirical assessments of what information is truly essential.

Through these contributions, a deeper understanding emerges of how interface design and information availability affect operator effectiveness in HSI scenarios. The remainder of the paper is structured as follows: Sect. 2 presents the background and related work; Sect. 3 describes the interface design and outlines the user study procedure; Sect. 4 details the experimental results; Sect. 5 provides a discussion of key insights; and Sect. 6 concludes the paper and suggests directions for future work.

2 Background and Related Work

This section presents the background and related work on HSI and SA, establishing the need for a comprehensive study of information degradation in HSI compared to previous research.

2.1 Human-Robot Interaction and Human-Swarm Interaction

When a human operator controls or guides a robot, the interaction is referred to as Human-Robot Interaction (HRI) [4]. However, when working with multiple robots, it is often beneficial to use a RS and integrate human input through teleoperation, resulting in HSI [5]. A RS is a group of autonomous robots that follow individual rules but collectively exhibit emergent behaviours to accomplish a shared goal [2, 6]. Due to limitations such as restricted sensing capabilities and battery life [7], human intervention can become critical for the safety of the RS. Through HSI, operators can assist and guide the swarm without compromising its autonomy [2].

Humans can influence a RS either through direct mechanisms—such as controlling leaders [2, 8], selecting algorithms [9], using avatars [10] or through indirect interaction, by modifying swarm parameters like radius, inter-robot distance, or environmental cues (e.g., pheromones [11], beacons [12], breadcrumbs [13]). In the study presented in this paper, users interact with the swarm via a tablet (to increase the portability [8]) using both direct (swiping to guide nearby robots) and indirect (marking cells to avoid) methods through a remote interface.

2.2 Situation Awareness in Human-Swarm Interaction

SA was originally introduced in the context of aviation, particularly in cockpit design for pilots. It is defined as a human's ability to perceive elements in the environment (Level 1), comprehend their meaning (Level 2), and project future states (Level 3) from available information [14]. Over time, SA has been studied in both HRI and HSI contexts. Several studies have employed techniques such as the Situation Awareness Global Assessment Technique (SAGAT) for objective SA [15] and the Situation Awareness Rating Technique (SART) for self-reported/subjective SA [16]. Some researchers have also explored physiological methods, such as EEG, claiming they provide a more objective measure than SAGAT [17]. While each method has its advantages and limitations, SAGAT and SART remain among the most widely used techniques and were incorporated into the study presented in this paper.

In some previous work, such as an urban search-and-rescue task using an Unmanned Ground Vehicle (UGV), SA was evaluated by asking participants to draw maps showing where they believed discovered items were located [18]. Although such methods are not standardised, they aim to capture users' situational understanding. Some studies have assumed that reducing available information results in reduced SA, and instead of measuring SA directly, they only evaluated TP. For example, one study limited access to global information, allowing participants to rely solely on sparse local data (such as sensor readings or a single image from the robot's camera) during a swarm aggregation task with obstacle avoidance [9]. However, reducing information does not necessarily equate to reduced SA. Users may compensate by engaging more deeply in comprehension or projection when perception is limited. Therefore, in this study, we explicitly measure SA while systematically degrading specific types of information one at a time. This approach allowed us to assess whether and how different types of degraded information impacted the user's SA and TP.

2.3 Task Performance and Situation Awareness in Human-Swarm Interaction

Various studies have investigated the relationship between SA and TP. While some suggest that lower SA leads to lower TP [9], others have shown that this is not always the case. TP may still be poor even if the operators' SA is high, but they are unable to effectively utilise their SA [15].

Previous work [3] has studied how different levels of SA correlate with TP in a target search task using a RS. In this study, participants teleoperated a swarm via a tablet-based interface. Results indicated that SA components related to perception and projection were positively correlated with TP. Furthermore, SA tied to specific information elements—such as the location of the robots and the target—also showed strong positive correlations with performance outcomes.

Building on these findings, the study in this paper explores how systematic degradation of specific information types affects SA and, in turn, impacts TP. This allows us to assess not only whether SA is diminished under degraded conditions, but also how this influences TP and, in turn, whether we can identify the most essential information elements based on their impact on SA.

2.4 Regulations for Remotely Controlling Robots

Currently, remote pilots are permitted to operate only one drone at a time unless they obtain specific approval for multi-drone operation [1]. This regulatory constraint parallels challenges in HSI, where a single operator may be expected to manage multiple robots simultaneously. In countries like Australia, the regulatory environment is actively evolving to support more advanced forms of drone operation. Australia is one of the nations listed as having experimental Beyond Visual Line of Sight (BVLOS) operations, which allows exceptions to the standard Visual Line of Sight (VLOS) requirements under certain conditions [19]. This experimental framework provides a testbed for exploring multi-drone teleoperation, potentially paving the way for regulations that support managing multiple drones by a single pilot in the future.

Despite these advancements, many jurisdictions—including Vietnam [20] and others [19]—lack detailed operational guidelines for teleoperation, indicating a significant regulatory gap. Australia's ongoing research and regulatory experimentation highlight a recognition of the need for comprehensive standards that address operational safety, capacity, and interoperability for multi-drone management [19]. Developing such standards based on empirical evidence is essential to transform these experimental practices into widely accepted, safe, and effective operational procedures. Empirical studies such as the one presented in this paper are therefore essential for informing policy and establishing evidence-based guidelines for safe and effective HSI practices.

3 Interface Design and the User Study

This section describes the interface design approach and introduces the HSI task used, followed by an explanation of the user study procedure.

3.1 Interface Design and Information Degradation

The framework introduced in [3] was applied to develop a tablet-based user interface for a single-target search task involving 20 robots operating in an open environment (Fig. 1). The robots autonomous operation is controlled by Particle Swarm Optimization (PSO) with the LBEST model [21], operate under limited communication range, and include a basic obstacle avoidance algorithm. The scenario simulates a spreading fire, with periodic messages alerting the operator to hazardous grid cells. Robots entering these cells become deactivated, making it the operator's responsibility to mark them for avoidance. Operators can also assist both active and trapped robots by swiping near them, causing nearby robots to follow the swipe direction briefly.

Based on the framework from [3], six information dimensions were identified: TargetInfo (positions of robots and the target), MoveInfo (robot movement and spatial dynamics), TimeInfo (task time tracking), MsgInfo (hazard alerts), LossInfo (status of deactivated robots), and TrappedInfo (robots unable to progress). To create degraded versions from the full-information state (which is shown in Fig. 1), the following changes were applied each time: A and B were removed to degrade TargetInfo; C was removed (while retaining D) to degrade TimeInfo; E was modified to reference ranges instead of specific danger cells to degrade MsgInfo; and F and G were shown in dark blue (similar to active robots) to degrade LossInfo and TrappedInfo, respectively. MoveInfo was not degraded as it reflects emergent swarm behaviour. In total, seven interface versions were created: one full-information, five with individual dimension degradations, and one with AllInfo degraded.

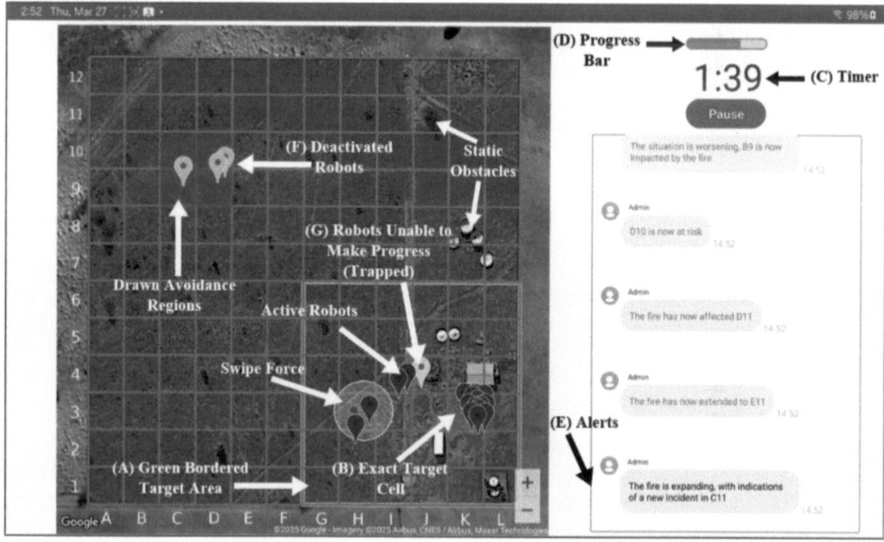

Fig. 1. Tablet interface with all the information.

3.2 User Study Procedure

The user study was carried out using the developed interface to evaluate SA and TP, following approval from the Human Research Ethics Panel at UNSW Canberra. 40 participants (21 male, 19 female) took part, ranging in age from 18 to 54, with the majority (26) between 25 and 34 years old. Most were postgraduate or undergraduate students, and 35 were non-native English speakers. Participants' prior knowledge of robotic swarms was normally distributed across the group. SA was assessed using both SAGAT queries [22] and the standard SART questionnaire [16], while TP was measured using multiple metrics: centroid active robot position, nearest and farthest active robot positions, and nearest quartile positions (Q1–Q3). The study was conducted individually with each participant. After providing consent, participants watched a 10-min instructional video and completed a pre-task questionnaire on demographics. They then performed a 5-min task under seven interface variants (see Sect. 3.1). Each task included two pauses where 14 SAGAT questions (Multiple-Choice Questions (MCQ) and Cell-Marking Questions (CMQ)) were asked (28 in total) to assess objective SA. After each task, participants completed a post-task questionnaire using SART to measure subjective SA. Each session lasted about two hours per participant.

4 Results

This section presents the results and analysis from two experiments. The first experiment further explores the relationship between SA and TP under conditions of information degradation. The second experiment highlights that not all information equally affects SA, and identifies the most critical information dimensions.

4.1 Experiment to Identify Correlations Between Situation Awareness and Task Performance Across Degraded Information Dimensions

Aim: This experiment investigates whether TP in a HSI can be inferred from users' objective SA, under various degraded information conditions. Specifically, it explores whether SAGAT scores correlate with TP, allowing us to assess performance based on objective SA and vice versa, and what the insights are for policy making for single operator, multi-robot control.

Hypothesis: There is a positive correlation between specific objective SA elements and TP metrics. Higher objective SA should lead to better performance.

Method: Participants completed HSI tasks under seven different interface conditions, each designed by degrading specific types of information. During these tasks, TP was measured using various robot proximity metrics such as centroid of active robots, nearest robot distance, and quartile distances (Q1, Q2, Q3) based on the positions of the robots by the end of the task. Objective SA was measured using the SAGAT technique, capturing scores at the overall, SA level-wise, and dimension-wise levels. Each question carried a maximum of 100 points. For MCQs, scoring was straightforward: 100 points for a correct answer and 0 for an incorrect one. CMQs also carried 100 points each, with partial credit awarded based on the proximity of the marked cell to the correct one—the closer the mark, the higher the score. Average scores were computed for each SA level and each information dimension (e.g., a separate score for TargetInfo-related questions). An overall SAGAT score (S_{SAGAT}) was calculated by averaging all responses, each out of 100. For each degradation condition, Spearman correlations were calculated to examine the relationships between objective SA and TP. The strength of correlations was interpreted as strong ($\geq$0.70), moderate (0.50–0.69), or weak (<0.50). Better TP is when robots are closer to the target, which corresponds to lower TP metric scores (i.e., distance). Therefore, lower TP metric values indicate better TP. All the correlations discussed here are negative when considering TP metrics (distance) and SA scores. This implies that the relationship between TP and SA is positive.

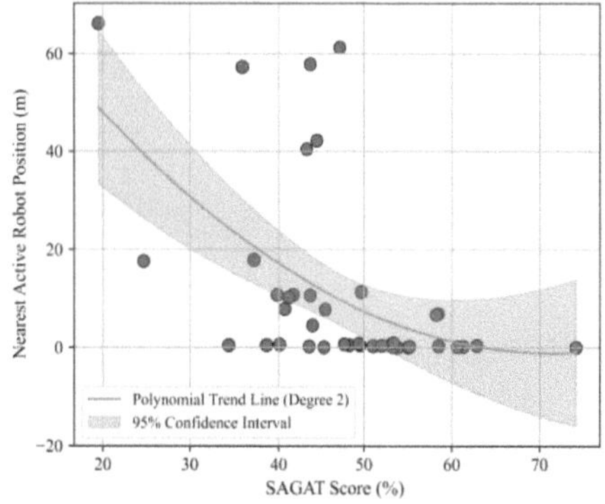

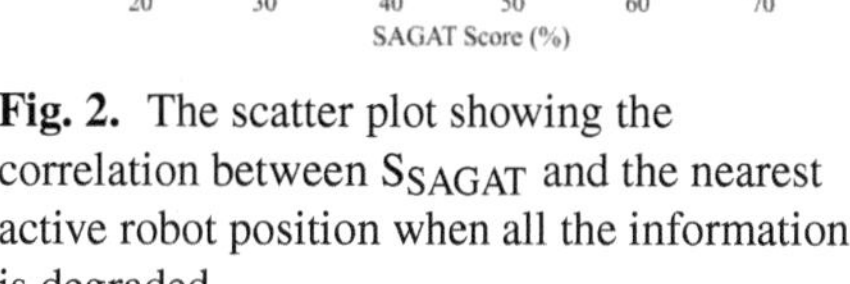

Fig. 2. The scatter plot showing the correlation between S_{SAGAT} and the nearest active robot position when all the information is degraded.

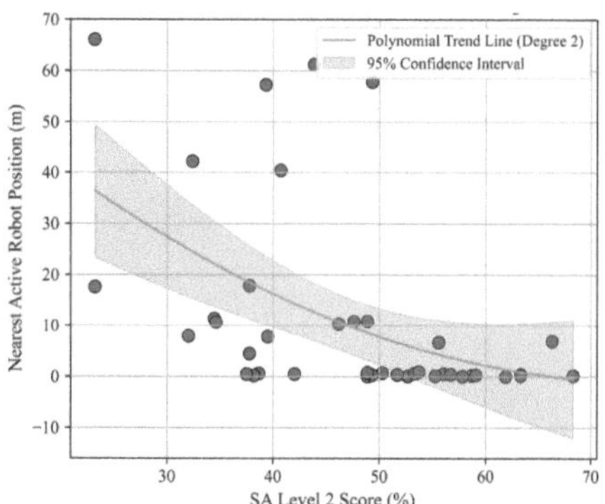

Fig. 3. The scatter plot showing the correlation between SA Level 2 and the nearest active robot position when all the information is degraded.

Results: For objective SA and TP, moderate negative correlations were observed under the AllInfo degraded condition. Specifically, SAGAT scores and SA Level 2 scores showed correlations of around −0.62 to −0.65 with the nearest active robot position as shown in Fig. 2 and Fig. 3. This suggests that under extreme degradation, operators concentrate their effort on bringing at least one robot close to the target, as a compensatory strategy to cope with uncertainty. Also, when AllInfo was degraded, participants' SA score related to TargetInfo moderately correlated with nearest (−0.65), nearest Q1 (−0.52) and Q2 (−0.56) active robot positions as shown in Fig. 4, Fig. 5 and Fig. 6, which means when they have high knowledge about the location information and the target, they were able bring at least 50% (Q2) of the robots closer to the target, which makes TargetInfo an important information dimension for TP. This supports H1. This finding highlights that maintaining SA about key task-relevant information dimensions is essential for effective TP, even when the rest of the interface is limited.

Conclusion: The results of this experiment reveal several important findings regarding the relationships between SA and TP under degraded interface conditions. First, TP related to the nearest active robot was meaningfully inferred from objective SA when information is severely degraded. This implies that when information was degraded, the operator narrowed their focus to bringing at least one robot closer to the target. This suggests that having an additional operator could help maintain awareness and control over the entire swarm.

During the study design, when information was degraded, it was hypothesised that users would compensate for missing cues by shifting from SA Level 1 (perception) to SA Level 2 (comprehension). The observed correlation between SA Level 2 scores and the nearest active robot position under the AllInfo degraded condition suggest that when all information was removed, if the users relied more on their comprehension to interpret the situation and guide the swarm, their ability to bring at least one robot closer to the target has been increased, indicating that users' Level 2 SA was directly linked to effective TP in the absence of clear perceptual cues. This suggests that incorporating decision-support tools that help synthesise minimal data into meaningful patterns, especially when full information is unavailable, could help operators.

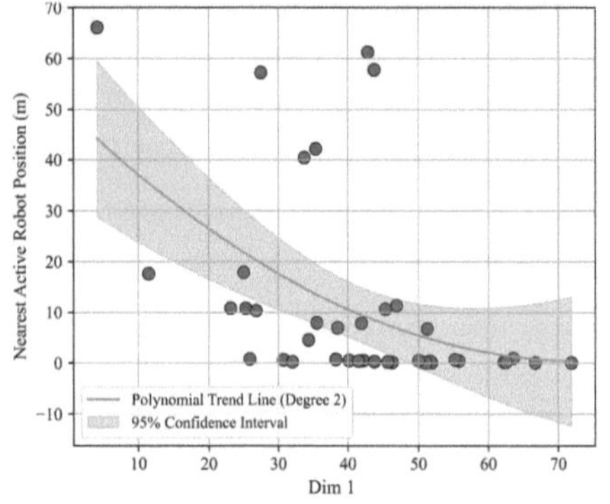

Fig. 4. The scatter plot showing the correlation between Dim 1 (TargetInfo) related objective SA and the nearest active robot position when all the information is degraded.

Fig. 5. The scatter plot showing the correlation between Dim 1 (TargetInfo) related objective SA and the nearest Q1 active robot position when all the information is degraded.

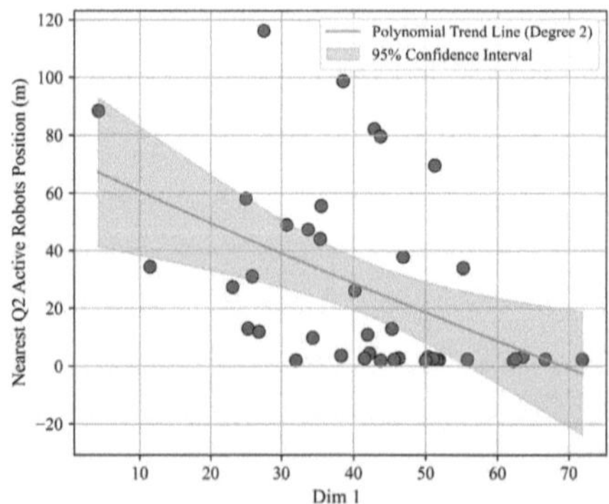

Fig. 6. The scatter plot showing the correlation between Dim 1 (TargetInfo) related objective SA and the nearest Q2 active robot position when all the information is degraded.

Further, it showed that when all the information is degraded and participants have high knowledge about the location information and the target, they were able to bring at least 50% (Q2) of the robots closer to the target, which helped in identifying the location information and the target as an important dimension in this task. This finding highlights that maintaining SA about key task-relevant information dimensions is essential for effective performance, even when the rest of the interface is limited.

4.2 Experiment to Compare Objective Situation Awareness Related to Each Individual Information Dimension Degradation with Full Information

Aim: To examine whether degrading a specific information dimension results in a measurable reduction in users' objective SA related to that dimension. This helps determine which information dimensions are most critical for maintaining SA during HSI.

Hypothesis: Degrading a particular information dimension will lead to a significant decrease in users' objective SA scores related to that specific dimension when compared to the condition where full information was available.

Method: The calculation of the SAGAT scores was similar to Experiment 4.1. In this experiment, for each information dimension, participants' objective SA scores of each dimension were compared between two conditions: (i) when that particular dimension was degraded, and (ii) when all information was available (i.e., the full information condition). The median and interquartile ranges (IQR) of objective SA scores were computed for each condition. Statistical significance was assessed using the Wilcoxon paired test, and p-values were reported to evaluate differences in medians.

Results: The results are summarised in Table 1. Only TargetInfo showed a statistically significant reduction in objective SA when degraded ($p = 0.000$), indicating a clear decline in awareness for that dimension when the relevant information was removed.

Conclusion: The findings indicate that degrading TargetInfo had a significant negative impact on users' objective SA, suggesting that this type of information provides critical perceptual cues that participants could not compensate for through higher-level reasoning, highlighting that compensation has limits. In contrast, other types of information showed no significant reduction in SA when removed, indicating that not all information is equal in its contribution to effective swarm guidance. This reinforces the need to prioritise task-specific SA needs when designing interfaces or operational protocols for HSI. Specifically, this underscores the importance of preserving high-value information dimensions in operator interfaces and tailoring interface requirements based on the task to maintain safe and effective control in real-world multi-robot scenarios.

5 Discussion

The study explored the relationship between SA and TP. TargetInfo was identified as a key factor. Even when all information was degraded, participants who had a strong understanding of TargetInfo were able to guide at least 50% of the robots closer to the target. These results suggest that TargetInfo is essential for effective TP in this context. This finding reveals that degrading TargetInfo not only reduced SA but also impaired task effectiveness. The implications for future regulatory or operational guidelines for operator control of robotic swarms could include the definition of minimum information requirements for interfaces that support multi-robot control.

Table 1. Comparison of objective SA scores of each information dimension when that dimension was degraded and all the information was available.

Dimension	Median [IQR] (Degraded)	Median [IQR] (Full Info)	p-value
TargetInfo	41.9 [33.9 - 53.4]	59.5 [50.5 - 68.2]	0.00
TimeInfo	66.7 [66.7 -100.0]	66.7 [66.7 - 66.7]	0.28
MsgInfo	66.5 [54.8 - 78.5]	76.5 [56.0 - 92.3]	0.07
LossInfo	57.3 [43.8 - 75.0]	62.5 [50.0 - 75.0]	0.39
TrappedInfo	25.0 [25.0 - 50.0]	25.0 [25.0 - 65.6]	0.24

Furthermore, it was found that TP concerning the nearest active robot position could be meaningfully inferred from objective SA when AllInfo was degraded. This behaviour mirrors a narrowing of focus, which has an effect with real implications for future multi-robot operations. If regulatory frameworks evolve to permit one pilot to operate multiple robots or drones, this result suggests that in degraded-information scenarios, the operator may prioritise only one robot—typically the one closest to completing the task—while potentially neglecting the rest of the swarm. This poses serious challenges to operational safety and effectiveness in multi-robot (swarm) deployments. Regulators and policymakers should consider whether support systems (e.g., secondary operators) are necessary to ensure full swarm coverage under challenging conditions.

Moreover, the findings from the second experiment revealed that not all types of information have such an impact on SA. In particular, degrading TargetInfo significantly reduced users' objective SA, while removing other types of information did not show such effects. This challenges the assumption made in several previous studies that a reduction in information automatically leads to a degradation in SA. Instead, our results highlight the importance of explicitly measuring SA rather than assuming it declines with information loss. In this case, only certain information elements (most notably TargetInfo) were shown to have a critical influence. This suggests that, while operators could compensate for some degraded information through reasoning or experience, they could not compensate for the absence of important information like that in TargetInfo. This suggests the need for redundancy or support systems (e.g., a secondary operator or AI assistant) in scenarios where full information may not be guaranteed. The results reinforce the idea that policy cannot be one-size-fits-all—what counts as critical information (like TargetInfo here) may change with task type. Thus, regulatory frameworks should allow flexibility informed by empirical assessments like this one.

6 Conclusions and Future Work

In this study, the effects of systematic information degradation on SA and TP in HSI were investigated. Five key information dimensions were removed one by one, and interface variations were created for each. Another variation was created by removing all information at once, along with a final version where all information was presented—resulting in seven variations in total. Then, a user study was conducted, where 40 participants interacted with a RS through a tablet-based interface under seven different information conditions. Objective SA was measured using SAGAT and subjective SA through SART, while TP was assessed using multiple robot distance metrics.

The findings demonstrated that some information types—particularly TargetInfo—had a high impact on both SA and TP, highlighting the importance of identifying critical information elements in interface design and regulatory planning for multi-robot control. These insights have broader implications beyond academic findings. As regulations evolve to permit single-operator control over multiple robots, the results suggest that under degraded conditions, operators may tend to narrow their focus to a single or a few robots, potentially compromising the control and awareness of the rest of the swarm. Also, this highlights the need for explicit measurement of SA and suggests that some dimensions of information are more critical than others, depending on the task. Moreover, since the importance of specific information elements can vary by task, regulatory policies should avoid one-size-fits-all standards and instead be flexible, evidence-based, and task-informed.

Ultimately, this study contributes both empirically and practically to the growing field of HSI, offering a foundation for human-centred policies that ensure safe and effective multi-robot control in real-world operations. These insights are expected to inform future interface design and support more robust HSI strategies in information-constrained environments. Building on this work, several future directions were identified. First, further experiments could explore different task contexts to determine whether critical information elements like TargetInfo remain consistently important or vary by scenario. Second, studies could examine the role of adaptive or AI-driven interfaces that selectively restore or prioritise critical information in degraded conditions. Moreover, studies could evaluate the effect of real-world operational constraints—such as end-to-end communication delays or lag in the robots—on SA and task performance.

Overall, this research contributes foundational evidence to guide flexible, task-informed policy development and the design of resilient human-swarm systems under degraded information conditions.

References

1. Australian Government Civil Aviation Authority (2024). https://www.casa.gov.au/drones/flight-authorisations/flying-multiple-drones-once
2. Kolling, A., Walker, P., Chakraborty, N., Sycara, K., Lewis, M.: Human interaction with robot swarms: a survey. IEEE Trans. Hum.-Mach. Syst. **46**, 9–26 (2016)
3. Wattearachchi, W.D., Lakshika, E., Kasmarik, K., Barlow, M.: A Study on Human-Swarm Interaction: A Framework for Assessing Situation Awareness and Task Performance. arXiv preprint arXiv:2503.14810 (2025)
4. Riley, J.M., Strater, L.D., Chappell, S.L., Connors, E.S., Endsley, M.R.: Situation awareness in human-robot interaction: challenges and user interface requirements. In: Human-Robot Interactions in Future Military Operations, pp. 171–192 (2010)
5. Hussein, A., et al.: Characterization of indicators for adaptive human-swarm teaming. Front. Robot. AI **9**, 16 (2022)
6. Schranz, M., Umlauft, M., Sende, M., Elmenreich, W.: Swarm robotic behaviors and current applications. Front. Robot. AI **7**, 36 (2020)

7. Arvin, F., Watson, S., Turgut, A.E., Espinosa, J.A., Krajník, T., Lennox, B.: Perpetual robot swarm: long-term autonomy of mobile robots using on-the-fly inductive charging. J. Intell. Rob. Syst. **92**, 395–412 (2017)
8. Hoang, M.-T.O., Berkel, N.V., Skov, M.B., Merritt, T.R.: Challenges and requirements in multi-drone interfaces. In: Extended Abstracts of the 2023 CHI Conference on Human Factors in Computing Systems, pp. Article 65. Association for Computing Machinery, Hamburg, Germany (2023)
9. Kapellmann-Zafra, G., Salomons, N., Kolling, A., Groß, R.: Human-robot swarm interaction with limited situational awareness. In: Swarm Intelligence: 10th International Conference, ANTS 2016, Brussels, Belgium, September 7–9, 2016, Proceedings 10, pp. 125–136. Springer (2016)
10. Bashyal, S., Venayagamoorthy, G.K.: Human swarm interaction for radiation source search and localization. In: 2008 IEEE Swarm Intelligence Symposium, pp. 1–8 (2008)
11. Daily, M., Youngkwan, C., Martin, K., Payton, D.: World embedded interfaces for human-robot interaction. In: 36th Annual Hawaii International Conference on System Sciences, 2003. Proceedings of the, p. 6 (2003)
12. McLurkin, J., Smith, J., Frankel, J., Sotkowitz, D., Blau, D., Schmidt, B.: Speaking swarmish: human-robot interface design for large swarms of autonomous mobile robots. In: AAAI spring symposium: to boldly go where no human-robot team has gone before, pp. 72–75. Palo Alto, CA, (2006)
13. Bowley, S.J., Merrick, K.: A 'Breadcrumbs' Model for Controlling an Intrinsically Motivated Swarm Using a Handheld Device, pp. 157–168. Springer (2017)
14. Endsley, M.R.: Toward a theory of situation awareness in dynamic systems. Hum. Factors **37**, 32–64 (1995)
15. Endsley, M.R.: Direct measurement of situation awareness: validity and use of SAGAT. Situation Awareness Anal. Measur. **10**, 147–173 (2000)
16. Taylor, R.M.: Situational awareness rating technique (SART): the development of a tool for aircrew systems design. In: Situational Awareness, pp. 111–128. Routledge (2017)
17. Fernandez Rojas, R., et al.: Encephalographic assessment of situation awareness in teleoperation of human-swarm teaming, pp. 530–539. Springer (2019)
18. Larochelle, B., Kruijff, G.J.M., Smets, N., Mioch, T., Groenewegen, P.: Establishing human situation awareness using a multi-modal operator control unit in an urban search & rescue human-robot team. In: 2011 RO-MAN, pp. 229–234 (2011)
19. Jones, T.: International commercial drone regulation and drone delivery services. RAND Santa Monica, CA, USA (2017)
20. Tran, T.-H., Nguyen, D.-D.: Management and regulation of drone operation in urban environment: a case study. Soc. Sci. **11**, 474 (2022)
21. Eberhart, R., Kennedy, J.: A new optimizer using particle swarm theory. In: MHS 1995. Proceedings of the Sixth International Symposium on Micro Machine and Human Science, pp. 39–43 (1995)
22. Endsley, M.R.: Situation awareness global assessment technique (SAGAT). In: Proceedings of the IEEE 1988 National Aerospace and Electronics Conference, vol. 783, pp. 789–795 (1988)

Learning Preferences in Additive Separable Hedonic Project Games

Jaber Valizadeh(✉), Dongmo Zhang, and Omar Mubin

School of Computing, Data and Mathematical Sciences, Western Sydney University, Sydney, Australia
{J.valizadeh,D.Zhang,O.Mubin}@westernsydney.edu.au

Abstract. We study Additive Separable Hedonic Project Games (ASHPGs), which model coalition formation considering both agents' subjective preferences over their coalition members and objective incentives for project rewards. In this setting, each autonomous agent selects a single project from a given set to maximize its own utility. We first formalize ASHPGs in a single-shot scenario with known preferences, where each agent has sufficient information to evaluate her preferences before project selection. We then extend the model to a learning setting where preferences are initially unknown, and agents gradually learn them through repeated feedback as they iteratively select projects and form coalitions. Motivated by this, we propose an Upper Confidence Bound (UCB)-based online learning algorithm with semi-bandit feedback, in which each agent observes the feedback received from all individual members of their coalition. Our theoretical analysis demonstrates that ASHPGs with symmetric preferences possess an exact potential function that guarantees the existence of a Nash stable outcome. Furthermore, we prove that the proposed algorithm achieves sublinear Nash regret and converges to an ε-approximate Nash equilibrium over time. Experiments on real and synthetic data validate these theoretical results.

Keywords: Multi-Agent Systems · Hedonic Games · Project Games · Online Learning

1 Introduction

In many multi-agent systems, individuals must form coalitions to undertake shared tasks or projects [30]. Examples include collaborative scientific research teams [1], crowdsourcing platforms [33], and decentralized marketplaces [29]. In these settings, agents are motivated by two interdependent considerations: their subjective preferences over *whom* they work with (as in additively separable hedonic games [10]), and *what* they work on (as in project games [8]). For instance, a researcher may choose a lower-impact project to work with trusted colleagues. Classical models often address only one of these dimensions, either the social preference for teammates or task-based rewards, but real-world scenarios typically involve both.

M. Liu et al. (Eds.): AI 2025, LNAI 16371, pp. 491–505, 2026.
https://doi.org/10.1007/978-981-95-4972-6_38

While classical models of coalition formation, such as hedonic games [22], focus primarily on agents' preferences over the composition of the group, they typically assume that the value of a coalition depends only on who belongs to it, without considering any external rewards or benefits arising from the completion of a specific task or project. Conversely, models of project selection and task allocation often adopt the opposite perspective: they assume that agents' utilities are determined entirely by the outcomes or rewards produced by projects, regardless of the identities of the other participants.

Although classical hedonic games typically assume static preferences and full information [3], recent research has begun to explore dynamic and more realistic environments [9,12]. For example, learning in coalition formation has been studied under the assumption that agents know their preferences in advance [5,16] or within restricted action spaces [15]. However, much of this prior work either presumes that agents are fully aware of their preferences or focuses exclusively on cooperative settings without considering strategic deviations.

In this paper, we introduce *Additive Separable Hedonic Project Games* (ASHPGs), a strategic game framework in which agents select projects while implicitly revealing their preferences over other agents. The utility of each agent depends on both the co-participants (assuming additively separable) and the reward for completing the project. We first formalize ASHPGs as single-shot games with known preferences and then extend the model to repeated interactions with initially unknown preferences, where agents learn their preferences based on observed interactions. To address this learning problem, we develop a reinforcement learning approach based on semi-bandit feedback and the *optimism principle.*

2 Related Work

Originating from economic theory [22], hedonic games have been extensively studied in works such as [6,10,13]. A comprehensive survey of this literature is provided by Aziz et al. [4]. Several subclasses have been proposed to capture different preference structures among agents, including Additively Separable Hedonic Games (ASHGs) [10], Fractional Hedonic Games (FHGs) [2], and Hedonic Coalition Nets [23]. Extensive work has examined expressive preference representations [4], existence of stable outcomes [7], and algorithmic aspects [11]. However, classical hedonic games neglect external factors like task rewards, focusing only on who is in the group and not what the group achieves. Extensions such as ASHGs [10] and FHGs [2] still largely ignore objective project rewards. ASHPGs directly address this shortcoming by introducing a dual-dependency model where agents care about both coalition composition and project payoffs.

Project games extend traditional coalition formation by incorporating strategic interactions in which agents select projects and contribute to them with predefined weights [8]. These games are related to valid utility games [31] and have strong connections to load balancing games [32], congestion games [28], and hedonic coalition formation [4,22]. Project games accurately reflects outcome-oriented incentives but inherently treats agents as indifferent to their collaborators, assuming that all coalitions on a project are equally acceptable. In contrast, many real-world settings (e.g., team-based research, collaborative crowdsourcing) show that interpersonal dynamics and preferences over teammates critically affect outcomes. ASHPGs fill this gap by incorporating coalition preferences alongside project rewards, thus blending social and task-based incentives into a unified decision model.

Bandit algorithms and reinforcement learning techniques have been applied to decentralized matching and coalition formation problems [26,27], providing regret bounds under various feedback settings. Online learning hedonic coalition formation games builds upon these foundations by integrating them with dynamic coalition formation [18,19]. It also draws from general online learning theory [14] and regret minimization in potential games [21], enabling convergence guarantees under bandit feedback with unknown agent preferences.

3 Definitions and Notations

This section defines the core concepts and notations for our ASHPG framework, blending agents' preferences with project rewards for realistic coalition modeling. We start with the single-shot game, assuming known preferences, which allows for direct computation of utilities and stability assessments. This is then extended to a repeated game setting with initially unknown preferences, where agents iteratively learn through stochastic interactions, reflecting practical applications such as crowdsourcing platforms or decentralized collaborative systems. We focus on individual deviations, using Nash equilibrium (and approximations) as the stability metric. Regret minimization, specifically average Nash regret, evaluates learning dynamics over time. Finally, we detail the semi-bandit feedback model, enabling efficient exploration-exploitation in our UCB-based algorithm, supporting guarantees like potential functions and sublinear regret.

3.1 Model

Let $M = \{1, 2, \dots, m\}$ denote a finite set of projects, and let $N = \{1, 2, \dots, n\}$ be a finite set of agents. Each agent can participate in any project in M. Every project $j \in M$ has an associated positive reward, formally defined by the reward function $r : M \to \mathbb{R}_+$, where $r(j)$ denotes the reward of project j. Each agent must select one of the m projects. Let $s_i \in M$ denote the project chosen by agent i as her *strategy*. A *strategy profile* is a vector $\mathbf{s} = (s_1, \dots, s_n)$, and the set of all strategy profiles is $\mathcal{S} = M^n$. Given a strategy profile $\mathbf{s} \in \mathcal{S}$, the *coalition* assigned to each project $j \in M$ is denoted by $C_j(\mathbf{s}) = \{i \in N \mid s_i = j\}$, and we write $|C_j(\mathbf{s})|$ for its size.

Each agent i has a *preference function* $p_i : N \to \mathbb{R}$, where $p_i(k)$ represents the value that agent i assigns to being in the same coalition as agent k, such that $p_i(i) = 0$. Positive values represent attraction (*friends*), negative values represent aversion (*enemies*). Preferences are said to be *symmetric* if $p_i(k) = p_k(i)$ for all $i, k \in N$. For each agent $i \in N$, let $\mathcal{N}_i = \{C \subseteq N \mid i \in C\}$ denote the set of coalitions containing i. Preferences are *additively separable*, i.e., the preference of agent i for coalition $C \in \mathcal{N}_i$ is $P_i(C) = \sum_{k \in C} p_i(k)$.

An additive separable hedonic project game is formally defined as follows:

Definition 1. *Let $\mathcal{G} = \langle N, M, r, \mathcal{S}, (p_i)_{i \in N}, (u_i)_{i \in N} \rangle$, be an additive separable hedonic project game (ASHPG), where:*

- *$N = \{1, 2, \ldots, n\}$ is the set of n players.*
- *$M = \{1, 2, \ldots, m\}$ is a finite set of m projects.*
- *$r : M \to \mathbb{R}_+$ is a reward function.*
- *$\mathcal{S} = M^n$ is the space of strategy profiles.*
- *$p_i : N \to \mathbb{R}$ is the valuation function for each player i with $p_i(i) = 0$.*
- *$u_i : \mathcal{S} \mapsto \mathbb{R}$ denotes the utility function for each player $i \in N$.*

Project rewards are assumed to be divisible and are shared equally among all agents who choose the same project. To reconcile the units of the reward (e.g., dollars) and the preference score (e.g., utility points), we use a *normalizing coefficient* γ. The *utility* of agent i under strategy profile $\mathbf{s}$ is given by:

$$u_i(\mathbf{s}) = \frac{r(s_i)}{|C_{s_i}(\mathbf{s})|} + \gamma \cdot P_i(C_{s_i}(\mathbf{s})) \tag{1}$$

where the first term represents agent i's share of the reward from project s_i, and the second term is the total preference that agent i assigns to the other agents participating in the same project as i under $\mathbf{s}$. Since each agent selects a project, for any agent i, $|C_{s_i}(\mathbf{s})| \geq 1$, as $i \in C_{s_i}(\mathbf{s})$.

Special Cases. ASHPGs generalize two well-known classes of games:

- If $p_i(k) = 0$ for all $i, k \in N$, the model reduces to a *singleton congestion game* [28], where utilities depend only on project rewards and coalition sizes.
- If $r(j) = 0$ for all $j \in M$, the model reduces to an *additively separable hedonic game* [10].

Mixed Strategies. For each player i, the *pure* strategy space is M. The *mixed strategy space* of the player i is denoted by $\Delta(M)$, representing the set of probability distributions on M. A *mixed strategy* for player i is then defined as a probability vector $\boldsymbol{x}_i = (x_i^1, x_i^2, \ldots, x_i^m) \in \Delta(M)$, where each probability $x_i^j \in [0, 1]$ represents the probability that player i selects project j, and $\sum_{j \in M} x_i^j = 1$. A *mixed strategy profile* $\mathbf{x} = (\boldsymbol{x}_1, \boldsymbol{x}_2, \ldots, \boldsymbol{x}_n) \in \bar{\Delta}$, specifies a mixed strategy for every player, where $\bar{\Delta} = \times_{i \in N} \Delta(M)$ denotes the mixed strategy profile space. This profile induces a probability distribution over the set of all possible project selections $\mathcal{S}$. The *expected utility* of player i under a given mixed strategy profile

$\mathbf{x} \in \bar{\Delta}$ is computed by taking the expectation over all possible strategy realizations $\mathbf{s} \sim \mathbf{x}$:

$$\mathbb{E}[u_i(\mathbf{x})] = \sum_{\mathbf{s} \in \mathcal{S}} \left(\prod_{i \in N} x_i^{s_i} \right) u_i(\mathbf{s}) \tag{2}$$

In many practical scenarios, such as crowdsourcing platforms or distributed multi-agent systems, agents may lack prior knowledge about others' identities or preferences. These preferences are gradually learned through repeated interactions. In the next section, we extend this model to a repeated game setting where agents initially lack information about their preferences regarding other participants.

3.2 Repeated ASHPGs with Unknown Preferences

We now extend to a repeated setting where agents initially lack knowledge of their preferences over others. Time proceeds in discrete rounds $t = 1, 2, \ldots, T$. At each round, every agent selects a project and forms a coalition.

For each pair $i, k \in N, i \neq k$, agent i has an unknown and fixed distribution $\mathcal{D}_i(k)$ supported on $[p_{\min}, p_{\max}]$, with mean $\bar{p}_i(k)$. At each round t, when agents i and k join the same coalition, agent i observes a realization $p_i^t(k)$ drawn independently from $\mathcal{D}_i(k)$. Otherwise, $p_i^t(k)$ is unobserved. Thus, learning occurs only through co-participation. We assume symmetry at the level of expectations, i.e., $\bar{p}_i(k) = \bar{p}_k(i)$ for all $i, k \in N$, while individual draws $p_i^t(k)$ may be asymmetric.

In each round t, every agent $i \in N$ receives a stochastic utility based on the strategy profile $\mathbf{s}$, defined as: $u_i^t(\mathbf{s}) = \frac{r(s_i)}{|C_{s_i}(\mathbf{s})|} + \gamma \cdot \sum_{k \in C_{s_i}(\mathbf{s})} p_i^t(k)$, where $p_i^t(k) \sim \mathcal{D}_i(k)$. The agents aim to maximize their expected utility, given by: $\bar{u}_i(\mathbf{s}) = \frac{r(s_i)}{|C_{s_i}(\mathbf{s})|} + \gamma \cdot \sum_{k \in C_{s_i}(\mathbf{s})} \bar{p}_i(k)$, through repeated interactions, each agent learns $\bar{p}_i(k)$ by observing the realized preferences $p_i^t(k)$ for all $k \in C_{s_i}(\mathbf{s}) \setminus \{i\}$ after each round.

In round t, each agent i independently selects a project $s_i \in M$, resulting in a strategy profile $\mathbf{s} \in \mathcal{S}$. The coalition $C_{s_i}(\mathbf{s})$ is revealed after all selections, and agent i observes $p_i^t(k)$ for all $k \in C_{s_i}(\mathbf{s})$. The estimated utility of agent i with respect to the mixed strategy profile $\mathbf{x}$ sampled at time t is $\hat{u}_i^t(\mathbf{s}) = \frac{r(s_i)}{|C_{s_i}(\mathbf{s})|} + \gamma \cdot \sum_{k \in C_{s_i}(\mathbf{s})} \hat{p}_i^t(k)$, where $\hat{p}_i^t(k)$ denotes the empirical mean preference of agent i for agent k up to time t (see more details in Sect. 3.3).

A central question in this setting is whether the project selection process converges to a stable outcome, that is, a *Nash equilibrium*. Given a pure strategy profile $\mathbf{s}$, agent i is said to have a *profitable deviation* if switching from her current project s_i to an alternative project $j \neq s_i$, while keeping others' strategies fixed, improves her utility; formally, this is a *Nash deviation* if there exists $j \in M$ with $j \neq s_i$, such that $u_i(j, \mathbf{s}_{-i}) > u_i(s_i, \mathbf{s}_{-i})$.

When working with mixed strategies, we define another notion of Nash deviations. At time step t, let $\mathbf{x}_{-i}^t = (\boldsymbol{x}_k^t)_{k \neq i}$ be the mixed strategy profile of all players except i. At each time t, for each player i, $\boldsymbol{x}_i^{\star,t} \in \Delta(M)$ best response to the other players' mixed strategies satisfying $\mathbb{E}[u_i(\boldsymbol{x}_i^{\star,t}, \mathbf{x}_{-i}^t)] =$

$\max_{\boldsymbol{x}_i \in \Delta(M)} \mathbb{E}[u_i(\boldsymbol{x}_i, \mathbf{x}_{-i}^t)]$. Thus, for any $\varepsilon \geq 0$, the mixed strategy profile $\mathbf{x}^t$ sampled at time t is ε-approximately Nash equilibrium (ε-NE) if

$$\max_{i \in N} \left(\mathbb{E}[u_i(\boldsymbol{x}_i^{\star,t}, \mathbf{x}_{-i}^t)] - \mathbb{E}[u_i(\mathbf{x}^t)] \right) \leq \varepsilon$$

Regret Minimization. To evaluate the learning performance, we adopt the standard notion of *regret* [25]. Regret quantifies the average difference between the expected utility an agent could have achieved by always playing her best possible strategy (i.e., the best response to others' strategies based on current estimates) and the expected utility she actually obtained by following her adaptive strategy over all rounds of the game. Formally, given a sequence of mixed strategy profiles $\mathbf{x}$, the *average Nash regret* [27,34] of agent i after T rounds is defined as:

$$\mathcal{R}_i^T = \frac{1}{T} \sum_{t=1}^{T} \left(\mathbb{E}[u_i(\boldsymbol{x}_i^{\star,t}, \mathbf{x}_{-i}^t)] - \mathbb{E}[u_i(\mathbf{x}^t)] \right)$$

We define the *total average Nash regret* over all agents as $\mathcal{R}^T = \sum_{i \in N} \mathcal{R}_i^T$. A learning algorithm is said to achieve *sublinear regret* if $\mathcal{R}_i^T = o(1)$ as $T \to \infty$, implying that the per-round regret converges to zero over time [25].

3.3 Optimism Principle and Feedback Settings

In our setting, agents select projects in an *uncertain* environment where their preferences over their coalition members are initially *unknown*. To navigate this uncertainty, we introduce a UCB-based algorithm according to the principle of *optimism in the face of uncertainty*[1], which states that one should act as if the environment is as nice as *plausibly possible* [25]. At each time step t, and for any strategy profile $\mathbf{s}$, we assume that each agent i can receive *semi-bandit* feedback, as commonly studied in multi-agent online learning [19,20,24].

Semi-Bandit Feedback. In the semi-bandit feedback model, each agent receives partial feedback that allows her to estimate the value of participating in her coalition. Specifically, agents learn their preferences for other agents based on observed interactions over time. According to the optimism principle, at the time step $t = 0$, each agent i initializes her preference $\hat{p}_i^0(k) = 0$ for all $i \in N$. In each round t, agent i selects a project $s_i \in M$ that maximizes her UCB estimate, leading to a strategy profile $\mathbf{s} = (s_1, \ldots, s_n)$. Once the coalition $C_{s_i}(\mathbf{s})$ is formed (i.e., the set of agents who selected the same project as i), agent i observes individual feedback values $p_i^t(k)$ for each member $k \in C_{s_i}(\mathbf{s}) \setminus \{i\}$.

Agent i maintains an empirical estimate of her preference for each agent $k \in N \setminus \{i\}$ by averaging the feedback she has received in past co-occurrences. Let $M_{i,k}^t(\mathbf{s}) = \sum_{\tau=1}^{t} \mathbf{1}\{s_i^\tau = s_k^\tau\}$ denote the number of times that agent i and

[1] For bandits, the *optimism principle* means using the data observed so far to assign to each project a value, called the *upper confidence bound*, that with high probability overestimates the unknown expected utility.

agent k selected the same project up to time t, where, for any time $\tau \in [t]$, $\mathbf{1}\{s_i^\tau = s_k^\tau\}$ equals 1 if $s_i^\tau = s_k^\tau$, and 0 otherwise. Then, the empirical mean preference of agent i for agent k up to time t is

$$\hat{p}_i^t(k) = \frac{1}{M_{i,k}^t(\mathbf{s}) \vee 1} \sum_{\tau=1}^{t} p_i^\tau(k) \cdot \mathbf{1}\{s_i^\tau = s_k^\tau\}$$

where the denominator is set to 1 whenever agent i and agent k have not co-selected any project up to time t, and otherwise equals $M_{i,k}^t(\mathbf{s})$. Using these estimates, the total empirical preference of agent i over her coalition at time t is $\hat{P}_i^t(C_{s_i}(\mathbf{s})) = \sum_{k \in C_{s_i}(\mathbf{s})} \hat{p}_i^t(k)$, such that $\hat{p}_i^t(k) = 0$, and the corresponding estimated utility is:

$$\hat{u}_i^t(\mathbf{s}) = \frac{r(s_i)}{|C_{s_i}(\mathbf{s})|} + \gamma \cdot \hat{P}_i^t(C_{s_i}(\mathbf{s})) \tag{3}$$

Agents use this feedback to guide future exploration and exploitation. To balance these, each agent constructs a UCB estimate that combines the empirical utility with an exploration bonus. Then the UCB estimate for agent i under profile $\mathbf{s}$ is $\hat{\mathcal{U}}_i^t(\mathbf{s}) = \hat{u}_i^t(\mathbf{s}) + \hat{b}_i^t(\mathbf{s})$, where $\hat{b}_i^t(\mathbf{s})$ is the exploration bonus that is given by:

$$\hat{b}_i^t(\mathbf{s}) = \lambda \cdot \sum_{k \in C_{s_i}(\mathbf{s}) \setminus \{i\}} \sqrt{\frac{2\log(4mnT/\delta)}{M_{i,k}^t(\mathbf{s}) \vee 1}} \tag{4}$$

Here, $\delta \in (0, 1]$ is a confidence parameter that controls the trade-off between exploration and exploitation (see [25] for more discussion), and $\lambda > 0$ is a normalizing coefficient. The bonus term decreases as more information about a project is collected, encouraging agents to explore unfamiliar projects while favoring those with historically higher observed utility.

3.4 Existence of a Nash Equilibrium

In this section, we establish structural properties of additive separable hedonic project games that enable convergence guarantees. We first show that both single-shot and repeated ASHPGs admit exact potential functions, which in turn imply the existence of Nash equilibria. Then, we analyze the convergence of utility estimates, cumulative Nash regret, and equilibrium properties of the agents' strategies.

Lemma 1 demonstrates that in the case of a single shot (i.e., $T = 1$), and known preferences (no uncertainty), a Nash equilibrium is guaranteed.

Lemma 1. *Every additively separable hedonic project game with known and symmetric preferences admits an exact potential function.*

In the case of a repeated game, where the symmetric preferences are unknown at any time t, we show that the game associated with each time t also admits a potential function:

Theorem 1. *At any time t, let $\hat{\mathcal{G}}^t$ be an additive separable hedonic project game with unknown preferences. Let* $\mathbf{s}$ *a strategy profile for $\hat{\mathcal{G}}^t$, and $\hat{\mathcal{U}}_i^t(\mathbf{s}) = \hat{u}_i^t(\mathbf{s}) + \hat{b}_i^t(\mathbf{s})$ be the estimated utility of each agent $i \in N$. Assume that the true underlying pairwise preferences are symmetric. Then, the function $\hat{\Phi}^t(\mathbf{s}) = \sum_{i \in N} \hat{\mathcal{U}}_i^t(\mathbf{s})$ is an exact potential function for $\hat{\mathcal{G}}^t$ during time t.*

Theorem 2 establishes that the `UCB-NE` algorithm in ASHPGs with unknown preferences achieves sublinear Nash regret. This ensures convergence to an ε-approximate Nash equilibrium, balancing exploration and exploitation effectively.

Algorithm 1. `UCB-NE`

Input: Number of projects m, agents n, time horizon T, confidence $\delta \in (0, 1]$, rewards r, bounds $p_{\min}, p_{\max}$, normalizing γ, and $\varepsilon \in (0, 1]$
Output: Mixed strategy profile $\mathbf{x}^t$ that is an ε-NE

```
Initialize p̂_i^0(k) = 0 for all i ∈ N, k ∈ N \ {i} (optimistic initialization); p̂_i^0(i) = 0
Initialize co-occurrence counts M_{i,k}^0 = 0 for all i, k ∈ N
Initialize arbitrary mixed strategy profile x^1 ∈ Δ
for each round t = 1 to T do
   for each agent i ∈ N do
      Sample project s_i ~ x_i^t
      Observe p_i^t(k) ~ D_i(k) for all k ∈ C_{s_i}(s) \ {i}
      Update empirical mean for each k ≠ i:
```

$$\hat{p}_i^t(k) = \frac{1}{M_{i,k}^t(\mathbf{s}) \vee 1} \sum_{\tau=1}^{t} p_i^\tau(k) \cdot \mathbf{1}\{s_i^\tau = s_k^\tau\}$$

- Compute empirical preference: $\hat{P}_i^t(C_{s_i}(\mathbf{s})) = \sum_{k \in C_{s_i}(\mathbf{s}) \setminus \{i\}} \hat{p}_i^t(k)$
- Compute estimated utility: $\hat{u}_i^t(\mathbf{s}) = \frac{r(s_i)}{|C_{s_i}(\mathbf{s})|} + \gamma \cdot \hat{P}_i^t(C_{s_i}(\mathbf{s}))$
- Compute exploration bonus: $\hat{b}_i^t(\mathbf{s}) = \lambda \cdot \sum_{k \in C_{s_i}(\mathbf{s}) \setminus \{i\}} \sqrt{\frac{2 \log(4mnT/\delta)}{M_{i,k}^t(\mathbf{s}) \vee 1}}$
- Compute UCB estimate: $\hat{\mathcal{U}}_i^t(\mathbf{s}) = \hat{u}_i^t(\mathbf{s}) + \hat{b}_i^t(\mathbf{s})$

end for

Update counts: $M_{i,k}^t = M_{i,k}^{t-1} + \mathbf{1}\{s_i^t = s_k^t\}$ for all $i, k \in N$

Update next strategies: For each i, set $\boldsymbol{x}_i^{t+1} \in \arg\max_{\boldsymbol{x}_i \in \Delta(M)} \mathbb{E}_{\mathbf{s} \sim (\boldsymbol{x}_i, \mathbf{x}_{-i}^t)}[\hat{\mathcal{U}}_i^t(\mathbf{s})]$

Check approximate Nash equilibrium:

for each agent $i \in N$ **do**

- Select best response: $\boldsymbol{x}_i^{\star,t} \in \arg\max_{\boldsymbol{x}_i \in \Delta(M)} \mathbb{E}[u_i(\boldsymbol{x}_i^t, \mathbf{x}_{-i}^t)]$
- Compute deviation gain: $\Delta_i^t = \mathbb{E}[u_i(\boldsymbol{x}_i^{\star,t}, \mathbf{x}_{-i}^t)] - \mathbb{E}[u_i(\mathbf{x}^t)]$

end for

Set current $\Delta = \max_i \Delta_i^t$

if current $\Delta \leq \varepsilon$ **then**

- **return** $\mathbf{x}^t$ as an ε-NE

end if

end for

Theorem 2. *At any time* t*, let* $\hat{\mathcal{G}}^t$ *be an additive separable hedonic project game with unknown preferences. Let* $U_{\max} = \max_{i \in N, \mathbf{s} \in \mathcal{S}} \hat{\mathcal{U}}_i^t(\mathbf{s})$*. Assume that the true underlying pairwise preferences are symmetric. Then, for any* $\varepsilon \in (0, 1]$ *there exists a time step* $T_0 \leq \lceil \frac{nU_{\max}}{\varepsilon} \rceil$ *such that*

$$\max_{i \in N} \left(\max_{\boldsymbol{x}_i \in \Delta(M)} \mathbb{E}[u_i(\boldsymbol{x}_i, \mathbf{x}_{-i}^{T_0})] - \mathbb{E}[u_i(\mathbf{x}^{T_0})] \right) \leq \varepsilon$$

To complement our theoretical analysis, we provide an empirical evaluation of the proposed framework in practical settings. The goal is to investigate whether the theoretical guarantees of sublinear regret and convergence to approximate equilibria translate effectively to real and synthetic environments. The following section outlines our experimental design, datasets, and key findings.

4 Experiments

To empirically assess the effectiveness of the proposed `UCB-NE` algorithm in ASHPGs with unknown preferences, we conduct an experimental analysis to support and validate the theoretical results presented in the preceding sections using real-world and synthetic datasets. For a real-world dataset, we have used a publicly available dataset from *gMission*, a research-oriented and open-source crowdsourcing platform [17], which comprises 1,200 tasks collected from real-world crowdsourcing activities, each characterised by publication time, geographic location, and associated rewards [33].

Synthetic instances are generated by sampling the number of players $n \sim \mathcal{U}(5, 100)$ and projects $m \sim \mathcal{U}(10, 200)$, in three scenarios in different scales: small with five agents and 10 projects, medium with 20 agents and 50 projects, and large scale with 100 agents and 200 projects. The preferences have been drawn from a uniform distribution in the range $[-1, 1]$. Each experiment was conducted for 300 to 500 rounds, using a dimension coefficient of 1.0 and an exploration parameter $\varepsilon = 0.1$ for the ε-greedy algorithm. We use epsilon-greedy as a baseline to evaluate the `UCB-NE` algorithm.

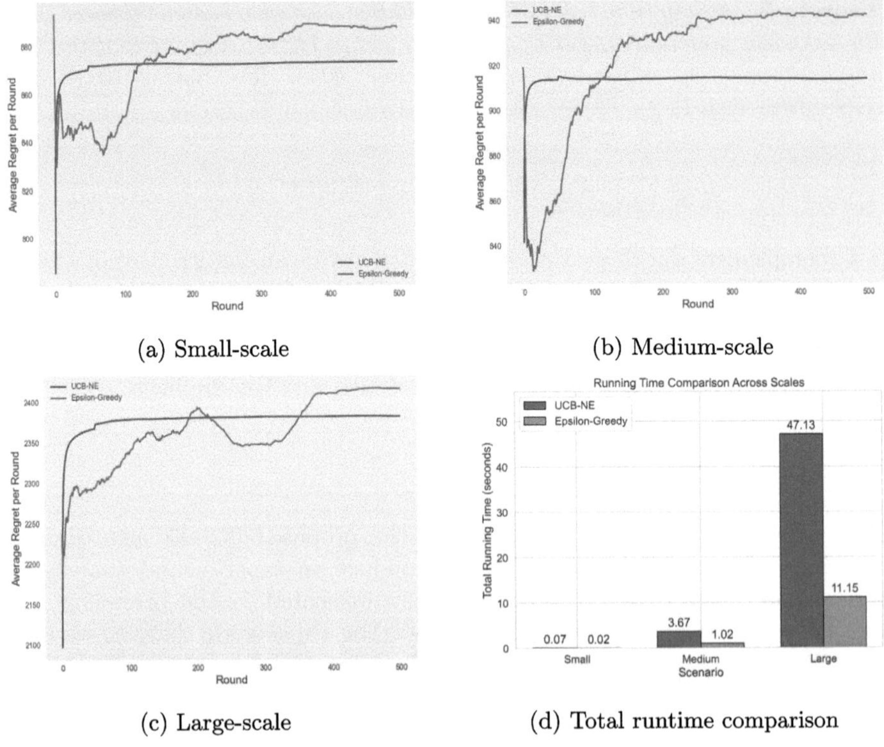

(a) Small-scale

(b) Medium-scale

(c) Large-scale

(d) Total runtime comparison

Fig. 1. Performance comparison of algorithms across different problem scales.

4.1 Results

Figures 1a to 1c present the comparative performances of the evaluated algorithms across various experimental settings. As evidenced by the results, the proposed `UCB-NE` algorithm consistently outperformed the baseline approaches, particularly in terms of convergence speed and cumulative regret. Agents utilizing the `UCB-NE` algorithm were able to form stable outcomes significantly faster than those using the epsilon-greedy baseline. Moreover, cumulative regret remained substantially lower across all time horizons and experimental scales. These results also demonstrate that the semi-bandit feedback model is efficient, allowing agents to learn individualized preferences within coalitions without requiring full information about the environment.

To assess the computational efficiency of `UCB-NE`, we compared its runtime with that of the Epsilon-Greedy baseline across small, medium, and large-scale settings. For each case, we recorded the total execution time over a fixed number of rounds, including initialization and learning. Figure 1d shows that both algorithms scale well with increasing number of agents, projects, and rounds. In small instances, the runtimes are nearly identical owing to similar coordination and update costs. In larger settings, `UCB-NE` incurs a slightly higher overhead in computing confidence bounds and managing exploration bonuses. However, this overhead is modest, and `UCB-NE` remains competitive in terms of runtime while offering improved learning performance and regret. Overall, the timing results confirm the practicality of `UCB-NE` for large-scale use, effectively balancing computation and solution quality.

4.2 Conclusion and Discussions

In this study, we introduced ASHPGs, a non-cooperative framework that combines agents' preferences over coalition partners with project-based rewards. This model bridges additively separable hedonic games and project games, thereby capturing both subjective (partner preferences) and objective (project rewards) incentives in multi-agent environments. We formalized the framework in both a single-shot setting with perfect information and a repeated setting with imperfect information. To handle cases where agents' preferences are initially unknown, we proposed a UCB-based online learning algorithm with semi-bandit feedback, enabling agents to adapt their strategies through experience within coalitions.

Computing Nash equilibria in coalition formation games is generally hard, but the potential function in ASHPGs with symmetric preferences enables equilibrium search via local improvements. Each update in UCB-NE requires $\mathcal{O}(nm)$ operations per round, with modest extra cost for confidence bounds. Our experiments show that the algorithm scales well up to 100 agents and 200 projects, indicating practical runtime despite theoretical challenges; future work may explore faster approximations and distributed implementations for larger settings.

Our results rely on symmetric preferences, which enable the potential-game structure and guarantee Nash equilibria. With asymmetric preferences, this structure breaks down: exact potential functions no longer exist, and only weaker notions such as ordinal or approximate equilibria may apply. Asymmetry also raises the risk of manipulation, since agents could misreport or exaggerate preferences to influence coalition outcomes. While UCB-NE handles stochastic uncertainty, it does not address adversarial behavior. Extending the model to asymmetric settings and identifying what stability concepts remain attainable is an important direction for future work.

Our semi-bandit model assumes full, noise-free per-member signals; however, in real-world platforms, feedback may be partial, noisy, or biased. Such occlusion reduces effective sample size, noise inflates variance, and non-random missingness can introduce bias. Addressing these challenges (e.g., via inverse-probability weighting, hierarchical pooling, or explicit selection models) and adapting UCB-NE to maintain robustness and regret guarantees remains an important avenue for future work.

A Appendix

A.1 Proof of Lemma 1

Proof. Define the potential function:

$$\Phi(\mathbf{s}) = \sum_{j \in M} \sum_{i \in C_j(\mathbf{s})} \frac{r(j)}{|C_j(\mathbf{s})|} + \frac{\gamma}{2} \sum_{j \in M} \sum_{\substack{i,k \in C_j(\mathbf{s}) \\ i \neq k}} p_i(k)$$

The first term sums the reward shares for all agents across all projects, and the second term accounts for pairwise preferences within each coalition, scaled by $\gamma/2$ to adjust for symmetry.

Suppose agent i changes their strategy from $s_i = j$ to $s_i' = j'$, yielding $\mathbf{s}'$. The utility change is:

$$u_i(\mathbf{s}') - u_i(\mathbf{s}) = \left(\frac{r(j')}{|C_{j'}(\mathbf{s}')|} - \frac{r(j)}{|C_j(\mathbf{s})|} \right) + \gamma \left(\sum_{k \in C_{j'}(\mathbf{s}')} p_i(k) - \sum_{k \in C_j(\mathbf{s})} p_i(k) \right).$$

Note that $C_{j'}(\mathbf{s}') = C_{j'}(\mathbf{s}) \cup \{i\}$ (since i joins project j') and $C_j(\mathbf{s}') = C_j(\mathbf{s}) \setminus \{i\}$ (since i leaves project j).

Now, compute the change in the potential function Φ. The reward term in $\Phi(\mathbf{s})$ is:

$$\sum_{j \in M} \sum_{i \in C_j(\mathbf{s})} \frac{r(j)}{|C_j(\mathbf{s})|} = \sum_{j \in M : C_j(\mathbf{s}) \neq \emptyset} r(j)$$

since $\sum_{i \in C_j(\mathbf{s})} \frac{r(j)}{|C_j(\mathbf{s})|} = r(j)$ for non-empty coalitions. For project j, if $C_j(\mathbf{s}') \neq \emptyset$, the contribution remains $r(j)$. For project j', the contribution is $r(j')$. The change in the reward term due to agent i's move is:

$$\frac{r(j')}{|C_{j'}(\mathbf{s}')|} - \frac{r(j)}{|C_j(\mathbf{s})|}$$

reflecting i's new reward share minus the old one.

For the preference term, consider the sum $\sum_{j \in M} \sum_{\substack{i,k \in C_j(\mathbf{s}) \\ i \neq k}} p_i(k)$. For project j, in $\mathbf{s}$, the terms involving i are $\sum_{k \in C_j(\mathbf{s}) \setminus \{i\}} (p_i(k) + p_k(i)) = 2 \sum_{k \in C_j(\mathbf{s}) \setminus \{i\}} p_i(k)$ (using symmetry $p_i(k) = p_k(i)$). In $\mathbf{s}'$, these terms are removed. For project j', in $\mathbf{s}'$, new terms are $\sum_{k \in C_{j'}(\mathbf{s}) \setminus \{i\}} (p_i(k) + p_k(i)) = 2 \sum_{k \in C_{j'}(\mathbf{s}) \setminus \{i\}} p_i(k)$. Thus, the change in the preference term is:

$$\frac{\gamma}{2} \cdot 2 \left(\sum_{k \in C_{j'}(\mathbf{s}) \setminus \{i\}} p_i(k) - \sum_{k \in C_j(\mathbf{s}) \setminus \{i\}} p_i(k) \right) = \gamma \left(\sum_{k \in C_{j'}(\mathbf{s}')} p_i(k) - \sum_{k \in C_j(\mathbf{s})} p_i(k) \right)$$

since $p_i(i) = 0$.

Combining both terms, the change in Φ is:

$$\Phi(\mathbf{s}') - \Phi(\mathbf{s}) = \left(\frac{r(j')}{|C_{j'}(\mathbf{s}')|} - \frac{r(j)}{|C_j(\mathbf{s})|}\right) + \gamma\left(\sum_{k\in C_{j'}(\mathbf{s}')} p_i(k) - \sum_{k\in C_j(\mathbf{s})} p_i(k)\right)$$

which equals $u_i(\mathbf{s}') - u_i(\mathbf{s})$. Thus, Φ is an exact potential function.

A.2 Proof of Theorem 1

Proof. The estimated utility is $\hat{\mathcal{U}}_i^t(\mathbf{s}) = \hat{u}_i^t(\mathbf{s}) + \hat{b}_i^t(\mathbf{s})$. Consider agent i deviating from project s_i to s_i'. The coalitions change as follows: $C_{s_i'}(\mathbf{s}') = C_{s_i'}(\mathbf{s}) \cup \{i\}$, $C_{s_i}(\mathbf{s}') = C_{s_i}(\mathbf{s}) \setminus \{i\}$, and $C_i(\mathbf{s}') = C_i(\mathbf{s})$ for $j \neq s_i, s_i'$. The change in agent i's utility is:

$$\hat{\mathcal{U}}_i^t(\mathbf{s}') - \hat{\mathcal{U}}_i^t(\mathbf{s}) = \left(\frac{r(s_i')}{|C_{s_i'}(\mathbf{s})|+1} - \frac{r(s_i)}{|C_{s_i}(\mathbf{s})|}\right) + \gamma\left(\sum_{k\in C_{s_i'}(\mathbf{s})} \hat{p}_i^t(k) - \sum_{k\in C_{s_i}(\mathbf{s})\setminus\{i\}} \hat{p}_i^t(k)\right)$$
$$+\lambda\left(\sum_{k\in C_{s_i'}(\mathbf{s})} \sqrt{\frac{2\log(4mnT/\delta)}{M_{i,k}^t(\mathbf{s})\vee 1}} - \sum_{k\in C_{s_i}(\mathbf{s})\setminus\{i\}} \sqrt{\frac{2\log(4mnT/\delta)}{M_{i,k}^t(\mathbf{s})\vee 1}}\right)$$

The potential function change is:

$$\hat{\Phi}^t(\mathbf{s}') - \hat{\Phi}^t(\mathbf{s}) = \sum_{i\in N}\left(\hat{\mathcal{U}}_i^t(\mathbf{s}') - \hat{\mathcal{U}}_i^t(\mathbf{s})\right)$$

For coalition members, reward terms shift as in congestion games, while preference and bonus terms add symmetrically and cancel across pairs. Hence,

$$\hat{\Phi}^t(\mathbf{s}') - \hat{\Phi}^t(\mathbf{s}) = \hat{\mathcal{U}}_i^t(\mathbf{s}') - \hat{\mathcal{U}}_i^t(\mathbf{s})$$

A.3 Proof of Theorem 2

Proof. From Theorem 1, $\hat{\mathcal{G}}^t$ is a potential game with exact potential $\hat{\Phi}^t(\mathbf{s}) = \sum_i \hat{\mathcal{U}}_i^t(\mathbf{s})$. Each agent updates via `UCB-NE`, maximizing expected $\hat{\mathcal{U}}_i^t$. Since $\hat{\Phi}^t$ is bounded by $nU_{\max}$, each deviation increases it by the deviator's utility gain.

By potential game theory, these updates yield no-regret dynamics, converging to an ε-approximate Nash equilibrium once $\hat{\Phi}^t$ is within ε of its maximum. Because the total possible increase is at most $nU_{\max}$, the number of steps needed is at most $\lceil nU_{\max}/\varepsilon\rceil$.

Thus, there exists $T_0 \leq \lceil nU_{\max}/\varepsilon\rceil$ such that $\mathbf{x}^{T_0}$ is an ε-approximate Nash equilibrium.

References

1. Alcalde, J., Revilla, P.: Researching with whom? Stability and manipulation. J. Math. Econ. **40**(8), 869–887 (2004)
2. Aziz, H., Brandl, F., Brandt, F., Harrenstein, P., Olsen, M., Peters, D.: Fractional hedonic games. ACM Trans. Econ. Comput. (TEAC) **7**(2), 1–29 (2019)
3. Aziz, H., Bredereck, R., Elkind, E., Walsh, T.: Stable group activity selection. Theoret. Comput. Sci. **754**, 50–64 (2019)
4. Aziz, H., Savani, R.: Hedonic games (2016)
5. Balcan, M.F., Procaccia, A.D., Zick, Y.: Learning cooperative games. In: Proceedings of the 24th International Joint Conference on Artificial Intelligence (IJCAI), pp. 475–481 (2015)
6. Banerjee, S., Konishi, H., Sönmez, T.: Core in a simple coalition formation game. Soc. Choice Welfare **18**(1), 135–153 (2001)
7. Bilò, V., Fanelli, A., Flammini, M., Monaco, G., Moscardelli, L.: Nash stable outcomes in fractional hedonic games: existence, efficiency and computation. J.Artif. Intell. Res. **62**, 315–371 (2018)
8. Bilò, V., Gourvès, L., Monnot, J.: Project games. Theoret. Comput. Sci. **940**, 97–111 (2023)
9. Boehmer, N., Bullinger, M., Kerkmann, A.M.: Causes of stability in dynamic coalition formation. ACM Transactions on Economics and Computation (2023)
10. Bogomolnaia, A., Jackson, M.O.: Stability in hedonic coalition formation games. Games Econom. Behav. **38**(2), 201–230 (2002)
11. Brandt, F., Bullinger, M., Wilczynski, A.: Reaching individually stable coalition structures in hedonic games. In: Proceedings of the AAAI Conference on Artificial Intelligence, vol. 35, pp. 5211–5218 (2021)
12. Bullinger, M., Romen, R.: Stability in online coalition formation. In: Proceedings of the AAAI Conference on Artificial Intelligence, vol. 38, pp. 9537–9545 (2024)
13. Cechlárová, K.n., Romero-Medina, A.: Stability in coalition formation games. Int. J. Game Theory **29**, 487–494 (2001)
14. Cesa-Bianchi, N., Lugosi, G.: Prediction, Learning, and Games. Cambridge university Press (2006)
15. Chalkiadakis, G., Elkind, E., Wooldridge, M.: Computational and gametheoretic aspects of coalition formation. Synthesis Lect. Artif. Intell. Mach. Learn. **5**(6), 1–168 (2011)
16. Chalkiadakis, G., Jennings, N.R., Boutilier, C.: Learning, coalition formation and coordination. In: AAAI (2008)
17. Chen, Z., et al.: gMission: a general spatial crowdsourcing platform. Proc. VLDB Endow. **7**(13), 1629–1632 (2014)
18. Cohen, S., Agmon, N.: Online friends partitioning under uncertainty. In: ECAI 2024, pp. 3332–3339. IOS Press (2024)
19. Cohen, S., Agmon, N.: Online learning of coalition structures by selfish agents. In: Proceedings of the AAAI Conference on Artificial Intelligence, vol. 39, pp. 13709–13717 (2025)
20. Cui, Q., Xiong, Z., Fazel, M., Du, S.S.: Learning in congestion games with bandit feedback. Adv. Neural. Inf. Process. Syst. **35**, 11009–11022 (2022)
21. Ding, D., Wei, C.Y., Zhang, K., Jovanovic, M.: Independent policy gradient for large-scale Markov potential games: Sharper rates, function approximation, and game-agnostic convergence. In: International Conference on Machine Learning, pp. 5166–5220. PMLR (2022)

22. Dreze, J.H., Greenberg, J.: Hedonic coalitions: optimality and stability. Econom.: J. Econom. Soc. 987–1003 (1980)
23. Elkind, E., Wooldridge, M.J.: Hedonic coalition nets. In: AAMAS (1), pp. 417–424. Citeseer (2009)
24. Jones, M., Nguyen, H., Nguyen, T.: An efficient algorithm for fair multi-agent multi-armed bandit with low regret. In: Proceedings of the AAAI Conference on Artificial Intelligence, vol. 37, pp. 8159–8167 (2023)
25. Lattimore, T., Szepesvári, C.: Bandit Algorithms. Cambridge University Press, Cambridge (2020)
26. Liu, L.T., Ruan, F., Mania, H., Jordan, M.I.: Bandit learning in decentralized matching markets. J. Mach. Learn. Res. **22**(211), 1–34 (2021)
27. Liu, Q., Yu, T., Bai, Y., Jin, C.: A sharp analysis of model-based reinforcement learning with self-play. In: International Conference on Machine Learning, pp. 7001–7010. PMLR (2021)
28. Rosenthal, R.W.: A class of games possessing pure-strategy nash equilibria. Internat. J. Game Theory **2**, 65–67 (1973)
29. Subramanian, H.: Decentralized blockchain-based electronic marketplaces. Commun. ACM **61**(1), 78–84 (2017)
30. Valizadeh, J., Zhang, D., Mubin, O.: Enhancing the efficiency of systems with overlapping coalition formation. In: Pacific Rim International Conference on Artificial Intelligence, pp. 284–290. Springer (2024)
31. Vetta, A.: Nash equilibria in competitive societies, with applications to facility location, traffic routing and auctions. In: The 43rd Annual IEEE Symposium on Foundations of Computer Science, 2002. Proceedings, pp. 416–425. IEEE (2002)
32. Vöcking, B.: Selfish load balancing. Algorithmic Game Theory **20**, 517–542 (2007)
33. Xia, J., Zhao, Y., Liu, G., Xu, J., Zhang, M., Zheng, K.: Profit-driven task assignment in spatial crowdsourcing. In: IJCAI, pp. 1914–1920 (2019)
34. Zinkevich, M., Johanson, M., Bowling, M., Piccione, C.: Regret minimization in games with incomplete information. Adv. Neural Inf. Process. Syst. **20** (2007)

Policy Gradient–Based Reinforcement Weighted Aggregation for Efficient Federated Learning on Heterogeneous Data

Mahmudul Hasan[1], Md Palash Uddin[1](✉), Yong Xiang[1], John Yearwood[1], and Longxiang Gao[2,3]

[1] School of Information Technology, Deakin University, Geelong, VIC 3220, Australia
{mahmudul.hasan,m.uddin,yong.xiang,john.yearwood}@deakin.edu.au

[2] Qilu University of Technology (Shandong Academy of Sciences), Jinan 250202, Shandong, China
gaolx@sdas.org

[3] Shandong Computer Science Center (National Supercomputer Center in Jinan), Jinan 250202, Shandong, China

Abstract. Federated Learning (FL) is a decentralized paradigm that enables collaborative model training without sharing raw data. However, traditional FL methods often struggle with slow and unstable convergence, especially under non-Independent and non-Identically Distributed (non-IID) data, which degrades model performance and training stability. Most existing aggregation strategies rely on static, uniform, or heuristic-based weighting schemes that fail to adapt dynamic client behavior and overlook variations in local training quality. Although some recent approaches explore client selection or value-based reinforcement processes (e.g., Q-learning), they fall short of directly optimizing the aggregation process in an end-to-end fashion. To address these limitations, we propose Federated Reinforcement Weighted Aggregation (FedRWA), a policy gradient-based aggregation approach that dynamically assigns client-specific weights based on local training characteristics on the server side. Each client generates a feature vector consisting of training loss and accuracy, which is fed into a central policy network. This network outputs aggregation weights and is trained via a reinforcement process, using the global model's validation performance as the reward. Extensive experiments on the UNSW-NB15 and NSL-KDD datasets, under both IID and non-IID conditions, show that FedRWA consistently outperforms state-of-the-art baselines in terms of classification metrics and communication efficiency. In addition, the learning curves demonstrate the superior convergence speed and training stability of FedRWA, validating its effectiveness as a robust solution for adaptive aggregation in heterogeneous FL environments.

Keywords: Federated Learning · Reinforcement Weighted Aggregation · Policy Network · Data Heterogeneity · Adaptive Aggregation · Non-IID Data

M. Liu et al. (Eds.): AI 2025, LNAI 16371, pp. 506–518, 2026.
https://doi.org/10.1007/978-981-95-4972-6_39

1 Introduction

Federated Learning (FL) has emerged as a powerful decentralized machine learning paradigm that enables multiple clients to collaboratively train a shared global model without exchanging their local data [1]. This privacy-preserving approach is particularly beneficial in sensitive domains such as healthcare, finance, and personalized services, where data centralization is restricted due to privacy concerns [2]. In FL, each client trains the global model locally on its private data and transmits the updates to a central server, which aggregates them to form an updated global model. Among existing aggregation strategies, FedAvg [1] is widely adopted due to its simplicity and scalability. It computes a weighted average of local model updates, typically using client data sizes as weights. However, in real-world scenarios, FedAvg and similar uniform aggregation methods are fundamentally limited, as they treat all client updates equally. This assumption is problematic under non-Independent and non-Identically Distributed (non-IID) data conditions, where it leads to biased updates and unstable convergence behavior [3]. Non-IID settings pose significant challenges because locally optimized models may diverge from one another, resulting in degraded generalization, slower convergence, and poor global model performance. Furthermore, under non-IID distributions, FL requires a larger number of communication rounds (CRs) to reach convergence. To address these challenges, aggregation mechanisms must become more flexible, performance-aware, and sensitive to client-specific attributes.

Recent research has explored adaptive aggregation strategies that assign different weights to clients based on factors such as data volume, model divergence, or local performance [4]. However, most of these approaches rely on static or heuristic-based weighting schemes that are not robust across training rounds or generalizable to diverse conditions. A more promising but understudied direction involves employing a learning-based decision-making process to assign dynamic aggregation weights based on observable client parameters and model performance. Some studies have introduced a reinforcement process into FL to optimize communication efficiency, client selection, or resource management [5,6]. However, few have focused on directly learning the aggregation function in an end-to-end manner. Existing RL-based methods often use value-based approaches (e.g., Q-learning) and overlook critical client-side training characteristics such as local accuracy, training loss. More importantly, they typically lack a direct feedback loop from the global model performance, which is crucial for guiding the learning of effective aggregation strategies.

To address these limitations, we propose a novel approach called Federated Reinforcement Weighted Aggregation (FedRWA), which dynamically learns client-specific aggregation weights based on local training dynamics and global model feedback. Each client generates a feature vector summarizing its local training status, which is processed by a central policy network in the central server. The network outputs a scalar weight for each client, which is normalized and used to compute a weighted average of local models. The policy network is trained using a reinforcement process framework, with the global model's validation performance serving as the reward. This design ensures that aggregation decisions are directly optimized to enhance overall model performance. As the reinforcement process is executed entirely on the server side, it preserves client privacy by relying only on high-level aggregated statistics (such as training loss and accuracy) rather than raw data. This process introduces minimal computational and communication overhead, as the policy network is lightweight and operates solely on summarized client features. The key contributions of this work are as follows:

- We propose FedRWA, a novel aggregation approach that dynamically assigns client weights based on local training characteristics and global model feedback.
- We design a lightweight policy network in FedRWA that processes client-specific feature vectors and employs a reinforcement process to enable an end-to-end optimization loop.
- We conduct extensive experiments on two benchmark intrusion detection datasets (UNSW-NB15 and NSL-KDD) under both IID and non-IID settings, demonstrating that FedRWA achieves superior performance and convergence compared to existing aggregation methods.

The remainder of this paper is organized as follows: Sect. 2 reviews related work. Section 3 details the proposed method. Section 4 presents the experimental setup and results. Section 5 concludes the paper and discusses future directions.

2 Related Work

Traditional FL aggregation schemes, such as FedAvg [1], assume uniform data quality and client contribution, computing a simple average of local updates. This assumption fails in non-IID scenarios, where it results in biased global updates and degraded model performance. To address this, several adaptive aggregation techniques have been proposed. For instance, FedDynamic [7] adjusts aggregation weights based on model divergence, data quality, and local precision. While it accounts for heterogeneity, it depends on predefined rules rather than trainable models, limiting its adaptability. SelfFed [8] employs a game-theoretic framework to assign client weights based on expected contribution quality but lacks a learning mechanism to adapt dynamically during training.

Recent efforts have integrated the reinforcement process into FL. FedRQ [9], for example, uses a reinforcement process to select clients based on their historical contributions. However, this method focuses solely on client selection and does not learn aggregation weights. The FedDreinforcement process [10] uses a deep reinforcement process to adjust client impact factors in cluster-skewed environments, employing Q-learning without leveraging local performance features or global feedback. Similarly, the study in [11] uses Q-learning for aggregation weight assignment but does not utilize policy gradients or client-specific training dynamics. FedAA [12] uses the Deep Deterministic Policy Gradient (DDPG) algorithm to adjust aggregation weights, focusing primarily on robustness against adversarial attacks rather than addressing data heterogeneity. FedAWA [13] assigns weights based on cosine similarity between local and global updates, capturing directional alignment but lacking adaptability and feedback-driven learning. FedLaw [14] introduces a consistency-based mechanism for weight learning across training rounds. FedBE [4] applies Bayesian ensemble averaging to enhance performance under non-IID settings but incurs high computational costs and lacks dynamic weight learning capabilities. In contrast to these methods, our proposed FedRWA leverages a policy gradient-based reinforcement process to dynamically learn aggregation weights, incorporating both local training features and global performance feedback in a unified, end-to-end manner.

3 Methodology

3.1 Approach Overview

The proposed FedRWA method builds upon the standard FL framework. Initially, the dataset is divided into training and testing subsets. The training data is then distributed among multiple clients according to both IID and non-IID partitioning strategies. A fixed number of CRs is performed, during which the complete FL process is executed in each round. On the server side, various aggregation algorithms, including the proposed FedRWA, are applied sequentially, and their performance is assessed using multiple evaluation metrics. Additionally, the number of CRs required by each method to reach a predefined target accuracy is recorded to measure convergence efficiency. After conducting experiments across all IID and non-IID configurations, the performance of each technique is systematically compared to identify the most effective aggregation strategy. An overview of the entire process is depicted in Fig. 1.

3.2 Objective Formulation

The primary goal of FedRWA is to build a global model θ_g that minimizes global aggregate loss between all clients. Formally, this is defined by Eq. 1.

$$\min_{\theta_g} \mathcal{L}_{global}(\theta_g) = \sum_{i=1}^{N} p_i \cdot \mathcal{L}_i(\theta_g), \tag{1}$$

where $\mathcal{L}_i(\theta_g)$ is the local loss in the client i, and p_i is the weight proportional to the contribution of the client (typically $p_i = \frac{|\mathcal{D}_i|}{\sum_j |\mathcal{D}_j|}$) in traditional settings. In our FedRWA, we replace the fixed p_i with adaptive weights w_i from a policy network π_ϕ, and w_i is calculated based on client-specific features. Therefore, the aggregation objective becomes Eq. 2.

$$\theta_g = \sum_{i \in \mathcal{C}_t} w_i \cdot \theta_i \quad \text{where} \quad w_i = \pi_\phi(f_i), \quad \sum_{i \in \mathcal{C}_t} w_i = 1. \tag{2}$$

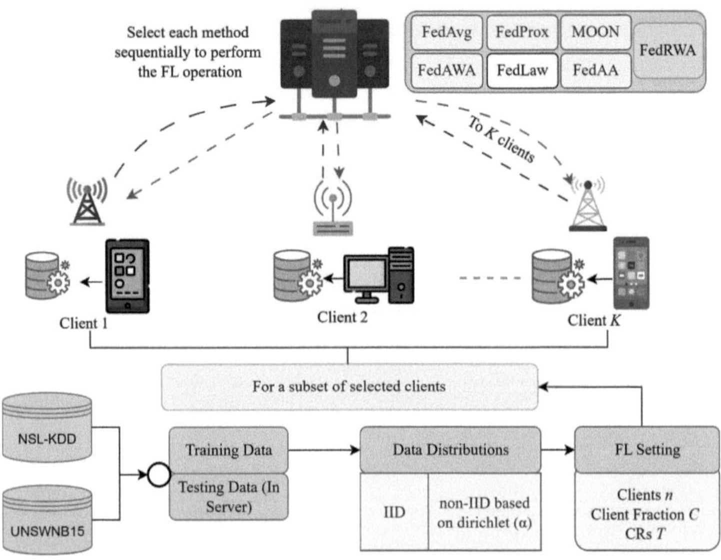

Fig. 1. Overview of the proposed FedRWA framework.

The policy network maximizes the performance of the global model by generating adaptive aggregation weights. We formulate the policy optimization as a reinforcement process problem, where the reward R is a function of the validation accuracy of θ_g, like $\max_\phi \mathbb{E}_{\pi_\phi}[R(\theta_g)]$. Using the reinforcement process, the corresponding loss for the policy network becomes Eq. 3.

$$\mathcal{L}_{\text{policy}} = -R(\theta_g) \cdot \sum_{i \in \mathcal{C}_t} \log w_i. \tag{3}$$

This equation indicates that FedRWA jointly solves two objectives: it optimizes the global model θ_g via performance-aware weighted aggregation and optimizes the policy network π_ϕ to dynamically generate weights w_i that maximize the global model's performance.

3.3 FedRWA Algorithm Formulation

The FedRWA algorithm begins by initializing the global model $\theta_g^{(0)}$ and the policy network parameters ϕ. For each CR $t = 1$ to T, the server selects a subset of clients $\mathcal{C}_t \subseteq \{1, \ldots, N\}$, where each client $i \in \mathcal{C}_t$ receives the global model $\theta_g^{(t-1)}$, performs local training, and obtains an updated local model $\theta_i^{(t)}$. In addition, each client computes a feature vector $f_i = [\ell_i, a_i]$, where ℓ_i is the training loss, and a_i is the training accuracy. Each client sends both $(\theta_i^{(t)}, f_i)$ back to the server. On the server side, a shared policy network π_ϕ is used to calculate the weights of the aggregation for all participating clients. For each client $i \in \mathcal{C}_t$, the feature vector f_i is first passed through a hidden layer to compute

Algorithm 1. The FedRWA algorithm

Require: Number of clients N, total CRs T, local epochs E, policy network π_ϕ, and validation dataset $\mathcal{D}_{val}$

Initialize the global model $\theta_g^{(0)}$ and policy parameters ϕ
for each round $t = 1$ to T **do**
    The server selects a subset of clients $\mathcal{C}_t \subseteq \{1, \ldots, N\}$
    for each client $i \in \mathcal{C}_t$ **in parallel do**
        Receive global model $\theta_g^{(t-1)}$
        Train locally for E epochs to obtain $\theta_i^{(t)}$
        Compute local features $f_i = [\ell_i, a_i]$
        Send $(\theta_i^{(t)}, f_i)$ to the server
    end for
    for each client $i \in \mathcal{C}_t$ **do** ▷ Server-side: Compute aggregation weights
        $h_i \leftarrow \text{ReLU}(W_1 f_i)$
        $\tilde{w}_i \leftarrow \sigma(W_2 h_i)$
    end for
    Normalize: $w_i = \frac{\tilde{w}_i}{\sum_{j \in \mathcal{C}_t} \tilde{w}_j}$ for all i
    $\theta_g^{(t)} = \sum_{i \in \mathcal{C}_t} w_i \cdot \theta_i^{(t)}$ ▷ Weighted aggregation
    Evaluate $\theta_g^{(t)}$ on $\mathcal{D}_{val}$ to obtain the reward R
    Compute policy gradient: $\nabla_\phi \mathcal{L}_{\text{policy}} = -R \cdot \nabla_\phi \log w_i$
    Update the policy parameters $\phi \leftarrow \phi - \eta \cdot \nabla_\phi \mathcal{L}_{\text{policy}}$
end for
return the final global model $\theta_g^{(T)}$

the intermediate representation as $h_i = \text{ReLU}(W_1 f_i)$. Then the output layer computes a normalized score using the sigmoid activation as $\tilde{w}_i = \sigma(W_2 h_i)$. These scores are normalized across all selected clients to produce valid aggregation weights as $w_i = \frac{\tilde{w}_i}{\sum_{j \in \mathcal{C}_t} \tilde{w}_j}$ and then the server aggregates the local models using a weighted average as $\theta_g^{(t)} = \sum_{i \in \mathcal{C}_t} w_i \cdot \theta_i^{(t)}$. To evaluate the quality of the aggregation, the server tests the updated global model $\theta_g^{(t)}$ on the testing dataset $\mathcal{D}_{val}$, obtaining a reward signal R, typically based on testing accuracy. This reward is used to update the policy network using the reinforcement process. The policy objective is defined as $\mathcal{L}_{\text{policy}} = -R \cdot \log w_i$. The corresponding gradient is used for updating the policy network like $\nabla_\phi \mathcal{L}_{\text{policy}} = -R \cdot \nabla_\phi \log w_i$. Finally, the policy parameters are updated using a gradient descent process as $\phi \leftarrow \phi - \eta \cdot \nabla_\phi \mathcal{L}_{\text{policy}}$. This iterative process continues for T rounds. The policy network learns to assign higher weights to clients that positively impact global performance. The complete pseudocode of our proposed FedRWA is provided in Algorithm 1.

4 Experimental Result Analysis

4.1 Setup

We evaluate our proposed method using two widely used intrusion detection datasets, namely UNSW-NB15 and NSL-KDD. These datasets are selected due to their diversity in attack types, realistic traffic patterns, and widespread use in cybersecurity and FL research [15]. The UNSW-NB15 dataset comprises 257,673 samples with 49 features and 10 classes (1 normal and 9 attack types). The NSL-KDD dataset contains 125,973 samples with 41 features and 5 classes (Normal, DoS, Probe, R2L, and U2R), and is a cleaned and balanced version of the KDD-Cup'99 dataset. Both datasets are partitioned across 100 clients. For IID settings, we employ uniform partitioning following McMahan et al. [1]. For non-IID settings, we use a Dirichlet distribution-based partitioning strategy with concentration parameters $\alpha = 1.0$ and 0.5 to control the level of heterogeneity. We assess performance using the training loss, testing accuracy, precision, recall, F1 score, and the number of CRs required to reach a target accuracy. We used loss and accuracy as state features since they offer stable and privacy-preserving indicators of client reliability, and adopted a lightweight two-layer MLP policy network as it provided the best balance between accuracy, stability, and efficiency compared to deeper or more complex alternatives. The input vector is passed through a fully connected layer with 64 hidden units, followed by a ReLU activation function to introduce non-linearity. The resulting hidden representation is then passed through a second fully connected layer that maps it to the number of output classes. A softmax activation function is applied at the output layer to produce a probability distribution over the class labels. In our experiments, we compare the proposed FedRWA against six baselines and state-of-the-art FL methods, including FedAvg [1], FedProx [16], MOON [17], FewAWA [13], FedLaw [14], and FedAA [12].

4.2 Performance Analysis

Performance Under IID Data Distributions. The results on both datasets under IID settings are presented in Table 1. We evaluate all methods using Accuracy (Acc), Precision (P), Recall (R), F1-score (F1), and the number of CRs required to reach a target accuracy. FedRWA consistently outperforms all baseline methods in terms of both predictive performance and communication efficiency. On the UNSW-NB15 dataset, FedRWA achieves the best results in both partial and full client participation. Under partial participation, it reaches an accuracy and F1-score of 0.973 in just 8 CRs, demonstrating a 4.5× speedup over FedAvg. In the full client setting, FedRWA improves further, reaching an accuracy and F1-score of 0.983 in only 6 CRs, reflecting a 6.33× faster convergence than FedAvg. On the NSL-KDD dataset, FedRWA achieves the highest accuracy and F1-score of 0.985 in only 8 CRs under partial participation, significantly outperforming FedAA and FedLaw, which require 97 and 94 rounds, respectively. Even in the full client setting, where FedRWA takes more rounds (157), it delivers the best F1-score of 0.942, outperforming all other baselines. These results confirm FedRWA's ability to achieve superior model performance with substantially fewer CRs.

Performance Under Non-IID Data Distributions. Results under non-IID conditions using $\alpha = 1.0$ and $\alpha = 0.5$ are shown in Table 2 (UNSW-NB15) and Table 3 (NSL-KDD). For the UNSW-NB15 dataset at $\alpha = 1.0$, FedRWA achieves

Table 1. Performance of FedRWA and baseline methods on the UNSW-NB15 and NSL-KDD datasets under IID data distribution.

UNSW-NB15 IID										
Model	Partial Client ($C = 0.2$)					Full client ($C = 1.0$)				
	CRs	Acc	P	R	F1	CRs	Acc	P	R	F1
FedAvg	36	0.951	0.954	0.951	0.951	38	0.941	0.946	0.941	0.941
FedProx	18 (2×)	0.968	0.969	0.968	0.968	17 (2.23×)	0.974	0.974	0.974	0.974
Moon	17 (2.12×)	0.956	0.957	0.956	0.956	42 (0.91×)	0.962	0.964	0.962	0.962
FedAWA	20 (1.80×)	0.961	0.963	0.961	0.961	18 (211×)	0.960	0.962	0.960	0.960
FedLaw	15 (2.40×)	0.962	0.963	0.962	0.962	25 (1.52×)	0.971	0.971	0.971	0.971
FedAA	10 (3.60×)	0.966	0.966	0.966	0.966	13 (2.92×)	0.969	0.970	0.969	0.969
FedRWA	08 (4.50×)	0.973	0.973	0.973	0.973	06 (6.33×)	0.983	0.983	0.983	0.983
NSL-KDD IID										
FedAvg	–	0.925	0.859	0.925	0.890	–	0.925	0.872	0.925	0.893
FedProx	–	0.946	0.901	0.946	0.923	–	0.945	0.900	0.945	0.922
Moon	–	0.923	0.876	0.923	0.892	–	0.936	0.921	0.936	0.910
FedAWA	–	0.925	0.858	0.925	0.890	–	0.925	0.858	0.925	0.890
FedLaw	94 (Imm)	0.968	0.950	0.968	0.959	–	0.948	0.930	0.948	0.933
FedAA	97 (Imm)	0.967	0.949	0.967	0.957	–	0.946	0.900	0.946	0.923
FedRWA	08 (Imm)	0.985	0.985	0.985	0.985	157 (Imm)	0.956	0.938	0.956	0.942

Table 2. Model performance on the UNSW-NB15 dataset under non-IID settings, evaluated across different levels of non-IID ($\alpha = 1.0$ and $\alpha = 0.5$) with both partial and full client participation.

UNSW Alpha = 1.0										
Model	Partial Client ($C = 0.2$)					Full client ($C = 1.0$)				
	CRs	Acc	P	R	F1	CRs	Acc	P	R	F1
FedAvg	26	0.959	0.961	0.959	0.959	45	0.971	0.972	0.971	0.971
FedProx	20 (1.3×)	0.969	0.965	0.968	0.968	26 (1.73×)	0.974	0.975	0.974	0.974
Moon	28 (0.93×)	0.962	0.963	0.962	0.962	32 (1.41×)	0.962	0.963	0.962	0.961
FedAWA	37 (0.70×)	0.968	0.968	0.968	0.968	96 (0.47×)	0.961	0.962	0.961	0.960
FedLaw	72 (0.36×)	0.962	0.963	0.962	0.962	42 (1.07×)	0.971	0.972	0.971	0.971
FedAA	18 (1.44×)	0.971	0.972	0.971	0.971	13 (3.46×)	0.968	0.970	0.968	0.968
FedRWA	12 (2.17×)	0.973	0.974	0.973	0.973	05 (9.00×)	0.979	0.980	0.979	0.979
UNSW Alpha = 0.5										
FedAvg	69	0.969	0.970	0.969	0.969	49	0.969	0.969	0.969	0.969
FedProx	24 (2.47×)	0.963	0.963	0.963	0.963	21 (2.33×)	0.962	0.963	0.962	0.962
Moon	55 (1.25×)	0.950	0.952	0.950	0.950	–	0.948	0.949	0.948	0.948
FedAWA	113 (0.61×)	0.959	0.960	0.959	0.958	–	0.945	0.947	0.945	0.945
FedLaw	38 (1.82×)	0.963	0.963	0.963	0.963	41 (1.20×)	0.962	0.963	0.962	0.962
FedAA	21 (3.29×)	0.969	0.970	0.969	0.969	24 (2.04×)	0.961	0.962	0.961	0.961
FedRWA	19 (3.63×)	0.970	0.971	0.972	0.972	10 (4.90×)	0.972	0.973	0.972	0.972

Table 3. Model performance on the NSL-KDD dataset under non-IID settings, evaluated across varying data heterogeneity levels ($\alpha = 1.0$ and $\alpha = 0.5$) with both partial and full client participation.

NSLKDD non-IID Alpha = 1.0										
Model	Partial Client ($C = 0.2$)					Full client ($C = 1.0$)				
	CRs	Acc	P	R	F1	CRs	Acc	P	R	F1
FedAvg	–	0.926	0.861	0.926	0.891	96	0.966	0.948	0.966	0.956
FedProx	–	0.946	0.901	0.946	0.923	–	0.925	0.858	0.925	0.890
Moon	–	0.939	0.923	0.939	0.921	–	0.945	0.928	0.945	0.929
FedAWA	158 (Imm)	0.955	0.927	0.955	0.941	–	0.908	0.837	0.908	0.870
FedLaw	195 (Imm)	0.957	0.938	0.957	0.944	149 (0.64 ×)	0.965	0.953	0.967	0.962
FedAA	–	0.925	0.861	0.925	0.891	–	0.947	0.901	0.947	0.923
FedRWA	37 (Imm)	0.965	0.946	0.955	0.955	40 (2.40 ×)	0.970	0.972	0.970	0.971
NSLKDD non-IID Alpha = 0.5										
FedAvg	–	0.947	0.901	0.947	0.923	–	0.925	0.858	0.925	0.890
FedProx	–	0.925	0.883	0.925	0.890	–	0.915	0.889	0.924	0.895
Moon	–	0.924	0.881	0.924	0.890	–	0.936	0.892	0.936	0.913
FedAWA	–	0.941	0.896	0.941	0.918	–	0.927	0.884	0.927	0.901
FedLaw	198 (Imm)	0.952	0.914	0.952	0.932	–	0.937	0.891	0.937	0.912
FedAA	54 (Imm)	0.959	0.951	0.969	0.960	–	0.948	0.932	0.948	0.934
FedRWA	45 (Imm)	0.961	0.953	0.971	0.961	67 (Imm)	0.963	0.965	0.963	0.964

the highest F1-scores in both partial and full client participation. In the partial participation case, it reaches an F1-score of 0.973 in just 12 CRs, outperforming FedAA (0.971 in 18 CRs) and significantly reducing CRs compared to FedAvg (26 CRs). In the full client setting, FedRWA achieves an F1-score of 0.979 in only 5 CRs, demonstrating a 9× speedup over FedAvg (45 CRs), while surpassing FedProx and FedLaw. At the more challenging $\alpha = 0.5$ level, FedRWA maintains its dominance. It achieves F1-scores of 0.970 in 19 CRs (partial) and 0.972 in 10 CRs (full), whereas FedAWA requires 113 CRs with lower performance, and FedProx and FedAA require 24 and 21 CRs, respectively, to converge. For the NSL-KDD dataset with $\alpha = 1.0$, FedRWA again leads. It reaches an F1-score of 0.955 in 37 CRs under partial participation and 0.971 in 40 CRs under full participation, outperforming FedLaw and FedAA, which require 195 and 149 CRs, respectively. Other methods, such as MOON and FedAWA, lag behind in both convergence speed and accuracy. At $\alpha = 0.5$, FedRWA achieves an F1-score of 0.961 in 45 CRs (partial) and 0.964 in the full setting, once again outperforming all baselines. Note that some inherent variability exists in the performance of different FL methods due to factors such as random client selection, data heterogeneity, and hyperparameter configurations. This performance variability is a well-known and commonly observed phenomenon in empirical studies on FL [3,16].

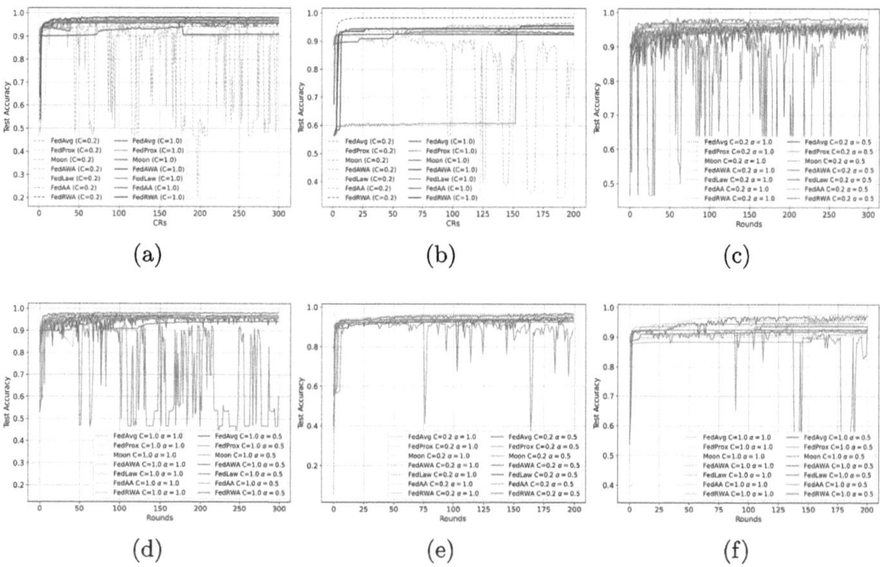

Fig. 2. Testing accuracy vs. CRs for: (a) UNSW-NB15 under IID, (b) NSL-KDD under IID, (c) UNSW-NB15 under non-IID ($C = 0.2$), (d) UNSW-NB15 under non-IID ($C = 1.0$), (e) NSL-KDD under non-IID ($C = 0.2$), (f) NSL-KDD under non-IID ($C = 1.0$).

Impact of Data Heterogeneity. Different non-IID levels ($\alpha = 1.0$ and $\alpha = 0.5$) show that increasing heterogeneity (i.e., lower α) degrades global model performance across most methods. However, FedRWA maintains high performance and requires significantly fewer CRs. On the UNSW-NB15 dataset, as α decreases, baseline methods exhibit lower F1-scores and increased CRs. For instance, FedAWA's F1-score drops from 0.960 to 0.945 and requires over 100 CRs in full participation. FedRWA achieves F1-scores of 0.970 (partial) and 0.972 (full) at $\alpha = 0.5$, in just 19 and 10 CRs, respectively. As non-IIDness increases in NSL-KDD, most baselines suffer in performance and require more rounds. For example, FedLaw takes 195 rounds to reach an F1-score of 0.944 at $\alpha = 1.0$ under partial participation, while FedRWA achieves a higher F1-score of 0.955 in only 37 CRs. At $\alpha = 0.5$, FedRWA again outperforms all methods, reaching F1-scores of 0.961 (partial) and 0.964 (full) with minimal communication. These findings confirm that while most FL methods struggle under increasing non-IIDness, FedRWA consistently delivers high accuracy, faster convergence, and lower communication overhead.

Training and Testing Progress Visualization. Figures 2 and 3 visualize the training and testing progression for FedRWA and other methods across IID and non-IID settings. In IID scenarios, FedRWA shows the highest accuracy and lowest training loss, with smooth and stable curves, which indicate strong convergence behavior. In contrast, methods like MOON and FedAWA exhibit notice-

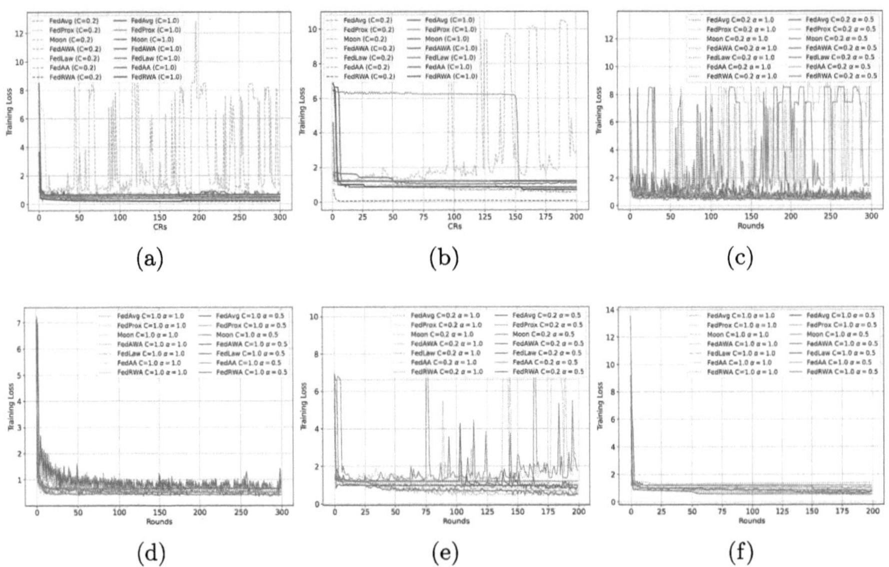

Fig. 3. Training loss vs. CRs for: (a) UNSW-NB15 under IID, (b) NSL-KDD under IID, (c) UNSW-NB15 under non-IID ($C = 0.2$), (d) UNSW-NB15 under non-IID ($C = 1.0$), (e) NSL-KDD under non-IID ($C = 0.2$), (f) NSL-KDD under non-IID ($C = 1.0$).

able instability and fluctuation. FedRWA's training loss decreases rapidly and stabilizes, while other methods either plateau early or oscillate unpredictably. Under non-IID conditions, FedRWA consistently outperforms all baselines. As heterogeneity increases, methods like MOON and FedAWA demonstrate erratic accuracy and loss behavior. FedRWA, on the other hand, maintains smooth and stable trends in both $\alpha = 1.0$ (dotted lines) and $\alpha = 0.5$ (solid lines). Its consistently low loss and high accuracy reinforce the effectiveness of the proposed aggregation strategy. These visual results further confirm FedRWA's superiority.

5 Conclusion and Future Work

In this study, we proposed FedRWA, a novel RL-based aggregation strategy for FL that dynamically assigns client weights based on local training characteristics and global feedback. Unlike existing methods that rely on static, uniform, or heuristic-based strategies, FedRWA leverages a policy gradient approach to learn client-specific aggregation weights using a lightweight policy network. The network is trained end-to-end using reinforcement process, with the global model's validation performance serving as the reward signal, ensuring that the learned aggregation policy aligns with the overall learning objective. Comprehensive experiments conducted on two benchmark intrusion detection datasets, UNSW-NB15 and NSL-KDD, under both IID and non-IID conditions, validate the effectiveness of FedRWA. The proposed method consistently outperforms state-of-the-art baselines in terms of accuracy, F1-score, and communication efficiency. It demonstrates faster convergence and greater stability, particularly in highly heterogeneous data settings.

Our evaluation is limited to two benchmark intrusion detection datasets (UNSW-NB15 and NSL-KDD), which ensures reproducibility but leaves room for validation on more diverse and larger-scale datasets to demonstrate broader applicability. In addition, while FedRWA shows strong empirical performance, a formal theoretical convergence analysis is not yet provided, and future work will address this to strengthen the method's foundations. We plan to extend the FedRWA framework to support personalized federated models, incorporate mechanisms to handle adversarial or unreliable clients, and explore multi-objective reinforcement processes to balance fairness, robustness, and performance.

Acknowledgment. This work was supported in part by the Australian Research Council under grant DP220100983.

References

1. McMahan, B., Moore, E., Ramage, D., Hampson, S., y Arcas, B.A.: Communication-efficient learning of deep networks from decentralized data. In: Artificial intelligence and statistics, pp. 1273–1282. PMLR (2017)
2. Uddin, M.P., Xiang, Y., Hasan, M., Bai, J., Zhao, Y., Gao, L.: A systematic literature review of robust federated learning: issues, solutions, and future research directions. ACM Comput. Surv. **57**(10), 1–62 (2025)
3. Kairouz, P., et al.: Advances and open problems in federated learning. Found. Trends® Mach. Learn. **14**(1–2), 1–210 (2021)
4. Chen, H.-Y., Chao, W.-L.: FedBE: making Bayesian model ensemble applicable to federated learning. In: International Conference on Learning Representations (2021)
5. Hu, M., et al.: AutoFL: a Bayesian game approach for autonomous client participation in federated edge learning. IEEE Trans. Mob. Comput. **23**(1), 194–208 (2022)
6. Qi, J., Zhou, Q., Lei, L., Zheng, K.: Federated reinforcement learning: techniques, applications, and open challenges. Intell. Robot. **1**(1), 18–57 (2021)
7. Zeng, Y., et al.: Adaptive federated learning with non-IID data. Comput. J. **66**(11), 2758–2772 (2023)
8. Singh, N., Adhikari, M.: SelfFed: self-adaptive federated learning with non-IID data on heterogeneous edge devices for bias mitigation and enhance training efficiency. Information Fusion **118**, 102932 (2025)
9. Li, M., Zeng, J., Xiong, Y.: Federated learning on non-IID data with reinforcement learning and quantifying historical contribution. In: 4th International Conference on Internet of Things and Smart City (IoTSC 2024), vol. 13224, pp. 320–325. SPIE (2024)
10. Nguyen, N.H., et al.: FedDRL: deep reinforcement learning-based adaptive aggregation for non-IID data in federated learning. In: Proceedings of the 51st International Conference on Parallel Processing, pp. 1–11 (2022)

11. Imani, H., Anderson, J., Farid, S., Amirany, A., El-Ghazawi, T.: RLFL: a reinforcement learning aggregation approach for hybrid federated learning systems using full and ternary precision. IEEE J. Emerg. Sel. Top. Circuits Syst. (2024)
12. He, J., Chen, W., Zhang, X.: FedAA: a reinforcement learning perspective on adaptive aggregation for fair and robust federated learning. In: Proceedings of the AAAI Conference on Artificial Intelligence, vol. 39, pp. 17085–17093 (2025)
13. Shi, C., Zhao, H., Zhang, B., Zhou, M., Guo, D., Chang, Y.: FedAWA: adaptive optimization of aggregation weights in federated learning using client vectors. In: Proceedings of the Computer Vision and Pattern Recognition Conference, pp. 30651–30660 (2025)
14. Lu, J., Zhang, H., Zhou, P., Wang, X., Wang, C., Wu, D.O.: FedLaw: value-aware federated learning with individual fairness and coalition stability. IEEE Trans. Emerg. Top. Comput. Intell. (2024)
15. Ghimire, B., Rawat, D.B.: Recent advances on federated learning for cybersecurity and cybersecurity for federated learning for internet of things. IEEE Internet Things J. **9**(11), 8229–8249 (2022)
16. Li, T., Sahu, A.K., Zaheer, M., Sanjabi, M., Talwalkar, A., Smith, V.: Federated optimization in heterogeneous networks. In: Proceedings of Machine Learning and Systems, vol. 2, pp. 429–450 (2020)
17. Li, Q., He, B., Song, D.: Model-contrastive federated learning. In: Proceedings of the IEEE/CVF Conference on Computer Vision and Pattern Recognition, pp. 10713–10722 (2021)

Correction to: PLNet-12: A Vision-Language Benchmark for Zero-Shot Physical Literacy Analysis Across 12 Fundamental Movements

Tianchen Guo, Peter Anthony Logan, Thomas Wackwitz, and David Martin

Correction to: Chapter 19 in: M. Liu et al. (Eds.): *AI 2025: Advances in Artificial Intelligence*, LNAI 16371, https://doi.org/10.1007/978-981-95-4972-6_19

The chapter 19 of this book was published with error. The figure 2 of this chapter has beeen revised. The erroneous affiliation for co-author David Martin has been removed.

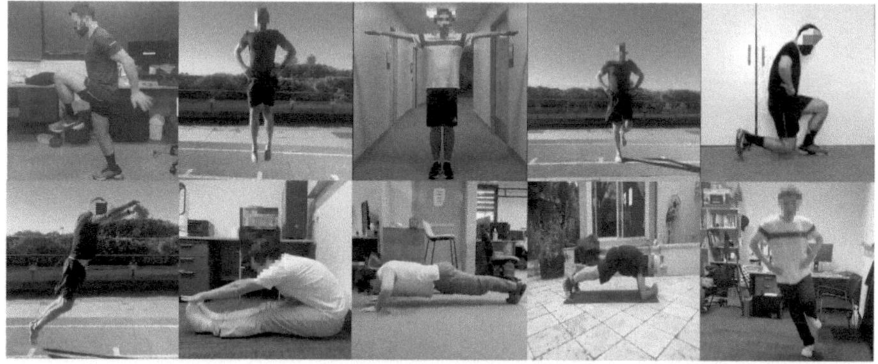

Fig. 1. Visualization examples from our dataset, showcasing diverse environments including both indoor and outdoor settings, as well as participants from different age groups (young, adults, and elderly). Some images also depict subjects wearing the Vicon system for motion capture.

The updated version of this chapter can be found at https://doi.org/10.1007/978-981-95-4972-6_19

M. Liu et al. (Eds.): AI 2025, LNAI 16371, p. C1, 2026.
https://doi.org/10.1007/978-981-95-4972-6_40

Author Index

M. Liu et al. (Eds.): AI 2025, LNAI 16371, pp. 519–521, 2026.
https://doi.org/10.1007/978-981-95-4972-6

MIX
Papier aus verantwortungsvollen Quellen
Paper from responsible sources
FSC® C105338

If you have any concerns about our products,
you can contact us on
ProductSafety@springernature.com

In case Publisher is established outside the EU,
the EU authorized representative is:
Springer Nature Customer Service Center GmbH
Europaplatz 3, 69115 Heidelberg, Germany

Printed by Libri Plureos GmbH
in Hamburg, Germany